AF770612

Ecological Silviculture

Ecological Silviculture

Foundations and Applications

BRIAN J. PALIK
USDA Forest Service-Northern Research Station

ANTHONY W. D'AMATO
University of Vermont

JERRY F. FRANKLIN
University of Washington

K. NORMAN JOHNSON
Oregon State University

WAVELAND PRESS, INC.
Long Grove, Illinois

For information about this book, contact:
Waveland Press, Inc.
4180 IL Route 83, Suite 101
Long Grove, IL 60047-9580
(847) 634-0081
info@waveland.com
www.waveland.com

10-digit ISBN 1-4786-3847-8
13-digit ISBN 978-1-4786-3847-6

Printed in the United States of America

7 6 5 4 3 2 1

DEDICATION

This book is dedicated to our friend and colleague, the late Robert (Bob) Mitchell. Bob was there at the beginning when we started to develop our ideas and approach. Had he lived, he would have been our coauthor on this book. Even now his thoughts and words are found throughout. Our sincerest hope is that Bob would consider our book to be "great stuff."

Dr. Robert J. (Bob) Mitchell II *1955–2013*
(Photo credit: Mark Hainds. Taken at an ecological forestry workshop at the J. W. Jones Ecological Research Center in 2008.)

Table of Contents

Preface

The approach to silviculture presented in this text is one based on natural forest models, that is, models that emulate natural disturbances, sustain biological legacies, emulate development processes, and allow time to take its course in shaping stands, all to the benefit of ecosystems and the people who use them. Our perspective for a natural models approach has been decades in the making, representing an accumulation and integration of a wealth of ideas and experiences. While we work in different regions, in different types of forests, and in different sociopolitical climates, we share several motivating experiences that have led to our collective perspective on silviculture and, ultimately, this book. First, we have spent our careers learning how forests work in terms of their disturbance regimes, stand development, and structure and function, and we all have scratched our heads over the often entrenched view that classical silvicultural approaches somehow emulate these natural forest dynamics. It is rare that this emulation is actually the case with classical silviculture, and consequently, forests managed using a timber-focused model are fundamentally different in structure, composition, and function, from the natural models we have come to understand. This book embodies this understanding and realization.

Another shared motivation is that we have had the privilege of working with organizations and entities that have put a natural models approach into practice. Often they have done this for many decades and without a knowledge of (or concern for) the academic debate surrounding the issue. These examples, many of which you will be introduced to later in this text, reflect silviculture motivated by the need to protect threatened and rare species, to sustain biological diversity, to provide habitat for game and nongame species, to create forest conditions that are socially acceptable to a broad range of landowners and the public, and importantly, to manage for timber in profitable ways. The foresters involved in these efforts all came to understand that these diverse objectives could be pursued together, but that timber-focused silviculture was not a good model for doing so, at least not everywhere across the forest estate. This text presents the scientific basis for why these foresters have been successful.

There are other important motivations for undertaking this project. First, we have been actively involved in presenting workshops on ecological forestry over many years and have seen the need (as have workshop attendees) for a comprehensive text that can serve as a reference that more fully describes the ideas and applications of our instruction; until now, such a text did not exist. Secondly, several of us have taught silviculture and forest dynamics courses at major universities in the US and have firsthand experience with the difficulty of trying to provide students with appropriate, comprehensive text material on ecological silviculture. Once again, such a text did not exist until now. In addition, all of us have been actively engaged in forest management and restoration projects for both public and private ownerships in North America and beyond throughout our careers. Through this work, we have gained an appreciation of the need practicing foresters have for the approach we present, but pointing them to a comprehensive source of this information has not been possible until now.

Finally, all of us, in one way or another, have been accused of presenting old (and perfectly good) ideas as new and have been reminded that the source of the comprehensive approach we seek is in fact the classical textbooks for the field. We do appreciate the rich history of silviculture instruction and application emanating from Europe and North America reflected in these textbooks. In our estimation, however, the problem students—both new and old—now face is that silviculture instruction as reflected in these texts is too steeped in classical theory and methods, particularly those based on production objectives using an agricultural model, which has been a major focus of silviculture since its inception. Students of forestry in the twenty-first century must deal with contemporary and emerging issues not addressed well by existing textbooks, including climate change mitigation and adaptation, forest certification, invasive species, and ecosystem restoration. These and other ecologically oriented objectives have risen to prominence since the late twentieth century, yet remain largely ignored by silviculture texts. Our text is intended to serve the need for addressing these issues by providing a comprehensive approach to silviculture that is truly based on a natural models approach to understanding and managing forests.

We also note that our text is a companion volume to another Waveland publication, one that takes a broad view of forest management, also from an ecological perspective: *Ecological Forest Management* (Franklin, Johnson, and Johnson 2018). The reader will notice some similarity of thought and material, as the two texts share some authors, and Franklin et al. 2018 discusses natural forest development and some silvicultural applications of this information. However, our text delves much deeper into the translation of natural disturbance and development models into silvicultural practice and prescription than does Franklin et al. and conversely, our text does not address economic, social, or policy issues in forest management. The reader will be well served by having both volumes.

Acknowledgements

Many influential researchers and teachers in the fields of forest ecology, forest management, and silviculture have helped shape our ideas, through both their writing and mentorship, but more often through thoughtful discussions. They include, in no particular order, Bob Seymour, Mac Hunter, Kurt Pregitzer, Lena Gustafsson, Linda Nagel, Shawn Fraver, Charles Goebel, John Bradford, Lisa Schulte-Moore, Craig Lorimer, Jim Long, Ellen Macdonald, Patricia Raymond, Dan Kneeshaw, David Foster, Klaus Puettmann, Steve Jack, Kay Kirkman, William K. Ferrell, David Lindenmayer, Steve Bédard, David Kittredge, Paul Catanzaro, Tom Spies, Mark Harmon, Mark Swanson, Fred Swanson, Dale Thornburgh (deceased), James Freund, Matthew Betts, James Lutz, Robert Van Pelt, Bill Beese, Robert Williams, Abe Wheeler, and Andrew Larson, among others. Several forward-thinking foresters from Minnesota opened our eyes to the real-world possibilities of ecological silviculture, including Jack Rajala, who was listening to the forest well before we came up with our ideas, and Mark Jacobs, who has been championing profitable ecological forestry for decades.

Our home institutions provided the time and financial support needed to complete this project, including the Northern Research Station of the USDA Forest Service (Palik), the University of Vermont's Rubenstein School of Environment and Natural Resources (D'Amato), the University of Washington College of Forest Resources (Franklin), and Oregon State University's Department of Forest Resources (Johnson). Tony D'Amato received support from the Harvard Forest, Charles Bullard Fellowship in Forest Research for a portion of his contribution. We thank them for this support.

Several people contributed to this book through writing sections or providing reviews. Notably, Chapter 12 and parts of Chapters 6 and 8 were written by Derek Churchill, providing us with a much needed perspective on frequent fire forests and managing for heterogeneity. Individual chapters or sections of the book were reviewed by Greg Corace, Lisa-Schulte-Moore, Matt Powers, Nicole Rogers, Mark Jacobs, John Bailey, John Lhotka, Beth Jacqmain, Michael Snyder, and Chris Woodall. We thank them for their time, thoughtful advice, and critical reflection on how to best translate our ideas into the classroom and field. Finally, thanks to Waveland Press for giving us the opportunity to pursue this book, especially Don Rosso, Senior Editor, and Jessica d'Arbonne, who did the heavy lifting of initial editing.

Brian Palik
Anthony D'Amato
Jerry Franklin
Norm Johnson

About the Authors

Dr. Brian Palik is science leader for applied forest ecology with the USDA Forest Service-Northern Research Station, in Grand Rapids MN, working broadly on questions related to the ecological sustainability of managed forests through use of operational-scale and long-term silviculture research in the western Great Lakes region. He has Ph.D. and M.S. degrees in forestry and ecology from Michigan State University and a B.S. in biology from Alma College. Prior to working for the USFS, he was a research ecologist at the Joseph W. Jones Research Center in Georgia, working on coastal plain forest ecology and management. He has authored over 230 scientific articles and book chapters on topics related to riparian forest management, ecological silviculture based on natural forest models, and more recently, on adapting forests to an uncertain climate future. During his tenure with the USFS, he has consulted with and sat on numerous advisory panels for federal, state, county, NGO, and tribal groups on issues related to riparian forestry and ecological forest management.

Dr. Anthony D'Amato is a professor of silviculture and applied forest ecology and director of the Forestry Program in the Rubenstein School of Environment and Natural Resources at the University of Vermont where he teaches silviculture and forest conservation courses. He received his BS in Forest Ecosystem Science from the University of Maine, MS in Forest Science from Oregon State University, and his PhD in Forest Resources from the University of Massachusetts. Prior to joining the faculty at the University of Vermont, he was an associate professor in the Department of Forest Resources at the University of Minnesota and a Charles Bullard Fellow at Harvard University's Harvard Forest. He has authored over 150 scientific articles and landowner outreach pamphlets on topics related to forest stand dynamics, ecological silviculture, forest carbon dynamics, and adaptation strategies. Throughout his career, a primary area of focus has been on training and advising students, landowners, and practitioners to develop silvicultural strategies that meet a wide range of ecological, adaptation, and economic objectives.

Dr. Jerry Franklin is professor emeritus for forest ecosystems in the School of Environmental and Forest Sciences at the University of Washington where he has taught ecosystem management and science, silviculture, landscape ecology, and forest ecology. He received his BS and MS in Forest Management from Oregon State University and his PhD in Botany and Soils from Washington State University. He began his career as a research forester at the US Forest Service's Pacific Northwest Research Station where he worked on silvicultural topics, eventually becoming project leader and Chief Plant Ecologist. During this time he was leader of a scientific team that characterized the ecological nature and importance of old-growth forest ecosystems. He has been an author on over 500 scientific articles and 12 books. Over his career, much of this time has been spent advising federal, state, Native American, and private landowners on ecological forest management.

Dr. K. Norman Johnson is professor emeritus of forest resources in the College of Forestry at Oregon State University where he has taught forest policy, forest management, forest economics, and collaboration. He received his BS in Forestry from the University of California at Berkeley and his PhD in Forest Management from Oregon State University, emphasizing natural resource economics. He spent the early part of his career on the theory and practice of timber harvest scheduling and on building forest planning models. He has published numerous articles on these topics and co-authored two books on forest management planning. Outside of teaching and policy analysis, much of his time has been spent advising Oregon's Governors and Congressional Delegation on federal forestry and working with Indian tribes on forest planning. Most recently, he has worked with Dr. Franklin to demonstrate ecological forestry principles in forests of the Pacific Northwest.

PART I

A Background on Ecological Forestry and Silviculture

CHAPTER 1

Forestry in the Twenty-First Century

When most people look at a forest and see trees, they are content with what they see, assuming that the forest is as it should be: vigorously growing and full of desirable plants and animals (Figure 1.1a). A deeper perception of the tree component, including spatial arrangement of open and closed canopy, crown structures of individual trees, relative abundance of different tree species, age and size structures of stands, and the amount and characteristics of dead trees, are less of a concern (Figure 1.1b). This perspective is not exclusive to the lay public; until recently it has also been the perspective of many practicing foresters, who have been taught to tend trees and forests as an agricultural crop, with a primary emphasis on growing trees for timber production. The end result, although likely unintended, is of forests devoid of the rich structural variety and natural diversity that is the hallmark of natural ecosystems (Puettmann et al. 2008).

While a forest is certainly defined by its trees (Palik and Engstrom 1999), ecologists have come to understand that there is much more going on in a forest, and in need of attention by foresters, than the number and size of its trees. When broader aspects of ecosystem function and health are held in high value, then it becomes apparent that a narrow focus on growing trees is not a sufficient management philosophy. This deeper understanding of the structure and function of forests has evolved out of a long-running and diverse research effort into **natural disturbance** and **stand development** of different types of forests around the world, and how this understanding can be used to shape a new approach to forestry, i.e., an ecological approach (e.g., Franklin et al. 2002). The tools to carry out on-the-ground forest management are of course what silviculture is all about, but the silvicultural toolbox for a natural models approach to **ecological forestry** (see Chapter 2) must contain different tools than classical silviculture.

This is the subject of our book: to provide the scientific foundation and tools to develop an ecological approach to silviculture, i.e., *ecological silviculture*. We define ecological silviculture as an approach for managing forest ecosystems, including trees and associated organisms and ecological functions, based on emulation of natural models of development and that explicitly incorporates principles of continuity (legacy management), complexity/diversity (variety and heterogeneity of structure; diversity of species), timing (basing actions on ecological queues) and context (considering the landscape context) into prescriptions. You will

Figure 1.1 *Contrast between a plantation of red pine trees (a) and a natural red pine ecosystem (b). Photo credit: Shawn Fraver (b).*

learn much more about the details of this definition beginning in Chapter 3 and in subsequent chapters. In the meantime, to understand the need for ecological silviculture, it is helpful to review how forestry has gotten to where it is today, that is, what is the context of forestry in the twenty-first century?

A Context for Forestry in the Twenty-First Century

A proper context is needed to appreciate the need for ecological silviculture in forest management. To gain this contextual understanding, it is necessary to consider several things, including the distribution of different forest conditions (plantations and production forests, multiuse forests, reserves) in the global forest. Such a perspective will lead to a better appreciation of where ecological forestry fits into the global forest landscape. It is also useful to review how the objectives for managed forests have evolved from the mid-twentieth to the twenty-first century, and why ecological silviculture is needed to meet these objectives. The global drivers of forest health have also changed (and continue to do so) over this same time period, so it is useful to review these changes to better understand why ecological silviculture is needed, especially if we are to sustain healthy ecosystems in a rapidly changing global environment. Finally, it is useful to consider how forest science has evolved to become much more **ecosystem**-focused and, as a consequence, how our understanding of forests has expanded. The inevitable consequence of this expansion of knowledge is the need for ecological silviculture.

The Global Forest

The global forest can be examined in several ways including its origin, that is, plantation or natural. As of 2015 there were approximately 0.30 billion hectares of **plantation forest** (i.e., forest established by planting or seeding) and 3.7 billion hectares of **natural origin forests** (FAO 2015). For our discussion (and following Lindenmayer et al. 2012a), we assume that plantation forests are managed more intensively for wood and fiber than other types of forest and while they do provide some ecosystem services, for the most part ecological benefits are eliminated or marginalized in plantation settings (Cao 2008).

The global forest can also be divided into categories of use, including **intensive production forests,** (mostly plantation), **multiuse forests,** and various types of **reserve forests** (forests excluded by law or policy from timber harvest). As of 2015, there were approximately 0.65 billion hectares of legally protected **forest reserves.** For our discussion, we assume that reserves, where managing for timber is not an objective, come from the natural forest category (although this may not be entirely the case). As important as these forests are, they can never support adequate and well distributed populations of many forest-dependent species (Mascia and Pailler 2011), largely because the total area of reserve forest is unlikely to increase substantially in the future due to landownership patterns and economic factors (Schmitt et al. 2009; Bauhus et al. 2010). However, other types of forest may be designated for conservation of ecosystem services, including biodiversity, watershed and soil protection, or cultural and spiritual values, but with different legal status than reserve forests.

Our intent in presenting these approximations is to point out that the bulk of the global forest, approximately 3.05 billion ha of *natural-origin* forest not formally protected, is the stage where multiple environmental, economic, and social objectives are pursued (Edwards et al. 2010, Lindenmayer et al. 2012a) and where ecological silviculture is the tool most appropriate for realizing diverse objectives. Given the size of this forest area, management for wood products and economic objectives will continue to be important. In fact, the vast majority of the world's timber and fiber come from non-plantation forests. Sustaining economic output through forest management will be critical for providing financial incentives for governments, private landowners, and trust managers to use ecological approaches in these forests (Forest Stewardship Council 2009).

Forestry After World War II

In the United States, forestry soon after World War II had as a primary objective the provision of wood to supply the booming housing market fueled by returning veterans (Jackson 1985). While watershed protection, habitat management, and providing recreational opportunities were ostensibly objectives, it was the need for plentiful cheap lumber that was a major driver for how forest lands were managed in the bulk of the forest landscape. The operating paradigm for this type of management was command and control (see Chapter 3) of forest development, very much reflecting the desire to view trees as agricultural crops and to manage forests as if they were agricultural fields (Puettmann et al. 2008). Much like row crop production, this type of management emphasizes crop tree development early in stand or patch development, it attempts to manipulate stand structure such that crop trees all have equal access to resources, it eliminates nondesirable tree species as they may be competitors with crop trees, it attempts to minimize damage and mortality from insects and disease, and it views less-than-healthy individuals as nondesirable, or non-merchantable trees as they are called.

The outcome of decades of application of an agricultural management model on much of the forest estate has been a loss of **structural complexity** and **heterogeneity** in forest stands and a reduction in native tree species diversity (Figure 1.2; Franklin and Johnson 2013), particularly when that management occurs in a plantation setting (Box 1.1). We will discuss much more about forest structure in Chapters 4 and 5, but suffice it to say for now that ecological silviculture has as its core objectives sustaining or often restoring structural complexity and heterogeneity of forest stands and sustaining populations of all native species, especially trees. An agricultural model of forestry does not share these objectives. In later chapters, you will gain an appreciation for the importance of tree-derived forest attributes for sustaining species diversity and ecosystem processes, but simply put, ecological silviculture is very much focused on maintaining ecosystem structure and function broadly. Silviculture based on an agricultural model in contrast has a primary objective of sustaining the productivity of merchantable trees (Puettmann et al. 2008).

Figure 1.2 *Examples of structurally simple and compositionally depauperate forests. Photo credits: (c) John Elioff.*

The Changing Drivers of Forest Management

Over the latter part of the twentieth century and continuing today, the drivers of forest management policy and practice have changed fundamentally from what they were soon after World War II. For instance, non-traditional forest owners and stewards have become key players in the forest management realm. For instance, The Nature Conservancy (TNC), one of the largest owners of forest land in the United States, not only conserves forests but often manages them for ecosystem services broadly (Cornett and White 2013). Needless to say, TNC and other conservation organizations that manage forest land have markedly different objectives than those with an intensive production or timber-management focus and embrace the kind of ecological approaches to forestry that this book presents.

Even the objectives of longstanding forest managers have evolved over time to be much more inclusive of sustaining a wide array of ecosystem services, often using approaches that better emulate the natural structural and compositional dynamics of forests. The United States Department of Agriculture National Forest system is a prime example of this evolution, where today there is greater emphasis than ever on restoration of forests to reduce fire risks, improve tree health, and protect species and ecosystem functions (Vilsack 2009; USDA 2012). The ecological silviculture approach we present is meant to address the needs of these users.

Another driver of change has been the rise of third party **forest certification**, which has led many forest stewards, both public and private, to adopt more ecological approaches to management (Boyle et al. 2016). For instance, various indicators in Forest Stewardship Council certification speak directly to managing ecologically using a natural model approach (Forest Stewardship Council 2010; Box 1.2). The Sustainable Forestry Initiative (SFI) is another organization that supports certification especially in North America (www.sfiprogram.org). The science behind an ecological approach and practical tools for implementing it on the ground to meet these certification requirements are exactly what we present in our approach to ecological silviculture.

Box 1.1 **The role of plantations in an ecological forestry approach**

Plantations represent about 7.5% of global forest area and annually provide 33–46% of the world's industrial roundwood (Jürgensen et al. 2014; Payn et al. 2015). The ability of these ecosystems to produce at such high levels reflects the highly efficient agricultural model associated with plantation forestry. This often includes high up-front investments in site preparation, competition control, fertilization, and improved nursery stock that result in the development of merchantable-sized crop trees at timescales that are much shorter than those observed in natural forests (Figure 1.3; Fox et al. 2007). The simplified nature of plantations (i.e., uniform spacing of a single cohort and species), although ecologically undesirable, allows for highly efficient production and harvesting regimes that are calibrated to local site conditions as part of what has been termed "precision forestry" (Dyck 2003). As with other agricultural crops, plantations focus exclusively on production of commercial tree species, with loblolly pine, radiata pine, Douglas-fir, teak, spruce, and eucalyptus being some of the more common species and genera planted around the globe (Brown 2000).

Figure 1.3 *An intensively managed slash pine plantation on site in northern Florida that has been mounded and treated with herbicides to enhance early seedling survival and growth. The site will be fertilized at age 6, thinned between ages 12–14, and clear-cut harvested between ages 23–25.*

Although much of this text is focused on silvicultural techniques that foster and restore ecological complexity and diversity in areas managed for wood products, including former plantations, it is important to recognize that plantation forestry plays an important, complementary role to ecological silviculture. In particular, the high levels of production realized from plantations generally reduce the levels of demand for wood products from natural forests (Pirard et al. 2016). This reduction in demand provides an opportunity for flexibility and creativity in applying ecological silvicultural systems that, although also focused on generating wood products, are generally guided by ecological timeframes and feature high levels of permanent retention of living and dead trees. Our emphasis on silvicultural systems guided by ecological models is not meant to diminish the critical role of plantation silviculture in meeting global wood demands, but instead is intended to provide alternative approaches that function in a complementary fashion at landscape and regional scales in balancing ecological and economic objectives (cf. Seymour and Hunter 1992).

In a similar vein as certification standards, many states in the USA have developed **best management practices** (BMP) designed to protect ecosystem services, such as water quality, aquatic ecosystems, and wildlife habitats (Aust and Blinn 2004; Blinn and Kilgore 2001). Whether mandatory or voluntary, BMPs are widely implemented and have had a significant influence on how forests are managed. Typically, these guidelines include provisions to retain structural and compositional elements in a harvested forest to the benefit of plant and animal species. Retention of snags, tree patches, and cavity trees are typical BMP elements (e.g., Minnesota Forest Resources Council 2013) that can be in alignment with an ecological silviculture approach.

Another driver of significance is the shifting percentages of consumption and imports of wood and fiber over the last several decades of the twentieth century and into the twenty-first century. In some parts of world, the United States included, sources of wood for different forest product sectors have increasingly come from imports. The United States, for example, increasingly relies on imports to meet consumption, with imports as a percentage of consumption increasing from 13% to 21% from 1965 to 2011 (USDA 2013). Much of the increase in imports is from softwoods, with their imports rising from 15% to 27% of consumption between 1965 and 2011 (USDA 2013). Hardwood imports as a percentage of consumption increased from 3% to 5% during the same time period. A large percentage of the increase in softwood imports come from Canada, but in the future, an increasing percentage may come from intensive tree farms in the Southern Hemisphere (Franklin and Johnson 2004). Of note, wood consumption in the United States has been declining since the late 1980s (USDA 2013), so the increase in imports may represent a significant shift in how the US domestic forest is being managed.

The upshot of this trend in the United States at least is that an increasing percentage of the forest landscape may no longer have the demands of softwood production placed on it, and consequently is ripe for application of ecological silviculture approaches that keeps the forest producing

Box 1.2 **Forest Stewardship Council (FSC) certification terminology that addresses elements of managing forests based on natural models of ecological forestry**

FSC is an international non-profit organization founded in 1993 to promote sustainable management of the world's forests. One of the main tools used by FSC to achieve its goals is the setting of standards that must be met by organizations wishing to have their forests and forest products certified as sustainably produced. Standards of sustainability are met by adhering to principles and associated criteria, as measured by indicators. An example of a principle and criteria associated with ecological forestry follow.

PRINCIPLE 6: ENVIRONMENTAL IMPACT

Forest management shall conserve biological diversity and its associated values, water resources, soils, and unique and fragile ecosystems and landscapes, and, by so doing, maintain the ecological functions and the integrity of the forest.

Criterion 6.3: Ecological functions and values shall be maintained intact, enhanced, or restored . . .

Indicator 6.3.a.1: . . . maintains, enhances, and/or restores underrepresented successional stages . . . that would naturally occur on the types of sites found . . . where old growth of different community types that would naturally occur . . . are underrepresented in the landscape relative to natural conditions, a portion of the forest is managed to enhance and/or restore old growth characteristics.

Indicator 6.3.d: Management practices maintain or enhance plant species composition, distribution and frequency of occurrence similar to those that would naturally occur on the site.

Indicator 6.3.f: Management maintains, enhances, or restores habitat components and associated stand structures, in abundance and distribution that could be expected from naturally occurring processes. These components include:

a) large live trees, live trees with decay or declining health, snags, and well-distributed coarse down and dead woody material. Legacy trees where present are not harvested.
b) vertical and horizontal complexity.

Indicator 6.3.g.1: . . . when even-aged silvicultural systems are employed, and during salvage harvests, live trees and other native vegetation are retained within the harvest unit in a proportion and configuration that is consistent with the characteristic natural disturbance regime . . .

more diverse wood products, along with various other ecosystem goods and services that are part of an ecological management approach. For example, as softwood production declines in the US as a result of globalization there may be increased demand to value (monetarily) other ecosystem services, such as watershed protection and carbon sequestration, as well as development of markets for higher-quality forest products that are more resilient in global market fluctuations (Franklin and Johnson 2004).

Finally, the health threats facing forest are evolving rapidly and some are unprecedented in recorded history. These threats are interrelated to some degree and include climate change, catastrophic wildfires, and invasive exotic species. Projected changes in precipitation and temperature in many forested regions will alter habitat suitability for many tree and plant species (Allen et al. 2010; Gillson et al. 2013), greatly altering forest composition and structure from historical conditions. Wildfire risk is increasing in dry forest regions due to fuel buildup, with the potential for catastrophic fires to increase under a warmer, drier climate (Flannigan et al. 2009). Nonnative invasive species continue to be a large threat to forests in many regions, with often devastating results in terms of loss of species and altered forest structure (Gandhi and Herms 2010; Pyšek and Richardson 2010). A wise course of action in the face of all of these threats is to start with a species rich forest, relative to the natural condition, and structurally similar to the conditions generated by natural disturbance and development. The rationale being that a compositionally diverse forest is more likely to contain at least one species that is successful under a changing climate, or is resistant or resilient in the face of a new insect or pathogen threat. Moreover, a forest that is structurally similar to its natural condition in terms of age cohorts, tree densities, and dead wood condition may be less likely to suffer major change from a catastrophic wildfire or wind disturbance (Bauhus et al. 2013). Confronting each of these threats requires an approach to management that restores the structural and functional integrity of simplified forests. Ecological silviculture, as we present it, is such an approach.

Ecosystem Science and Ecological Silviculture

A key development that has propelled ecological silviculture to the forefront of forest management is a profound increase in basic understanding of **ecosystem structure and function**. This knowledge has emerged out of large-scale and long-term research efforts, most notably as part of the US National Science Foundation's Long-Term Ecological Research program (Knapp et al. 2012). As a result of this and similar efforts, we simply know exponentially more about natural disturbances, forest dynamics, and ecosystem function than we did during the heyday of the agricultural forestry era. Examples of this expanded knowledge include recognizing the importance of large **deadwood** as habitat for a wide variety of organisms (Figure 1.4a,b; Harmon et al. 2004), as well as the importance of very large and old trees for ecosystem function (Figure 1.4c, d; Lindenmayer et al. 2012b). In this context, it is instructive to note that Aldo Leopold (1949) viewed forests as ecosystems long before it was popular to do so, but his thinking was far ahead of ecosystem science. In a very real sense, the science has finally caught up to Leopold's concept of managing forests as ecosystems.

Somewhat coincident with the increase in understanding of ecosystem science has been the emergence of landscape ecology as a unique discipline that has offered valuable insight into how forests work across multiple scales of space (Forman 1995). An early revelation of landscape ecology was how silvicultural actions at stand scales, when repeated across the landscape, profoundly influence that structure and function of the forest landscape, for instance, through

Figure 1.4 *Ecosystem research in various forests around the world has uncovered the functional role of stand structural elements, including large deadwood (a, b) and very large trees (c, d).*

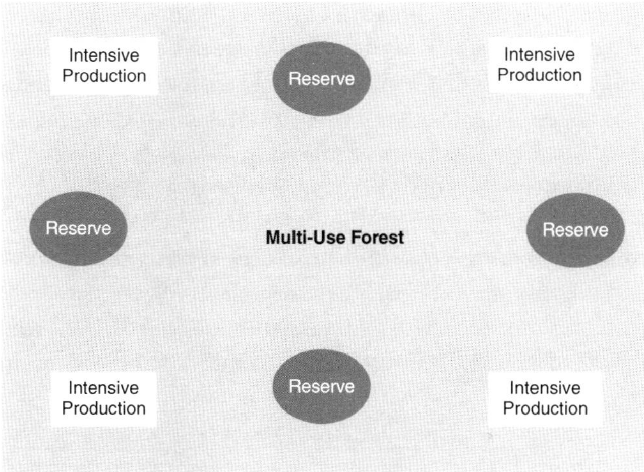

Figure 1.5 *Conceptual representation of a forest landscape consisting of intensive production forests (plantation for our purposes) and reserve forests that are protected by law for purposes other than timber management. The two categories are embedded in a landscape of multiuse forest, consisting of natural origin timber-focused forests, multiuse forests, and other types of conservation forests. The multiuse forest is arguably the place where ecological silviculture will be applied.*

edge effects and patch size distributions (Franklin and Forman 1987). In many regions, this larger-scale context perspective has fundamentally altered how stand-scale actions are implemented in a landscape. Another insight from landscape ecology has been appreciation for the role of the larger forest landscape, i.e., the bulk of the forest estate, for the sustainability of biodiversity broadly defined (Lindenmayer and Franklin 2002). We now recognize that species cannot be sustained in the landscape when the only habitat provision for them is reserve forest; the provision of habitat elements (e.g., large snags, den trees, etc.), as well as landscape heterogeneity and connectivity, within the multiuse landscape are critical for sustainability of many taxa (Hunter 2005; Lindenmayer et al. 2006). Armed with this understanding, the multiuse landscape emerges as the operating ground for ecological silviculture (Seymour and Hunter 1992) and it is where the principles we present are best applied (Figure 1.5).

The Greening of Forestry

In total, the various drivers, pressures, and changes facing forestry have converged into what might be thought of as a *greening of forestry*. This greening is a multi-scale phenomenon that reflects the interjection of ecosystem understanding into management of the forest landscape, as opposed to basing this management on long-held agricultural paradigms (Kohm and Franklin 1997). A general idea is that management of the multiuse landscape can and should better emulate natural models, particularly as we learn more about how ecosystems work, and apply this knowledge to forest management (Franklin et al. 2007). This green forestry proclivity is most evident in the recent generation of forestry graduates and newer stakeholders of forest management, such as The Nature Conservancy and consumers who base their purchases of wood products on sustainable certification labels. Their need for an ecological approach to silviculture is what we address in this book.

Our Beliefs for Ecological Silviculture

Given the context of forestry in the twenty-first century and the greening of forestry that has emerged from this context, what do we believe regarding the role and formulation of ecological silviculture as the toolbox for contemporary management? As a fundamental starting principle, we believe, as did Leopold (1949), that "to keep every cog and wheel is the first precaution of intelligent tinkering" and by extension, intelligent silviculture. While Leopold was referring to species richness specifically, his concept can be broadened to include all ecosystem structures, species, and processes. Moreover, in our contemporary forest landscape, intelligent silviculture may often focus on restoration of these parts before the focus can shift to "keeping" them in the forest. The ecological silviculture approach we present does just this, that is, it provides a foundation for both restoration and sustainability of structurally complex forests.

Secondly, we believe that sustaining or restoring ecosystems requires an approach based on a fundamental understanding of natural disturbance and forest development. Native species and ecosystem functions are dependent on the structural and compositional characteristics that develop in response to natural disturbances and stand development processes, so the closer these conditions can be emulated, the better the chances of sustaining whole ecosystems. Our conceptualization and application of ecological silviculture is based on this premise.

Third, we believe that wide application of ecological silviculture requires a vibrant forest industry. If ecological forestry is to be widely adopted there has to be a market for timber from natural forests; without this market, and with an ever increasing reliance on plantations for timber and fiber, there may be limited justification, and a financial ability, for adopting the approach we present.

Finally, we believe that a comprehensive rethinking of silviculture is needed to address the societal needs and objectives that have emerged as a result of changing drivers of forestry; a tweaking of the agricultural model of forestry is not sufficient to address these needs (Puettmann et al. 2013). Our book provides this comprehensive rethinking of silviculture. Specifically, we provide readers with a perspective on silviculture that is grounded in a comprehensive understanding for natural forest disturbance, dynamics, and stand development, and that recognizes the importance of structural complexity and stand heterogeneity for sustaining ecosystem function and biodiversity. This perspective is fundamentally different from that of other textbooks in the field.

The Tenets of Ecological Silviculture

Our beliefs can be codified into a set of *tenets* for ecological silviculture (Box 1.3). A tenet by definition is *any opinion, principle, doctrine, dogma, etc., especially one held as true by members of a profession, group, or movement.* In our case, the tenets of ecological silviculture are beliefs held closely by foresters who have gained a deep understanding of natural forest dynamics and who have devised and implemented practices that reflect a natural models approach to managing their forests. When accepted by a forester, the tenets provide the license to practice silviculture in a way that is in line with societal expectations for sustainability of a broad array of ecosystem services and is compatible with ecological principles. As you read this book, you will learn much more about the foundations of these tenets; we present them now so that you will have a sense of the direction this book will take you.

What You Will Learn from This Book

Our book is organized into five sections and 17 chapters, including this introduction (Chapter 1). In Section I, Chapter 2, we define ecological forestry and we show how silviculture based on natural models is fundamentally important for achieving the goals of ecological forestry. We also compare our approach to other ecological approaches, e.g., **close-to-nature** forest management (Bauhus et al. 2013), to illustrate similarities and differences. After reading this chapter, you will be able to defend the merits and objectives of ecological forestry to your agency, organization, or stakeholder and articulate an approach based on natural models. In Chapter 3, we review the fundamental concepts of silviculture and then recast classical silvicultural practices within the context of an ecological approach. After reading this chapter, you will start thinking about ecological silviculture within the context of classical silvicultural actions and systems, appreciating the antecedents of ecological silviculture, but recognizing that a tinkering of the classical school of thought is not sufficient to meet the objectives of ecological forestry.

Section II is organized into a series of chapters that describe the fundamental aspects of natural disturbance regimes and **biological legacies** (Chapter 4) and the developmental processes and patterns that shape established stands (Chapter 5), while Chapter 6 presents details on stand development and dynamics organized around several general but highly repeatable disturbance **archetypes**, using well-studied forest type as case studies. After reading this section, you will have a deeper understanding of the science behind an ecological approach to silviculture and you will be able to assess what aspects of this approach are missing from current practices.

Section III is the first applied portion of our book in which we address the four primary principles of an ecological approach: **continuity**, **complexity/diversity**, **timing**, **and context**. Specifically, in Chapter 7 you will learn how to manage biological legacies during regeneration harvests. In Chapter 8 you will learn about intermediate treatments in establishing and established stands that are designed to create heterogeneity of structure and you will gain greater appreciation for ecologically determined timing of silvicultural interventions. Finally, Chapter 9 will help you better

Box 1.3 **The tenets of ecological silviculture**

1. **Sustain ecosystems, rather than timber (or other singular objectives)**
   This tenet recognizes that ecosystem structure and function is the engine of growth and yield; by sustaining the former, the latter is better safeguarded.
2. **Seek complexity and diversity**
   This tenet acknowledges that structural complexity and heterogeneity, as well as native species diversity, are the hallmarks of healthy forests and that it takes time for these features to develop.
3. **Seek forest continuity rather than discontinuity**
   By this tenet, the ecological forester emphasizes what is left behind during a harvest as much or more than what is removed, thus recognizing the importance of legacies.
4. **Maintain options for the future**
   This tenet pays heed to Leopold by acknowledging the importance of keeping all the parts of a forest so as to maintain options for an uncertain future.
5. **Practice silviculture in the context of landscapes**
   With this tenet, the ecological forester recognizes that stand actions must be considered in a landscape context, where connectivity between adjacent ecosystems is considered and cumulative impacts are recognized.
6. **Let the ecosystem speak to you**
   The emphasis of this tenet is on the stand as an ecosystem, not just a collection of crop trees; the ecological forester considers the condition of the rich array of ecosystem components when developing objectives and actions.

appreciate the larger-scale context of your stand-scale silviculture. After studying these chapters, you will be armed with the applied knowledge and tools you need to clearly articulate your objectives for ecological silviculture and develop prescriptions for your forest.

Part IV is the second applied section of our book. In Chapter 10, we review the concept of a silvicultural system, but from an ecological perspective. In Chapters 11–14, we present in some detail a background on classical silviculture for the major forest-disturbance archetypes we introduced in Part III, as well as outline ecological silvicultural systems for these archetypes. As a whole, this section will give you some real world examples of ecological silviculture systems that you can adapt to your specific needs.

Finally, in Section V we examine some additional issues likely faced by the ecological forester unique enough to warrant special consideration. These include riparian silviculture (Chapter 15) and silviculture in the face of climate uncertainty (Chapter 16). Finally, in Chapter 17 we reiterate the major conceptual themes associated with ecological silviculture and finish by highlighting the rightful role and place of ecological silviculture in the global forest landscape.

CHAPTER 2

What Is Ecological Forestry?

This book is about ecological silviculture, that is, an approach to on-the-ground actions of managing a forest that is based on natural development models and ecological principles (see Chapters 1 and 3) and for ecological objectives. That is, ecological silviculture is the toolbox for meeting objectives of *ecological forestry*. To better appreciate the need for such a toolbox, it is helpful to more fully understand what we mean by ecological forestry as a management paradigm. We present this discussion largely by way of a contrast with timber-focused forestry as it has been (and continues to be) practiced by many organizations and forest owners, and in many regions. We also discuss the contrasts between ecological forestry and some other disciplines or approaches that may be viewed as ecological, for instance, close-to-nature forestry.

Broadly, our definition of ecological forestry is that of an approach that applies an understanding of the structure, function, and dynamics of natural forest ecosystems to achieve integrated environmental, economic, and social outcomes (Franklin et al. 2007; Franklin and Johnson 2012; Franklin et al. 2018). In practice, ecological forestry is a framework for management that rests on an ecological foundation. Ecological forestry is not a cookbook for management. Rather, it is a way of thinking about forest ecosystems and their management in a fundamentally different way than timber-focused forestry, which has a history of addressing ecological goals as secondary objectives.

Ecological forestry addresses management from the standpoint of conserving native **biodiversity** and ecological productivity. It is distinguished from timber-focused forestry by its emphasis on emulating, to varying degrees, natural patterns and processes of forests (Figure 2.1) understanding these patterns and processes, working in harmony with them, and maintaining their integrity, even when it becomes financially difficult or inconvenient to do so (Seymour and Hunter 1999).

Quoting Seymour and Hunter (1999), two early influential thinkers on ecological forestry, a "central axiom of ecological forestry is that manipulation of a forest ecosystem should work within the limits established by natural disturbance patterns prior to extensive human alteration of the landscape. The key assumption here is that native species evolved under these circumstances, and that maintaining a full range of similar conditions with management offers the best assurance against losses of biodiversity." While we augment this statement with the need to recognize that natural disturbance patterns are not static and may be affected by climate change and invasive species, it is still true that considering the circumstances under which native species evolved provides an invaluable context for conserving ecosystems and the biodiversity within them. This is at the heart of ecological forestry as we address it in this book.

Key elements of ecological forestry that stem from using natural models as guides (Franklin et al. 2007; Franklin and Johnson 2012; Franklin et al. 2018) include: 1) retaining structural and compositional elements of the preharvest stand in the new stand; 2) manipulating established stands based on understanding stand development processes; 3) employing return intervals for silvicultural activities consistent with recovery of desired structures and processes; and 4) planning management activities in the context of larger spatial scales, in accordance with knowledge of spatial patterns and ecological functions in natural landscapes.

Development of **silvicultural prescriptions** for ecological forestry must start with an understanding and recognition of natural processes and with a thorough understanding of the ecology of the tree species in relation to disturbance and developmental processes, as well as the other native biota in the forest. This recognition and consideration of all biota and ecological functions is important, since we are interested in managing forests as ecosystems and not simply as collections of trees. This understanding informs the development of forest management actions and, more specifically, silvicultural prescriptions, as we shall see in later chapters. (Note that much of what follows is adapted from Chapter 1 in Franklin et al. 2018).

Ecological Forestry and Timber-Focused Forestry: A Comparison

We add to our discussion of ecological forestry by identifying some of its principles and derived outcomes and

Figure 2.1 *Ecological forestry is grounded in an understanding and emulation of natural patterns and processes including: (a) natural disturbance, such as this bark beetle mortality in a Norway spruce forest (Sumava National Park, Czech Republic); (b) small-scale mortality in established stands, like individual tree death in a red pine forest (Chippewa National Forest, Minnesota, USA); and (c) the development of trees and forests over time, like in this old-growth eastern hemlock forest (Massachusetts, USA).*

contrasting these with a timber-focused management model. A timber-focused model can certainly include sustaining other ecosystem services, but this is generally pursued as secondary to timber production and often addressed through regulatory means, such as forest practice acts or best management guidelines. In the timber-focused model, management for timber is the template used to address additional objectives, whereas in an ecological model, ecosystem sustainability, broadly defined, will share importance with or have priority over timber management.

In its extreme, a timber-focused model may evolve into an intensive production model (e.g., fiber farms) designed to optimize production of wood for economic gain with little direct concern for other ecosystem services. Generally this is not the type of forestry we are concerned with, as there is little opportunity or incentive to use ecological silviculture in these settings. Intensive production forestry is and may remain (depending on public tolerance) an important source of wood globally and is often included as part of broad strategies for balancing wood production with conservation goals across large forested areas (Seymour and Hunter 1992; Messier et al. 2009).

Getting back to timber-focused forestry (as it is more broadly practiced), it is not our intent to give this term a pejorative label, but rather, we use it to provide a contrast with management designed to sustain or restore multiple ecosystem benefits using models based on natural forests. Moreover, our intent is to demonstrate that much of the global forest estate outside of reserves and true production forests can be more explicitly based on a natural model while still including timber production as a shared goal.

We start our comparison of ecological and timber-focused forestry with the broad conceptual foundations of the two approaches (as adapted from Franklin et al. 2018):

-*Ecological forestry* utilizes models from natural forest ecosystems as a basis for management systems. It incorporates principles of natural forest development, including the role of natural disturbances, in the initiation, development, and maintenance of stand structures and landscape mosaics.

- *Timber-focused forestry* utilizes crop tree growth or economic models as a basis for management systems. To varying degrees, it combines agricultural farming principles with either crop tree growth analysis (with a goal of maximizing cumulative mean annual increment) or with rate of return analysis (to find the amount and spatial organization of capital that will achieve the desired economic outcomes, including achieving an acceptable rate of return on capital).

We note that not all timber-focused forestry utilizes these principles to the same degree, but most applications share a heritage of doing so by virtue of a focus on crop tree management as a primary objective. For example, much commercial forestry may actually be passive in application, often involving nothing more than prescribing harvests to create conditions favorable for tree regeneration (potentially followed by some level of intermediate treatment) so as to guide development toward economically desirable compositional and structural conditions. However, even passive forestry tends not to be explicitly ecological, hence a contrast with a natural model is still valid.

These different foundations between ecological and timber-focused forestry, including passive forestry, have profound implications for forest management. In fact, we argue that they lead to fundamentally different ways of thinking about what constitutes good forest management. We present these differences below to better illustrate the distinction between approaches.

Contrasting Fundamental Principles

The fundamental principles of ecological forestry and timber-focused forestry differ markedly (Table 2.1). Ecological forestry has as a core focus on the management and sustainability of whole ecosystems, including their diverse array of species, structures, and functions. In contrast, timber-focused forestry has, as a deliberate or primary focus, only a subset of these components, notably those consistent with the economic goals for management, namely the commercial tree component and its productivity, and may also include provisions to sustain additional ecosystem functions (sediment flows, riparian habitat) when required by regulatory statutes.

Given its focus on ecosystems, ecological forestry has an emphasis on management that emulates, to varying degrees, the outcomes of natural disturbance and development, as reflected in structure, composition, and processes. Timber-focused forestry, by contrast, is at its heart based

Table 2.1 *Fundamental principles for ecological forestry and timber-focused forestry. (Adapted from Franklin et al. 2018).*

Ecological Forestry	Timber-Focused Forestry
Maintains ecosystems and their array of structures, functions (processes), and biota	Maintains a subset of ecosystem structures, functions, and biota consistent with economic goals
Uses natural stand development models, including effects of disturbances, as the basis for silvicultural prescriptions	Based on agronomic models, e.g., plant spacing, weeding, fertilization, as the bases for silvicultural prescriptions
Values complexity and heterogeneity of ecosystem attributes	Values simplicity and homogeneity of structure and composition
Emphasizes ecosystem diversity and resilience to reduce major disruption risks	Emphasizes optimizing growth of crop species to reduce risks

on an agricultural model (Puettmann et al. 2008), with an emphasis on controlling plant spacing to optimize access to limiting resources, **weeding** and **thinning** to maintain desired spacing, reduction in abundance of **non-commercial** species, and perhaps application of fertilizer to increase **crop-tree** growth.

Timber-focused forestry emphasizes, to varying degrees, simplicity and uniformity of composition and structure as an outcome of using an agricultural system model, and because of the economic motivation for managing commercial tree species. It is unusual for landowners who manage forests (either actively or passively) primarily for economic reasons to retain an ecologically complex forest. Rather, timber-focused forestry emphasizes that part of the forest ecosystem that contributes most directly to achieving the desired economic objective, while trying to minimize the resources devoted to the part of the ecosystem that may detract from those returns (e.g., noncommercial tree species). This is not to say that non-commodity components of the ecosystem are deliberately ignored, they are just not usually the driver for determining the management (or silvicultural) approach. In contrast, ecological forestry places high value on and manages for complex and spatially heterogeneous distribution of ecosystem structures and composition within units of land, i.e., stands to landscapes, often without constraints on the return on investment associated with doing so.

Finally, by focusing on ecosystems, ecological forestry assumes that risks from disturbances are minimized by maintaining diverse species assemblages with their inherent resiliency of function. By contrast, timber-focused forestry reduces risk through avoidance of disruption with efforts to achieve rapid growth and shorten rotations; that is, it tries to minimize the amount of time the capital is exposed to risk.

Contrasting Ecological Implications

Ecological forestry and timber-focused forestry have different ecological implications resulting from their differing fundamental principles (Table 2.2). Among the most significant difference is how the two approaches view sustaining attributes over time; ecological forestry seeks continuity of structure, composition, and function across forest generations, while timber-focused forestry results in discontinuity of these attributes between the old and new forest, given the cyclical focus of often complete (or near-complete) harvest of the economically mature forest with rapid replacement by a new, vigorously growing crop (Figure 2.2).

By valuing complexity and heterogeneity as fundamental principles, and through emulation of natural models, ecological forestry results in greater complexity, heterogeneity, and species diversity in stands compared to timber-focused forestry. Perhaps not so obvious is that an end result of the ecological model is a greater diversity of age structures and successional stages (see Chapter 5) across the landscape, including old forest, but also early successional (pre-forest) stages. In contrast, timber-focused forestry results in a narrow focus on the young forest stage (Chapter 5), with few old trees or stands or early successional habitat on the landscape.

Table 2.2 *Ecological implications of ecological forestry and timber-focused forestry. (Adapted from Franklin et al. 2018).*

Ecological Forestry	Timber-Focused Forestry
Results in continuity of structure, function, and biota across forest generations	Results in discontinuity of most structures, functions, and biota across forest generations
Results in stand-level structural complexity and species diversity	Results in stand-level structural simplicity and reduced species diversity
Results in spatial heterogeneity of structure and composition in stands	Results in spatial uniformity in stands
Maintains a more diverse array of successional stages at larger spatial scales, including older trees and forests, as well as early successional ecosystems	Maintains age variants of young and mature stages across a landscape; may not include old trees and stands or extended duration of early successional stages, namely the preforest stage, as management goals

Figure 2.2 *Ecological and timber-focused forestry contrast in temporal dynamics of forest attributes. (a) Ecological forestry seeks to maintain structural and compositional attributes across generations, such as patches of trees (aggregated retention) in this trembling aspen harvest in Minnesota, USA. (b) Timber-focused forestry in its extreme lacks this kind of continuity, as in this harvest in a trembling aspen forest where no legacy structures or trees were maintained (Minnesota, USA).*

Contrasting Economic Implications

Ecological forestry and timber-focused forestry, in general, utilize different measures of economic performance, with ecological forestry focused on income (cash flow) over time and the preservation of natural capital, and timber-focused forestry focused on an acceptable rate of return (Table 2.3). Higher rates of return on timber capital will generally be calculated for timber-focused forestry than for ecological forestry, in part because of the higher levels of forest inventory maintained in ecological forestry and the retention of some slower growing trees and stands. Additionally, ecological forestry generally requires greater care and more involvement from professional foresters, and may have higher treatment costs due to its more complex

Table 2.3 *Economic implications of ecological forestry and timber-focused forestry. (Adapted from Franklin et al. 2018).*

Ecological Forestry	Timber-Focused Forestry
Generally seeks financial return from management, especially income over time, but return on capital is not the primary performance measure	Moderate to high return on capital is the primary performance measure
Leaves substantial recoverable values on site at time of harvest; some recoverable values retained have low rates of return	Minimizes recoverable values on sites at time of harvest
Higher cost per unit of output due to complexity of prescriptions and technical expertise needed for management	Lower cost per unit of output due to simplicity of prescriptions and management
May produce significant periodic income from the forest over time, but maintains higher investment level	Producing a periodic income from the forest over time usually a secondary consideration to return on capital, considering all investment opportunities
Lower risk due to emphasis on structural complexity, compositional diversity, and resilience	Higher risk due to its emphasis on structural and compositional simplicity and uniformity

Table 2.4 *Social implications of ecological forestry and timber-focused forestry. (Adapted from Franklin et al. 2018).*

Ecological Forestry	Timber-Focused Forestry
Retaining ecosystem components and processes contributes to effectiveness of collaborative efforts	Simplified forests limit potential effectiveness of collaborative efforts
Retaining ecosystem components and processes makes it easier to respond to changing social goals	Simplified forests limit ability to respond to changing social goals
Tends to increase management options in response to evolving needs and social expectations	Tends to reduce management options, limiting responsiveness to evolving needs and social expectations

prescriptions. However, ecological forestry should be less risky than timber-focused forestry because of its emphasis on diversity and resilience and it is generally better positioned to generate income from other ecosystem services, such as carbon storage.

Contrasting Social Implications

Ecological and timber-focused forestry have differing social implications (Table 2.4) stemming from their differing fundamental principles that in turn result in different ecological and economic implications. First, compared to timber-focused forestry, ecological forestry is inherently more amenable to collaborative processes for setting goals and for determining the actions needed to meet goals. This follows from an emphasis on retaining all parts of a forest—Leopold's precaution on intelligent tinkering again (Leopold 1949); in doing so, ecological forestry is inherently better able to accommodate diverse interests and perspectives and adapt to changing societal expectations. Secondly, this same focus on keeping all the parts, as well as creating complexity and heterogeneity of conditions, results in greater options for responding to uncertain futures. Put simply, with ecological forestry, there are more tree species and more structural conditions to choose from in stands and landscapes, enhancing the ability to respond to changing conditions as brought about by, for example, climate change, invasive species, or changing markets. Timber-focused forestry tends to reduce options by focusing on single species management in simplified, homogenous management settings.

Contrasting Silvicultural Considerations

More fundamental to our interests in this book is the fact that ecological forestry and timber-focused forestry lead to very different considerations for silviculture. These differences are summarized in Table 2.5 and addressed in more detail in subsequent chapters. Most fundamentally, ecological silviculture, with its focus on emulating natural forest development and the resulting complexity and heterogeneity of forest conditions, may require different and more complicated prescriptions than timber-focused forestry.

Ecological forestry may also require more professional expertise and more time to implement—keeping all of the parts takes extra effort. Implementing ecological forestry prescriptions may be a novel undertaking for many organizations. Collectively, there is much to learn about developing silvicultural approaches that address the objective of ecological forestry; this book is our contribution to that effort.

Relationship of Ecological Forestry to Other Ecological Approaches

Ecological forestry may share similar goals or principles with other conceptual approaches for forest management that are ecological in their intent, such as **close-to-nature**

Table 2.5 *General silvicultural considerations of ecological forestry and timber-focused forestry. (Adapted from Franklin et al. 2018).*

Ecological Forestry	Timber-Focused Forestry
Focuses on native species and genotypes that provide an array of ecological and other values	Focuses on species with desirable economic characteristics; may include tree improvement
Inclusive of uneven-aged, two-cohort, and single-cohort stands (with older residuals), as reflective of the natural model for a given forest type.	Inclusive of uneven-aged and even-aged stands, with narrowly even-aged stands common in plantation settings
Incorporates biological legacies into regeneration harvests, including in uneven-aged systems	Incorporating biological legacies is not a primary goal
Emphases development of complexity with intermediate treatments, e.g., thinning; intermediate treatments used to modify all structural layers	Emphasizes development of uniformity with intermediate treatments; actions focused on growing stock
Values decadence; retains defective trees and structures as habitat	Eliminates defective trees and structures
Utilizes longer rotations (with stand-replacement forest types) or long cutting cycles (with uneven-aged forest types)	Utilizes economic rotations on high-productivity sites or opportunistic removals on low-productivity sites
Return intervals for management (e.g., rotations, cutting cycles) are keyed to ecological criteria	Return intervals for management interventions (e.g., rotations) are keyed to growth or economic criteria
Landscape context of stand-level treatments focuses on maintenance of ecosystem processes and structures	Landscape context of stand-level treatments focuses on efficiency of harvest patch, road, and logging designs

forestry of central Europe (Bauhus et al. 2013), or the **complex adaptive systems** approach (Messier et al. 2013). While it is true that these utilize an understanding of natural models for silvicultural guidance, there are key differences for each compared to ecological forestry, as we define it in this book.

Close-To-Nature Forestry

Close-to-nature forestry is an approach that developed in Bavaria and has spread widely throughout central Europe. Key principles of close-to-nature forestry include: maintaining continuous forest cover through avoidance of clearcutting; use of medium to long rotations (>80 years depending on tree species); use of natural regeneration; and avoidance of intensive soil cultivation, mineral fertilizers, and pesticides. There may also be an effort to increase tree species diversity, particularly of areas formerly dominated by broad-leaved trees but now in conifer-dominated forests. Often, the aim is to create uneven-aged and structurally diverse forests (Bauhus et al. 2013).

These principles and aims are certainly in line with aspects of ecological forestry as we present it in this book. However, a key distinction is that close-to-nature forestry has as a focus the mature stage of forest development and thus a general lack of consideration of managing forests across all developmental stages, the importance of which we will discuss in more detail in Chapter 5. The outcome of this is an avoidance of large openings and, as a result, a lack of pre-forest and young forest habitat in the landscape (see Chapter 5). For instance, in Bavaria, the distance between retained trees can be no greater than twice the average tree height, resulting in openings that are generally less than 0.3 ha (Hickey et al. 2015). This favors shade tolerant species and limits opportunities for regeneration of light-demanding species, such as birches, oaks, and pines. More recently, this limitation has been recognized and some call for more variation in silvicultural systems to facilitate regeneration of light-demanding species (Bauhus et al. 2013; Brang et al. 2014). Another key distinction is that close-to-nature forestry does not have a clear focus on management of legacy structures, especially decadent and declining trees and deadwood, as does ecological forestry as we define it. Perhaps it is fair to say that close-to-nature approaches, while ecological in origin and aim, are a subset of the full array of considerations that comprise our ecological approach.

Complex Adaptive Systems Approach

Another ecological approach first discussed early in the twenty-first century involves considering forests to be complex adaptive systems (Messier et al. 2013). This is a conceptual way of thinking about forests and their management that acknowledges the inherent complexity of ecosystems, including their structure, organization, and interactions. Complex adaptive systems share an array of properties that can render them difficult or unpredictable to manage, e.g. non-linearity of relationship among components wherein small actions may have large consequences and strong coupling among components that can result in cascading failures. On the other hand, complex systems are characterized by a high degree of adaptive capacity, giving them resilience in the face of disturbances and perturbations. Recognizing forests as complex adaptive systems can empower foresters with a conceptual perspective that allows them to manage forests in the face of uncertainty (see Chapter 16), for example, with climate change, and emphasizes many similar components as ecological forestry, including the importance of system memory (cf. legacies) for enhancing ecosystem resilience. In summary, we view the complex adaptive systems approach as less of an ecological forest management model and more of a conceptual way of thinking about forests that is applicable regardless of the management goals, i.e., ecological or timber-focused.

Ecological Restoration

It is also instructive to think about ecological forestry and how it compares to the ecological restoration of forests. Broadly, **ecological restoration** engages in activities that assist ecosystems in the recovery of functions when they have been degraded, damaged, or destroyed, and that enhance the capacity of an ecosystem to adapt to change. Certainly, there are aspects of ecological forestry that can be construed as restoration, such as legacy management actions that increase the abundance of large snags as an important habitat element in forests where deadwood has been depleted. By contrast, a program to reestablish trees in old fields that have been degraded by past agricultural practices is an example wholly within the realm of restoration, but not ecological forestry. In general, ecological restoration is less holistic than ecological forestry, having in its purest form, a focus on restoring key ecosystem functions in degraded forests or former forests, with less of a focus on integrated ecological, economic, and social goals. Referring back to the fundamental principles of ecological forestry (Table 2.1), the importance of emulating natural developmental processes and the outcomes of these processes is a defining feature of ecological forestry that likely is not shared with ecological restoration. In short, ecological forestry has a wider set of goals and is relevant over a much wider set of circumstances than ecological restoration.

What to Remember about Ecological Forestry

From this chapter, a forester can surmise that ecological forestry is a malleable concept that is adaptable to different settings, but one that can prove invaluable for basing forest management on an ecological foundation of emulating natural disturbance and stand development, as opposed to the agricultural foundation of traditional timber-focused forestry. Also remember, ecological forestry is a conceptual approach to forest management, and not a management prescription; an approach that is built on the assumption that emulating natural developmental processes of forests provides a framework for managing ecosystem for sustainability and adaptability in the face of uncertainty. While ecological forestry by definition puts high value on ecological goals for management, economic and social considerations must still be part of the mix, although as we have suggested, in different ways and with different outcomes than timber-focused forestry. Finally, the ecological forester can consider the similarities and differences between ecological forestry and other ecological forest management approaches. This way they will recognize that ecological forestry is a comprehensive conceptual approach that considers all stages of forest development and readily leads to the formulation of silvicultural prescriptions to achieve objectives, as we will explore later in this book.

CHAPTER 3

An Overview of Silviculture

What is Silviculture, Classically?

Silviculture is often viewed as the earliest forest science, reflecting the long history of human civilization's interaction with forests to derive raw wood materials for subsistence (Puettmann et al. 2009). These interactions, along with shortages of wood at various times, forced an early recognition of the importance of understanding tree biology and its relationship to the environment when trying to perpetuate a "forest crop." As such, the field of silviculture evolved as a discipline focused on understanding forest tree and **stand** (Box 3.1) behavior in order to anticipate the outcomes of harvesting practices. Early definitions of silviculture reflected these origins, with the "production of wood crops" and "growing of trees for the greatest production of wood products" as early examples (Hawley 1921; Cheyney 1942). What these early definitions underscore is the objective-driven nature of the field of silviculture, with the early, prevailing emphasis on forests as a supply of raw materials leading to methods and techniques revolving around sustained wood production. This objective-driven focus has allowed silviculture to evolve in response to changing societal values and policies related to forests, including those related to wildlife habitat, recreation, and aesthetics. This evolution is best typified by modern definitions of silviculture, which generally describe the discipline as an art and science focused on controlling aspects of forest development, composition, and health to meet broad societal needs. Fittingly, the advancement of ecological sciences and our understanding of forest responses to natural disturbance has led to the genesis of ecological silviculture as an approach to meet current societal objectives linked with biodiversity conservation and ecosystem resilience, while also satisfying traditional objectives related to producing wood products. As objectives continue to evolve, so too will the art and science of silviculture.

Historical Operating Paradigms

The application of silviculture at its core has historically revolved around sustaining a flow of wood to meet societal demands. Reflecting this underlying doctrine have been four operating paradigms that generally apply to classical silvicultural approaches: *command and control*, *agricultural systems model*, *simplify and homogenize*, and *manage for removals*. These interrelated paradigms largely signify a longstanding focus on using silvicultural treatments to increase predictability of commodity outputs from the forest and associated approaches for doing so. While more natural systems approaches have long existed (see *Silviculture: An Ecological Approach* below), the abovementioned paradigms have been most prevalent in the application of silviculture around the globe.

Command and control represents an overarching operating paradigm focused on reducing ecosystem variability, including the impacts of natural disturbances, in an attempt to increase predictability or stability in commodity outputs (Holling and Meffe 1996). This paradigm is certainly not exclusive to the classical applications of silviculture and has been applied to numerous ecosystem settings from which predictable levels of resource extraction are desired (e.g., agricultural systems or game populations) and where deviations from stable conditions have negative economic impacts (e.g., flood regulation).

An outgrowth of the command and control paradigm is the application of an **agricultural systems model** in the design and implementation of silvicultural treatments. This paradigm involves a largely singular focus on the production of a wood crop through the design of silvicultural interventions and is particularly apparent when observing the inputs and approaches associated with forest plantation systems in production forestry. As with the production of annual crops for food production, plantation silviculture relies on uniform spacing of trees to accommodate planting operations and optimize efficiency of harvesting equipment (Figure 3.2). Rotation ages are based on maximizing returns on the high number of upfront investments for plantation establishment and tending, falling well short of any measures of biological maturity (Bettinger et al. 2009). While plantation silviculture typifies the agricultural systems model, even silvicultural systems based on natural regeneration retain elements of this paradigm through a largely

Box 3.1 The stand concept in classical and ecological forestry contexts

The **stand** concept has long been central to the practice of silviculture and has traditionally been defined as a group of trees that are relatively homogenous in age-class distribution, composition, and structure growing on a site of uniform quality (Helms 1998). Stands, as defined in this context, have served as the primary unit of forest management around the globe with stand-by-stand application of silvicultural treatments often serving to homogenize a given area and further differentiate it from adjoining areas on the landscape. Delineating landscapes into stands has largely been justified by the operational efficiencies it presents for organizing treatments and harvest scheduling, particularly in areas managed with even-aged systems; however, this has largely ignored the role of natural disturbances in shaping patterns of vegetation at fine to large scales. Stand management in this fashion has profoundly changed the natural patterns of fine-scale, within-stand heterogeneity, the dynamics of natural stand boundaries, and the multiscale nature of ecosystem properties and wildlife dynamics across a landscape (O'Hara and Nagel 2013).

Given these shortcomings, the application of the stand concept, as traditionally defined, is quite incongruous with the practice of ecological silviculture and its associated principles. In particular, ecological silviculture in practice embraces and encourages heterogeneity within and across populations of stands and relies on biophysical templates (e.g., soils, geomorphology) and, in some types of ecosystems, natural disturbance patterns to define the size and configuration of silvicultural treatments, particularly harvesting, across a landscape. As with classical silvicultural approaches, the delineation of areas as foci for scheduling and applying ecological silviculture treatments is critical for achieving long-term ecological forestry objectives; however, the ***use of the term "stand" to describe these units requires a broader interpretation to include forest areas that incorporate significant heterogeneity in structure, age, composition, and spatial patterns of trees.***

A good example of this broader interpretation put into practice comes by way of prescriptions used to generate within-stand heterogeneity reflective of natural woodland structure in Ponderosa pine forests, an ecosystem that was historically maintained by frequent surface fire (Figure 3.1). In the figure, fine-scale delineations represent differences in silvicultural prescriptions within larger stand boundaries, including different thinning intensities, reserve areas, and regeneration openings. For implementation, georeferenced prescription maps are displayed on handheld tablets and used by harvest equipment operators to efficiently execute these spatially complex stand-level prescriptions. Boundaries of the stand and the specific silvicultural treatments within it are viewed as dynamic and future treatments will be applied to enhance structural and compositional complexity across the stand, building off of previous treatments.

Ecological silviculture, where possible, should rely primarily on ecological criteria to delineate stands and, correspondingly, will often be applied in areas much larger than those delineated based on classical criteria that focus on identifying homogenous units. Admittedly, delineating stand boundaries using ecological criteria, such as underlying biophysical conditions or natural disturbance patterns, can be more challenging than classical approaches. This is particularly true where clear topographic or edaphic boundaries do not exist or historic stand-level management or past land-use has created homogenous structural and compositional conditions that obfuscate natural, underlying patterns in ecological conditions. At a minimum, boundaries should be large enough to allow for multiscale interactions from local tree neighborhoods to broader spatial extents and should be viewed as diffuse in nature in recognition of the importance of multiscale processes and population dynamics across stands and landscapes. In addition, stand boundaries should be considered as flexible and dynamic over time to account for natural and treatment-related changes in the spatial distribution of forest conditions (O'Hara and Nagel 2013).

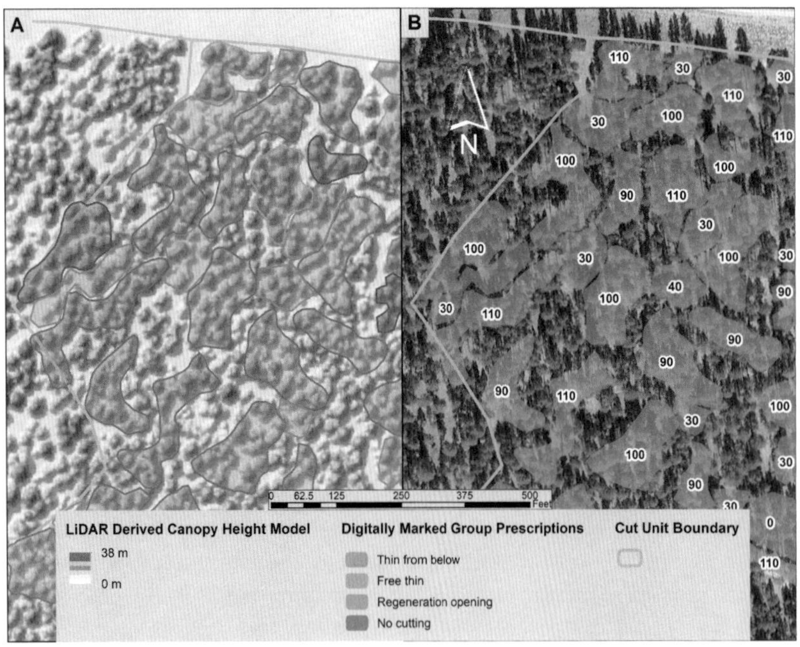

Figure 3.1 *Map depicting harvesting unit in an ecological forest restoration project for the Four Forest Restoration Initiative - Collaborative Forest Landscape Restoration Program on the Coconino National Forest, AZ, USA. A) Prescriptions for individual groups and interspace have been digitally "marked" based on a LiDAR-derived canopy height model to create spatially complex forest structural prescriptions that can be used by harvesting operators. B) In-cab harvesting operator map translating spatial pattern of interspace (area between groups where all trees are harvested except oaks and old-growth ponderosa pine trees), prescriptions (e.g., thin from below), and basal area targets (numbers inside marked groups, ft^2/acre) for individual groups. Image courtesy of Travis Woolley - The Nature Conservancy - Arizona.*

Figure 3.2 *(a) Plantation of longleaf pine in Georgia, (b) red pine plantation following the first thinning entry in Minnesota, (c) mixed red and white pine stand following shelterwood establishment cutting in Minnesota, and (d) northern hardwood forest following removal of canopy sugar maple as part of single-tree selection system. For (a) and (b), the agricultural systems model is reflected both in terms of uniformity of tree arrangement and crop-tree focus. For (c) and (d), the agricultural model is emphasized through singular focus on crop trees with competing hardwood species removed via selective herbicides in (c) and emphasis on crop-tree silvics in (d) through the creation of a limited range of gap sizes.*

crop-tree focus and associated treatments applied to selectively eliminate non-merchantable tree species or forms (Figure 3.2). Although optimal from a commodity-output standpoint, this crop-tree focus has been less than ideal for sustaining habitat elements such as cavity trees (Kenefic and Nyland 2007), and for sustaining the full array of canopy species in managed systems (Webster and Jensen 2007). For example, work examining long-term application of single-tree selection methods to sustainably produce sugar maples in northeastern North America have demonstrated that the consequences of this crop-tree focus has been a dramatic reduction in representation of other, less-tolerant canopy tree species, as well as important wildlife elements, such as snags, hollows, and downed woody material (Kenefic and Nyland 2007; Gronewold et al. 2010).

Another operating paradigm closely linked with the agricultural systems model, is **simplify and homogenize,** which underscores the focus of most timber-focused silvicultural treatments on creating conditions that generate uniformity in tree growth, quality, and value through simplification of naturally complex ecosystems. These structurally simple, homogenous conditions are generally sustained through top-down control of ecosystem attributes of economic interest, including favored species (crop trees), their lifespan (rotation), and spatial arrangement and density (to optimize access to resources). The long-held desire for simplification and uniformity in silviculture is best typified by the adjectives used in silvicultural terminology to describe stands that do not conform to standards of homogeneity in terms of size and age-class distributions, including "unbalanced" and "irregular" (Smith et al. 1997). Justifications for the use of such descriptors include pretenses on the inability of such irregular structures to produce a predictable and sustainable flow of volume; however, recent arguments have been made for the value of managing for irregular structures to achieve a diversity of economic and ecological objectives in managed areas (Raymond et al. 2009; O'Hara 2014).

The final paradigm, **manage for removals**, has been a logical focus of silvicultural systems given the early and continued emphasis on achieving societal objectives related to procurement of wood. With this paradigm, the primary focus of silvicultural treatments has been on encouraging the development of forest trees with the ultimate end goal of a final removal. Even with thinning treatments or regeneration methods using extended regeneration periods or continuous forest cover, an initial focus on how much is left behind (e.g., residual basal area following thinning or amount of overwood cover in shelterwood treatments) is accompanied by a final emphasis on when and how many mature trees will ultimately be removed. Although elements of classical silvicultural regeneration methods are based on regeneration responses to natural disturbance (Hawley 1921), the focus on managing for removals has ignored a primary element of disturbance processes, namely the abundance and importance of biological legacies following disturbance and their critical role in sustaining and enriching ecosystem structure and function (see Chapter 4). The advent of site-level management guidelines in many regions of the globe has forced the integration of retention trees as part of harvest operations historically focused solely on removals (Gustafson et al. 2012); however, even the individual trees selected to serve as legacies to satisfy these harvest guidelines reflect the prevailing emphasis on removals. In particular, non-merchantable species and smaller individuals are commonly selected as leave trees in intensively managed areas (D'Amato et al. 2009), reflecting an emphasis on maximizing profitability of removals while placing little attention on managing for what is being left behind (Figure 3.3).

The Major Silvicultural Systems and Associated Methods and Treatments

The general organizing structure classically used for applying silvicultural treatments are **silvicultural systems**, which are a planned program of treatments over the life of a forest stand to meet long-term objectives. Silvicultural systems are designed to reflect specific stand- and forest-level management objectives, as well as site-level conditions that may influence the degree to which a given silvicultural treatment may need to be applied (e.g., need for release treatments on more productive sites). The classification of silvicultural systems has historically been based on the primary source of regeneration relied upon for crop trees (seedling versus vegetative origin) and the age structures created and perpetuated through regeneration methods employed (even-aged, two-aged, uneven-aged). Although silvicultural systems are often presented as rigid plans of treatments over the life of a forest, silviculturists have long emphasized the notion that systems be treated as a working hypothesis with the intent of application in an adaptive management framework (Smith 1962). Unfortunately, this emphasis has often been disregarded in favor of "cookbook" silvicultural guides for managing a given forest type across a region (e.g., Benzie 1977), with corresponding negative results on the range of forest conditions maintained and the willingness of practitioners to step outside of the box to experiment with alternative methods for a given ecosystem.

Figure 3.3 *Leave trees following clearcut harvests in (a) paper birch and (b) white spruce-dominated forests in northern Minnesota. Note emphasis on smaller and non-merchantable individuals.*

Silvicultural systems generally include three main phases and associated treatments: 1) regeneration, 2) tending, and 3) harvest (Matthews 1989; Nyland 2007), with a strong interdependence among these phases, particularly the regeneration and harvest phases. There are several excellent texts on classical application of silvicultural systems and associated methods (Smith et al. 1997; Matthews 1989; Nyland 2007) and the remainder of this section is meant to provide a general background as a foundation and point of contrast to ecological silviculture.

Harvests within the context of silvicultural systems represent the application of **regeneration methods** designed to initiate a new cohort or cohorts of trees. In contrast with a strictly commercial view of timber harvesting, harvests associated with regeneration methods—while accounting for economic practicality—are designed to create favorable environmental conditions for the reproduction of desired tree species. Beyond exerting control over future species composition, these methods also dictate forest age structure and range from **even-aged methods** such as clearcutting, seed-tree, and shelterwood systems to **uneven-aged methods** like single-tree and group selection. **Two-aged** variants of even-aged methods, in which a portion of the initial cohort is retained through the next rotation, have also long been recognized, giving rise to systems reflecting an extension of even-aged approaches (e.g., clearcutting, seed-tree, and shelterwood with reserves; coppice with standards).

Historically, the primary factors affecting the selection of regeneration methods have included ecological characteristics of the desired tree species (e.g., shade and exposure tolerance) and economic rotation of desired wood products (Matthews 1989). **Clearcutting methods** are the most widely applied approach and remove the entire mature stand in a single entry to regenerate shade intolerant, exposure tolerant species, such as southern pines and Douglas-fir. Although variants of this method, such as strip clearcutting, rely on adjacent, uncut areas to provide seed for harvested areas, this method is primarily regenerated with artificial planting of tree seedlings given the difficulties with securing adequate natural regeneration across large open areas, particularly in regions such as the Pacific Northwest and southeastern U.S. Correspondingly, this approach is often highly reflective of an agricultural systems model in its application. **Seed-tree methods** favor a similar suite of species as clearcutting methods, but retain mature trees, either singly or in groups, across harvested areas to serve as a seed source for the new, even-aged stand. While methods based on clearcutting have been suggested as analogs for natural stand-replacing disturbance, seed-tree methods actually provide a closer, albeit still distant, surrogate given their incorporation of a critical driver of post-disturbance cohort establishment: surviving seed-tree legacies (Figure 3.4).

The broadest, most diverse group of regeneration methods are **shelterwood methods**, which range from conventional, even-aged shelterwoods to two-aged and uneven-aged irregular shelterwood variants, such as extended and continuous cover irregular shelterwood methods

Figure 3.4 *Jack pine-dominated forests following (a) stand-replacing fire, (b) clearcutting (and planting), and (c) establishment cutting as part of the seed-tree regeneration method.*

(Raymond et al. 2009). This diversity of terms speaks to the flexibility of this approach in accommodating both a wide range of species regeneration requirements and site conditions. The common element of these variants is that overstory trees are partially or completely removed in a series of harvests in a manner that creates favorable environmental conditions in the understory for the establishment of desired tree species. Species generally favored with these methods are mid-tolerant to tolerant tree species whose primary regeneration mechanism is reliance on advance regeneration. As such, the partial, successive removals of the canopy with shelterwood methods is intended to mimic "top-down" disturbances, such as windstorms and insect defoliation, that release previously established regeneration.

Selection methods are uneven-aged regeneration methods in which scattered individuals or groups of mature trees are removed to initiate a new cohort. These selection cuttings are applied every five to twenty-five years, depending on residual structure and species composition objectives, to maintain an uneven-aged forest condition at the stand scale. Selection methods are most commonly practiced with shade tolerant species; however, a range of species tolerances can be maintained with these methods through the use of larger group openings and appropriate stocking control. The distribution of group selection cuttings generally resembles area regulation at the stand scale, whereas harvest levels and patterns associated with single-tree selection are generally guided by target diameter distributions (see *BDq method* in Chapter 8).

The final group of regeneration methods are **coppice methods**, which rely primarily on vegetative reproduction for regeneration of a new cohort. These methods represent the simplest and oldest regeneration method and are primarily used for producing fuelwood given the poor stem form of coppice-origin trees. Although the application of coppice methods is generally less common relative to other regeneration methods for most regions, large areas of the boreal and sub-boreal portions of North America are managed with this approach to favor aspen species for the production of pulp, paper, and manufactured wood products. Often, two-aged variants of the coppice approach are used (**coppice with standards**) in which individuals or groups of trees are maintained through multiple coppice rotations to serve as a source of habitat or higher quality products.

Site Preparation Treatments

The successful establishment of regeneration depends not only on availability of propagules and creation of suitable environmental conditions, but also on the presence of favorable seedbeds (Roe et al. 1970). Most temperate and boreal species require competition-free, stably-moist seedbed conditions for germination and establishment. In natural systems, these seedbeds are provided either by well-decomposed logs on the forest floor (Franklin et al. 1987; Bolton and D'Amato 2011) or through exposure of the mineral soil via natural disturbances such as windthrow and fire (Hutnick 1952). In order to enhance the effectiveness of regeneration methods, silvicultural systems often include the application of **site preparation** treatments during the harvest and regeneration phases to increase the availability of favorable seedbed conditions for the desired tree species and facilitate planting efforts when artificial regeneration is being used. Common site preparation techniques include mechanical methods in which large machinery is used to create desired seedbed conditions or move logging slash to facilitate planting efforts, chemical methods where herbicides are applied to reduce the abundance of interfering vegetation, and prescribed fire (Figure 3.5).

Tending Treatments

After the regeneration phase, additional treatments are often applied to encourage the development of desired regeneration and enhance the growth, quality, and health of crop trees over the course of stand development. These treatments are collectively referred to as **tending** or **intermediate treatments**, and in their typical application are critical elements of the *command and control* and *simplify and homogenize* paradigms. Included in this broad group of treatments are **release treatments**, which focus on reducing or eliminating the effects of competing vegetation (including less desirable tree species) on crop trees early in their development. These treatments are used to control forest successional patterns after harvest disturbance in a manner that shifts a large share of site resources toward desired species.

As stands develop and levels of inter-tree competition increase, **thinning treatments** are often applied to reduce overall stand density and increase resources available to better-formed, often larger trees. The primary goals of thinning include accelerating growth on crop trees and increasing stand-level wood production through the removal of individuals with a high mortality risk. Thinning treatments primarily target smaller and poorer-formed individuals for removal and often serve to increase levels of homogeneity in tree size, form, and spatial distribution in a stand. The intensity and frequency of thinning entries are based on regional, species- or forest type-specific **stocking guides** or **density management diagrams** which draw upon plant population biology principles related to maximum size-density relationships in self-thinning populations (Silvertown and Lovett Doust 1993). Such diagrams are used to maintain forest stocking conditions low enough to provide adequate resources to accelerate individual crop tree growth and reduce production losses to self-thinning

Figure 3.5 *Mechanical and chemical site preparation techniques and the seedbed conditions they aim to emulate to enhance regeneration success. Exposed mineral soil seedbeds naturally created by (a) windthrow and (b) surface fires are generally created using mechanical site preparation techniques such as (c) blading, trenching, or scarifying, while seedbed conditions free of competition naturally created by surface fires and wind are typically created using (d) broadcast or spot herbicide applications. Photo credit (d): Rick Klevorn.*

mortality, yet dense enough to fully utilize site potential for stand-level growth (Long 1985). Chapter 8 includes more details on thinning approaches.

A final suite of intermediate treatments, improvement, sanitation, and salvage cuttings, focus primarily on reducing biological legacy elements in managed stands through embodying the *manage for removals* paradigm. **Improvement cuttings** are applied during the same period of stand development as thinning treatments; however, the primary focus of improvement cuttings is removing non-merchantable tree species or poorly formed stems to increase overall stand quality from an economic standpoint. Although these treatments are viewed as an "improvement" from a commercial growing stock perspective, these removals serve to reduce and degrade key elements associated with biodiversity conservation in managed forests, including diversity of overstory trees (Palik and Engstrom 1999) and dying and dead trees (McComb and Lindenmayer 1999). **Salvage cutting**, which removes dead or dying trees resulting from non-harvest-related disturbance to recover economic value, also reduces detrital legacy elements in a similar manner to improvement cuttings; however, these treatments often extend to removals of logs recruited into downed woody material pools. Finally, **sanitation cuttings** focus on reducing the impacts of injurious insects and diseases on crop trees either through anticipatory harvests that remove potential host trees or deliberate removal of infested individuals to reduce spread. The singular focus of salvage cutting and sanitation cutting on economic objectives generally results in these operations occurring with little regard for long-term ecological objectives, including retaining some dead and dying trees on site (Box 3.2).

Box 3.2 **Salvage logging in a twenty-first century context**

The desire to command and control forest development through active forest management has been a longstanding central operating paradigm associated with the practice of silviculture. Despite this intent, natural disturbances invariably affect managed forest ecosystems, killing mature trees and presenting a near-term economic loss for landowners. As a result, salvage logging of dead trees following natural disturbance to recover economic value has been a common practice for centuries. Recent studies of the near- and long-term ecological impacts of this practice have highlighted that salvage operations often negatively affect forest regeneration, native biodiversity, and forest carbon stores due in large part to the compounding effects of logging disturbance following the original natural disturbance (Donato et al. 2006; Lindenmayer and Noss 2006; Bradford et al. 2012). The primary target of salvage operations are the biological legacy elements, namely standing and downed dead trees, which are critical for sustaining innumerable forest ecosystem processes. Although considerable work around the globe has been devoted to developing best management practices for retaining these legacy elements when live-tree harvests are occurring, these same guidelines are generally not followed during salvage and sanitation cuttings (Schmiegelow et al. 2006). Given the recognized importance of these legacies to ecosystem sustainability, there is a need for guidelines and policies that weigh the often marginal economic returns from salvage logging (Smith et al. 1997) with the long-term wealth of ecological benefits provided by retaining a portion if not all of these legacies on site. Within an ecological silviculture context, such retention systems could be approached in a similar manner as variable retention harvests to provide for detrital legacies at varying spatial scales across a site undergoing salvage logging.

Silviculture: An Ecological Approach

Historical Origins

Applying ecological principles to guide the development of silvicultural prescriptions is not a new advancement in forestry, as many early silviculturists recognized the importance of developing silvicultural treatments and prescriptions reflective of the general ecology of a given species or forest community (Hawley 1921; Toumey 1927). In most cases, the focus of these early "ecological silviculturists" was on forest vegetation development as it related to site factors and the environment. For example, the earliest advocates for a more nature-based, ecological approach were silviculturists in Germany and Switzerland who during the mid- to late-nineteenth and early twentieth centuries began advocating for managing locally adapted species and mixed-species stands using selection-based silvicultural regimes as an alternative to widespread application of even-aged monocultures of conifers, such as Norway spruce (Gayer 1886).

As the fields of ecology and silviculture developed in the early twentieth century, principles associated with plant succession and climax theory became a common guiding element to nature-based silvicultural thinking. In particular, one of the earliest recognized forms of nature-based forestry was the **Dauerwald** approach, which reflected a strong emphasis on permanently maintaining mature forest cover in areas being managed for wood products (Troup 1927). Although this approach did not advocate for the use of any particular silvicultural system, its rejection of systems based on clearcutting and focus on retaining continuous forest conditions often led to the application of selection and irregular shelterwood systems to achieve these objectives. Consistent with the theories of plant succession being developed by Frederic Clements and other plant ecologists during this time period (Clements 1916), advocates for the Dauerwald approach justified this management philosophy on the premise that the forest was a living, highly interconnected organism (Möller 1922; Bauhus et al. 2013). Successional theories have been refined since these early formulations, but the ideas underlying much of the Dauerwald approach reflect an early example of viewing silviculture from an ecosystem, rather than crop-tree, management perspective.

The development of forestry practices and silvicultural systems in North America during the late nineteenth and early twentieth centuries drew heavily from European practices. As such, early North American silviculture was largely concerned with converting historically unmanaged systems into highly regulated and predictable forests (in relation to wood production) (Hawley 1921), a condition common across Europe at the time. Interestingly, a silvicultural philosophy not widely imported from Europe was the Dauerwald approach, with the exception of its application in southern New England (Box 3.3).

Ecological Silviculture in a Twenty-First Century Context

While there have been elements of ecological silviculture scattered throughout the evolution of silviculture as a practice, this historical antecedent has generally been restricted to a given region or management system (e.g., multi-aged methods and the Dauerwald approach). The remainder of this book is devoted to outlining both the overarching principles of ecological silviculture in its modern context and its application to a diversity of forest types and management

Box 3.3 **Early roots of ecological silviculture in North America at the Harvard Forest**

The Harvard Forest in Petersham, MA, USA, was established as Harvard University's forestry school in 1907 under the leadership of Dr. Richard Fisher. Like other American forestry schools, early research and teaching efforts at the Harvard Forest were focused on refining European practices to develop sustainable silvicultural systems for forests in southern New England. After years of unsuccessful attempts to establish conifer plantations in areas that historically supported hardwood-dominated forests, a unique perspective on ecological silviculture was developed and formalized here in the early 1930s that drew heavily from the Dauerwald principles developed in Europe decades earlier. In what was labeled a "naturalistic" approach to silviculture by other forestry schools in the region (Steen 1990), foresters and scientists at Harvard Forest focused on studying natural processes, particularly those found in the few remaining old-growth forests in southern New England to serve as a means for informing the development of silvicultural practices. This is clearly reflected in the writings of forest scientists at the Harvard Forest during this period, including in Al Cline and Stephen Spurr's (1942) monograph on old-growth eastern hemlock and white pine forests:

> *"the soundest basis for the development of the art of silviculture lay in an understanding of the 'natural' forests of the region—the complex of trees, shrubs and herbaceous plants which would occupy a given piece of ground when free from interference from man."*

An outgrowth of this period of silvicultural exploration at Harvard Forest was the first known application of the term "ecological forestry" in the forestry literature in the early 1940s (Spurr and Cline 1942), over four decades before its global popularization (Franklin 1989). The emphasis of this approach was similar to the Dauerwald approach and advocated for selection methods to protect the site and perpetuate multi-aged, mixed-species stands composed of shade-tolerant species; however, this ecological silviculture also included recommendations to vary the intensity and frequency of disturbance to perpetuate a mixture of less tolerant species on a site using an understanding of forest succession patterns and natural disturbance regimes. Correspondingly, many of the scientific methods used for studying and documenting natural stand developmental patterns and dynamics (namely forest reconstruction techniques) were developed and refined at the Harvard Forest from the 1920s through the 1950s (e.g., Marshall 1927; Stephens 1955; Henry and Swan 1974). An important distinction between this early practice of ecological silviculture and the contemporary approach described in this text and elsewhere is that the focus of early ecological silviculturists was on natural models of regeneration to achieve desired tree composition goals (Spurr and Cline 1942). Although this remains a core component of modern ecological silviculture, an equal level of emphasis is placed upon integrating other elements of natural disturbance regimes into silvicultural prescriptions, namely structural legacies, to fully reflect natural disturbance models and sustain core ecosystem functions and biodiversity.

settings across North America and beyond. Given the commonalities in the ecological processes that influence natural forest development around the globe, the intent of the remainder of this text is to provide a level of understanding that allows for the application of ecological silviculture in any geographic locale and forest type. In light of the broad suite of objectives and complexity of stressors affecting modern forest ecosystems, ecological silviculture provides a framework and path forward for continuing to produce commodity outputs from forest ecosystems, while balancing diverse social objectives and reinforcing ecosystem resilience in an adaptive framework.

Relationships to Classical Silviculture

Practicing ecological silviculture does not imply an abandonment of the centuries of silvicultural research and experience that have come before this text, as many of the same terms and tools developed from classical silviculture are often applied to achieve ecological silvicultural objectives. For example, crown thinning and other thinning methods have been used to meet objectives related to accelerating the development of large trees so as to increase representation of a live-tree structural condition characteristic of old-growth forests (Singer and Lorimer 1997). Similarly, site preparation treatments, including prescribed burning and mechanical scarification, are often used in conjunction with followup release treatments to favor the establishment and development of ecologically and economically important tree species (Prevost et al. 2010).

Silviculture as a discipline has always called for the artful application of these classical tools to meet new objectives and adapt to new ecological and economic conditions. At the same time, there has been a countervailing and inflexible doctrine in the field regarding the need for rigid adherence to standard terminology and practices. This rigidity of terminology has recently been criticized as constraining how truly artful silviculture as a discipline can be in responding to new objectives and challenges when adhering to inelastic classifications of silvicultural methods (O'Hara 2014). Correspondingly, there are several terms and approaches associated with ecological silviculture that, although they contain elements of a classical silvicultural approach, differ greatly in both their application and objectives. These include **legacy trees**, **variable retention harvests**, **variable density thinning**, and **recovery periods** (Table 3.1).

Table 3.1 *Crosswalk between commonly used terms in ecological silviculture and classical silviculture*

Ecological Term	Definition	Classical analogues	Similarities between terms and their intent	Differences between terms and their intent
Retention tree	Canopy tree retained on site during a regeneration harvest, often to live out its full lifespan, to serve as a source of mature forest structure and associated functions (i.e., seed sources, mycorrhizae)	Reserve trees, standards	All are examples of mature canopy trees retained during regeneration harvests	Classical analogues are typically left as part of a deferred harvest and ultimately may be harvested. When reserved for economically motivated objectives, trees with the greatest potential to increase in financial value are left, as opposed to retention trees which are primarily left for their ecological value and may include "overmature", decadent, and declining trees.
Variable retention harvest	Retention of living and dead trees (retention trees) during regeneration harvest in a range of spatial patterns *(dispersed and aggregated)* to serve as a source of mature forest structure and function in the post-harvest community	Irregular shelterwood	Both contain a range of overstory tree retention patterns following initial regeneration entries	The traditional purpose of these spatial arrangements are to influence the regeneration process of midtolerant and tolerant species. These retained trees are ultimately harvested after successful regeneration is achieved. The primary objective of variable retention harvests is to retain structural legacies to sustain ecological processes and biological diversity from the previous stand when regenerating a new cohort.
Variable density thinning (VDT)	A thinning approach in which the density of tree removals is varied across a stand or management unit to increase heterogeneity in horizontal and vertical structure. With VDT, thinning levels often range from small group removal, to standard thinning with individual tree removals, to patches with no removal. Applications of VDT have been referred to as the "skips and gaps" or "Individuals, Clumps, and Openings" (ICO; Churchill et al., 2013) approach.	(a) Thinning (crown, low, dominant, free, row) (b) Single, group, and patch selection; irregular shelterwood	(a) Both accelerate development of large tree structure (b) All represent multi-age regeneration methods	(a) The term "thinning" in a classical context indicates no attention is being directed toward regenerating a new cohort, whereas VDT explicitly focuses on increasing structural heterogeneity, including recruiting new cohorts in areas containing group removals. In addition, classical thinning strives to create homogeneity in mature tree structure and density across a site, whereas the focus of VDT is to create spatial heterogeneity to influence development of canopy trees, as well as understory and midstory components (b) For selection methods, the primary focus is on the uniform application of these methods across a stand to generate regulated conditions and sustained timber flows. Irregular shelterwoods, another multi-age system, may also generate stand-level heterogeneity in mature tree structure similar to VDT, particularly with continuous cover variants (Raymond et al. 2009); however, permanent reserve patches are not traditionally a component of these systems, as they may be with VDT
Recovery period	Period of time between regeneration harvests of sufficient duration to allow for development of mature forest structure at tree, patch, stand, and management unit scales	Rotation; cutting cycle	All represent intervals of time between regeneration harvests in a forest	The criteria for establishing rotation and cutting cycle length differs from recovery period. With even-aged systems, rotation lengths are based on the age that maximizes economic return on investment or average levels of stand productivity (MAI). For uneven-aged systems, cutting cycle lengths are generally associated with the severity of harvest entries (residual basal area retained) and stand growth rates needed to meet an operable harvest. Both rotation and cutting cycle length are generally fixed for a landscape (even-aged systems) or stand (uneven-aged systems), whereas recovery periods vary to reflect the range in natural disturbance return rates across landscapes and over time and on forest development processes. All of these will vary based on site productivity and associated rates of tree growth and forest development.

Ecological Operating Principles

As with the differences in terminology and objectives described above, so too are there differences in operating principles for ecological silviculture relative to the operating paradigms that largely guide classical silviculture. These *foundational* principles are: 1) continuity, 2) complexity/diversity, 3) timing, and 4) context. ***Continuity*** is the provision for *continuity in forest structure, function, and biota between pre- and post-harvest ecosystems* during regeneration harvests. In ecological silviculture, interventions in forests, such as those associated with regeneration harvests, are designed to provide for significant continuity in forest structure, function, and composition between the old and new forest. This emulates a key element of natural disturbance regimes by retaining biological legacies from the predisturbance forest (see Chapter 4) with the silvicultural goal of providing for significant continuity in structural, functional, and compositional elements between forest generations.

A common application of this principle in ecological silviculture is the selective retention of structures, organisms, and patterns at the time of forest harvest, sometimes referred to as variable retention harvesting (Franklin et al. 1997; Franklin et al. 2007). Retention can focus on individual structures, such as large live and dead trees (snags) and downed logs on the forest floor, or on small patches or islands of intact forest. It can also focus on retention of specific species in the overstory or understory or broad functional groups (e.g., conifers or hardwoods). Retention may also focus on protection of ground layer plant populations or patches of advance regeneration. Retention may be either distributed widely through the harvest unit (***dispersed retention***), concentrated in intact forest patches (***aggregated retention***), or at the gap-level (***gap-level retention***) in the context of uneven-aged systems (Figure 3.6). Each approach has its ecological, economic, and cultural advantages (see Table 8.4 in Lindenmayer and Franklin 2002), so silvicultural prescriptions will often include elements of both (Figure 3.7).

Figure 3.6 *(a) Dispersed and (b) aggregate retention following clearcut harvesting in aspen mixed-wood forests in northern Minnesota. Within-gap retention can be applied to group selection openings in uneven-aged systems to retain important structural elements at the gap-level, as in these 0.1 ha group selection openings in northern hardwood forests in (c) Massachusetts and d) Michigan.*

Figure 3.7 *(a) Variable retention harvest in red pine-dominated forests in Minnesota in which a combination of (b) dispersed and (c) aggregated retention patterns are being applied.*

Complexity/diversity reflects the need to *create and maintain structural complexity and species diversity, including spatial heterogeneity of conditions at multiple spatial scales* through all silvicultural interventions (Table 3.2). Ecological silvicultural interventions are designed to manipulate established forests to create, restore, and/or sustain desired conditions of structural complexity and species diversity. Snags, logs, and decadent features of living trees and old trees are critical elements of forest complexity that are included in ecological silviculture prescriptions. This is most often accomplished by creative application and modification of classical silvicultural approaches, including thinning and two-aged and uneven-aged regeneration methods, to accelerate and restore a diversity of forest structural conditions and provide regeneration opportunities for a wide range of tree species. Other tools that may be utilized include prescribed fire, planting, artificial creation of habitat (e.g., cavities or killing of trees to provide snags or logs), pruning, and selective use of herbicides to combat invasive or noxious species.

Table 3.2 *Attributes associated with complex and simple forest conditions. Promoting and maintaining complex forest conditions is a central principle in the design of ecological silviculture systems.*

Complex	Simple
Multiple tree species in canopy layer	Single species
Multiple age cohorts	Single cohort
Wider size and age ranges	Narrow ranges; young trees
More noncommercial species	Few species
Understory plant populations	Depleted plant populations
Abundant tree-derived structures	Few tree-derived structures
Abundance of coarse, woody material across range of decay states	Low abundance and representation of advanced decay states
Spatial heterogeneity	Spatial homogeneity

Creation and maintenance of heterogeneous (nonuniform) spatial patterns of structure and composition is an additional and important element in structurally complex forests. Forest ecosystems that exhibit both a diversity of individual structures and complex spatial arrangements of those structures can be viewed as "niche rich" as they provide the diversity of conditions, such as structures and microclimates, needed to sustain high species and functional diversity. This was historically recognized in the context of tree species through the development of irregular shelterwood systems (Troup 1928); however, the principle of *complexity/diversity* extends beyond the regeneration layer and is more broadly focused on restoring and sustaining ecosystem complexity and species diversity. Correspondingly, creation of spatially heterogeneous outcomes is a common objective in ecological silviculture and is often achieved in developing stands using variable density thinning prescriptions (Table 3.1). Variable density thinning

(VDT) can take a variety of forms (see Chapter 8), such as skips and gaps VDT (Carey 2003; Harrington et al. 2005), but a common objective of all VDT approaches is to hasten the development of structural heterogeneity in simplified stands by altering the intensity of thinning across space, such that resulting tree densities vary spatially. Specific targets for the spatial pattern of thinning/density can come from reference systems (e.g., Churchill et al. 2013) or from empirical knowledge of the relationships between the pattern and desired functions, such as the habitat needs of key species, fire resistance, snow retention, understory development, tree regeneration, or other objectives. In many cases, VDT approaches can be combined with other prescriptive elements, such as crop tree release, to enhance rapid development of trees of larger diameter or with larger branch systems.

Timing refers to the importance of applying *silvicultural interventions at ecologically appropriate time intervals.* The use of ecologically based time intervals is a broad consideration for ecological silviculture that includes both the timeframes necessary for recovery of forest ecosystem structure and function following a major treatment (i.e., some minimal time interval that is required), as well as the maximum interval that should be allowed between an intervention (e.g., prescribed burning). Biological measures, such as the culmination of mean annual increment (CMAI) or economic criteria such as maximum net present value, have historically been used to dictate time intervals for silvicultural interventions, such as regeneration harvests; however, determining time intervals for ecological silviculture is often more complex because considerations are not limited to timber production or financial returns. For example, in setting intervals between regeneration harvests, ecological foresters have to consider the time needed for specific structural elements and functional attributes to recover and operate at desired levels, e.g., development of a population of larger diameter trees or large well-decayed snags (Figure 3.8).

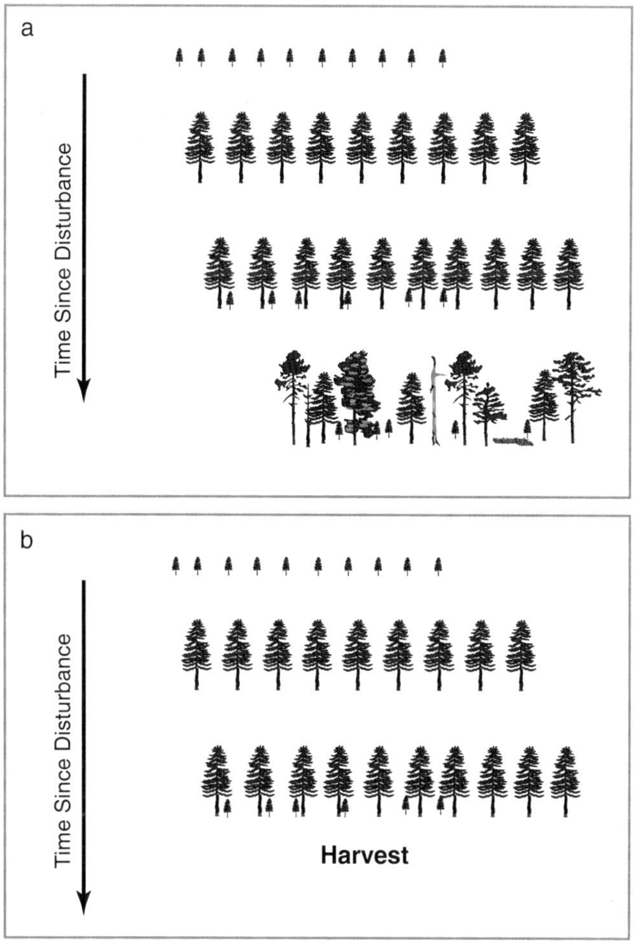

Figure 3.8 *Conceptual representation of the importance of timing for the development of structural complexity. (a) A young post-disturbance forest begins with simplified structure and composition (i.e., no biological legacies) and over time develops significant structural complexity due primarily to tree growth, mortality, and small-scale canopy disturbance. (b) In many classical forestry approaches, the forest is harvested before significant compositional and structural complexity has had time to develop.*

Retention harvesting is one ecological silviculture approach that allows for the retention and creation of mature forest structures or structural conditions for multiple rotations, rather than having to recruit these following each harvest, as would be necessary under classical even-aged silvicultural systems. In many respects, retention approaches greatly reduce ecological concerns over appropriate rotation age, since significant elements of the pre-harvest forest are partially or fully retained allowing for greater flexibility in setting rotations while sustaining desired structures or functions.

Finally, ***context*** underscores the importance of *planning and implementing silvicultural interventions in the context of how these actions accumulate to influence the structure and function of landscapes.* While the first three principles of ecological silviculture guide the development of silvicultural treatments and prescriptions at the stand or harvest-unit level, this final principle involves the integration of stand-level silvicultural activities at larger spatial scales. For ecological silviculture, landscape-level silvicultural planning emphasizes spatial heterogeneity through the creation of diverse stand conditions. Developing a better appreciation of the landscape-scale context of silviculture will be covered in Chapter 9. In short, the spatial heterogeneity generally sought on larger landscapes managed according to ecological silviculture incorporates a diversity of forest developmental stages from openings occupied by pre-forest or early successional vegetation to maturing and old forest ecosystems of diverse composition and structural complexity. Disturbances of either human or natural origin will be necessary to initiate the sequential developmental stages characteristic of forest ecosystems subject to partial or complete forest replacement disturbances (see Chapter 4).

Moving Forward

The principles outlined above and the processes they strive to emulate and sustain will be explored in greater detail throughout the remainder of the text, but you can see from examining them in the context of different management objectives (Table 3.3) that they are widely applicable, regardless of objective. What is critical to remember as you proceed through this book is that ecological silviculture, like the classical practice it has evolved from, is not meant to be viewed as a static set of practices that are to be generalized across forest types and regions. Ecological silviculture, by default, is about matching the prescription to the site, ecosystem, and objectives in an adaptive framework that can anticipate and accommodate biological and socioeconomic changes. This includes recognizing the biological constraints for some forest types in achieving objectives associated with developing and maintaining high degrees of structural and compositional complexity (e.g., lodgepole pine versus mixed mesophytic hardwood forests). Many of the emerging challenges and recommended management approaches associated with global change draw heavily from the principles of ecological silviculture (Puettmann et al. 2009), yet unforeseen future issues will require constant innovation and evolution of practice. The oft-repeated yet rarely internalized categorization of silviculture as an "art and science" reflects the long held intent for this discipline to embrace and encourage creative, adaptive approaches to meet management objectives. As the remainder of this text will detail, ecological silviculture provides an arena within which the artistic elements of silviculture can be fully explored and expressed in largely unprecedented ways.

Table 3.3 *Example outcomes of application of the four principles of ecological silviculture to priority objectives for management.*

Principle	Commodity productivity	Biodiversity conservation	Global change resilience/adaptation
1) Continuity	• Opportunities for natural regeneration of a range of species • Larger high-value products	• Life boating of species requiring mature forest conditions • Greater diversity of food/energy sources from canopy species • Large snags/deadwood for saproxylic and cavity nesting species	• Options for regeneration in face of uncertainty • Amelioration of harsh environmental conditions ○ regeneration safe sites (shaded understory, decomposed wood) • Conservation of genetic diversity

2) Complexity/ Diversity	• Opportunities for multiple entries (outputs)
	• Diverse product mix
	• High-quality products (resulting from natural pruning, training)
	• Multiple opportunities for natural regeneration of desired species
		• Diversity of habitat niches
		  ○ tree size classes
		  ○ deadwood decay classes
		  ○ live-tree spatial conditions
		  ○ tree, shrub, understory species
			• Reduced vulnerability to disturbance
			  ○ spatial variability in fuels
			  ○ heterogeneity in wind risk (diverse heights)
			  ○ heterogeneity in potential host species (insects/disease)
			  ○ heterogeneity of tree sizes (host preferences, stress tolerance)
			• Multiple Recovery and developmental pathways
			  ○ diversity of seed sources
			  ○ advance regeneration
			• High levels of onsite mitigation potential (carbon storage)

3) Timing	• Higher-value products • Multiple entries (outputs) • Seed source over extended periods • Multiple species and lifespans (diversity of products/harvests over time)	• Opportunity for multiple life cycles for species with slower development • Habitats for large tree specialists (live and dead trees)	• Long-term maintenance of options for adaptation from current overstory species • Long-term amelioration of extremes in understory conditions • Reduced likelihood for compounding influence of harvesting with other stressors/disturbance • Accumulation of large onsite carbon stores
4) Context	• Diverse portfolio of products and potential harvest entries • Lower risk from changing market conditions	• Connectivity across landscapes and habitat gradients (e.g., riparian to upland, travel corridors) • Refugia at multiple scales • Diversity of structures/ composition at landscape-scale	• Reduced risk from landscape-scale stressors (drought) and disturbance (insects, fire, wind) • Greater options for adaptation potential at broad scales • Greater range of regeneration conditions for new species due to localized and landscape-scale heterogeneity in structure

PART II

Forest Dynamics Relevant to Ecological Silviculture

CHAPTER 4

Natural Disturbances and Biological Legacies

Foresters use an understanding of **natural disturbances** as models for regeneration systems (see Chapter 3) in broad conceptual ways. For example, **even-aged** regeneration systems are often described as analogs for stand-replacement disturbances, such as severe fires or windstorms (Smith et al. 1996). Individual tree and group selection systems are modeled on patterns of disturbance that involve the death of individual or small groups of trees within otherwise intact stands. Disturbance regimes are thus the conceptual foundation for a core silvicultural activity, i.e., regeneration harvests.

However, silviculturists have more recently begun to look beyond the type, severity, and scale of disturbances (particularly from the standpoint of what is removed), to consider the specific ecological conditions left by natural disturbances and to more fully incorporate these conditions into silvicultural prescriptions for regeneration harvests (Kohm and Franklin 1997). Successfully implementing such an approach requires that prescriptions be founded on a conceptual basis that links disturbance to the complexity of stands after the disturbance, as expressed in structure and composition, and heterogeneity of these features in space and time. This approach requires an understanding of the importance of biological legacies created by a tree-regenerating disturbance. That is, the structures, organisms, and patterns left behind from the previous forest after disturbance, and incorporating legacy management into harvesting prescriptions.

Providing this understanding is the goal of this chapter. Specifically, 1) we define the concept of natural disturbance in the context of ecological silviculture, 2) we review the concept and importance of biological legacies, and 3) we present a model of natural disturbance regimes that focuses on the types and amounts of biological legacies they create. Armed with this understanding, you will be prepared to more fully delve into the development of silvicultural prescriptions that better emulate natural disturbance by deliberately incorporating biological legacies.

The Nature of Natural Disturbance

Disturbances are discrete events that free resources, such as light, soil moisture, and nutrients, by killing or damaging existing organisms (Pickett and White 1985). Forest disturbances come in a wide variety of types, sizes, severities, and effects, with spectacular large and infrequent disturbances coming most immediately to mind. However, small chronic disturbances that result in the death of individual trees or small groups of trees are overwhelmingly the most common type of disturbance in forests. In this book, we define intervals between disturbances as either frequent (*chronic disturbances*), which are expected at intervals of one year to a decade, or infrequent (*episodic disturbances*), which are generally expected at intervals of a decade or more. The entire spectrum of disturbance types and scales, from the small and frequent to the large and infrequent, provides models for ecological silviculture. (Figure 4.1).

Few natural disturbances, including those considered **stand-replacing** (e.g., Figure 4.1c), "wipe the site clean." Rather, disturbances kill some trees and other organisms in the ecosystem, leaving behind a seemingly chaotic residue of dead and damaged trees and altered environmental conditions. While fire, windstorm, insect epidemics, and other natural disturbances convert trees from a live to dead state, very little of the biomass contained in these trees is consumed or removed from a site. These biological legacies are significant, biologically generated elements of a pre-disturbance ecosystem that persist in some form through even severe natural disturbance events.

The conditions created by disturbance events—including the biological and physical legacies left behind and their influence on ecosystem function—are the most important outcome of disturbances from a silvicultural perspective. Although we can measure disturbances in many other ways, such as the level and duration of heat in the case of fire or

Figure 4.1 *The gradient of natural disturbance from (a) frequent individual tree events, to (b) less frequent patch disturbances, to (c) infrequent whole-stand events. Photo credit (c): Timo Kuuluvainen.*

wind strength in the case of windstorm, it is ultimately the outcome or effects of disturbances via legacy creation and resource release that will be the important determinants of what happens next in a forest ecosystem (Figure 4.2). The type and abundance of biological legacies that are created and left behind are among the most important of those effects. We discuss biological legacies in more detail in the following section.

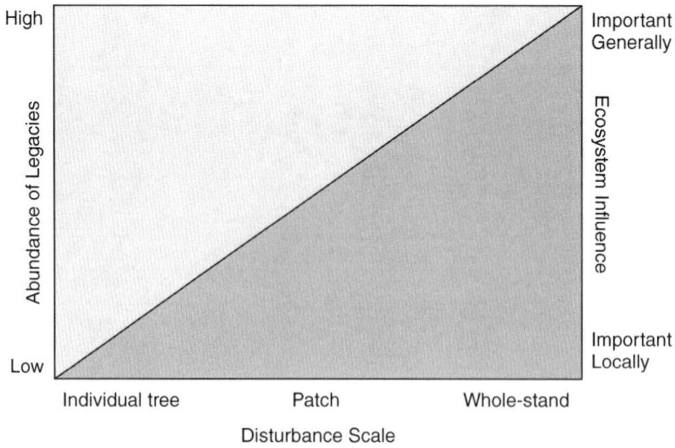

Figure 4.2 *Conceptual representation of a gradient in legacy characteristics and ecosystem consequences along the natural disturbance gradient. Note that although the legacies are localized and limited in abundance with small scale disturbance, the influence on ecosystem function is equally important at the local scale as it is at larger scales with stand-scale disturbance.*

Biological Legacies Defined

(Note that much of this section is drawn from Chapter 3 of Franklin et al. 2018) Biological legacies are defined as the organisms, organic materials, and organically derived patterns in soil and vegetation that persist from the pre-disturbance ecosystem into the post-disturbance environment (Franklin et al. 2002).

Biological legacies provide for significant continuity in ecosystem composition, structure, and function between forest generations (see Chapter 3). This perspective of continuity contrasts with the traditional view of disturbances as creating sharp discontinuities between forest generations. As a consequence of this continuity, biological legacies play important roles in ecosystem reorganization and recovery following disturbance (Franklin et al. 2000; Franklin and MacMahon 2000).

The biological legacy concept emerged out of research following several well-publicized major natural disturbances, most notably, the 1980 Mount St. Helens volcanic eruption. Post-disturbance conditions at Mount St. Helens were initially described as a "moonscape" (Franklin and MacMahon 2000), but subsequent research revealed a diverse array of legacies, including living plants and animals (Dale et al. 2005), revealing that even disturbances of this magnitude do not create the simplified and homogeneous environment previously imagined. Similar studies following other notable disturbance events, such as the Yellowstone fires of 1988 (Christensen et al. 1989; Wallace 2004) and Caribbean hurricanes (e.g., Walker et al. 1991; Pimm et al. 1994), confirmed the universality of biological legacies and their importance.

It is noteworthy that the concept of biological legacies is not wholly new; rather, it is a revision and elaboration of

Frederic Clements's concept of "*organic residuals*" proposed early in the twentieth century (Clements 1916). Clements's take on the concept focused primarily on the importance of residual organisms for secondary succession. However, since that time, scientists have added to our understanding of biological legacies, greatly expanding the definition of the concept to include i) living organisms, ii) organically derived structures, and iii) organically derived patterns.

Living Organisms

Organisms and propagules of organisms are biological legacies that directly perpetuate biota between a pre- and post-disturbance forest ecosystem. Living organisms include intact immature and mature plants and animals, perennating plant parts (such as rhizomes and plant parts with sprouting capacity), and propagules, such as seeds, spores, and eggs. Legacy organisms in the form of remnant trees are most easily visualized. For instance, legacies of disturbance commonly include significant numbers of live trees in the understory (Figure 4.3a) or overstory (Figure 4.3b and c).

Figure 4.3 *Legacy organisms include: (a) advance tree regeneration, such as these conifers, surviving a natural disturbance; (b) dispersed trees that survived a tornado (Brimfield, Massachusetts, USA); and (c) aggregated surviving trees after the Warner Creek Fire (Willamette National Forest, Oregon, USA).*

Organically Derived Structures

Legacies of organic matter that are not part of a living organism take many forms and play diverse roles in post-disturbance ecosystems. Such materials function as short- and long-term sources of energy and nutrients and enrich and bind soils. Much of this organic matter is in the form of particulates and fine litter.

Forest disturbances also leave larger legacies of dead organic matter, such as fully or partially intact snags, down dead trees, and large pieces of wood (Figure 4.4). Such structures can be abundant, massive, and persistent (Harmon et al. 2004). They also play many important roles in forests and associated aquatic ecosystems, in addition to functioning as long-term energy and nutrient sources. These roles include: influencing geomorphic and hydrologic processes, such as erosion, sediment deposition, and the development of channel or bank morphology in streams; providing critical habitat for a large array of living organisms; providing protective cover, nesting sites, and food sources; and ameliorating microclimatic conditions on disturbed sites (Harmon et al. 2004; Maser et al. 1988).

Organically Derived Patterns

Organically derived patterns are less obvious legacies that are most often encountered as patterns in soils or sediments or in understory plant communities. The presence of long-lived plants, such as trees, can generate distinctive patterns in soil chemical, physical, and microbiological properties as a result of the chemical and physical properties of their litter. When the trees are killed or removed, these patterns may persist for decades or even centuries. The disturbance itself can also generate some important

Figure 4.4 *Organically derived structural legacies include: (a) tree snags (Seney National Wildlife Refuge, Michigan, USA) and (b) dead and down trees (Sumava National Park, Czech Republic). Photo credit (a): Greg Corace.*

Figure 4.5 *Organically derived soil patterns in the form of pits and mounds resulting from tree blowdown (Sumava National Park, Czech Republic).*

patterns in soils and sediments, such as the mounds and pits created by tree uprooting (Figure 4.5).

Organically generated patterns in soil properties are common. For example, distinctive radiating patterns in soil properties may be associated with long-lived trees (e.g., Zinke 1962), such as old giant sequoia trees in the Sierra Nevada of California (Zinke and Crocker 1962). Persistent spatial patterns in soil properties are especially notable where long-lived tree species produce litter with strongly contrasting chemical or physical properties, such as western red cedar and western hemlock (Turner and Franz 1985) or eastern hemlock and sugar maple (Lovett et al. 2004). Trees or other plants with special functional or physiological capabilities may also create persistent and notable patterns in soil properties, such as tree species that host nitrogen-fixing organisms.

Similarly, forest overstories produce distinctive and persistent patterns in understory plant communities by controlling the availability of light and moisture. For example, strong patterning often occurs in forest understory plants in response to overstory tree density, with richer and better developed understories under canopy openings (canopy gaps) and little or no understory in areas with dense overstories (anti-gaps).

Functions of Biological Legacies

Biological legacies have three generalized functions that include lifeboating, structural enrichment, and softening the matrix. *Lifeboating* is the perpetuating of genotypes and species in situ. Specifically, mechanisms by which legacies "lifeboat" biological diversity include:

-*Perpetuating plant species*, as surviving immature or mature individuals or as reproductive structures, such as seeds, spores, or vegetative parts with sprouting capability,

-*Perpetuating biota* by providing habitat, supplying energy and nutrients, and modifying microclimatic conditions.

Secondly, legacies result in *structural enrichment* of the stand or patch (i.e., with snags, downed logs, and residual trees), thereby providing habitat for other organisms to recolonize the disturbance.

Lifeboating and structural enrichment functions are typically provided by larger biological structures that persist following a disturbance, such as live trees, snags, and downed boles. These structures sustain organisms in the

Figure 4.6 *Legacy Douglas-fir may be important for maintaining a source of mycorrhizal fungi inoculum in the post-disturbance forest (Gallatin National Forest, Montana, USA). Photo credit: Brian Harvey.*

Figure 4.7 *Post-disturbance forest displaying both lifeboating and structural enrichment functions through the large eastern white pine and snag legacies.*

Figure 4.8 *Legacy tree patches in the midground of this landscape help to soften the matrix of the disturbed forest, providing hiding and movement cover for some species.*

post-disturbance environment by providing necessary habitat (e.g., nesting sites and hiding cover), seed sources, and energy, especially right after the disturbance. Surviving green plants have particular importance in sustaining high-quality energy flows to belowground organisms and food webs, as well as to herbivores. For example, residual trees in Douglas-fir forests (Figure 4.6) are associated with increased diversity of ectomycorrhizal fungi, which are important for seedling nutrition and survival (Louma et al. 2006). Residual structures also modify microclimate, often bringing it within the acceptable environmental range for organisms to survive.

In practice, structural enrichment and lifeboating may be accomplished with the same legacies, such as surviving trees that maintain plant species in situ, but also provide habitat and resources for other biota both in situ and through recolonization. Similarly, legacy deadwood may lifeboat dependent species in situ, as well as structurally enrich the new forest to the benefit of recolonizing organisms (Figure 4.7).

Legacies also serve to *soften the matrix* of the forest at larger spatial scales, which improves connectivity across the landscape. For some organisms, the matrix softening function, by providing protective cover within the larger disturbed area, may allow for greater movement across the landscape (Figure 4.8).

The lifeboating, enrichment, and softening functions of legacies are particularly relevant in the context of stand-replacing disturbances (Lindenmayer and Franklin 2002). However, they can also be important to varying degrees with smaller disturbances, such as within midsize to large **canopy gaps** in otherwise intact stands. For example, surviving trees in a group opening provide seed, energy, and physical habitat to in situ and recolonizing organisms (Figure 4.9).

A General Model of Disturbance and Legacy Creation

Forest disturbances come in a broad array of sizes, severities, and frequencies. The majority of disturbance agents (e.g., wind, fire, and insects) are capable of generating disturbances across a wide range of scales depending on

Figure 4.9 *A live Norway spruce, along with abundant deadwood, in this gap caused by bark beetles in the Czech Republic.*

circumstances. For example, some forest types (e.g., ponderosa pine and longleaf pine) historically experienced frequent, low intensity surface fires that covered large areas, but these disturbances typically produced little mortality of mature trees, with the mortality that occurred consisting mostly of individuals or small clusters. Other forest types (e.g., Douglas-fir, western hemlock, and lodgepole pine) are primarily influenced by infrequent large disturbances, such as wildfire, that result in significant mortality of the dominant trees. Similar to wildfire, wind and insects can be responsible for either frequent small-scale or infrequent large-scale events.

For the ecological forester, emulating ecosystem responses to disturbances is best understood by considering not only patterns of destruction, but more importantly characteristics of what remains following the disturbance, i.e., biological legacies. Our general disturbance model incorporates scale and disturbance agent as the primary considerations in predicting the kind, quantity, and spatial pattern of biological legacies. Here we discuss these concepts generally, as they relate to creation of legacies. In subsequent chapters we delve more deeply into common disturbance "archetypes" and their characteristic biological legacies.

Disturbance Scale

Spatial scale of dominant tree mortality is one dimension of our model of legacy-generating disturbance (Table 4.1). Disturbances are scaled from low intensity, frequent disturbances that involve single or small groups of trees (and that maintain stands), to meso-scale, large-gap opening disturbances (which may be frequent to infrequent), to infrequent, intense, stand-replacing disturbances (Figure 4.10). This gradient in scale of individual disturbance events is correlated with the proportion of the remaining intact forest (Figure 4.10), as well as the characteristics of the legacies that are generated and the degree of heterogeneity of structure and composition.

Low severity, frequent stand-maintaining disturbances

Stand maintaining disturbances tend to be both frequent and of low severity. They include tree-scale and gap-scale mortality events that leave the forest largely intact such that the forest matrix still dominates the post disturbance environment (Figure 4.10b). This category also includes frequent low-severity surface fire regimes that may kill individual trees, but largely function to maintain open, low density stand conditions (Figure 4.11). Tree-scale and gap-scale events tend to create or perpetuate stand structural heterogeneity, although in greater degrees with gap-scale events. In latter successional forests, tree and gap mortality events do tend to perpetuate relatively stable canopy tree species composition by creating resource conditions that favor self-replacement of shade tolerant species.

Biological legacy characteristics of stand-maintaining disturbance regimes, by definition, include the live residual tree matrix of the stand, but also trees that have been damaged or killed by the disturbance, i.e., decadent trees and their derivatives, snags and down boles (Figure 4.12).

Stand-maintaining regimes that include multi-tree gap disturbances leave behind significant biological legacies within the opening, including snags or downed boles or both (Figure 4.13), and living legacies, such as intact understory layers (including regeneration) and living canopy trees.

Meso-scale disturbance

Meso-scale disturbances are less frequent than small-gap events, but more frequent than stand-replacing disturbance. Meso-scale disturbances result when multiple tree or gap events occur simultaneously (heavy, partial canopy disturbance), leaving a stand that contains similar amounts of both open and intact canopy (Figure 4.10c). Meso-scale disturbances often result in higher levels of retention than is characteristic of stand-replacement disturbances, but also remove more of the overstory in any one event than is characteristic of individual tree or gap disturbances. Meso-scale disturbance results in a high degree of structural heterogeneity and provides conditions for both new cohort establishment and the release of residual trees (Fajvan and Seymour 1993), often resulting in two-cohort stands (see Chapter 6, Archetype 4).

Meso-scale disturbances, with their resultant two- or three-cohort age structures, are characteristic of a wide

Table 4.1 *Biological legacies associated with scale[1] and agent of disturbance.*

	Disturbance Agent								
	Wind[2]			Fire[3]			Beetle		
Legacy	Tree	Gap	Stand	Tree	Gap	Stand	Tree	Gap	Stand
Live, mature trees	NA	Few/Absent	Few	NA	Few	Few	NA	Species dependent	Species dependent
Seedling bank	Possible	Possible	Possible	No	No/Rare	Rare	Possible	Possible	Possible
Intact understory	Possible	Yes	Yes	No	Rare	Rare	Possible	Yes	Yes
Snags	None/one	Few/many	Few/many	Yes	Abundant	Abundant	Yes	Abundant	Abundant
Logs	Yes	Abundant	Abundant	No	No	Common	No	No	No
Uprooted trees	One/none	Few/many	Few/many	No	No	No	No	No	No
Mineral seedbed	Yes	Yes	Yes	Yes	Yes	Abundant	No	No	No

[1]Within each agent, scale of disturbance varies from single tree, gap (multi-tree), or whole stands.
[2]Includes both tree-uprooting and bole-snapping events.
[3]Fire at the tree and gap scales is largely surface fire that spreads through a stand (with or without crowning) and may kill individual or small groups of trees.

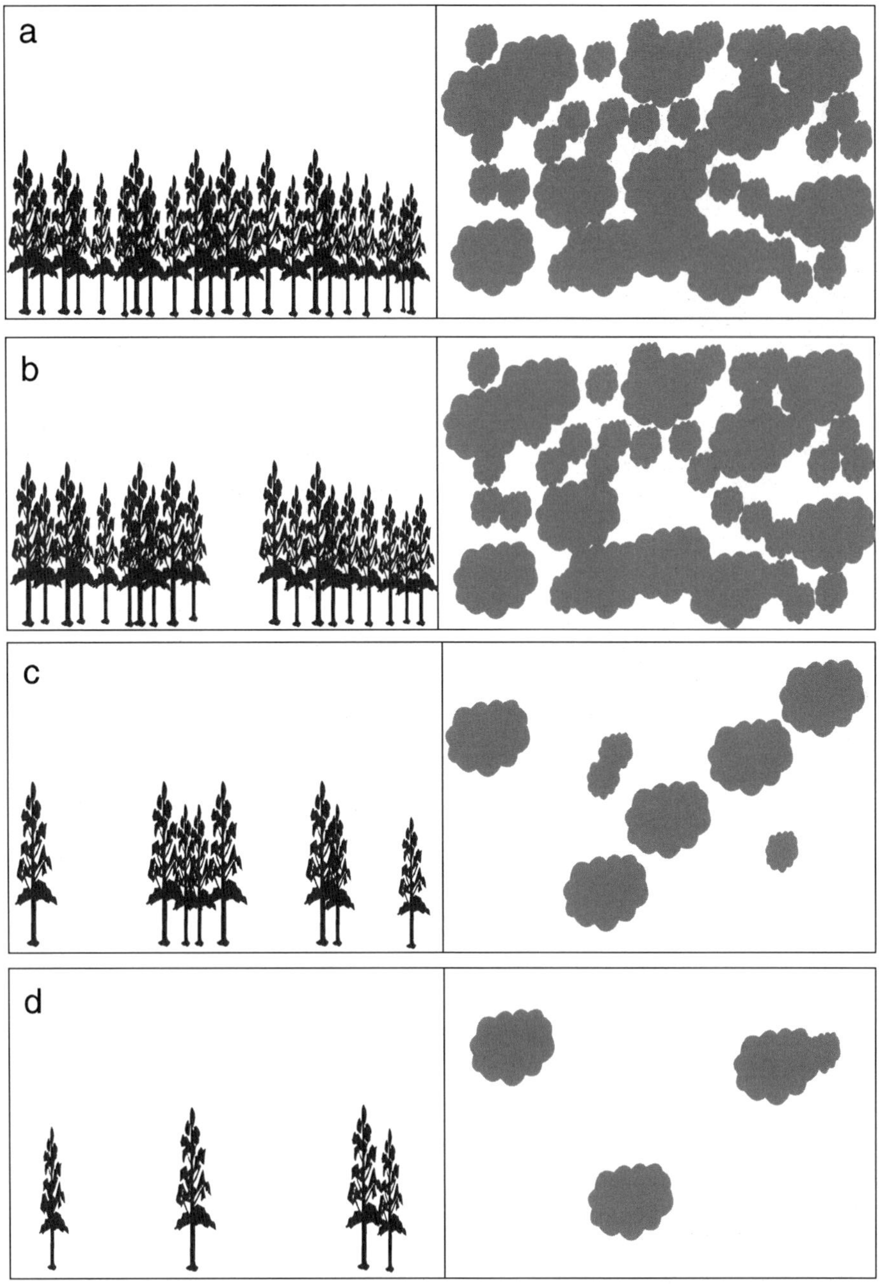

Figure 4.10 *Conceptual representation of spatial scale and distribution of overstory disturbance within a stand: (a) mature forest; (b) gap-scale disturbance; (c) meso-scale disturbance; (d) stand-replacement disturbance.*

array of forest types, including Acadian mixed-conifer forests (Fajvan and Seymour 1993; Seymour and Hunter 1999) and Great Lakes red pine-white pine forests (Heinselman 1981; Fraver and Palik 2012). Such disturbances result in substantial numbers of dead and down trees, often abundant live canopy trees, patches of understory plant populations, as well as spatial heterogeneity in stand structure.

Stand-replacing disturbances

Stand-replacing mortality events destroy the dominant forest cover and shift the matrix to an open post-disturbance environment (Figure 4.10d). Stand-replacing events will generally homogenize (in a relative sense) the post-disturbance tree population structure of a stand by providing conditions for establishment or release of a new tree cohort and generating abundant coarse, woody material.

Most prominent among the legacies of stand-replacement disturbance are the abundant snags and down boles, and often remnant live trees (Figure 4.14), with associated pits and mounds in the case of windthrow. Also, natural disturbances that occur at the stand-replacement scale typically generate substantial spatial heterogeneity of biological legacies (Eberhart and Woodard 1987; Fraver and Palik 2012).

Figure 4.11 *A stand-maintaining disturbance regime characterized by low severity, frequent fire is exemplified by longleaf pine woodlands in southwest Georgia, USA.*

Figure 4.12 *Snags and downed logs are left after small-gap disturbances in northern hardwoods (Dukes Experimental Forest, Michigan, USA).*

Figure 4.13 *A disturbance gap within a Norway spruce forest (Sumava National Park, Czech Republic). Note the biological legacies of snags and downed boles and advance regeneration.*

Figure 4.14 *Significant legacies in the form of snags, fallen trees, and live residual trees are left after an intense windstorm (Chequamegon-Nicolet National Forest, Wisconsin, USA).*

Disturbance Agent

Disturbance agent is the other important dimension of tree-regenerating disturbances that directly affects biological legacies. We use fire, wind, and bark beetles to exemplify the effect of disturbance agents on legacies, as these are common agents and relate to disturbance regimes in many well-known forest types, but we acknowledge that other agents of disturbance, while not discussed in detail here, also create legacies. Wind, fire, and bark beetles can kill trees across a range of spatial scales, from individual trees to entire stands, and these disturbance agents contrast greatly in their resultant biological legacies, irrespective of the scale of overstory tree mortality (Table 4.1).

Wind

Many forest types have a primary disturbance regime defined by wind-created gaps. Legacies within wind-generated gaps include boles on the forest floor and some large, live trees and snags in larger gaps (Table 4.1). Cohorts of tree seedlings and saplings or seedling banks are often present, sometimes in high densities. These regeneration banks are typically referred to as **advance regeneration** by foresters. Understory vegetation and seed banks are typically left intact, except around the uprooted area and where they are buried beneath boles and other material. Exposed mineral soil is typically limited to areas influenced by uprooting, i.e., tree-fall **pits and mounds**, which are also important structural legacies in wind-created gaps (Figure 4.15).

Wind-generated disturbances at larger spatial scales (e.g., hurricanes, typhoons, straight-line winds, and tornadoes) are similar to wind-gap disturbances in the nature of the biological legacies they produce (Table 4.1). Structural legacies are primarily downed boles with some snags and a few larger live trees (e.g., Foster and Boose 1992; Palik and Robl 1999; Peterson and Pickett 1995). Some snags often persist, probably because they have less surface area (i.e., no canopy), providing less purchase for winds. In many temperate and tropical hardwood forests, broken, bent, and partially uprooted trees have the capacity to re-sprout, even though they are no longer part of a dominant overstory (Foster et al. 1997; Walker et al. 1991). These damaged but live trees are an important structural legacy of the disturbance. The understory

Figure 4.15 *Localized soil disturbance in the form of a tipped-up root wad is characteristic of many tree blowdowns that involve uprooted trees, such as these Norway spruce (Sumava National Park, Czech Republic).*

Figure 4.16 Stand of jackstrawed (overlapping) old-growth trees after a severe blowdown.

of affected stands remains largely intact, except for areas disrupted by uprooting or located directly under boles; hence, seed and seedling banks and understory plants are a major legacy of stand-scale wind disturbance. A stand of windthrown or jackstrawed trees (Figure 4.16) also may protect pre-disturbance and post-disturbance seedlings and other understory plants from browsing animals (Franklin and Dyrness 1973; Sharpe 1956).

Fire

Wildfire tends to kill trees from below at either small (gap) or large (stand) scales, but produces comparable legacies at all scales (Table 4.1). Legacies of fire include a high density of snags and sometimes down boles on the forest floor, but relatively few large live trees, if it is truly a stand-replacement disturbance. However, some small number of large trees typically do survive even severe fires, as exemplified by such diverse forest types as jack pine in the Great Lakes region (Abrams 1984) and Douglas-fir in the western USA (Spies and Franklin 1991) (Figure 4.17). As with all disturbance events that leave standing dead trees, snags created by fire will eventually disintegrate to provide large woody material on the ground. Seedlings and seed banks, understory plants, and organic layers on the soil surface are all affected by fire; however, some understory trees and plants typically survive fire, either because of variable intensity or adaptations to fire, or both (Jacqmain et al. 1999). A tree seed bank may also persist as a result of ecological adaptations (e.g., **serotinous** cones) or fortuitous survival of portions of a current seed crop in the canopy (Larson and Franklin 2005). Fire also generates significant physical legacies of mineral soil and ash seedbeds and provides excellent conditions for regeneration of many shade-intolerant or pioneer tree species.

Fires that are frequent, but low severity, may cover a large area but kill few trees relative to the area burned.

Figure 4.17 *Scorched Douglas-fir trees that survived the Saddle Creek Fire (Idaho/Montana, USA). Photo credit: Brian Harvey.*

In southern longleaf pine woodlands, many understory and ground cover plants are adapted to regular burning (Kirkman et al. 2004), while many potential canopy species, such as oaks, survive fire, but do not grow tall (Jacqmain et al. 1999). In some of the forest types subject to frequent fires—such as longleaf pine—overstory tree mortality is caused primarily by lightning or bark beetles rather than by surface fires (Palik and Pederson 1996), but these lightning strikes are an important ignition source for fires, which are fueled by pine needles and flammable grasses (Williamson and Black 1981). Gap-scale stand openings may sometimes occur as part of a frequent, low to moderate intensity fire regime, as is characteristic of southeastern longleaf pine forests (Palik and Pederson 1996) and many ponderosa pine forests (Franklin and Van Pelt 2004). In these cases, surface fires occasionally crown, killing patches of overstory trees and creating gaps in the process.

Highly variable fire severities are characteristic of infrequent, stand replacing fire regimes; consequently, high levels of spatial heterogeneity in burn severity and type and density of biological legacies are common. This heterogeneity is found at all spatial scales, from within stands to landscapes, resulting in patches of varying burn severity, including unburned patches, i.e., **mixed-severity** fire regime (Agee 1993).

Bark beetles

Bark beetle disturbances can occur at a variety of spatial scales, with specifics depending upon the beetle and host species. For example, native bark beetles typically kill small groups of mature to old longleaf pine, producing a gap in an already open woodland (Palik and Peterson 1996) (Figure 4.18a). They may kill trees in a patchy, less than stand-replacement pattern, such as the Douglas-fir beetle (Figure 4.18b). Where extensive pure stands of host species

Figure 4.18 *Various scales of bark beetle mortality: (a) gap mortality of longleaf pine (Jones Ecological Research Center, Georgia, USA); (b) extensive, but patchy mortality of Douglas-fir from bark beetle (Big Horn Mountains, USA); (c) stand-level spruce beetle mortality (San Juan Mountains, Colorado, USA). Photo credits (b, c): Brian Harvey.*

Figure 4.19 *Snags left after a mountain pine beetle outbreak in lodgepole pine, Rocky Mountain National Park, Colorado, USA. Photo credit: Brian Harvey.*

exist, several species of bark beetles are capable of large-scale stand replacement disturbances, such as the spruce beetle in the western USA (Figure 4.18c). Outbreaks of bark beetles and other insects tend to kill from above, leaving and releasing understory communities, including seedlings and seed banks. They also create extensive structural legacies in the form of snags and, eventually, boles and other coarse, woody material on the forest floor (Figure 4.19) (Table 4.1). Undisturbed forest floors are also legacies of beetle disturbance.

Interaction between Disturbance Scale and Agent

Disturbance regimes for various forest types may not be easily categorized as to scale and agent. A notable example of this involves forests that are characterized by both stand-replacement fire events and gap-scale wind events. Developmental sequences initiated by catastrophic fire disturbances are invariably subject to gap-based disturbance processes that operate throughout stand development (Figure 4.20); the longer the developmental timespan, the more influential gap-based processes become (Franklin et al. 2002). In the context of understanding biological legacies in these forests, it is important for the forester to recognize the developmental stage being considered (see Chapter 5) and tailor their expectation and management of legacies accordingly. For example, in our hypothetical forest (Figure 4.20), there will be abundant snag legacies remaining after a stand-replacing fire, whereas in maturing forests characterized by gap-scale events, deadwood legacies will be scattered (among gaps) and likely consist of blowdown or snapped trees. We will delve more deeply into developmental patterns of major disturbance "archetypes" in Chapter 6.

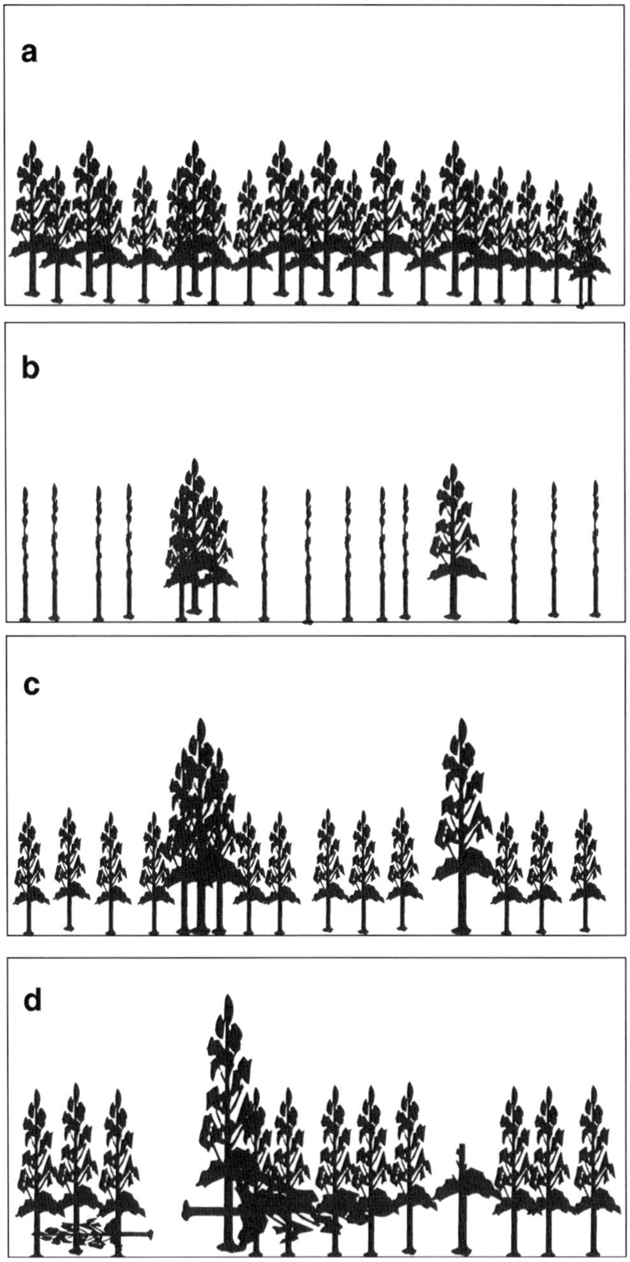

Figure 4.20 *Conceptual representation of a mature forest (a) that is disturbed by a stand-scale fire, or other stand-replacement event (b) with a few residual live trees that survive the disturbance. Over time, if the forest develops into mature and old forest stages (c) (see Chapter 5 for details), the forest is eventually impacted by smaller, gap-scale disturbances, likely from agents other than the stand-replacement agent (d).*

What to Remember about Natural Disturbance and Biological Legacies

A key principle to remember from this chapter is that natural disturbances are important as models for ecological silviculture primarily because of the biological legacies they create. It is not so much that a particular disturbance needs to be emulated; rather it is the outcome in the form of legacies that the forester seeks to emulate. Also remember that legacies exemplify a major pathway by which nature provides for continuity in composition, structure, and function across forest generations. Keeping this in mind allows the forester to remember why legacy conservation in managed forests is so important: through the mechanisms of lifeboating, structural enrichment, and softening the matrix, legacies conserve most pre-disturbance organic matter as future sources of energy and nutrients, they create large pulses of critical habitat structures (snags and logs), and they help maintain micro-environmental conditions that are tolerable for many forest-oriented organisms, allowing them to continue to persist on the disturbed site. Thinking beyond the forest itself, some biological legacies also play important roles in the function of adjacent wetland and aquatic ecosystems by influencing geophysical processes, such as erosion and the deposition of sediments. The ecological forester knows that these functions are equally as important in managed forests as they are in natural forests and that their management prescriptions will have a significant emphasis on legacy conservation for these functional roles. In later chapters we will return to the concept of legacy retention, specifically addressing how to incorporate legacy management into silvicultural prescriptions for different types of forests.

CHAPTER 5

Forest Development Patterns and Processes

Consideration of forest developmental processes, particularly those related to tree growth and mortality, is essential to the formulation of ecological silvicultural prescriptions. The complex structures and spatial heterogeneity that are distinctive features of old-growth forests are often the result of tree growth, decadence, and mortality that occur during stand development (Figure 5.1). Developmental processes and their structural outcomes are what the ecological silviculturist will seek to emulate with intermediate treatments in established (or establishing) stands and by setting ecologically appropriate time periods between regeneration harvests (topics we will consider in detail in Chapter 8), so having a comprehensive understanding of these processes is important.

Various conceptual models have been used to describe the processes and outcomes of tree and forest development following tree-regenerating (and legacy producing) disturbances. These include relatively simple four-stage models that focus largely on live tree establishment and development of vertical patterns (Oliver and Larson 1996) or those based on ecosystem processes, such as biomass accumulation and nutrient cycles (Bormann and Likens 1994). Here we present a more complex model of development (Franklin et al. 2002) that builds upon these prior models, but specifically considers the role of tree growth and decline, plant competition and competition-induced mortality, and small-scale (noncompetitive) disturbances during forest development, as generators of complexity and heterogeneity in structure and composition. In describing this model and its various stages, we discuss key attributes of forest ecosystems that vary over time, including tree sizes and size distributions, **large deadwood** and **resource availability**.

Forest Development Stages

Forest ecosystems (or cohorts of trees in these ecosystems) undergo continuous development after an initiating disturbance in response to internal and external processes. This developmental sequence can be usefully abstracted into a series of stages to better understand the dominant processes influencing their development. (Note that much of the discussion below is drawn from Chapter 3 in Franklin et al. 2018).

The complete sequence of stages or conditions following an initiating disturbance is called a **sere**. In the case of forests that are subjected to infrequent, high-severity disturbances (such as complete or partial stand-replacement wildfire or windstorm) these stages typically develop sequentially over time until the forest reaches an Old Forest Stage (OFS) after one to several centuries. The OFS is relatively stable at the level of a stand, but is quite dynamic at the level of tree populations due to small-scale disturbances that kill trees and create gaps (e.g., Franklin et al. 2002). In the case of forests subjected to frequent, low to moderate intensity disturbance, such as wind or surface fire, these developmental stages are typically represented as elements of a fine-scale spatial mosaic that collectively composes the complete forest ecosystem (Franklin and Van Pelt 2004).

The developmental sequence is primarily one involving gradual changes over time, so it can be represented as a temporal continuum; however, we segment the continuum into distinct stages that can have relevance to developing silvicultural prescriptions based on natural models (Table 5.1). The developmental sequence begins with a disturbance, which may occur at a relatively large scale (as in episodically disturbed forests) or at a patch-scale (as in forests subjected to chronic fire or wind). The disturbance increases resource availability by killing existing trees, but also leaves a legacy of organisms and structures from the pre-disturbance forest or patch, as discussed in Chapter 4.

The developmental stages that follow the initiating disturbance are referred to as the Preforest, Young Forest, Mature Forest, and Old Forest Stages (Franklin et al. 2018; Table 5.1). Although the exact nature and duration of each stage varies greatly—generically with different forest types and idiosyncratically with each sere—a generalized description of each stage and two of the key transitional events (Disturbance and Legacy Creation and Tree Canopy Closure) follows (Table 5.1). The developmental stages are labeled based on level of forest maturation (none, young, mature, and old), rather than

Figure 5.1 *Distinctive structural features of old forests include (a) canopy gaps, (b) large snags, (c) old, declining trees, and (d) deadwood on the ground, and are the result of processes that occur during forest development.*

Table 5.1 *Major developmental stages and transitional events in forest ecosystems as presented in this text, with a crosswalk to other classifications of forest developmental stages. (Source: Franklin et al. 2018.)*

Classification						
Franklin et al. (2018)	Franklin et al. (2002)	Oliver and Larson (1996)	Spies and Franklin (1996)	Carey and Curtis (1996)	Bormann and Likens (1994)	Frelich (2002)
Disturbance and Legacy Creation	Disturbance and legacy creation					
Pre-forest Stage	Cohort establishment	Stand initiation	Establishment phase	Ecosystem initiation	Reorganization phase	Initiation
Forest Canopy Closure	Canopy closure					
Young Forest Stage	Biomass accumulation/ com-petitive exclusion	Stem exclusion	Thinning phase	Competitive exclusion	Aggradation phase	Stem exclusion
Mature Forest Stage	Maturation	Understory reinitiation	Mature phase	Understory reinitiation	Transition phase	Demographic transition
				Botanically diverse		
Old Forest Stage	Vertical diversification	Old-growth	Transition phase (early)	Niche diversification	Steady-state	Multi-aged
				Old-growth		
	Horizontal diversification		Transition phase (late)			
	Pioneer cohort loss					

the dominant development processes (e.g., "biomass accumulation" or "competitive exclusion") (Table 5.1). We believe that a developmental sequence based on maturation is more easily adapted to a broad array of forest types than one based on processes, and is more easily translated into silvicultural application.

Disturbance and Legacy Creation Event

The Disturbance and Legacy Creation event (DLC) creates a sufficient opening in an existing forest, at a patch or whole-stand scale, to initiate a new sere. As discussed in Chapter 4, such disturbances will leave significant biological legacies from the previous forest, generally including standing and downed dead trees and live trees that are either dispersed or aggregated across space, which will be important structural elements in the post-disturbance ecosystem. The amounts and types of legacies will depend upon the type and severity of the disturbance (see Table 4.1).

Preforest Stage

The most fundamental structural feature of the Preforest Stage (PFS) is an absence of trees as the dominant form of plant life. The absence of a significant tree overstory in the PFS allows other plant life forms to flourish and new cohorts of trees, including light-demanding species, to become established (Greenberg et al. 2011; Swanson et al. 2011) (Figure 5.2). Labeling this stage as Preforest clearly identifies it as a non-tree-dominated ecosystem, but one that is occupying a site that will eventually be forested. Other labels applied to this stage, such as the ***stand initiation*** or ***cohort establishment*** stage (Table 5.1), focus solely on regeneration of a new cohort of trees and reflect a strongly tree-centric bias. In actuality, tree regeneration is only one of many important processes that occur in the PFS. While some live trees often survive from the pre-disturbance forest (Chapter 4), collectively their abundance is not enough to dominate the overall light environment of the stand or patch, although they do have localized influence. The PFS is also characterized by a rich structural legacy of snags and logs (Figure 5.2).

Following the disturbance, live plant cover develops from vegetative growth of surviving plants and plant parts (rhizomes, roots, stumps) and establishment of new plants from both off- and on-site seed sources. On many sites, annual and perennial herbaceous species are important initial dominants, but most are transient and replaced

Figure 5.2 *The Preforest Stage of a forest follows a disturbance and is distinguished by a lack of overstory tree dominance, plant communities composed of diverse life forms (e.g., mosses, herbs, shrubs, and individual trees), and often, significant biological legacies (e.g., snags and logs) from the previous forest, as in (a) a post-wildfire black spruce forest in northern Minnesota, USA, and (b) a Douglas-fir forest twelve years after the 2005 Biscuit fire in southwest Oregon, USA.*

(in abundance) in a few years by shrub species. Tree species may be present as legacy individuals or new individuals that establish from legacy vegetative parts or from seed.

There is relatively rapid recovery of many ecosystem functions during the PFS, including **primary productivity** and conservation of nutrients and soil. Net carbon balance is initially negative (the site is a net source of CO_2 released into the atmosphere), due to the abundance of decaying organic matter, warmer temperature, and reduced **leaf area** and rates of primary production. Accelerated rates of **nutrient cycling** and increased **nitrogen fixation** are characteristic of early seral habitats. The great diversity of litter inputs and warmer conditions are two factors that contribute to accelerated rates of nutrient release.

The PFS generally has the highest species diversity of any stage in a forest sere (Swanson et al. 2011). The floristic diversity is derived from a variety of sources, including legacies from the pre-disturbance forest (some initially present as seed banks) and a variety of weedy invasive species. Animal diversity includes many habitat specialists that find optimal conditions primarily or solely in the PFS.

The rich variety of habitat **niches** and food sources in the PFS are key factors in its high biotic diversity. Food sources are typically diverse in quality and quantity, which provides the opportunity for significant numbers and complexity of food webs based on herbivory and consumption of other plant materials, including reproductive parts (e.g., fruits, seeds, and nectar). Many species of vertebrate and invertebrate animals are commonly elements of these **foodwebs.** Warmer conditions found in openings are conducive to the development of cold-blooded animals, such as reptiles. Deer, elk, and other ungulates find the PFS critical for foraging. Structural legacies (such as snags and logs) also provide a rich array of foraging and resident habitats for vertebrate and invertebrate animals, including the larval stages of many invertebrates.

The duration of the PFS is highly variable depending primarily on the time required for trees to re-establish dominance and for canopy closure to occur (see next section). In some forest types, such as the mountain ash forests of southeastern Australia, this period is so ephemeral that it lacks any ecological significance (Lindenmayer 2009). In other forest types, such as the Douglas-fir-western hemlock forests of northwestern North America, elements of the PFS may persist for several decades, even when tree seed sources are abundant (Freund et al. 2014; Tepley et al. 2014). Many hardwood-dominated forest ecosystems experience a PFS of intermediate duration, but it still retains ecological significance (e.g., Greenberg et al. 2011). Duration of the PFS is also commonly much longer after some disturbance types, such as wildfire, which force non-sprouting vegetation to establish primarily from seed, and shorter following windstorms, where understories with abundant advance regeneration are often left largely intact and subsequent tree canopy recovery is relatively rapid.

Forest Canopy Closure

Forest Canopy Closure (FCC) results in dramatic changes in environmental conditions and biodiversity. With forest canopy closure, trees resume strong dominance of the site, significantly reducing the resources available to other plant life forms. The degree of dominance can vary greatly from nearly complete exclusion of other plant life forms to persistence of significant understory plant diversity and cover in less dense forests.

Details on the spatial pattern of canopy closure for most natural forests is limited, but it is likely that there is substantial spatial heterogeneity, particularly where the PFS persists for several decades (Donato et al. 2012). More rapid and uniform FCC may be more common in forest ecosystems dominated by hardwoods, especially when revegetation is primarily through sprouting from stumps and roots (Figure 5.3a). Gradual emergence of closed-tree-canopy patches, which eventually merge, might be a more common pattern than rapid uniform FCC when tree establishment is from seed (Figure 5.3b).

Young Forest Stage

The Young Forest Stage (YFS) is a period of strong or exclusive dominance by trees following establishment of a closed forest or tree patch. Such forests commonly have simple live tree structures composed of relatively uniformly sized trees (Figure 5.4); however, in most seres there will be structural legacies from the pre-disturbance forest in the form of some residual live trees, snags, logs, and other large woody material (Figure 5.4).

Biomass accumulation is a dominant developmental process in the YFS and has actually been labeled the biomass accumulation stage in other classifications (Table 5.1). High rates of primary productivity and biomass accumulation (including wood) are prominent, since essentially the site's light, nutrient, and moisture resources are fully utilized by trees. Dense forests (particularly of evergreen conifers) transpire large amounts of water, so annual water yields and summer low flows typically decline during the YFS.

Figure 5.3 *a) Forest canopy closure occurs rapidly (within a few years) after disturbance in forests having a substantial component of trees that establish vegetatively from sprouts, such as this trembling aspen forest in northern Minnesota USA. b) Canopy closure may take years to decades in forests where regeneration is from seed, such as in this pine-oak woodland in the New Jersey Pine Barrens, USA. Photo credit (b): Bob Williams.*

Figure 5.4 *The YFS is often densely dominated by trees with a relatively even-structured live tree component. A dense YFS may have very little understory, as in this spruce-fir forest in Maine, USA (a), whereas a less dense YFS may retain significant understories of shrubs and herbs, as in this jack pine forest in Minnesota, USA (b).*

High levels of nutrient conservation are characteristic of this stage.

Competition for light and soil resources is the other important process occurring during the YFS and, accordingly, this stage has been called the competitive exclusion phase in other classification systems (Table 5.1). Where tree stem densities are high, intense inter-tree competition develops and trees that fall behind in height growth are quickly suppressed and ultimately die. Competitive mortality is predictably concentrated in the smallest diameter classes. Mortality rates are also density-dependent, so the densest parts of the forest or patch experience the highest rates of competitive-based mortality and consequently this process is sometimes referred to as natural thinning. The net effect of this inter-tree competition is to move forests or patches toward a more structurally homogeneous state.

Although often viewed as a continuous process, competitive mortality may occur in pulses over the course of forest development (Horsley et al. 2000; Palik and Pregitzer 1993b). As trees grow in size, they occupy more growing space and compete for limited resources. Many trees in the stand may decline in growth as competition intensifies; however, inferior competitors decline in growth and vigor at a faster rate than superior competitors, and the former eventually die. This mortality frees growing space and resources, which in turn allows growth rates of vigorous trees to increase until the next period of competitive interaction and resultant mortality. Ultimately, the result of this process is the development of larger trees.

Competitive processes also often result in elimination of much or all of the understory vegetation as a result of low light levels. Branch pruning and the gradual elevation of the tree canopies result in the high (relative to the forest type), single layered canopies typical of the YFS. Competitive tree mortality also generates snags and deadwood on the ground, but this material is typically small and thus of limited value as habitat or as a modifier of microclimate (Figure 5.5).

In forests that initiated after stand-replacement disturbance, competitive tree mortality is increasingly concentrated in spatially distinct patches, as the stand transitions into the Mature Forest Stage (see next section). In forests characterized by gap-scale disturbance, competitive mortality occurs in discrete patches that have established or been released in canopy gaps (Figure 5.6). In natural forests, competitive tree mortality may not progress uniformly throughout a stand; different portions of the stand may thin at different rates and the residual trees will grow at different rates. The result is some degree of horizontal variation in stand density and tree sizes.

Where the YFS is composed of diverse tree species, competition and differential growth rates and life spans may result in significant compositional as well as structural changes over time. For example, slower growing species may be overtopped, although they may persist in intermediate canopy positions if they are at least moderately shade-tolerant (Oliver and Larson 1996). Short-lived tree species may play significant early roles in the YFS and subsequently drop out; their positions in the canopy are typically filled by canopy expansion by other canopy trees and by the emergence of slower growing trees from the initial cohort, which are often species of intermediate tolerance. Hence, compositional changes (in the dominant tree layer) may be significant in the YFS.

The YFS often has the lowest biodiversity of the developmental stages. Understories in densely shaded forests or patches are sparse with few vascular plant species

Figure 5.5 *Deadwood generated by competitive tree mortality (Cedar River Watershed, King County, Washington, USA). This material is typically small in size and not persistent and consequently has limited habitat value.*
Photo credit: James A. Lutz.

Figure 5.6 *Competitive mortality occurs within dense patches of regenerating trees in mature and old forests such as this patch of Norway spruce regeneration developing within a canopy gap (Šumava National Park, Czech Republic).*

Figure 5.7 *Young Forest Stage in a 45-year-old Douglas-fir plantation on the Gifford Pinchot National Forest, Washington USA. Photo credit: Andrew J. Larson.*

Figure 5.8 *Key features of the Mature Forest Stage, such as in this Norway spruce forest in the Czech Republic, include redevelopment of the understory, including shade-tolerant species; opening of the canopy from non-competitive mortality, such as the gap in the back right; and increased spatial heterogeneity of structure.*

(Figure 5.7), so herbivory-based food webs are poorly developed or absent, particularly in forests dominated by conifers. The majority of the biological diversity is associated with detritus-based food webs, since dead organic materials are abundant. Resident animals tend to be either participants in **detritus**-based foodwebs or species dependent upon reproductive structures produced by mycorrhiza-forming fungi (e.g., northern flying squirrels), or those that feed primarily on tree seeds (e.g., red squirrels). Higher levels of vertebrate diversity can be expected when large legacy structures, such as logs, are present, since vertebrates, such as salamanders and some voles, benefit from such structures. Vertebrates may also use dense, young forests for hiding and thermal cover.

Mature Forest Stage

Dominance by trees during the YFS eventually eases, as densities continue to decline from competition-induced mortality and the forest or patch transitions to the Mature Forest Stage (MFS). The forest is still dominated by the initial cohort of trees; however, light and moisture are becoming more available for the establishment or expansion of the understory, including reproduction and growth of shade-tolerant trees (Figure 5.8). The MFS is distinguished from the YFS by: 1) aspects of individual tree development, 2) shifts in causes of mortality in the initial cohort of trees, and 3) initiation of other important processes—such as increased decadence and spatial heterogeneity—that ultimately will characterize the Old Forest Stage (OFS), as discussed in the next section.

Overstory trees will achieve most of their height growth and crown spread during the MFS, although the crowns still tend to conform to classic forms, with well-defined apical dominance (see, e.g., Van Pelt and Sillett 2008). **Epicormic** or other **adventitious** branch systems may begin developing on many tree species, sometimes replacing canopy that was lost to natural branch pruning during the YFS. Decadent canopy and bole features, such as rots, brooms, and cavities, also become more evident. This is also a period when suppressed tree species may experience some release and enter the main canopy. Additional cohorts of shade-tolerant species are also likely to become established.

A major transition occurs in the pattern and dominant causes of tree mortality in the MFS, which significantly increases structural heterogeneity of the stand. Mortality shifts from primarily competitively based to agent-based (i.e., noncompetitive in nature). Agents of noncompetitive mortality include root rots, wind, bark beetles, lightning, ice damage, and surface fire (Franklin et al. 1987; Franklin and DeBell 1988; Kneeshaw and Bergeron 1998; Palik and Pederson 1996). Many of these same agents could be part of a stand or patch initiating regeneration disturbance (see Chapter 4), but mortality from these agents in the MFS occurs infrequently enough and at smaller spatial scale (i.e., single tree and small patch), such that a significant amount of regeneration of future overstory trees is not necessarily an outcome. In reality, small-scale mortality events grade into regeneration disturbances.

Noncompetitive mortality also is often spatially aggregated. The combination of mortality of larger (overstory) trees and contagion generates canopy gaps. Creation of gaps is a key process in development of spatial heterogeneity in structure. This heterogeneity in canopy conditions results in spatial variability of many attributes, including abundances of tree, shrub, and herbaceous plants, snag and downed wood loadings, tree size distributions, forest floor conditions, and mineral soil exposure. The development of structural heterogeneity counteracts the tendency toward structural homogenization produced by the competitively based mortality in the YFS.

The development of canopy gaps in the MFS leads to increased light and moisture availability locally. Gap-neighboring trees may capture these resources and increase growth accordingly (similar to competition-induced mortality), closing the gap laterally (Parsons et al. 1994). More importantly, from the standpoint of structural complexity, gaps may result in new establishment or proliferation of understory plant populations, including herbs, shrubs, and understory trees (McGuire et al. 2001). Depending upon the density of the YFS, some understory plants and tree reproduction may already be present and respond to gaps. However, additional establishment of shade-tolerant trees is a common element during the MFS. One important consequence of the growth of understory trees is the development of an intermediate canopy between the ground and upper crown, which begins moving the canopy toward the vertically continuous canopy that often characterizes the OFS (see next section).

Forests maintain moderate to high levels of productivity in the MFS, even though foresters traditionally view such forests as being "over mature" because generally they are beyond **culmination of mean annual increment** (CMAI) (Curtis 1995). Although CMAI was and sometimes still is considered to represent biological maturity, forests reaching the MFS are only just arriving at maturity from a tree form or forest ecosystem perspective. In fact, high levels of primary productivity can continue in the MFS and result in additional accumulations of wood and carbon.

The mass of coarse woody material that traces its origin back to snag and log legacies from the initiating disturbance has been declining during the YFS and generally reaches its lowest levels early in the MFS. Inputs of small, sapwood-dominated trees dying during the YFS have not made up for these losses. As the MFS proceeds, stocks of deadwood begin rebuilding, as agent-based mortality in dominant and codominant trees generates larger snags and logs. Development of decadent features, such as mistletoe brooms, cavities, and heart rots, accelerates significantly in the MFS and can also contribute to mortality.

Transitions from the MFS to the Old Forest Stage (OFC) are gradual and reflect the continued and cumulative effects of processes initiated in the MFS. For most forests, only limited information is available on the patterns and rates of these processes, which are known to vary with forest productivity (see, e.g., Freund, et al. 2014 and Larson et al. 2008).

Old Forest Stage

The Old Forest Stage (OFS) is typically the most functionally and structurally complex forest stage (Figure 5.9). Structural complexity typically includes a diverse mixture of live and dead tree sizes and conditions, large old trees of one or more tree species, and high levels of spatial heterogeneity in the distribution of stems and canopy. Forests in the OFS typically remain highly functional in term of productivity and other attributes, such as their ability to influence hydrologic and other material processes, despite the casual observer's sense of disorder and decadence. The OFS can have a high level of species diversity (second only to the PFS among stages in our model) and this diversity includes species that require the unique habitat provided by old trees and forests. The OFS in chronically disturbed (e.g., frequent-fire) forests does take a different form than in episodically disturbed forest ecosystems (see Chapter 6), but still has attributes directly related to the presence of large, old trees (Figure 5.10). In fact, large old trees are a particularly prominent feature of the OFS, providing a set of unique niches in this stage (Box 5.1). With the inclusion of intermediate sized trees and tree reproduction, size-class distributions often display the classic reverse-J or rotated sigmoid shapes, provided one or more shade-tolerant tree species is present. The broad range of live tree sizes is often used as an identifying metric for the OFS.

Decadence is an important and highly evident feature of the OFS. Agent-generated mortality of overstory and intermediate-sized trees is an active process, providing continuing inputs of large snags and logs to the forest. These agents include root, butt, and bole rots (fungi), wind (uprooting), insects (e.g., bark beetles), and lightning (Figure 5.13). Competition-induced decline may also have resumed, primarily in dense patches of tree reproduction that develop in canopy gaps. Other decadent features, such as cavities and decay pockets in live trees, as well as brooms in the crown of trees, develop in response to parasites or pathogens and are typically well represented in the OFS.

Horizontal heterogeneity is high in the OFS primarily because of continuing gap generation as a result of overstory mortality. Clustered or clumped patterns of tree regeneration in these gaps are also characteristic of older forests. Canopies in the OFS are typically multi-layered or sometimes continuous from the ground to the top of the canopy. Additionally, canopy foliage in some OFS forests may be bottom-loaded—i.e., the bulk of the canopy is lower in the profile—rather than single-layered and top-loaded as was the case in the YFS (Figure 5.14).

Forest ecosystems in the OFS typically continue to have high levels of gross productivity, however much of the

Figure 5.9 *The Old Forest Stage is the most structurally and functionally diverse stage in the development of many types of forests. Note the diversity of tree sizes and conditions, abundance of deadwood structures, high level of spatial heterogeneity, and vertically continuous canopy in these old-growth forests: (a) eastern hemlock forest in Massachusetts, USA, (b, c) eastern hemlock forests in Vermont, USA, (d) Lenga forest in Argentina.*

Figure 5.10 *Old Forest Stage in frequently burned ponderosa pine woodland displaying distinctly different structural features than the episodically disturbed forest, but still inclusive of large, old (relative to the forest type) trees and tree-derived structures. Photo credit: Bureau of Land Management. This image is a work of a Bureau of Land Management employee, taken or made as part of that person's official duties. As a work of the US federal government, the image is in the public domain of the United States.*

Box 5.1 The importance of large old trees (Drawn from Chapter 2 of Franklin et al. 2018)

Old trees differ greatly from young trees in the roles that they play in forest ecosystems. This is not just related to their generally larger size; i.e., old trees are not simply enlarged versions of younger trees. Particularly in long-lived species, older trees accumulate significant idiosyncratic features as a result of numerous injuries and infections and responses to those injuries, and as a consequence of altered light and temperature conditions (Figure 5.11).

Large old trees have large branches of primary origin, as well as secondary origin, i.e., epicormic branches in species with that capacity. Large branches are important to many canopy organisms and processes. For example, in tropical and some temperate regions, very large branches can accumulate massive epiphytic communities (Nadkarni 1984; McCune 1993) and provide support for the nests of some large birds, such as eagles. Older trees have had more time to experience stimuli that generate secondary branch systems. Canopies of old trees may also be deep as a result of secondary (e.g., epicormic) branch development. In contrast, canopies of old trees in frequent-fire ecosystems are likely to be elevated as the result of past damage or loss from fire.

Old trees likely will have developed larger percentages of **heartwood** than younger trees of the same species, which will result in more durable snags and logs. Heartwood follows different decay pathways than sapwood, with the consequence that it plays very different roles as habitat. Heartwood also may be very persistent in litter and soil and it behaves differently as a fuel because of the large amount of resinous material it typically contains.

Older trees in many species develop thick, furrowed bark, or loose and stringy bark, both of which create niches for invertebrates that are potential food sources for other species (e.g., Carey 2009; Lindenmayer 2009). Thicker bark makes older trees more resistant to wildfire and increases their probability of surviving such events.

The decadent features of older trees are some of their most important features from the standpoint of habitats for biological diversity. Cavities and other pockets of decay are important for a wide array of cavity-dependent species (e.g., see Hunter 1988 and Lindenmayer 2009). Development of multiple tops provides complex canopies, as well as thick vertical branches, that act as stable platforms where thick mats of organic matter may accumulate. Eventually, old trees die and as such are the source of large deadwood in the forest. Large snags and logs on the ground are important habitat features in old forests (Figure 5.12).

Figure 5.11 *Large old trees display unique characteristics not found in younger trees, including: (a) buttressed roots, (b) deep furrowed bark, (c) flaking bark, (d) decay and cavity formation, and (e) large horizontal branches.*

Figure 5.11 *Continued*

Figure 5.12 *Large snag in an Old Stage pine forest provides a structural platform for this raptor nest.*

Finally, old trees also represent a distinctive genetic resource, particularly in landscapes now dominated by managed forests. These trees not only represent diverse gene banks, but also include genotypes that can be viewed as long-term "winners"—i.e., they are trees that have survived diverse climates, attacks by fungi and insects, and possibly intense storms and fires.

A not uncommon term in the parlance of classical silviculture is that of the *over-mature* tree, i.e., a tree that has reached that stage of development when it is declining in vigor and health and reaching the end of its natural life span or has begun to lessen in commercial value because of size, age, decay, or other factors (SAF 2008). It should be clear that from an ecological perspective, there is no such thing as an over-mature tree. Allowing some trees to develop to old age and live out their natural life span is exactly what the ecological forester should do; they should actually embrace the structural uniqueness and ultimate decline of old trees because of their ecological importance.

Figure 5.13 *Wound from a lightning strike on a large, old tree that initiates decay and potentially could result in mortality.*

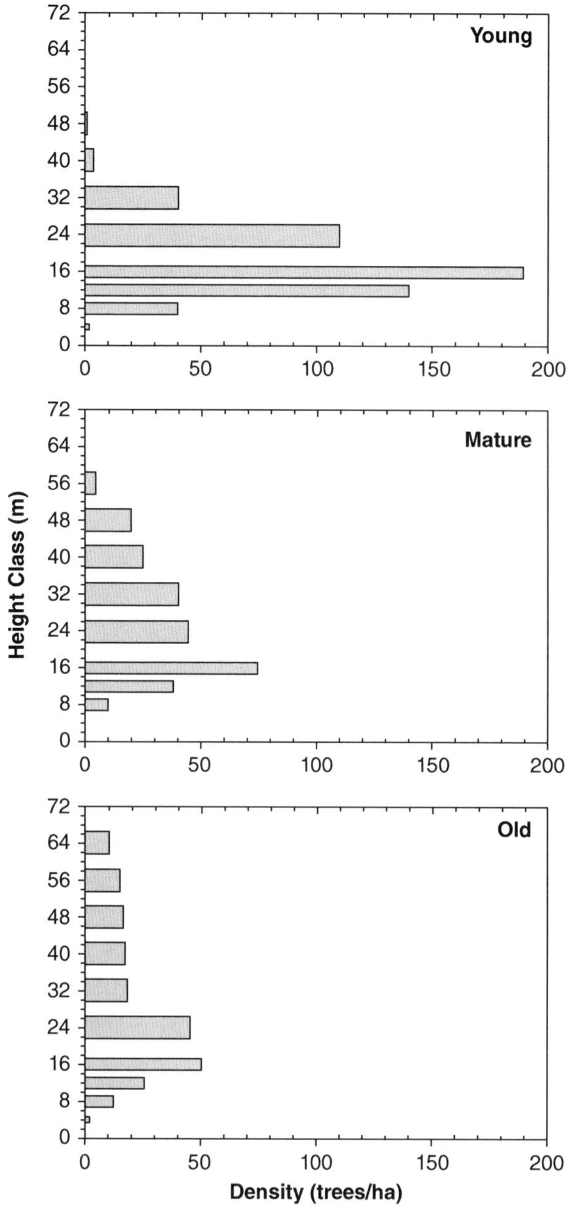

Figure 5.14 *Canopy distribution profiles for Douglas-fir forests based on tree height classes and densities in young, mature, and old stands. Only trees in codominant, dominant, or emergent canopy positions are included. Note that tree densities decrease and canopies become bottom-weighted with age. Redrawn from Bingham and Sawyer 1992.*

carbon uptake is balanced by losses to respiration and decomposition. Consequently, net productivity—additions to existing carbon stocks (organic matter)—may be relatively modest or even negative. Also, an increasing proportion of the carbon may be in large deadwood and not in live trees. In traditional forestry terms, the OFS typically has high gross growth but high mortality, and thus, little or no net growth from the standpoint of wood in live trees. Still, it is important to understand that: 1) many trees in old forests are still growing well, including some of the largest and oldest trees; and 2) that many old forests are still making small additions to their stock of accumulated organic matter—i.e., they are still sinks for atmospheric carbon. In fact, OFS ecosystems have the largest aboveground biomass pools of any developmental stage and are therefore critically important as terrestrial carbon stores.

Forests in the OFS continue to influence aspects of the hydrologic cycle, through i) interception of precipitation and snow in canopies and condensation of cloud and fog moisture on deep crowns, ii) transpiration, and iii) hydraulic redistribution of moisture in soils. Indeed, the earliest scientific studies of old forest function were those done by hydrologists who discovered that watersheds vegetated by old forests produced well-regulated flows of very high-quality water. Older forest ecosystems are generally highly retentive of nutrients and effective at stabilizing soils. Trees in the OFS typically have well-developed, deep root systems that are very effective at stabilizing soils and at capturing and holding soil nutrients. These are factors that contribute to the high quality of water flowing from watersheds dominated by the OFS.

In some forests, the OFS may have the highest tree species diversity of all the stages, particularly if representatives of the shade-intolerant pioneer species are still present (e.g., Aravena et al. 2002). In other forests, many of the plants, as well as vertebrate and invertebrate animals, that are present in the OFS may also be present in the YFS and MFS

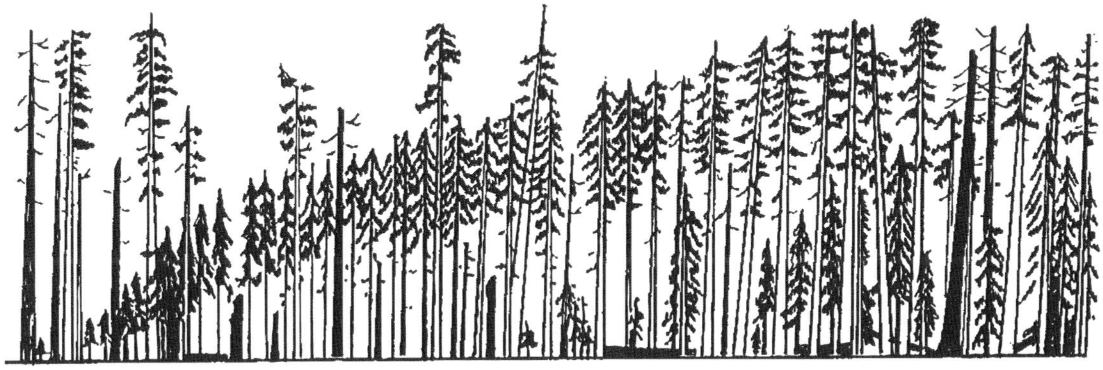

Figure 5.15 *Hypothetical forest development displaying increasing structural complexity over time as non-competitive tree mortality and tree growth processes accumulate.*

(e.g., Hansen et al. 1991). Still, there can be some animals that find conditions in the OFS to be optimal or essential for aspects of their life cycle.

Forest ecosystems that reach the OFS continue to change and evolve; that is they continue to undergo gradual directional changes in structure, function, and composition albeit at a much slower rate than in the previous stages. These are by no means static "climax" forests, as will be seen in subsequent discussions of specific developmental archetypes in later chapters.

Recovery Periods between Disturbance Events

Disturbance and Legacy Creation events (DLC) vary in agent, scale, frequency, intensity, and the legacies they produce. Often, the return interval between events is long enough that structural complexity, particularly as a result of noncompetitive mortality and tree growth processes, can develop (Figure 5.15). For example, the predicted return interval for Category 4 hurricanes in Coastal Plain pine forests in South Carolina is 260 years (Myers and van Lear 1998). Between these major disturbances, small-scale tree mortality events occur with increasing frequency as stands age (Palik and Pederson 1996), leading to development of complex heterogeneous stand conditions.

The importance of a recovery period for development of structural complexity is well illustrated by the dynamics of large deadwood in Pacific Northwest (USA) Douglas-fir forests. Spies et al. (1988) illustrate the temporal dynamics of large wood after stand-initiating fires (Figure 5.16). They show that initial high loadings of deadwood before disturbance increase dramatically after a fire, primarily due to fire mortality, with lesser contributions from carryover from the previous stand and, in the case of partial burns, contributions from the residual stand after the fire. Wood levels then decline dramatically within the first 150 to 200 years of stand development, falling well below initial post-disturbance values. Several centuries are required

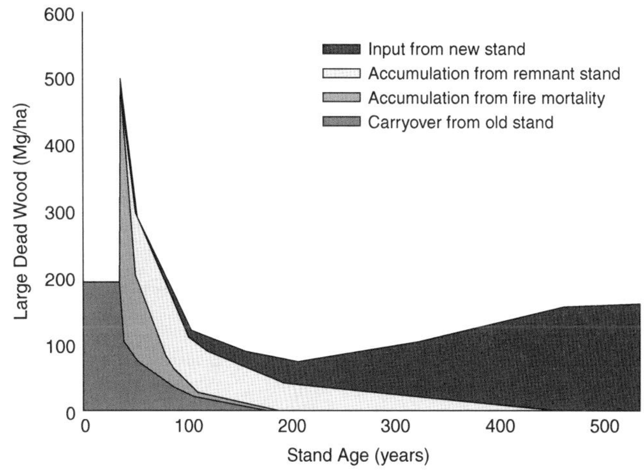

Figure 5.16 *The development of structural complexity over time as illustrated by large deadwood dynamics in Pacific Northwest (USA) Douglas-fir forests after partial stand-replacement fire. Deadwood is derived from four sources, including carryover from the pre-disturbance stand, mortality from the fire, mortality from the post-disturbance residual stand, and from the newly developing stand. Redrawn from Spies et al. 1988.*

before the amount of large, persistent deadwood increases to values approaching pre-disturbance levels, primarily as a result of small-scale canopy disturbances and mortality in the developing stand (Franklin et al. 2002). While the time period required to develop comparable levels of deadwood complexity is probably not so long in most forest types, the general premise still applies; it takes a significant period of time between major disturbances for small-scale mortality processes to create a structurally complex forest.

The concept of recovery period is clearly applicable to stand-replacement disturbance regimes, as illustrated above. However, there is value in considering this concept in the context of gap-replacement regimes as well. Forests that develop primarily as a result of gap-disturbance regimes—e.g., hardwood forest in eastern North America (Runkle 1982)—are spatially heterogeneous and structurally complex precisely because gaps occur infrequently in

Figure 5.17 *Gap scale development of structural complexity. Northern hardwood forests in (a) Wisconsin, USA and (b) Michigan, USA.*

space and time, creating a patch mosaic of forest in different stages of development. At the stand-scale, this mosaic of recovering patches includes recent canopy openings in which deadwood is recruited, resource availability is initially high, and new cohorts of trees and other plants are established. Although gaps might recur at or near the same location by chance, there is often a long recovery period between gap events at any one location. This allows for the development of large old trees, cavities, and ultimately a source of snags and deadwood on the ground (Figure 5.17).

What to Remember about Forest Development Patterns and Processes

A key point to remember from this chapter is that to varying degrees, natural forest ecosystems include a Preforest Stage (PFS) that is largely subverted in managed forests, but is an important element of biodiversity in the larger landscape. The ecological forester may deliberately choose to include the PFS in their management by not being so concerned with achieving full stocking of tree regeneration immediately after harvest (i.e., fast-forwarding to the Young Forest Stage), recognizing that full stocking may develop slowly over years to decades in many ecosystems in contrast to the often short (1–5 year) window used to judge success of regeneration establishment in timber-focused forestry. Moreover, the ecological forester should recognize that regeneration densities in many natural forests, even after "full stocking" is achieved, are often lower than what is used in timber-focused forestry, where full site occupancy and use of all available resources by trees is a driving consideration. A second point to remember is that developmental processes, particularly non-competitive mortality and, eventually, the development of unique attributes in old trees, are important generators of structural complexity and heterogeneity in natural forests that only begin to develop in the Mature Forest Stage. The ecological forester can use this knowledge to devise approaches for jump-starting the development of spatial heterogeneity and structural complexity by incorporating small gap cutting (not necessarily for the purposes of regenerating new trees, although this may still be an outcome) into thinning prescriptions (see Chapter 8: Variable Density Thinning). Ultimately, time is the only prescription for generating many of the structural characteristics of the Old Forest Stage, including large old trees and their unique structural attributes. Therefore, the ecological forester must consider appropriate recovery periods between regeneration harvests timed to ecological development, as much, or more so than economic or growth considerations. It follows that the ecological forester may allow a component of trees in a forest, even if it is a small proportion of all trees, to live out their natural life span, i.e., achieve large size, old age, and ultimate senesce, then die and transition into the woody material pool of the forest.

CHAPTER 6

Natural Disturbance Archetypes

In this chapter, we focus our attention on natural disturbance **archetypes**. The definition of an archetype is *the original pattern or model from which all things of the same kind are copied or on which they are based.* Forest archetypes then, as conceptualized in this chapter, are meant to provide the reader with model systems for major types of forest ecosystems, including their disturbance regimes, stand dynamics, structural and compositional features, and, ultimately, appropriate ecological silviculture practices. The archetypes also illustrate how the developmental stages presented in the previous chapter (Chapter 5) play out in different forest types. Note that references to the developmental stages in the different archetypes is based on the developmental model description in Chapter 3 of Franklin et al. (2018).

Many forests will not fit perfectly into the models, but elements of the archetypes will be apparent in almost all forests. Our goal is to help the reader better understand the developmental processes that go on in natural forests, which has rarely been emphasized in the forest management literature because of the historical emphasis on early stages of development. The archetypes that follow should not be viewed as models that must be directly emulated in management, but rather as sources of information to consider when developing ecological silviculture prescriptions. We focus on four archetypes, as distinguished by their natural disturbance regimes:

1. *forests initiated by infrequent severe disturbance;*
2. *forests characterized by frequent low-severity disturbance, primarily fire;*
3. *forests characterized by gap disturbance, notably from wind;*
4. *forests characterized by mixed-severity disturbance regimes*

Archetype 1 is initiated by severe forest- or landscape-scale disturbance at infrequent intervals (one hundred years or more), thereby initiating new developmental sequences. Development progresses through the Preforest Stage (PFS), Young Forest Stage (YFS), and Mature Forest Stage (MFS), and eventually arrives at a quasi-equilibrium Old Forest Stage (OFS). The OFS is sustained by frequent, small-scale (tree- and gap-level) disturbances until a new disturbance begins the sere anew.

Archetype 2 applies to forests that experience low severity disturbance, largely wildfire, which occurs at frequent intervals (e.g., one year to several decades). In this archetype, severe disturbances that eliminate overstory cover over large areas are rare. Tree mortality occurs primarily to individuals, and in small patches, creating openings for regeneration and subsequent stand development. In this archetype, developmental stages are present at small spatial scales as elements of a forest mosaic.

Archetype 3 is one in which gap-scale disturbances, primarily from wind, drive the dynamics of the ecosystem. It has similarities to Archetype 1 in that the *OFS* is also in a quasi-equilibrium of structure and composition defined by a shifting mosaic of gap disturbance. It differs from Archetype 1 in that gap disturbances define the ecosystem at earlier developmental stages and can be indefinite in tenure, with severe disturbance being extremely rare. Archetype 3 also has similarities to Archetype 2 in that a mosaic of development stages, albeit at the gap scale, is present in the forest.

Finally, Archetype 4, is best characterized by a mixed-severity disturbance regime that includes variation in severity among events and variation in the agent of disturbance. Variable amounts of canopy disturbance are an outcome among stands (i.e., sometimes, but not always, stand-replacing). This disturbance regime is not easily pigeonholed into any of the other archetypes, since it includes chronic, low-severity disturbance such as surface fire, disturbances that can be gap-based, and disturbances that infrequently are severe enough to result in stand-replacement. This disturbance archetype also may contain all developmental stages simultaneously as elements of a forest mosaic.

In the following sections, we discuss the archetypes in greater detail, including how our forest development model (Chapter 5) applies to each. We use a case study ecosystem to illustrate our ideas for each archetype, but we note that the generalities are meant to apply to the variety of similar forests in each archetype.

Archetype 1: Forests Initiated by Infrequent Severe Disturbance

Many temperate and boreal forests are characterized by natural disturbance regimes that include infrequent but severe disturbance resulting in significant areas of complete or near-complete stand-replacement. Attributes of this archetype can include: 1) initial dominance by a **shade-intolerant pioneer** species, which is gradually replaced by more **shade-tolerant** associates; 2) high levels of structural legacies (e.g., dead trees and logs) left by the initiating disturbance; 3) an extended and ecologically important Preforest Stage (PFS); 4) high levels of structural complexity and accumulations of live and dead biomass in the Mature Forest Stage (MFS) and Old Forest Stage (OFS), and 5) an often extended period of time needed for full development to the OFS.

Example: Douglas-Fir-Western Hemlock Forests of Western North America

The Douglas-fir-western hemlock (DF-WH) forests of the Pacific northwestern USA and southwestern Canada and the mountainous interior western US is our primary example of archetype 1 (Figure 6.1; Color Plate 6.1a), with the description below drawn from Chapter 3 of Franklin et al. (2018). Our description of the archetype refers to the variant of this forest that experiences severe (stand-replacement) events (see next section).

The DF-WH forest is dominated early in succession by Douglas-fir, with western hemlock increasing in importance over time. Other associates include western white pine, western redcedar, grand fir, and Pacific silver fir. Variants of this forest that occur adjacent to the Pacific Ocean support Sitka spruce (and have a greater abundance of western hemlock), and in northern California (US), coast redwood. Moving north into British Columbia (Canada), Douglas-fir eventually drops out and Alaska cedar becomes a common element. Additional associates in montane (mid-elevation) habitats of the Cascade Range include noble fir and sometimes mountain hemlock. Inland occurrences of these forests (e.g., northern Idaho, western Montana, and interior British Columbia) may also include western larch and even ponderosa pine.

Hardwood trees, while not dominant in Mature and Old Forest Stages, are ecologically important, particularly early in succession and on drier sites. These include bigleaf maple, red alder, black cottonwood, bitter cherry, cascara, Pacific dogwood, golden chinkapin, Pacific madrone, and in some parts of the range, Oregon white oak, California tanoak, California black oak, and canyon live oak.

Disturbance Regime

Large and severe wildfire initiates the DF-WH sere. Most wildfires of this type occur during a synoptic weather pattern in which east winds bring hot and dry air from interior continental regions during summer and early fall. The aftermaths of these wildfires are complex mosaics of fire severity, as they burn through diverse topographic and forest conditions over many days or weeks. Hence, these fires are complex events that invariably include significant areas of high tree mortality, but also result in large areas of intermediate and low overstory tree mortality, as well as unburned patches. This varied disturbance is sometimes described as a **mixed-severity** regime, but here we focus on its most important distinguishing element: areas of high-severity fire behavior.

The DF-WH forest can experience severe wind storms, which can remove the overstory. In fact, the largest forest disturbance event recorded in the Douglas-fir region was the Columbus Day windstorm of 1962. As reviewed in a later section, successional development following wind events follows different pathways than wildfire with

Figure 6.1 *Transect profile of an old-growth Douglas-fir-western hemlock forest. Cedar Flats Research Natural Area, Washington, USA. Photo credit: Robert Van Pelt.*

succession towards shade-tolerant species accelerated by such events.

Developmental Pathway

Disturbance and Legacy Creation

A distinctive attribute of the DF-WH forest is a Disturbance and Legacy Creation (DLC) event that leaves behind large amounts of dead trees and logs. These legacies play important short- and long-term roles in the post-disturbance forest (Figure 6.2). There are also legacies of living trees scattered through the burned landscape as individuals, in clusters, and in larger patches that persist for varying periods of time (Figure 6.3).

Biological legacies left following windstorms contrast with those from wildfires (Figure 6.4). These differences include wood legacies predominantly in the form of fallen boles (rather than snags), uprooting and creation of root wads and depressions, and a largely undisturbed understory, which often includes significant tree seedling banks.

Preforest Stage

A prolonged Preforest Stage (PFS) is a notable attribute of the DF-WH forest. It is ecologically important because this is the only period in a potentially multicentury sere that is not dominated by evergreen conifers. The PFS in the DF-WH forest typically persists for 40 to 60 years (Freund et al. 2014; Tepley et al. 2014) and has the highest biological diversity of any stage in part because of the richness of available habitats and food sources. The PFS persists, in part, because of the relatively slow rate of conifer tree establishment and growth, which must be from seed provided by trees within or near the burn. Readily dispersed annual and perennial herbaceous plants dominate initially and are subsequently replaced by shrubs, many of which were present in the pre-disturbance forest either as established plants or in the seedbank (Figure 6.5). Full development and maturation of shrub species requires several decades. The PFS is generally of much shorter duration following windstorms than it is

Figure 6.3 *Legacies of living Douglas-fir and western larch after severe wildfire. Colville National Forest, Washington, USA.*

Figure 6.4 *Intense windstorms leave massive legacies of downed trees, snags, root wads, and associated soil depressions when mature and old moist temperate forests are blown down. Photographed in the Bull Run drainage of the Mount Hood National Forest, Oregon.*

Figure 6.2 *Legacies of standing dead trees and logs in severely burned Douglas-fir-western hemlock forest. Willamette National Forest, Oregon, USA.*

Figure 6.5 *Shrubs and perennial herbaceous plants typically assume dominance in the Preforest Stage of the Douglas-fir-western hemlock forest. Several different species of shrub are visible in this photo, including Pacific rhododendron and snowbrush, as well as the perennial herb beargrass. H. J. Andrews Experimental Forest, Willamette National Forest, Oregon, USA.*

Figure 6.6 *The Young Forest Stage (YFS) in Douglas-fir-western hemlock forest is often a dense, tree dominated ecosystem with a relatively even-structured live tree component. (a) Dense young forest may have very little understory. (b) Lower density stands have understories of shrubs and herbs.*

following wildfires due to the tree seedling banks that survive and respond rapidly to the increased solar radiation.

Young Forest Stage

The Young Forest Stage (YFS) in the DF-WH forest is a period of near exclusive dominance by Douglas-fir and other coniferous tree species, which form dense stands of relatively even age and size (Figure 6.6a). Some short-lived hardwood trees, such as bitter cherry and cascara, are often present but die out after 40 to 80 years. Occasional longer-lived hardwood trees, such as bigleaf maple and Pacific dogwood also may be present. The understory in the YFS can be extremely sparse or have significant herb and shrub cover depending upon tree density. Many naturally regenerated DF-WH forests appear to have established at relatively low densities so that significant understories are retained (Figure 6.6b) and the amount of competitively based tree mortality is relatively low in comparison with densely stocked stands. The YFS can have high levels of structural complexity because of legacies of snags and downed logs left after disturbance.

Mature Forest Stage

The Mature Forest Stage (MFS) of the DF-WH forest begins the transformation of the young, competitively driven forest into a more complex older forest (Figure 6.7). Douglas-fir continues to dominate the overstory and undergoes additional growth during its second century; e.g., Douglas-fir typically adds 40 percent of its total height after an age of one hundred years. Agent-based mortality from wind, root and butt rots, and bark beetles becomes significant during the MFS, resulting in canopy gaps. This, in turn, stimulates additional regeneration of shade-tolerant trees in the understory, including western hemlock, which begins the process of creating a mid-level tree canopy. An additional understory tree species, Pacific yew, may become more evident beginning in this stage. The mortality of larger trees begins the process of restoring larger masses of deadwoody material, as the large original legacy of deadwood has largely decayed away.

Figure 6.7 *Mature Forest Stage in a Douglas-fir-western hemlock forest in which shade-tolerant western hemlocks have formed an intermediate canopy layer and dominant Douglas-fir are achieving most of their height growth and crown spread. Photo credit: H. J. Andrews Experimental Forest, Oregon, USA.*

Old Forest Stage

The Old Forest Stage (OFS) of the DF-WH forest achieves great structural complexity (Figure 6.8; Color Plate 6.1b). It is also the most diverse forest stage of this sere. Notable attributes include: 1) the presence of large old trees, which often have major epicormic branch systems; 2) diverse tree sizes; 3) abundant large deadwood, including large old tree-derived snags and downed logs and other decadent features; 4) vertically continuous canopies with the majority of the foliage in the lowest one-third of the canopy; and 5) substantial spatial heterogeneity in the distribution of structures.

The OFS can be viewed as a dynamic fine-scale structural patchwork that effectively recapitulates all of the earlier stages in forest development (i.e., DLC, PFS, YFS, and MFS). This dynamic patchwork is sustained by chronic small-scale disturbance events (e.g., insects, disease, and wind) and tree population processes. However, an important exception is that the gaps present in the OFS are usually not big enough for successful recruitment of the sere's pioneer species-Douglas-fir. Consequently, the forest gradually transitions to dominance by western hemlock during this stage. The eventual loss of Douglas-fir would have major consequences for the structure and function of the ecosystem, particularly when the primary species replacing it was western hemlock, a species of smaller dimension and greatly lesser decay resistance; few stands reach such an advanced state, however, before experiencing a major disturbance event that provides the opportunity for Douglas-fir regeneration.

Figure 6.8 *Old-growth Douglas-fir-western hemlock forest.*

Variations on the Archetype

The large wildfires that initiate the DF-WH forest always include a high severity forest-replacement component, but large portions of such burns experience low to moderate severity fire as a result of variable burning conditions. The developmental conditions produced by these less intensively burned areas give rise to a large set of possible developmental pathways that are alternatives to the primary pathway presented above (Rapp and Spies 2003; Spies 2009; Tepley et al. 2013). For example, less severe burns may leave most of the dominant fire-resistant Douglas-firs alive but provide suitable conditions for abundant reproduction of shade-tolerant species, such as western hemlock, resulting in two-aged stands (Tepley et al. 2013).

Successional development following a severe windstorm follows a very different pathway from that following wildfire. With a windstorm, most of the existing understory, including tree reproduction, survives the disturbance and is released by the removal of the overstory. The PFS after wind disturbance is much shorter than that following wildfire, because advance tree regeneration is present. Furthermore, advance tree regeneration is generally composed of shade-tolerant species, such as western hemlock, western redcedar, and Pacific silver fir. Consequently, the new stand that develops will generally be dominated by shade-tolerant species and not Douglas-fir.

Other Examples of the Archetype

There are a number of conifer-dominated forest ecosystems that are initiated by stand-replacement wildfire, although

Figure 6.9 *(a) Lodgepole pine stand, Gifford Pinchot National Forest, Washington, USA. (b) Jack pine stand, Superior National Forest, Minnesota, USA.*

few of these achieve the same level of age and complexity characteristic of the Douglas-fir-western hemlock forest. Jack pine forest in the Great Lakes region of North America and lodgepole pine-subalpine fir-Engelmann spruce forests of western North America exemplify much shorter-lived and simplified versions of Archetype 1 (Figure 6.9).

Here we focus on the lodgepole pine example. Generally these are even-aged forests regenerated following complete or partial forest-replacement disturbances by wildfire, mountain pine beetle epidemics, or, occasionally, windstorms. The PFS can be relatively short because initial growth of lodgepole pine is rapid, although severe site conditions may result in a PFS of longer duration. Where dense forests regenerate, development may occur slowly through an intensely competitive YFS. As these forests mature and thin after a century, lodgepole pine typically begins reproducing in the understory along with other tree species, such as Engelmann spruce, subalpine fir, or mountain hemlock. Over time, another major disturbance event (fire, beetles, or storm) will initiate a new sere. While all three of these disturbances kill the overstory of larger old lodgepole pines, wildfire has an additional impact because it kills the understory as well as the overstory. Bark beetles and windstorms, on the other hand, kill only the larger of the trees so that most of the tree reproduction survives the disturbance and is released. Where lodgepole pine dominates the regeneration, it will again dominate the new stand, but where the reproduction is primarily composed of more shade-tolerant species, such as subalpine fir, Engelmann spruce, and mountain hemlock, then these species are likely to dominate the new stand.

Archetype 2: Forest Characterized by Frequent Low-Severity Fire Disturbance

(Contributed by Derek Churchill)

This archetype characterizes forests that experience frequent, predominantly **low-severity** fires. Frequent fire forests are fundamentally different from forests that experience mixed- to high-severity disturbance (Franklin and Johnson 2012). Instead of progressing through a sequence of developmental stages after a major disturbance, these forests persist as dynamic, fine scale mosaics of all developmental stages (Chapter 5) over small areas (e.g., within stands). In this sense, they are similar to the Old Forest Stage in other ecosystems that are maintained by chronic wind, insect, and pathogenic disturbances. However, fire is such a dominant force in shaping these ecosystems that frequent fire forests display a distinct and common set of structural attributes, despite having major differences in productivity, species composition, and climate. We define frequent fire forests as having a mean fire **return interval** of 1 to about 20 years and predominantly low severity fires (less than 20 percent mortality of basal area) (Agee 1993). However, return intervals can be upwards of 35 years in some cases.

Frequent fire forests are found in regions around the world that share four characteristics. First, the climate must include a warm, dry period sufficient for surface fuels to dry and become flammable and for favorable fire weather conditions to occur. Second, an ignition source must occur reliably during the dry period. Lightning is the primary ignition source, although aboriginal burning was historically a major source in many places (Boyd 1999). The third characteristic is a plant community dominated by fire resistant tree species that are able to survive most low to moderate severity fires. Conifers with thick, fire resistant bark occur in almost all frequent fire forests. Oak species that have fire resistant bark and the ability to sprout are also common. Finally, the ecosystem must be sufficiently productive for surface fuels to rebuild quickly after a fire in order to carry the next fire. Surface fuels are typically a mixture of fire adapted herbaceous plants (mostly grasses), leaf litter, and in some cases, shrubs.

Example: Ponderosa Pine and Dry Mixed-Conifer Forests of Western North America

Our primary example of this archetype is the pure ponderosa pine and dry mixed-conifer forests that extend over most of the western intermountain region of North America from interior southern British Columbia, Canada south along the eastern slopes of the Cascade Range and the Rocky Mountains into northern Mexico (Figures 6.10; Color Plate 6.2). Annual droughts are a defining feature of this region and forest type. Across most of the range, most moisture falls as snow during winter with a drought occurring during the summer. In parts of the Central and Southern Rocky mountains, most of the precipitation results from summer storms and a drought period can occur in both the spring and fall (Reynolds et al. 2013). The historical record (including the twentieth century) commonly shows evidence of prolonged, multiyear droughts (Westerling and Swetnam 2003).

Ponderosa pine is the most common tree species in this forest type (e.g. Hagmann et al. 2014). As one moves from hot-dry to cooler-moister sites, composition shifts from nearly pure forests of ponderosa pine to forests that include greater mixtures of Douglas-fir and western larch (Graham and Jain 2005). All three of these species are fire and drought resistant, although Douglas-fir does not develop thick, fire resistant bark as quickly as the other two species. Ponderosa pine and western larch are shade intolerant and need significant light to recruit into the overstory, whereas Douglas-fir is moderately shade tolerant. Lodgepole pine can be a significant component in cold and dry locations or in unusual topo-edaphic environments such as the tephra deposition zone in central Oregon, USA. Other minor species include western white pine, sugar pine, Jeffrey pine, and incense-cedar. Both shrub and herbaceous dominated understory communities occur and combine with pyrogenic tree litter to provide the fine fuel necessary to carry frequent fire (see next section). Understory plants

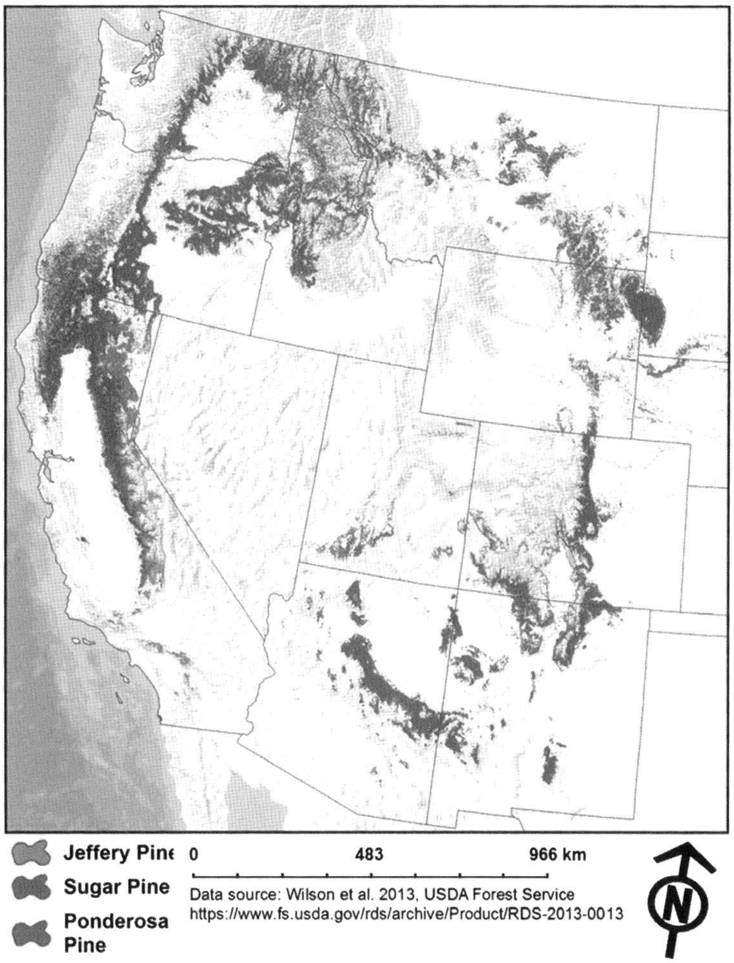

Figure 6.10 *Approximate distribution of historical frequent fire forests in the western US. The distribution of ponderosa pine, sugar pine, and Jeffrey pine are used as surrogates for historical presence of a frequent fire regime. Other fire-resistant species, such as Douglas-fir or western larch, are also common. Note that these species can occur in moist, historically mixed-severity forest types as a minor component; these areas were removed from the map. Frequent fire forests in Mexico are not shown. Map courtesy of Kerry Kemp, the Nature Conservancy of Oregon. Data adapted from Wilson et al. (2013).*

also suppress tree regeneration by competing for soil moisture (Rummel 1951).

Disturbance Regime and Forest Structure

Surface fire is the defining feature of natural disturbance regimes in these forests. Extensive fire history studies across the western US show that fires burned in these forests every 5 to around 20 years prior to Euro-American settlement and that these fires were often very large (Heyerdahl et al. 2008; Reynolds et al. 2013; Addington et al. 2018). Fire return intervals fluctuated within this range based on decadal shifts in climate patterns with more frequent fire in drier, warmer periods and less fire in cooler, wetter periods (Heyerdahl et al. 2002). Return intervals occasionally exceeded 20 years, but longer intervals tended to result in more mixed-severity fire. In mountainous regions with more complex topography, higher fuel moisture levels on northerly aspects prevented some fires from spreading, resulting in longer intervals between fires (Hessburg et al. 2007). Lightning was the primary ignition source in most areas as thunderstorms occur most years during the dry months. Aboriginal burning prior to the mid- to late-1800s was also common in some areas (Boyd 1999).

A defining characteristic of the forest is an open canopy dominated by a low density (and basal area) of large and old trees (Figures 6.11; Color Plate 6.3a). Canopy cover is generally less than 30 percent, although patches of higher cover and density do occur. Frequent fire maintains this open canopy by thinning out smaller trees (and occasional larger trees), resulting in stocking levels far below maximum carrying capacity. Understory vegetation and much of the downed woody material are periodically consumed by fire as well. This prevents a buildup of surface fuels and the development of horizontally and vertically continuous canopies that would otherwise lead to high severity fire.

In addition to large, old trees, a full range of tree diameters are generally present in patches and diameter distributions are often relatively flat. The forests tend to be multi-aged, with age classes spanning multiple centuries. The different canopy strata tend to be spatially segregated in a low-contrast, fine-scale mosaic (Franklin and Van Pelt 2004; Figure 6.12; Color Plate 6.3b). Multistoried patches with canopy layers that extend close to the ground are less common because they are prone to crown fire torching.

Developmental Dynamics

At the stand and larger spatial scales, frequent fire forests do not typically progress through the four developmental stages of development more characteristic of the other archetypes (Chapter 5), because they are rarely subjected to stand-replacement disturbances. Instead, old trees tend to be present throughout much of the forest. In this sense, they are similar to the Old Forest Stage in other systems where the frequent disturbances are provided by chronic low severity winds, insects, and pathogens. Still, most of the structural conditions and processes associated with the developmental stages and events, as described in Chapter 5, are present, but within small patches or clumps (0.004 ~ 0.4 ha) (North et al. 2016), as reviewed below.

Figure 6.11 *Frequent fire forests in the western US: a) Klamath, California, b) Montana.*

Disturbance and Legacy Creation

Large tree mortality occurs frequently, but at low levels for each disturbance event (0–20 percent mortality). Agents can include fire, bark beetles and other insects, windthrow, structural failure due to basal fire scars, lightning, root rots, and mistletoe. Individual trees can die (both isolated trees and members of clumps), or partial to complete mortality of overstory clumps can occur (Figure 6.13). Frequent fire accelerates the transition of standing dead trees into downed logs, which are then consumed by a combination of insects and fire. Legacy structures persist for a much

Figure 6.12 *Profile of frequent-fire forest showing different canopy strata and patch types. Note the presence of regeneration thickets, old trees, and young to mature trees all within a small area. Illustration credit: Robert Van Pelt.*

Figure 6.13 *Death of overstory trees from fire.*

shorter timeframe than in other archetypes and deadwood levels are generally lower. They are created much more frequently, however.

Preforest Openings

Preforest openings are a critical component of the spatial mosaics of frequent fire forests. While large stand-scale patches of Preforest are rare, openings still make up the majority of area cumulatively (Reynolds et al. 2013). Due to low overall canopy cover and variable arrangement of trees, openings in these forests are not canopy gaps in the typical sense. Instead, they tend to be amorphous, often sinuous and linear in shape, and generally hard to delineate and quantify. Opening sizes range from 0.05 ha to a hectare, with infrequent larger openings. They contribute to the resilience of the forests by impeding the spread of high severity crown fire, insect, and pathogenic disturbances. Openings also enhance shrub and herbaceous plant diversity and abundance, and thus provide important habitat elements for many birds, insects, and mammals.

Given adequate seed supply and suitable climatic conditions, tree seedlings preferentially establish and survive on exposed mineral soil within fire-created openings (Figure 6.14), leading to an overall aggregated distribution of tree seedling establishment. **Belowground competition** with herbaceous plants is reduced in these regeneration patches for several years. Exposed mineral soil also provides a more favorable moisture regime than tree litter. Slow accumulation of surface fuels in regeneration clumps growing in openings provides seedlings some protection from fire. Microclimatic and fuel moisture conditions within dense seedling and sapling clumps may also limit fire spread, conferring some additional protection from surface fires.

Not all regeneration is aggregated in openings, however, as some isolated individuals or small groups of seedlings establish and survive in openings or in proximity to overstory trees. Finally, not all openings fill in with trees. Openings can persist through time on sites with poor soils, root rot centers, or on sites with deep organic soil layers that favor herbaceous species.

Figure 6.14 *Regeneration of trees in a fire-created opening.*

Figure 6.15 *Density reduction in a young clump from surface fire that killed some of the regenerating trees.*

Young Forest Clumps

Clumps of trees that establish in openings develop into small patches where many of the processes associated with the Young Forest Stage occur. As seedlings within the regeneration clump develop into saplings, surface fuels begin to accumulate from the litter of the young trees themselves and facilitate surface fires burning through the clump, which kill a portion of the young trees (Figure 6.15). Competition also leads to crown recession and density dependent mortality within the clump. The process of thinning is accelerated and more severe when regeneration clumps are intermingled with residual overstory trees, which suppress seedling growth rates while increasing surface fuel accumulation rates.

Figure 6.16 *Tree maturation, resulting in a range of tree sizes.*

Mature Forest Clumps

Over time, tree neighborhood dynamics lead to a range of tree sizes in both even and uneven-aged clumps (Figure 6.16), paralleling developmental processes associated with the Mature Forest Stage. Trees on the perimeter of a clump maintain higher diameter growth rates and attain larger sizes than trees in the interior. Position within the clump causes differentiation of tree crown attributes: trees growing on the edge of a clump may develop asymmetrical crowns with large limbs and tend to lean away from the clump center, while trees in the center of the clump tend to have relatively smaller symmetrical crowns. Mortality from fire and competition slows, but is still ongoing. Depending on the severity of intra-patch mortality, new cohorts of seedlings can establish, leading to the development or exaggeration of an uneven-aged structure in some tree clumps, while preserving the spatially aggregated tree distribution.

Old Tree Clumps

As clumps age, intra-clump competition and mortality stabilize most clumps in arrangements of 1 to 6 trees that can live for centuries (Figure 6.17a). Some large overstory tree clumps composed of as many as 10 to 30 trees can persist and develop into old tree patches as well (Figure 6.17b). Widely spaced individual old trees are also common (Figure 6.17c). Individual clumps are often even-aged, but can also contain multiple-age cohorts due to partial overstory mortality and subsequent regeneration within or adjacent to clumps (White 1985). Although some crown and moisture competition may still be occurring, old trees within clumps have been shown to share water and carbohydrates across root grafts and **mycorrhiza** networks (Simard et al. 2012). This can facilitate survival and growth of regeneration growing close to old trees, especially on harsh sites. Maximum tree ages are generally around 300–500 years in most ponderosa pine and mixed conifer forests, although 600–800 year old trees do occur. The death of old trees leads to local accumulations of coarse surface fuels, setting the stage for the creation of new openings and safe sites for tree regeneration and the restarting of the cycle.

Old Forest Stage

As discussed earlier, ponderosa pine forests persist for centuries in the Old Forest Stage across a range of spatial extents

Figure 6.17 *a) Old trees in clumps of 2 to 6 trees; b) old trees in dense clumps of 10 to 30 trees; c) widely spaced individual old trees.*

(tens to thousands of hectares). The developmental stages discussed above exist at fine-grained (0.004 to 0.4+ ha) spatial scales within these old forests. Spatial patterns are characterized by widely spaced individual trees, clumps ranging from 2 to more than 20 trees, and openings (Graham and Jain 2005; Larson and Churchill 2012; Reynolds et al. 2013; Matonis and Binkley 2018). Regeneration is relatively continuous over time and new cohorts generally recruit in dense thickets within openings. The cyclic repetition of fires, tree regeneration, and tree mortality within the old-forest matrix perpetuates a shifting mosaic of tree clumps and openings in different stages of development (Figure 6.18; Color Plate 6.4).

Variations on the Archetype

Ponderosa pine and dry mixed-conifer forests can display a wide range of structure and pattern and tolerate a wide range of fire return intervals. If fire is absent for more than a few decades, they can shift towards a mixed-severity regime (Archetype 4), especially on more productive sites where understory tree layers recruit and grow more quickly. High severity fires and insect disturbances, although relatively rare, did occur historically. These infrequent, extreme climate driven events would put forests onto the four-stage development pathway like Archetype 1, but subsequent frequent fire would nudge them back toward the frequent fire archetype over time (Weaver 1961).

Ponderosa pine and mixed conifer forests thus have the capacity to follow multiple developmental pathways, which is largely a function of fire return intervals (Schoennagel et al. 2004), although structure and function under alternative pathways differs substantially. These forests correspondingly provide the ecological forester with a range of natural development pathways to emulate when facing different management challenges.

Other Examples of the Archetype

The Ponderosa pine and mixed conifer forests described in the example above are similar to other frequent fire forests in western North America. The Sierran mixed conifer and Jeffery pine forests found in California and northern Mexico are key examples (Stephens et al. 2008; North et al. 2015). Oak-pine forests are also found through the western US and typically grade into mixed conifer forest, sharing similar low-severity, frequent fire disturbance regimes. Oak and oak-hickory forests of the central US are another important example of this archetype.

Frequent fire forests are also common in the southeastern USA. The most iconic examples are longleaf pine ecosystems of the southeastern Coastal Plain. Longleaf pine savannas and woodlands have one of the shortest fire return intervals of ecosystems globally, ranging from 1 to 10 years (Mitchell et al. 2006). Longleaf pine needles, along with abundant bunch grasses, provide ideal fine fuels to support

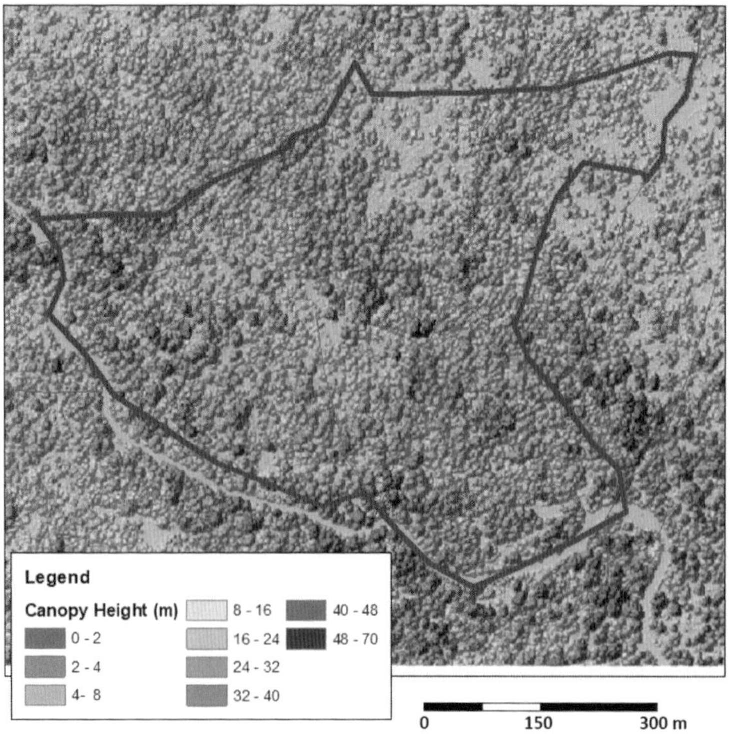

Figure 6.18 *Overhead LiDAR image of a frequent fire forest with an active fire regime in Yosemite National Park, USA. Note the small patch sizes of different structural conditions, ranging from clumps of old trees, to patches of mid-sized trees, to thickets of regeneration, to openings with few or no trees.*

Figure 6.19 *Clumped regenerating longleaf pines in an opening within a patchy matrix of mature and old trees. Jones Ecological Research Center, Georgia USA.*

frequent fire (Hendricks et al. 2002; Myers 1990). Surface fires interact with canopy disturbance to control the structure and dynamics of longleaf pine ecosystems. Tree densities and canopy cover tend to be low and spatially patchy in frequently burned longleaf pine ecosystems (Platt et al. 1988; Palik and Pederson 1996; Palik et al. 2003). Within this sparse canopy matrix, larger canopy openings develop, often resulting from the coalescing of single tree disturbances from lightning and windthrow (Palik and Pederson 1996; Myers and Van Lear 1998), or from larger-scale disturbances from hurricanes or downbursts (Palik et al. 2002). Longleaf pine regeneration generally occurs in distinct clumps in these openings (Platt et al. 1988; Palik et al. 1997) (Figure 6.19). Over time, dense patches of regeneration naturally thin as a result of competition and fire. Eventually, a small number of individuals from these clumps reach dominant positions in the overstory, reinforcing the open, sparse canopy structure of savannas or woodlands.

Archetype 3: Forests Characterized by Gap-Scale Wind Disturbance

Many temperate forests are characterized by natural disturbance regimes dominated by frequent, low severity wind disturbance resulting in complex, multi-aged conditions across broad spatial scales. Attributes of this archetype can include: (1) live-tree legacies from the pre-disturbance stand, including broken and partially uprooted trees, advance regeneration, and **soil seed banks**; 2) a general absence of the PFS due to rapid site capture by advance regeneration, vegetative growth from surviving trees, and additional seedling establishment; 3) dominance by shade-tolerant species across all developmental stages; (4) eventual emergence of uneven-aged, structurally complex OFS of high tree species diversity due, in part, to frequent formation of canopy gaps spanning a range of sizes; and (5) duration of the OFS far exceeding lifespan of canopy trees given the rarity of severe disturbance.

Example: Northern Hardwood Forests of Northeastern North America

Our example of Archetype 3 is northern hardwood forests occurring over a large portion of the cool, moist, temperate forest landscapes of northeastern North America. An important attribute of these ecosystems, relative to fire-driven ecosystems, is the shade tolerance of most tree species, including sugar maple, American beech, and eastern hemlock. This condition is highly advantageous given the disturbance-development model for this forest is one in which gap-scale disturbances driven by wind, ice, insects, and disease play a prominent role. These gap-scale events occur frequently, generally removing 0.5–2 percent of the canopy annually, with more moderate-severity disturbances (30–50 percent canopy removal) such as thunderstorm microbursts or ice storms occurring every 300–400 years (Frelich and Lorimer 1991). Stand-replacing events are rare with return intervals exceeding the maximum longevity of tree species found in these forests (>1000–1800 years; Lorimer and Frelich 1994; Seymour et al. 2002). Despite the general absence of fire from these forests, fire-adapted species, including eastern white pine and northern red oak, are occasionally minor components of northern hardwood forests, with their occurrence associated with physiographic settings and disturbance events that allow for their successful recruitment in these mesophytic systems (Frelich 2002; Fahey and Lorimer 2014; Figures 6.20, 6.21; Color Plate 6.5a).

Developmental Pathway

In contrast to Archetype 1, and similar to Archetype 2, the application of sere concepts to describe structural development in Archetype 3 forests is less useful given the return interval for stand-replacing events is 4 to 5 times greater than the average longevity of dominant canopy trees. As a result, a point of focus for Archetype 3 is the predominance of the OFS and the range of compositional and structural conditions created and maintained through recurring low severity disturbances and less frequent, mesoscale events. The predominance of the OFS in this archetype is similar to Archetype 2, with a key distinction being the marked difference in structures between the two types (closed forest in this archetype versus an open woodland structure in Archetype 2) that results from the different primary disturbance agents.

Disturbance and Legacy Creation

In Archetype 3 forests, wind disturbance generates a mosaic of gap-scale conditions across the landscape that span

Figure 6.20 *Profile of an old-growth northern hardwood forest with emergent eastern white pine, Porcupine Mountains, Michigan, USA. Illustration credit: Robert Van Pelt.*

Figure 6.21 *Old-growth northern hardwood forest in Richmond, New Hampshire, USA, dominated by sugar maple and eastern hemlock with a minor component of northern red oak (right foreground). Photo credit: A.C. Cline, Harvard Forest Archives.*

Figure 6.22 *Multiple-tree fall gap in old-growth northern hardwood forest in Sylvania Wilderness, Michigan USA, representing YFS at the gap scale. Photo credit: Craig Lorimer.*

the general developmental stages described for Archetype 1 (Figure 6.22) and creates a fine-grained mixture of tree sizes and ages and abundance of structural legacies (Frelich and Graumlich 1994; Després et al. 2014). For consistency with the other archetypes, the following description of developmental stages is based on patterns observed following severe, stand-replacing windstorm events to illustrate key developmental processes (Figure 6.23); however, it is important to recognize that such events and stand conditions rarely affect more than 15% of the landscape (Frelich and Lorimer 1991; Lorimer and Halpin 2014). As such, the following developmental model should largely be translated and applied to the gap or patch scale, with the exception of the OFS, which occurs at stand to landscape scales.

Figure 6.23 *Post-disturbance conditions following severe windstorm event in old-growth northern hardwood forest in northern Wisconsin. Photo credit: Craig Lorimer.*

Preforest Stage

The Preforest Stage (PFS) in this archetype lacks the significance of the PFS in the fire-generated, conifer-dominated archetypes and is generally not a common feature of northern hardwood forest systems. The transience and insignificance of the PFS is due to the rapid site capture following canopy disturbance through release of advance tree regeneration, vegetative response of broken but surviving trees, and the rapid establishment of new seedlings from soil seedbanks or windblown seed. There may be a short period of 5 to 10 years during which a significant shrub component dominated by raspberries (*Rubus* spp.) is present on site following disturbance, serving as the dominant source of litter inputs and conserving nutrient losses (Covington and Aber 1980; Johnson et al. 1997). Nonetheless, with the rapid reestablishment of tree cover, the cover and contribution of these species to ecosystem processes declines rapidly (Donoso and Nyland 2006). With rapid reestablishment of tree cover, the productivity of these ecosystems quickly rebounds with litterfall rates returning to pre-disturbance levels in less than a decade post-disturbance (Plotkin et al. 2013).

Young Forest Stage

Composition of the Young Forest Stage (YFS) reflects a range of regeneration mechanisms, including long-lived, shade-tolerant trees released from advance regeneration, including sugar maple and American beech, and short-lived pioneer species originating from soil seedbanks and wind-dispersed seed, such as pin cherry and paper birch, respectively (Figure 6.24; Merrens and Peart 1992). Scattered throughout this cohort are older residual trees that survived the windstorm, including mid-tolerant yellow birch, which has the lowest susceptibility to windthrow of canopy trees in these forests (Canham et al. 2001; Hanson and Lorimer 2007).

Figure 6.24 *Young forest dominated by early successional bigtooth aspen and pin cherry (large stems in center) and late-successional sugar maple and American beech in the White Mountains, New Hampshire USA, fifteen years after severe disturbance.*

The live-tree structure and composition of the YFS is initially a reflection of differences in early height growth rates and longevity across species forming the new post-disturbance cohort. Early on, the YFS is generally a stratified stand with rapidly growing, short-lived species, such as pin cherry, trembling and bigtooth aspen, and paper birch forming upper canopy strata and the slower growing mid and shade-tolerant species occupying lower canopy positions. Senescence of most of the short-lived, early successional tree species occurs within 50–80 years, resulting in a shift in composition towards species ultimately dominating all canopy layers in the OFS. Canopy structure also shifts toward a largely single-layered stand with intense, density-dependent competitive thinning becoming the major cause of tree mortality (Lorimer et al. 2001). Canopy openings created via tree mortality are generally filled by lateral crown expansion of existing trees, given the small crowns of overstory trees dying during this stage (Dahir et al. 1996). This stage of development may last for 100 to 150 years (Frelich 2002; Bormann and Liken 1994).

During the YFS, downed log and snag legacies originating from the windstorm largely disappear by 50–80 years owing to the comparatively higher decomposition and snag-fall rates of the dominant hardwood species in these forests relative to other conifer-dominated archetypes (Harmon et al. 1986). An exception are northern hardwood ecosystems containing a significant eastern hemlock component; downed wood from this species may persist for over a century, serving as important substrates for the establishment of yellow birch and hemlock seedlings as they advance in decay (Figure 6.25; Marx and Walters 2008). Despite the disappearance of windthrown logs, the pit and mound microtopography created by these events is persistent (Lang et al. 2009), representing key microsites for establishment of small-seeded tree species, including yellow birch, during the PFS and ultimately driving spatial heterogeneity in understory herbaceous plant communities throughout forest development (Miller et al. 2002).

Mature Forest Stage

The Mature Forest Stage (MFS) represents a transitional condition between YFS and OFS and has been referred to as the *demographic transition* stage (Frelich 2002) and *transition* state (Bormann and Likens 1994) in previous studies describing northern hardwood forest development. Much of the compositional transition in these ecosystems occurs in the *YFS*, whereas the MFS is a period of structural transition in which the death of larger canopy trees results in

Figure 6.25 *a) An eastern hemlock log serving as a microsite for yellow birch establishment in an old-growth northern hardwood forest. b) A row of eastern hemlocks that established on a hemlock nurse log in the Upper Peninsula of Michigan.*

significant canopy openings recruiting new age classes and the development of large snags and downed logs. Competitive thinning remains an important mortality agent during this period; however, localized windthrow of canopy dominants also becomes an important process owing to greater susceptibility with increases in height and crown area (Canham et al. 2001; Lorimer et al. 2001). Owing to these increasing levels of mortality, this stage represents the peak in both the abundance of large trees (Lorimer and Halpin 2014) as well as in aboveground biomass in northern hardwood ecosystems (Bormann and Likens 1994; Halpin 2014). Correspondingly, **sapling** and **pole**-sized shade-tolerant trees make up an increasing proportion of the live-tree structure as the MFS progresses (Figure 6.26), with this stage generally lasting 250–300 years after the initial disturbance (Lorimer and Halpin 2014).

Old Forest Stage

Compared to other archetypes, the OFS has disproportionate importance as the dominant forest condition across northern hardwood-dominated landscapes (Frelich 2002; Bormann and Likens 1994; Lorimer and Halpin 2014). Long-term, chronic gap-disturbance results in high levels of complexity in the OFS, with shade-tolerant sugar maple, eastern hemlock, and American beech found across a broad range of tree ages and represented from seedlings and saplings through mature to old trees (Figure 6.27; Color Plate 6.5b; Leak 1975; Frelich and Graumlich 1994). Species diversity in the canopy layer is largely maintained through large, multiple tree-fall gaps and less frequent mesoscale disturbances like microbursts that generate legacies such as tip-up mounds and sufficient light conditions for yellow

Figure 6.26 *Mature Forest Stage in northern hardwood forests with (a) death of canopy trees resulting in a shift toward a (b) multi-aged condition with an increasing proportion of sapling and pole-sized trees.*

Figure 6.27 *Old Forest Stage northern hardwood forest in the Dukes Research Natural Area, Michigan USA. Note the range of tree sizes at a fine spatial grain.*

birch and other less tolerant species to establish and ascend to canopy positions (Webster and Lorimer 2005; Hanson and Lorimer 2007). The interplay of this disturbance regime with the range of species tolerances present in northern hardwood forests typically results in live-tree diameter distributions that are descending monotonic or rotated sigmoid in shape, reflecting a U-shaped mortality function in relation to tree size (Leak 1996; Lorimer et al. 2001).

Abundance of large woody material is also at its greatest level since the early PFS with hollow snags and logs becoming increasingly common owing to the increasing levels of heartwood decay on large diameter canopy trees (Hale et al. 1999). There is a high level of spatial heterogeneity associated with these legacies and live-tree conditions resulting from a history of gap development, and filling that very much resembles the fine-scale mosaic characteristic of the OFS in the fire-generated Douglas-fir-dominated archetype. One major difference is that the openings created by tree and gap-based mortality agents in the Douglas-fir-western hemlock example are generally not sufficient to allow for regeneration of the major pioneer species, Douglas-fir.

Variations on the Archetype

The disturbance regime for northern hardwood forests is generally wind-dominated; however, there are portions of the range of this forest type, particularly approaching the prairie-forest border in the western Lake States and in the transition to boreal regions, where wildfire is an important disturbance element in some deciduous hardwood forests (Fahey and Lorimer 2014; Frelich 2002). In these cases, fire-dependent species, such as northern red oak, eastern white pine, and red pine, may constitute a greater component of the canopy; however, much of the developmental sequence described above would still apply to these systems.

Figure 6.28 *Old-growth European beech forest, Stužica, Slovakia. Photo credit: Miroslav Svoboda.*

Other Examples of the Archetype

The European beech forests of western Europe are an important example of this archetype (Figure 6.28), with disturbance regimes quite similar to those documented for northern hardwood ecosystems of North America. As a result, these forests are also dominated by shade-tolerant species, including European beech and silver fir, that are favored by a disturbance regime consisting primarily of frequent, low severity disturbances from windthrow and disease, that generally remove less than 10 percent of the canopy each decade (Kenderes et al. 2008; Rugani et al. 2013; Nagel et al. 2014). As with North American northern hardwood forests, the OFS is the dominant condition for European beech ecosystems, with similarities in live-tree and age distribution forms to those described for northern hardwoods (Westphal et al. 2006; Nagel et al. 2014). The canopies of these forests in the OFS also contain an admixture of less tolerant tree species, such as sycamore maple, European ash, and Scotch elm that establish primarily in mesoscale disturbances, namely windstorms that remove 20–50 percent of the canopy every 100–500 years (Nagel et al. 2014). In addition to providing the levels of light availability necessary for canopy ascension, these more severe events generate microsites that lack beech and fir advance regeneration, allowing for the successful establishment of a diversity of species in these forests over time (Nagel et al. 2014).

Archetype 4: Forest Development with Mixed-Severity Disturbances

Here we describe forests having a development model not easily characterized by the natural disturbance regimes that fit neatly into the other archetypes discussed previously in this chapter. These are forests that sometimes experience stand-replacement events, like Archetype 1, but do not typically follow a similar sequential developmental pathway because the dominant late successional species establishes early and subsequent disturbances may be frequent enough and vary enough in severity that a shifting old-growth mosaic (e.g., Archetype 1) does not develop. These forests may experience frequent disturbances, largely surface fire like Archetype 2, but these fires may occasionally grade into canopy disturbance events, although not necessarily stand-replacing ones. Finally, these forests may be characterized by a mosaic of gap disturbances, from wind and other agents, similar to Archetype 3, but typically the gaps are at a larger patch scale, and the age structures of the forests are not nearly as complex as Archetype 3 forests, given the range of other disturbance processes affecting these systems beyond the low severity events predominating in Archetype 3. We place these forests into a fourth developmental model (Archetype 4), one characterized by mixed-severity and, potentially, mixed-agent disturbances.

Archetype 4 forests can be characterized by canopy disturbances that range from infrequent, near stand-replacing events to heavy, but often less than stand-replacing events, to large gap (patch) scale events. When fire is the source of this disturbance, it may originate from frequent surface fires that are of low severity, but that occasionally intensify into a canopy disturbance event. Key attributes of this archetype include 1) early and maintained dominance by shade-intolerant species, 2) important inclusions of additional tree species that span a range of shade tolerance and successional status, 3) the potential for an extended YFS, 4) broadly single cohort to distinctively two- or three-cohort age structures, and 5) high levels of structural complexity and spatial heterogeneity in the MFS and OFS.

The disturbance regime for this archetype does not neatly fit the classic definition of a (spatially) mixed-severity disturbance regime (Agee 1993), where individual disturbances are large but vary in severity and outcome across the landscape. Rather, the archetype includes stand-scale variation in disturbance severity among events across a landscape, as well as (sometimes) variation in the agent of disturbance. The most severe disturbance events in these forests tend to drive structure toward a broadly single or double-age cohort condition, with chronic and low-severity agents creating much of the complexity in developmental stages in the forest mosaic over time.

Example: Red Pine Forests of the Western Great Lakes Region, North America

The red pine dominated forest of the western Great Lakes region of North America is our primary example of Archetype 4 (Figure 6.29; Color Plate 6.6a). These forests can occur in the northern half of the Lower Peninsula of Michigan, the entire Upper Peninsula of Michigan, northern Wisconsin, north central Minnesota, which are all in the US, as well as portions of southern Ontario, Canada. Historically, a red pine dominated, mixed-species ecosystem was an important upland component on more xeric sites in this region, occupying over 4 million ha prior to significant Euro-American settlement (Frelich 1995). While characterized as mixed-species, red pine along with eastern white pine are the dominant tree species, with varying amounts of other conifers and hardwoods, including balsam fir, white spruce, jack pine, trembling and bigtooth aspens, red maple, northern red oak, bur oak, and paper birch.

The disturbance-development model for the red pine ecosystem is one in which fire plays a prominent role, both as surface and crown fire. In the classic interpretation, frequent, low severity surface fires, with return intervals on the order of 5 to 75 years, reduces hardwood competitors, thins populations of pines, and maintains an open woodland structure, with patchy, open (50–75 percent) canopy cover (MN DNR 2003; Figure 6.30). Infrequent stand-replacing crown fire occurred with a return interval of 150–250 years (Heinselman 1996), resulting in the regeneration of even-aged stands of red pine.

The actual disturbance/development model for red pine forests is more complex than the classic model (Box 6.1). Historically, fires (mostly surface) could be quite frequent, with mean return intervals ranging from 9 years and up to 50 or more years (MN DNR 2003; Guyette et al. 2016; Drobyshev et al. 2008a). It is suggested that this variation was related to Native American burning activity, with shorter return intervals associated with more active use areas. On top of anthropogenic ignition, lightning was certainly also an occasional source of ignition, but perhaps not the primary source (Loope and Anderton 1998).

Regardless of cause, surface fires periodically intensified, resulting in significant mortality of canopy trees, but often in a patchy, less-than-stand-replacing spatial arrangement (Fraver and Palik 2012; Bergeron and Brisson 1990; Drobyshev et al. 2008a, 2008b). The developmental attributes discussed below result largely from the heterogeneous nature of the fire regime. In addition to fire, there is evidence (largely unstudied) for gap or patch-scale tree mortality from wind and fungi in the *Armillaria* genus (see McLaughlin 2001; Silver et al. 2013). These may be important drivers of heterogeneous stand structures and composition

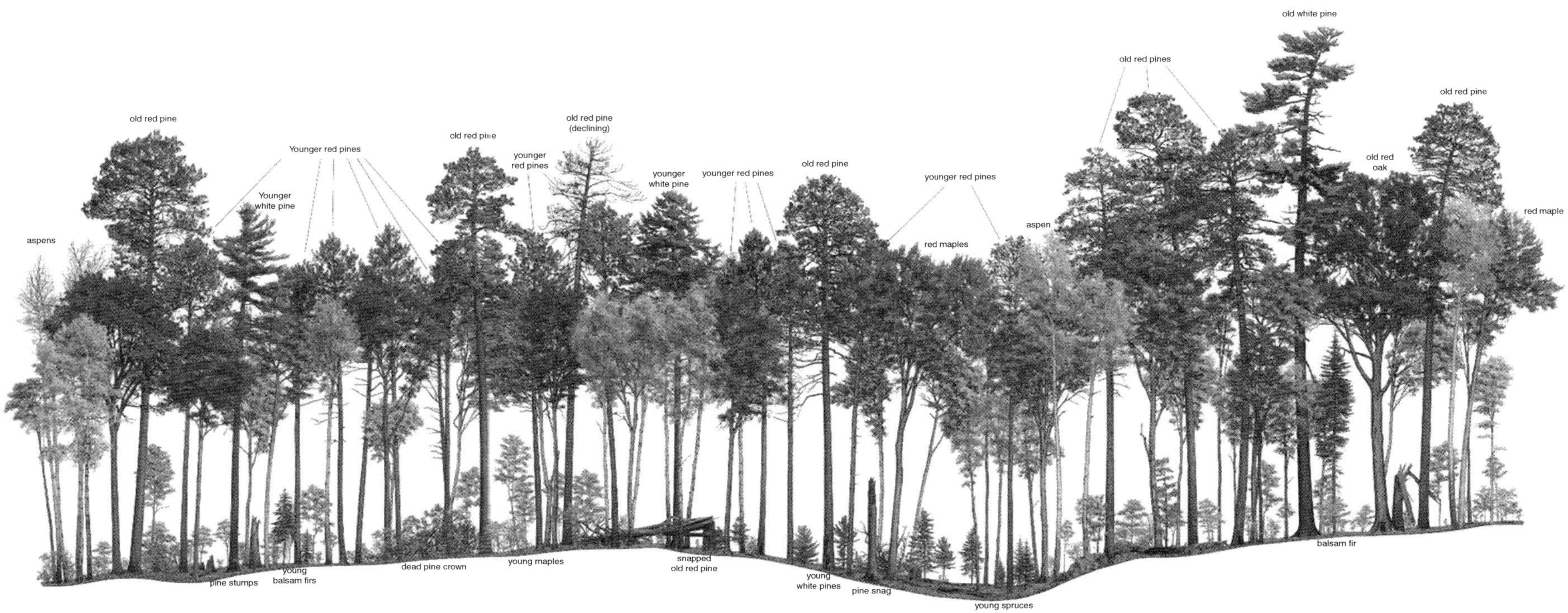

Figure 6.29 *Profile of an old-growth red pine forest. Illustration credit: Robert Van Pelt.*

Figure 6.30 *Surface fire in a red pine dominated forest in the Huron-Manistee National Forest in Michigan, USA. Note its open, woodland structure. Photo credit: Jack McGowan-Stinski.*

Box 6.1 **Evidence for complex age structures in red pine forests**

The classic disturbance/developmental model for red pine forests, in which infrequent stand-replacing fire results in the establishment of largely even-aged stands, has permeated the thinking of ecologists and foresters for decades. Interpretation of this model has served as justification for an even-aged silvicultural system, primarily clearcutting followed by planting, as an emulation of natural disturbance regimes. However, there is ample evidence for more complex age structures, and by inference, more complex disturbance regimes. This evidence comes from early published references, historical photographs, and more recent **dendroecological** research.

Early researchers often referenced the complex age structures of old red pine stands (Bergman 1924; Shirely 1932; Eyre and Zehngraff 1948), recognizing that these conditions likely developed from less-than-stand replacement fire and, quoting Shirely (1932), were *typical* of the forest. Evidence also comes by way of old photographs from the early nineteenth century that depict multi-cohort age structure of natural red pine forests, regeneration in gaps, and the patchy canopy structure of these stands (Figure 6.31).

Perhaps the best evidence for complex disturbance regimes and resultant structures comes by way of stand

Figure 6.31 *Photo documentation of gap regeneration of red pine (colloquially referred to as Norway pine) in what is now the Chippewa National Forest, Minnesota, USA. The caption for this photo, dated 1939, reads* nature's own group selection: red pine reproduction often became established in great abundance following a seed year, but because of the density of the overstory such reproduction was generally shaded out within a few years except in openings caused by tree mortality. *Photo credit: Forest History Society.*

reconstructions, such as these from northern Minnesota, USA (Figure 6.32; Fraver and Palik 2012). These demonstrate that disturbance dynamics had to be variable across the regional landscape, as reflected in the variable red pine age structures, and that at the stand-scale, partial canopy disturbances were common, an interpretation supported by research from other areas in the Lake States and Canada (Bergeron and Brisson 1990; Drobyshev et al. 2008a, 2008b).

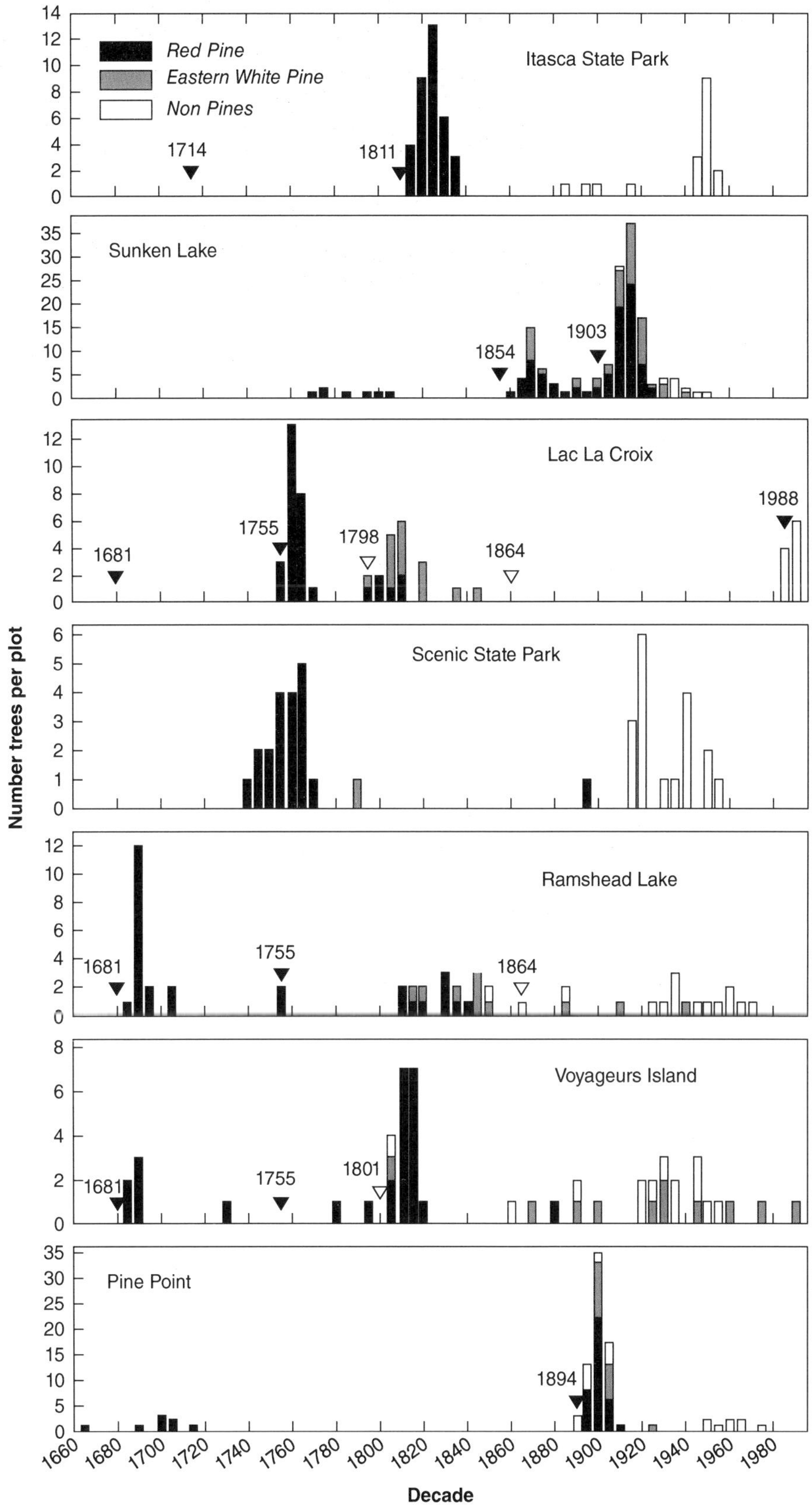

Figure 6.32 *Age structures for old-growth red pine stands in Minnesota, USA, showing a variety of forms, including broadly one-cohort, two-cohort, and three-cohort structures. Dates are known fire occurrences. (Redrawn from Fraver and Palik 2012.)*

(i.e., hardwoods versus conifers), essentially reinforcing the heterogeneous structure resulting from patchy fire.

Developmental Pathway

Disturbance and Legacy Creation

Similar to the Archetype 1 example of Douglas-fir-western hemlock, structural legacies, including residual live trees and deadwood structures, are an important attribute of red pine forests after a stand or patch initiating disturbance. The distinction from Archetype 1 is that red pine forests may be characterized by fewer numbers of deadwood legacies, greater numbers of live tree legacies, and greater spatial heterogeneity of arrangements, reflecting the often less-than-stand-replacing and patchy nature of disturbance (Figure 6.33). The spatial heterogeneity of disturbance and legacy creation can be surmised by examining the distribution of tree cohort establishment in an old-growth stand (Figure 6.34). From this it is easy to see how after the disturbances that initiated new cohorts, the distribution of residual trees, and the tree-derived structures that would have been present at the time, would have been spatially heterogeneous, due to varying spatial severity of disturbance. The key message here is that after a regeneration disturbance, spatial heterogeneity of legacy structures within stands is a hallmark of this ecosystem.

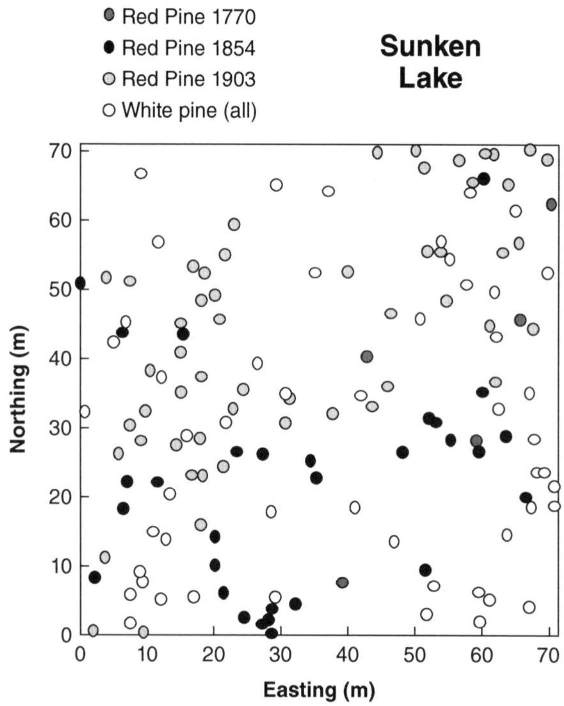

Figure 6.34 *Spatial arrangement of trees in an old-growth red pine stand, Sunken Lake Natural Area, Minnesota, USA. Red pine trees are listed according to cohort establishment dates; establishment dates for eastern white pines are pooled. (Redrawn from Fraver and Palik 2012.)*

Figure 6.33 *Old red pine stand after a partial canopy disturbance, leaving residual live tree legacies. Circa early twentieth century, Chippewa National Forest, Minnesota, USA. Photo credit: Forest History Society.*

Most legacy trees in this forest will be red pine, with some eastern white pine, reflecting the dominance of these species in the OFS. These trees will be equally structurally rich and as large as those in the OFS (see below), with the potential for additional branch and top breakage and wounding leading to cavities. Additionally, snag and dead tree legacies will be largely red and eastern white pines as well, and their sizes will be reflective of trees in the *OFS*.

Preforest Stage

There has not been much research on the Preforest Stage (PFS) in red pine forests. In some cases, it may be that it is not an important attribute of this ecosystem, as it is in Archetype 1. The reason for a short duration PFS in red pine forests results from the potential for aggressive vegetative sprouting after disturbance of shade-intolerant hardwoods, including bigtooth and trembling aspens and paper birch, species that can be important components of this ecosystem. When individuals of these species are present in the MFS or OFS, a sprout-origin matrix establishes from roots (aspens) and stumps (paper birch) within the first year after disturbance, forming dense populations of stems where they occur (Figure 6.35). When this dynamic occurs, the PFS is essentially bypassed. However, this forest did have

Figure 6.35 *Trembling aspen root suckers develop in an opening in a mature red pine forest, Chippewa National Forest, Minnesota, USA. Photo credit: Christel Kern.*

more of a woodland structure historically, with a patchy, interrupted canopy cover ranging from 50–75 percent (MN DNR 2003), suggesting a naturally lower density of trees than is typical in managed stands of this forest. As discussed below, if establishment of pine populations occurred over an extended period, a PFS may have been important, at least in some settings.

Young Forest Stage

An underappreciated aspect of the red pine forest, or at least an aspect that is often ignored during management, is its extended Young Forest Stage (YFS). This may occur at the whole-stand scale, in cases where stands experience a near-replacement disturbance, or it may occur at the patch scale, where cohorts of trees establish in large gaps. In either case, establishment periods and dominance by early successional species can be lengthy. For instance, establishment periods for red pine after disturbance can range from 40 to 70 years (Figure 6.32; Fraver and Palik 2012) and initial dominance by early successional hardwoods, as described above, can last similar lengths of time.

In many examples of this forest, there may be early shared dominance between red pine and pioneer species, including aspens and paper birch, but also jack pine, which establishes early after disturbance from seed sources in or near the disturbed area. As such, a distinctive attribute of this phase, compared to Archetype 1, is early establishment of a shade-intolerant tree species, i.e., red pine, which maintains its dominance into the MFS and OFS. Given this dynamic, there is not always a clear and prolonged successional replacement dynamic that regularly occurs with this forest, that is, one in which red pine replaces a dominant earlier successional species and a later successional species replaces red pine. The prolonged dominance of red pine over time is similar to the pattern in Archetype 2, where frequent surface fires maintain dominance of shade-intolerant species. By contrast, the establishment of red pine amongst shorter-lived species, such as aspen, paper birch, and jack pine, introduces a background template of patch-scale mortality occurring early in the development of these forests that is not observed in Archetype 2. The patches dominated by these shorter-lived co-associates become important nuclei for gap-scale processes, such as root disease pockets, wind gaps, and patchy crown fires as the red pine cohort develops.

Mature Forest Stage

As development continues into the Mature Forest Stage (MFS), more shade tolerant, later successional species, including eastern white pine, northern red oak, and red maple increase in dominance. These species may be present early after disturbance, often sprouting from stumps in the case of the hardwoods (Palik and Pregitzer 1993) or establishing from nearby seed sources for pine (Palik and Pregitzer 1994), but since they are more tolerant of shade, they gain dominance below the emergent red pine canopy later in development. However, under a typical disturbance frequency, these species rarely replace red pine. Still, the importance of co-occurring species is an important feature of the red pine forest, particularly as it develops into the MFS. In fact, historically, a suite of other tree species were often equally important as red pine in terms of cumulative importance (Whitney 1986; Nowacki et al. 1990; Palik and Pregitzer 1992; Frelich and Reich 1995).

Another important attribute of the MFS is the development of spatially heterogeneous structure resulting from infrequent, less than stand-replacing fire and other disturbances that kill mature trees in patches. It is in the MFS, and continuing into the OFS, that a new cohort of red pine may establish in these openings (Figure 6.36). Moreover, localized dominance of hardwoods may be facilitated by patch mortality of pines from *Armillaria* root-rot, enhancing compositional variation. Given the ability of this fungus to operate as a **saprotroph** and **biotroph**, these patches often originate in portions of the forest where jack pine formed a component of the YFS and begin affecting red pine during periods of stress, such as prolonged droughts.

Old Forest Stage

Similar to the other archetypes, it is in the Old Forest Stage (OFS) that the red pine forest achieves great structural complexity, notably including large old trees and diverse tree sizes, abundant large woody material, including large-tree derived snags and downed logs (Figure 6.37; Color Plate 6.6b), and

Figure 6.36 *After patchy disturbance in a MFS red pine stand (140 years old), red pine reproduction established in groups in openings. Chippewa National Forest, Minnesota, USA. Photo credit: Forest History Society.*

substantial spatial heterogeneity in the distribution of these structures. Again, this structure results from the patchy nature of disturbance, as described in the MFS, but also from the long life span and large stature achieved by red pines and co-occurring eastern white pines. Both species have potential life spans over 300 years (Burns and Honkala 1990). At these old ages, trees will have developed large diameters and branch systems, and sometimes broken tops and cavities. Snags and downed logs derived from these trees will be equally large.

The mixed-species composition of this forest is another attribute of the OFS. While the overstory, particularly the emergent canopy, continues to be dominated with red pine and often co-occurring eastern white pine, all the other species mentioned above will typically have a presence in the OFS. While the aspens and paper birch, as well as jack pine, may have greatly decreased in importance over time due to shorter life spans, these species may be able to maintain a presence in this forest with continued large gap-scale disturbances. Mid- to shade-tolerant eastern white pine, northern red oak, and red maple, species that increased in importance in the MFS, will continue to do so in the OFS, again due to the patchy nature of disturbance, which allows these more fire susceptible species to recruit into larger size classes in areas skipped by fire.

Another important compositional attribute of the OFS in this forest is that with continued infrequent larger patch disturbances, red pine continues to establish and maintain dominance in the forest. This is contrary to the OFS in the Douglas-fir-western hemlock example of Archetype 1, where gap disturbances in the OFS are not large enough for new Douglas-fir to establish and, consequently, the forest gradually transitions to dominance by later successional western hemlock. Again, such a successional replacement dynamic is not characteristic of the red pine forest, at least not when it experiences a natural disturbance regime.

Variations on the Archetype

The disturbance (fire) regime characteristic of the red pine forest—frequent surface fire punctuated by infrequent cohort establishing (but less than stand replacing) fire—certainly graded into a higher severity forest-replacement component at times and places in the landscape. Establishment histories from some old-growth stands demonstrate this (Fraver and Palik 2012; Silver et al. 2013). When this occurred, the variant on the archetype was more of an even-aged stand, although one with establishment windows that might be measured in decades as opposed to years (Figure 6.32). There are also examples of this forest type where hardwoods, particularly species of oak, are more abundant and of larger stature (Whitney 1986; Nowacki et al. 1990). This would occur where the surface fire interval is extended, or fires were particularly patchy, allowing these species to establish and grow into large sizes.

Other Examples of the Archetype

The Scots pine-dominated forest of southern Fennoscandia is a good example of an Archetype 4 ecosystem (Figure 6.38). The natural disturbance regime for this forest is characterized by infrequent severe, but patchy fires that result in establishment of pine cohorts. For instance, in a study in Finland, a Scots pine site experienced three regeneration-scale fires over a 260 year period (Wallenius et al. 2002). However, it is recognized that fire scar dating in Scots pine may grossly underestimate the frequency of low-severity surface fires, which were common historically in this ecosystem (Piha et al. 2013). That these fires varied in severity within the fire areas (i.e., stands) is reflected in the multi-cohort (2–3 cohorts) age structure noted in several studies (Volkov et al. 1997; Kuuluvainen et al. 2002; Wallenius et al. 2002). Moreover, it has been noted that even severe fires historically were not always stand replacing, with many of the older Scots pine surviving even several fires, i.e., live tree legacies (Kuuluvainen et al. 2002). Similar to our red pine example, periodic fire served to maintain Scots pine dominance by opening the canopy and creating suitable conditions for the establishment of new cohorts from seed provided by legacy trees, while arresting successional replacement by Norway spruce and deciduous trees.

Utility of the Forest Archetypes

What is the value of considering the four archetypes described above? The archetypes as presented are simplified models of how forest ecosystem development can proceed in unmanaged forest settings in response to primary disturbance regimes. As we have presented them, the episodically disturbed archetype is characterized by linear, segmented,

Figure 6.37 *Old Forest Stage features in red pine forests in Minnesota, USA including: (a) large trees and deadwood on the ground, Scenic State Park; (b) large snag, Voyageurs Island; (c) diverse tree sizes and hardwood inclusions, Itasca State Park; (d) large old pines and red pine cohort regeneration, Pike Bay, Chippewa National Forest circa 1939; (e) heterogeneous understory conditions, Norway Beach, Chippewa National Forest circa 1938; (f) emergent pine over more tolerant tree species, Voyageurs Island. Photo credits: (a–c, f) Shawn Fraver, (d, e) Forest History Society.*

Figure 6.38 *Old Forest Stage Scots pine stand in northern Finland displaying multi-cohort age structure, abundant deadwood on the ground, large snags, heterogeneous canopy, and charring from past fires. Photo credit: Timo Kuuluvainen.*

and predictable dynamics, whereas the frequent-fire archetype, and to some extent the gap-disturbed archetype, rarely experiences stand-replacement disturbances events, with the heavy-partial disturbance archetype representing something in between.

Of course, the natural world is never this simple. The complexity of the real world begins with the complexity and stochastic nature of the disturbances themselves. Stand initiating disturbances (Archetype 1) never act uniformly over very large areas. For example, large disturbances are rarely severe throughout, i.e., they do not kill all of the dominant trees of the pre-disturbance forest. The areas of intermediate disturbance exhibit highly varied levels of legacies and openness. The resulting forest landscapes are complex mosaics, including patches of diverse tree ages and structural conditions, which undergo developmental patterns very different from the archetypical model. Forests generated and maintained under frequent-fire regimes (Archetype 2) would appear to be extraordinarily stable as an ecosystem—at least as long as their fire regime is maintained. However, fire behaviors do vary and extensive areas of a frequent fire landscape may have fire regimes that are sufficiently variable that mixed behaviors are encountered, perhaps transitioning them into Archetype 4 where spatially variable severity of disturbance may occur. In turn, Archetype 4 may grade into the episodic disturbance archetype and follow that developmental pathway in some portions of the landscape.

Given the complexity of the real world, why are we presenting these simple archetypes? First of all, they were devised to help students of ecological silviculture see as clearly as possible how disturbance and developmental processes covered in Chapters 4 and 5 play out in real world examples. Just as the classical silvicultural practices and systems were developed based on an understanding of forest biology, production ecology, and succession (Chapter 3), these archetypes can serve as benchmarks for basing the development of ecological silviculture for different forest systems based on knowledge of their developmental processes. Equally important, the archetypes help to illustrate that the product of these developmental processes is not just structural or compositional changes, but changes in the functionality of the forest (as reviewed in Chapter 5). Different forest stages in the different archetypes have different functional capabilities.

The second major value of the archetypes, even in the context of a highly variable real world, is to help us understand how we can develop approaches to silviculture that will sustain arrays of ecosystem services similar to those provided by the natural forest. For example, biological legacies exemplify the ways in which structural, compositional, and, most importantly, functional attributes can be sustained through multiple generations of forest. Natural forests rarely start from scratch after a disturbance; rather, the legacies associated with even major disturbances demonstrate ways that an array of organisms and functions can be sustained in managed systems. The tree and gap-level dynamics of several of the archetypes demonstrate the tendency of forests—in the absence of major exogenous disturbances—to evolve toward species-rich and structurally complex ecosystems that are sustained over long periods of time by small-scale chronic disturbance regimes. Forest conditions dominated by chronic disturbances appear to be the most stable of forest states and, in many cases, the easiest to sustain by management that emulates these disturbances.

Finally, the intent of the archetype models is to examine the patterns and processes associated with natural forest development so as to provide us with a set of principles that we can use as the basis for silvicultural prescriptions for ecological management, which we begin to address in the next section. Our understanding of natural forest developmental processes has advanced considerably since the formalization of most silvicultural approaches around the globe and the presentation of these archetypes provides an opportunity to reflect on these models of development as we chart a path forward through the remainder of the text.

PART III

Applying Principles of Natural Models to Silviculture

CHAPTER 7

Incorporating Continuity into Silviculture: Legacy Retention in Regeneration Harvests

Recall from Chapter 3 that the first foundational principle for ecological silviculture is *continuity*, which we defined as the provision for continuity in forest structure, function, and biota between pre- and post-harvest ecosystems. In Chapter 4, we talked in more detail about biological legacies as the way continuity is maintained in forest ecosystems after a natural disturbance severe enough to result (ultimately) in tree regeneration. It follows then that an important application of the principle of continuity in ecological silviculture is the deliberate retention of structures and biota between the old and new forest as part of regeneration harvests. This is in contrast to many timber-focused regeneration systems, which intentionally or inadvertently eliminate much of the pre-harvest forest in stands (or at the gap-level), thereby creating significant discontinuities between forest generations (Franklin et al. 2018). Our objective in this chapter is to delve more deeply into the principle of continuity as implemented through legacy retention. For this, we will: 1) define retention harvesting and consider its importance in the practice of ecological silviculture; 2) contrast legacy retention (or the lack thereof) in classical regeneration systems with their (purported) natural disturbance counterparts; and 3) consider how legacy retention can be incorporated into any regeneration system. This chapter is meant to be a general overview, while in later chapters we will discuss legacy retention as part of silvicultural prescriptions for the four disturbance archetypes presented in Chapter 6.

Biological Legacies Revisited

We learned in Chapter 4 that legacy structures include large healthy trees, those declining in health (i.e., decadent), snags, and large deadwood on the forest floor, such as boles and branches. In the case of deadwood legacies, most will likely be created by the disturbance, but some may have carried over from the pre-disturbance forest, particularly forests in the Old Forest Stage, where large snags and logs in various stages of decay are most abundant. There are two features of structural legacies that a forester should keep in mind when considering their management. First, structural legacies endure after (and of course are created by) even the most intense stand-replacement disturbances, including fire, with the amount of live trees and deadwood generally being proportional, in opposite directions, to the intensity of disturbance (see Figure 4.2). Second, legacy structures play critical roles as habitat and modifiers of the physical environment for a variety of organisms (i.e., see *lifeboating, structural enrichment,* and *softening the matrix* in Chapter 4). Some biota may be associated with the pre-disturbance forest and are able to persist in the new forest because of habitat conditions created by legacy structures. Other organisms may be associated with the Preforest or Young Forest Stages, but still require structural legacies from the old-forest for habitat. Retention of *compositional legacies* (e.g., various tree species and other plants) are also an important consideration. Sometimes compositional legacies are important because they retain certain growth forms or life histories, such as hardwood trees in an otherwise conifer-dominated stand. Compositional legacies may also be important because they maintain special functional capabilities in the new forest, such as nitrogen fixation or hard mast production. Compositional legacies may include advance regeneration or seedbanks of tree species that will be important in the young forest stage.

Legacy Retention in Practice

Increasingly, legacy retention is being incorporated into regeneration prescriptions throughout many forest regions worldwide (Gustafson et al. 2012; Lindenmayer et al. 2012). Largely, this relates to the ability of retention to reduce ecological impacts of classical regeneration systems (e.g., Mori and Kitagawa 2014; Fedrowitz et al. 2014; Franklin et al. 2018). For example, species richness and abundance of song birds in different forest types have been shown to be higher in partial harvests with retention compared to clearcuts (Norton and Hannon 1997; Lance and Phinney 2001), although not necessarily as high as the uncut forest.

Legacy retention can be used to sustain structures and conditions that require time periods longer than a traditional **cutting cycle** or **rotation** to develop (Franklin et al. 2018).

Such legacies include large old trees, which provide habitat directly and are also the source (eventually) of snags and dead trees on the ground. Retention can be used to create structurally complex forests that are similar to the Old Forest Stage, but in stands managed on time periods much shorter than occur naturally (Franklin et al. 2018). In fact, by using retention approaches, old tree structures can be incorporated into managed stands without having to grow an entire old-forest stand because individual structures (such as old trees) can be retained or developed over multiple rotations of cutting cycles.

Legacy retention or management is often implemented using *variable retention harvesting* or VRH (Franklin et al. 1997). The Society of American Foresters defines VRH as:

> *". . . an approach to harvesting based on the retention of structural elements or biological legacies (trees, snags, logs, etc.) from the harvested stand for integration into the new stand to achieve various ecological objectives. . . . Major variables are types, densities, and spatial arrangements of retained structures" (Helms 1998).*

As this definition indicates, and as we discuss in more detail later, the variable modifier refers to the opportunity for the forester to alter what and how legacies are retained both within and among harvest units. Variable retention harvesting can be focused on individual structures, such as live trees, snags, and downed logs on the forest floor (Figure 7.1). It can also be focused on retention of particular species in the overstory or understory or on broad functional groups (e.g., conifers or hardwoods). Retention may be either distributed widely through the harvest unit (i.e., dispersed retention; see *Where to Retain Legacies* below) or concentrated in intact forest patches (i.e. aggregated retention) (Figure 7.2). As we will see later, the "variability" of VRH makes the approach readily adaptable to a broad array of forest types and management objectives. Before we address VRH in more detail however, we must first review the place (or lack thereof) retention has had in classical regeneration systems.

Figure 7.1 *Biological legacies that may be retained during a regeneration harvest can take the form of individual structures, such as (a) large trees, (b) large snags, and (c) logs on the ground.*

Legacy Retention in Classical Regeneration Systems

Biological legacies typically left after natural disturbance (see Chapter 4, Table 4.1) and those left after regeneration harvests based on classical models can be distinctly different in their type, condition, amount, and spatial pattern. The contrasts between legacy conditions in natural systems and managed counterparts can be significant, whether silvicultural approaches involve even-aged, two-aged, or uneven-aged regeneration systems (Table 7.1). In fact, none of the classical regeneration systems have purposely retained elements of harvested stands for ecological reasons, particularly where even-aged regeneration systems are being applied (Ashton and Kelty 2017), but this is also true at the scale of harvest gaps within stands.

Figure 7.2 *Biological legacies may be dispersed across a harvest unit (a) or aggregated into patches of trees (b), which may contain various legacy structures, as well as other conditions that are desirable to protect. Photo credit (a): Bob Williams.*

Table 7.1 *Biological legacies associated with classical regeneration systems as traditionally applied. (Adapted from Franklin et al. 2007.)*

	Regeneration Method					
	Even-Aged			Two-Aged	Uneven-Aged	
Legacy	Clearcut with site preparation	Seed tree with site preparation	Shelterwood with site preparation	Clearcut, Seed tree, or Shelterwood, *with reserves*,and site preparation	Group selection	Single-tree selection
Live, mature trees	No	No/Maybe[1]	No[2]	Yes	Few/No (in group)	n.a.
Seedling bank	No	No	Yes	Yes	Possible	Possible
Intact understory	No	No	No	Possible	Possible	Possible
Snags	No	No	No	No	No (in group)	Few/No[3]
Logs	Few/No	Few/No	Few/No	Few/No	Few/No (in group)	No
Uproots	No	No	No	No	No	No
Mineral seedbed[4]	Yes	Yes	Yes	Yes	Possible	Possible

[1]As traditionally applied, seed trees are removed, although logistics/economics often prevents this from happening.
[2]Following final removal of overstory.
[3]Declining, poor quality, and dead trees receive priority for removal under single-tree selection resulting in low numbers of snags.
[4]Assuming ground-based harvesting.

We review the legacy outcomes of classical regeneration systems below, referring to the parallel natural disturbance regime, so as to develop an appreciation of the desired legacy outcomes when emulating regenerating disturbances with silviculture. (Note that Franklin et al. 2007 and 2018 were sources for material in the following sections.)

Legacies in Even-Aged Regeneration Systems

Clearcut, seed tree, and shelterwood are the standard even-aged regeneration systems that, in concept, have been considered as emulative of stand-replacement natural disturbances (Chapter 3). In their purest forms, they do not

Figure 7.3 *The level of complexity in stands managed with even-aged regeneration systems can be substantially reduced compared to natural forests, as in this example of (a) jack pine forest after a stand-replacing fire and (b) a jack pine forest after a harvest. Photo credits: (a) Shawn Fraver, (b) Josh Kragthorpe.*

provide for long-term retention of live or dead trees in the harvested stand (Table 7.1). Typically, overstory trees retained during a seed tree cut and an initial shelterwood entry are eventually removed after successful establishment of regeneration (Ashton and Kelty 2017). The option of retaining live trees indefinitely is explicitly mentioned only in some forms of seed tree cutting (Ashton and Kelty 2017) and in modification of other types of even-aged systems, i.e., **clearcut with reserves** or **irregular shelterwood**.

The lack of live tree legacies is then a major difference between these classical even-aged regeneration methods and natural stand-replacement disturbances whether by fire, wind, insects, or some other agent. Other legacy structures will also typically be missing or at least low in abundance from classical even-aged regeneration systems, including large snags and large downed boles (Figure 7.3) and pit and mounds in the case of windthrown trees. Importantly, the lack of significant structural retention in even-aged regeneration systems by default results in homogeneous post-disturbance conditions (Bergeron et al. 1999), whereas natural disturbances that occur at the stand-scale typically generate substantial spatial heterogeneity of biological legacies (Eberhart and Woodard 1987; Fraver and Palik 2012).

Modifying clearcut or shelterwood prescriptions to incorporate wildlife trees, snags, and logs is widely practiced and represents a good step toward more effective integration of legacy management into siviculture. Moreover, even-aged harvesting with reserves is a more formalized approach to incorporating some legacies into classical even-aged regeneration systems. However, neither of these applications may recognize the numerous ecological roles played by reserved structures or the array of structural retention prescriptions possible, such as spatially variable patterns of retention. Moreover, the long-term functions of legacy structures often are not considered in the context of the overall silvicultural system, as little provision is made for how areas adjacent to reserved structures might be managed differently over time (Tappeiner et al. 2015). Also, confusion or misunderstanding can also arise from use of the term *with reserves*. When substantial structural legacies are retained in an area identified as a clearcut, observers may be confused because the area is clearly not what most observers visualize as a clearcut (Box 7.1). Moreover, using terms like reserve shelterwood or seed tree in the context of legacy retention is misleading when the primary goal for retaining trees is sustaining ecological services, rather than regeneration potential.

Legacies in Two-Cohort Regeneration Systems

Partial harvesting approaches that create stands with two cohorts of trees is a middle point along the gradient from even-aged to uneven-aged regeneration systems and in some ways are emulative of meso-scale or mixed-severity natural disturbances (see Chapters 4). In the nomenclature of classical silviculture, an irregular shelterwood (formally defined as retention time in years that exceeds 20 percent but less than 60 percent of the rotation age) may result in a two-cohort age structure, although multi-cohort age structures are also possible under this system (Raymond et al. 2009). There is retention of live trees by definition in such systems. However, irregular shelterwoods, particularly extended irregular shelterwoods, are often implemented in ways that create homogeneity in structure, resource availability, and competitive environments (Figure 7.5), and pay limited attention to retention of deadwood legacies (Table 7.1).

Box 7.1 **Clearcutting with reserves as an application of legacy retention**

A common approach to incorporating retention of biological legacies into practice is to modify a classical even-aged regeneration system so as to include retention guidelines. The following is an excerpt from an actual silvicultural prescription from an ownership in northern Minnesota, USA that takes this approach.

Existing stand condition: Mature trembling aspen stand (11 ha) with scattered balsam fir, red maple, paper birch, red pine, eastern white pine, and white spruce in the overstory. Aspen seedlings dominate the understory; red maple, white spruce, paper birch, and balsam fir seedlings are also present.

Desired condition: Conversion to a red pine dominated stand with a mixture of additional species present.

Harvest method: *Clearcut with reserves*

Timber marking guide:

1. Cut all aspen, balsam fir, paper birch, and white spruce.
2. Retain other species (red maple, red pine, eastern white pine).
3. Retain a total of at least 0.8 ha (2 ac) in reserve areas (legacy patches).

The problem with using language such as this (i.e., *cleacut with reserves*) is that even to the casual observer it is evident that the stand was not regenerated using a clearcutting regeneration system (Figure 7.4). In fact, the intent of this harvest was to create complex and spatially heterogeneous structural conditions in the developing stand using a variable retention harvest. Both the scattered retained trees and the legacy patches result in a two-cohort structure, an outcome not possible with clearcutting. Using the modifier *with reserves* also does not fully covey the intent and desired conditions. This incongruity between the name and the intent of the prescription highlights the need to apply silviculture terminology that best describes the objectives being accomplished as opposed to insisting on rigid adherence to the use of classical terms.

Figure 7.4 *Post-harvest view of a clearcut with reserves harvest (the actual prescription terminology that was used) in Minnesota, USA. Rather than calling this a clearcut, this is better defined as an exemplary application of variable retention harvesting used to create a structurally complex stand with a two-cohort age structure.*

Generally, residual trees in an irregular shelterwood are dispersed uniformly across the stand, creating fairly uniform resource and competitive environments. A primary consideration when retaining overstory trees in such systems has been their expected growth potential. There also may be interest in providing adequate seed sources for a new cohort of natural regeneration. For both reasons, a single tree species may be selected preferentially for retention, whereas natural disturbance may leave representatives of most of the major species that occurred in the stand (e.g., Palik and Robl 1999). Moreover, two-cohort systems may not explicitly consider the ecological importance of retaining dying and diseased trees; generally these are considered liabilities and are removed during harvesting. These trees, which can be plentiful after natural disturbance have great immediate and future value as wildlife habitat.

Legacies in Uneven-Aged Regeneration Systems

Single-tree and group selection are the classical uneven-aged regeneration systems (Ashton and Kelty 2017; Chapter 3). Theoretically, selection approaches are modeled after individual tree or gap-based natural disturbances. In practice, these approaches can be highly formalized, such as where

Figure 7.5 *An example of an irregular shelterwood that created homogeneous structural and competitive conditions across the stand, Menominee Reservation, Wisconsin, USA.*

selection of trees is driven by efforts to create and maintain balanced diameter distributions, even though most natural stands do not exhibit such balance or stability in form over time (Matthews 1989; O'Hara 1996; Chapter 8). Moreover, selection as classically implemented seeks to improve residual stand quality over time by removal of diseased, dying, and poorly formed trees. The problems then with selection systems from the standpoint of our discussion is that there is little attention given to retention of legacies in single-tree selection, including trees larger than a maximum prescribed diameter or those with poor form (i.e., habitat trees), or as legacies within openings created through group selection (Table 7.1). As such, uneven-aged regeneration systems, just like even-aged approaches, can also result in simplification of structure (Seymour and Hunter 1999). For example, selection as traditionally applied in northern hardwood ecosystems in the North American Great Lakes region, even when applied over long periods in old-growth stands, greatly reduce the abundance of coarse woody material in the stand (Gronewold et al. 2010).

Single-tree selection

Typical single-tree selection systems contrast significantly with tree-scale natural disturbances in terms of the types, amounts, and pattern of trees removed, as well as the nature of the resultant biological legacies (Table 7.1). Most selection systems involve the systematic removal of at least some of the large, old, or decadent trees (Moser et al. 2002; Kenefic and Nyland 2007). Some of these trees are the most valuable in the stand from a timber perspective, but are viewed as no longer contributing significantly to stand growth based on a presumption that large old trees are **over-mature** and at greater risk of dying in the near future. Thus, their removal is driven more by economic rather than ecological considerations. However, studies of tree demography in many forest types and for many tree species show that large trees actually have lower rates of mortality—i.e., greater probability of survival—than smaller trees, at least until individuals get very large (Lorimer et al. 2001; Monserud and Sterba 1999). Moreover, as discussed earlier in the book, large old trees have tremendous ecological value, as they provide habitat in the form of cavities, deeply fissured bark, and large limbs.

In practice, most single-tree selection approaches also do not sufficiently recognize the ecological importance of decadent trees or of snags and down boles. In fact, individual-tree selection as traditionally applied may prioritize the removal of decadent, diseased, or poorly formed trees; however, see Mitchell et al. (2000) for an outstanding example of a selection approach in longleaf pine woodlands that explicitly retains such trees for ecological purposes. Additionally, the fine-scale removal of canopy trees with single-tree selection rarely includes efforts to emulate localized soil disturbance generated from windthrow and soil tip-up mounds and pits.

Group selection

Classical group selection regeneration systems also contrast in biological legacies compared to multi-tree, gap-based natural disturbance regimes. Essentially, the differences are similar to those between even-aged regeneration systems and stand-replacement disturbance, albeit at a smaller spatial scale. Recall that natural multi-tree gap disturbances leave behind biological legacies in the form of snags or downed boles, understory plants, living canopy trees, and pits-and-mounds in the case of windthrown tree disturbance (Chapter 4). Few silvicultural approaches based upon group selection have explicitly incorporated such structural legacies into practice (Table 7.1; Figure 7.6).

Retaining Biological Legacies: What, How Much, Where, and Over Time

Providing for continuity between forest generations using legacy retention is applicable to all regeneration systems, including all the approaches discussed in the previous section. Incorporating biological legacies into regeneration systems that create large openings (i.e., even-aged systems) is the most obvious way to better approximate the outcomes of natural stand-replacing disturbances. In fact, VRH has most often been used to modify classical even-aged regeneration systems in forest types and sites characterized by stand-replacement disturbances.

However, retention practices are equally applicable to other regeneration systems, including two-cohort

Figure 7.6 *(a) Large harvest opening in red pine forest that is devoid of structural legacies, Minnesota, USA. (b) Opening in the same forest with a retained snag. Photo credit (a): Christel Kern.*

approaches and selection systems. Two-cohort regeneration systems provide excellent opportunities to incorporate retention with only minor adjustments (O'Hara 2001). A key adjustment is to incorporate greater spatial heterogeneity in structural and compositional conditions across the harvested stand and to deliberately retain deadwood legacy structures.

In individual tree selection, old and decadent trees, snags, and logs can be explicitly retained to achieve ecological goals. Classical approaches to controlling selection, such as the application of stocking control concepts like the *BDq* method (see Chapter 8), can be modified to allocate a portion of the residual stand structure to permanent legacies. Similarly, the effective area being regenerated in group selection entries could be reduced to account for the integration of within-gap retention at each harvest entry.

Regardless of the regeneration system, a forester incorporating legacy retention into a regeneration harvest must answer four fundamental questions: what to retain, how much to retain, the appropriate spatial pattern of retention, and how best to account for retained elements over time. The answers will vary with forest type, management objectives, and priorities among objectives. Given the various factors involved, a broad spectrum of prescriptions for retention are possible, but again all should address the four fundamental questions. We will address each of these questions below (with some material adapted from Franklin et al. 2007 and 2018).

What to Retain?

Structural Retention

The most obvious answer to this question and the way it is commonly answered, is retention of the big three legacy structures: live trees, snags, and downed logs. Generally, attention should be given to retaining individual structures that are large for the forest in question—that is, large diameter trees, large snags, and large logs on the ground (Figure 7.7). Beyond the obvious habitat value of large structures, there is the practical consideration that in most forests, large size generally equates to old age; that is, these structures often take decades to centuries to develop. Given the lengthy time required to develop large structures, the ecological forester is wise to deliberately leave some of the largest structures available at harvest, potentially in perpetuity in the case of live trees, i.e., let these trees live out their natural lifespan and ultimately transition into the snag and downed log pools.

Guidelines for retention of live trees have been incorporated into many organizational management practices (e.g., Minnesota Forest Resources Council 2013); however, even with these formalized approaches there may be a tendency to select smaller diameter trees, including those that have low commercial value (Figure 7.8). While not without some ecological value, such trees tend to meet the letter, but not intent, of the continuity principle, where larger generally

Figure 7.7 *The primary structural legacies of concern in retention harvests are (a) large trees, (b) snags, and (c) logs on the ground. Photo credits: (a) Bob Williams, (c) Shawn Fraver.*

Figure 7.8 *An example of trees selected for retention because they have little commercial value due to their small size; such trees also have limited ecological value.*

equates to greater ecological value. Natural disturbance is a useful guide for thinking about retention tree sizes. For example, in an aspen mixed-wood forest in northern Minnesota, USA, trees that survived a severe wind disturbance (Figure 7.9) included some from the upper end of the diameter distribution. These were long-lived pines and other conifers that predated in age the main cohort of the forest, as well as some large trembling aspen, which were part of the main cohort. Using this as an example, and if continuity after natural disturbance through legacy retention is a principle being addressed, it follows that some large conifers and aspens should be left during harvest of similar types of forests.

The ecological forester should also consider the condition of legacy structures, particularly tree quality and decay class of dead trees. In the case of snags and downed logs, maintaining a range of decay classes equates to maintaining

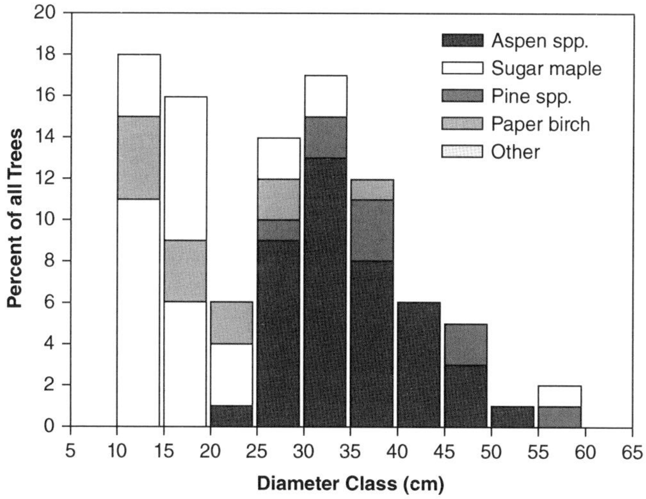

Figure 7.9 *Diameter distribution for trees surviving a severe windstorm in a mixed-wood forest, Minnesota, USA.*

Figure 7.10 *This well-decayed log provides important habitat for wood decay fungi and other organisms. The ecological forester attempts to avoid destruction of these legacies during harvesting and subsequent activities. Photo credit: Shawn Fraver.*

a range of habitat conditions for deadwood dependent organisms. Moreover, retaining a range of decay classes better ensures a more continuous supply of well decayed snags and logs (the most ecologically valuable condition) over time in a stand. Not all forests will have snags and downed logs across a range of decay states, but to the extent that different decay conditions exist, including well-decayed logs (Figure 7.10), these should be protected during harvest, so as to be represented in the legacy portfolio.

Live tree retention should also incorporate a range of live tree quality if possible, inclusive of healthy, vigorous trees, as well as decadent, decaying, and declining trees. There is a natural tendency in timber-focused forestry to remove damaged and decadent trees. Historically, these have been viewed as a liability because of safety concerns for workers and because of the growing space they usurp from healthy, vigorous trees. Moreover, they may have the potential to form the epicenter of forest pest outbreaks. To the ecological forester, the current and future habitat value of at least some portion of these trees will likely outweigh these risks. Trees with damaged boles, scars from fire, or large branch breakage (Figure 7.11), are likely to form cavities in the near future. Trees that are in a state of decline due to disease, pests, or disturbance provide habitat for insects that are, in turn, a food source for other organisms, such as songbirds (Figures 7.12).

Compositional Retention

Compositional retention is focused on three aspects of maintaining continuity of plant communities during regeneration harvesting: 1) the species composition of large legacy trees, 2) retention of non-tree vegetation, and 3) protection of advance tree regeneration. For large legacy trees, attention should be given to retaining a diversity of tree species, relative to the tree composition of the ecosystem in question (Figure 7.13). Not all forests are species rich, but most will contain at least two or three species, even if all of these are not considered commercially desirable. Most may have some unique value to wildlife that makes their retention important. As such, the practice of cleaning stands by eliminating noncommercial species may have significant negative impacts on biological diversity (Palik and Engstrom 1999) and should be avoided as inconsistent with the principle of continuity. For instance, there is a history of removing broadleaf hardwood tree species from conifer plantations in Finnoscandia, but these species are known to provide important habitat for wood decay fungi, epiphytic byrophytes, and cavity nesting birds (Markkanen and Halme 2012; Oldén et al. 2014; Holm 2015). Turning back to the wind-disturbed, aspen mixedwood forest in Minnesota, USA as a model for continuity management (see Figure 7.9), some or most of the tree species that survived disturbance might be considered for retention in a regeneration harvest, favoring larger individuals (in a relative sense).

Changes in understory plant communities are expected with regeneration harvesting and are part of normal successional processes. By design, regeneration harvests alter physical, chemical, and other environmental conditions of forests, and in so doing, change habitat characteristics for plant species that occur in the pre-harvest forest. Nevertheless, there may be opportunities to retain herbaceous or shrub species of unique interest, e.g., species of conservation concern or species that are rare in the regional landscape, by identifying these populations prior to harvesting and then associating retention aggregates (see below: *Where to Retain Legacies*) so that they include these populations, or avoiding selection harvesting around them in the case of uneven-aged regeneration methods.

Figure 7.11 *(a) Large tree with damaged bole forming a cavity. (b) Red pine with multiple fire scars. Photo credits: (a) Karin Fassnacht, (b) Robert Van Pelt.*

Figure 7.12 *Large decadent Norway spruce with bore holes from woodpeckers.*

Figure 7.13 *Retention trees in expanding gap irregular shelterwood in Acadian mixedwood forests in central Maine, USA. Trees reflect mature individuals spanning the species composition of original stand.*

By definition, legacy management is associated with a regeneration harvest, with a new cohort of trees establishing from seed, sprouting, planting, or advance regeneration. If regeneration is dependent on new establishment, then it is important to link legacy management with other silvicultural activities aimed at controlling composition of the new cohort (Box 7.2). If the new cohort is largely secured from advance regeneration, then consideration

Box 7.2 **Linking legacy retention with regeneration of trees**

Recall from Chapter 3 that silvicultural systems include treatments associated with both regeneration and harvesting and that these activities are closely linked—harvests are the application of regeneration methods designed to create favorable environmental conditions for the reproduction of desired tree species. There is nothing inconsistent with this interpretation in the context of ecological silviculture and the principle of continuity. The same attention to regeneration treatments, particularly site preparation and release treatments and planting or seeding of new individuals, may be needed to control composition of the new stand after a harvest that includes legacy retention. Establishment of regeneration from seed requires appropriate seedbed conditions, often including exposed mineral soil or minimal litter, or for some species, well-decayed logs. Mineral soil exposure can be achieved by a variety of methods, including mechanical treatments such as patch or row scarification, raking, disking, and bedding. Prescribed fire can also be effective at reducing litter. Regeneration from seed, sprouts, or planted stock may require reduction in competing vegetation to increase survival and early growth of the new trees. There are a variety of methods that might be used for competition control, including prescribed fire and mechanical treatments, such as spot or area brushing. Herbicides may be applied alone or in conjunction with mechanical site preparation to control vegetation. These methods, including herbicides, are not inconsistent with ecological silviculture. However, a distinction with their use in the former, compared to timber-focused silviculture, is that their use needs to be consistent with the objectives of legacy management. For example, area-wide raking of logging slash may remove deadwood legacies, or whole-stand application of herbicides or mechanical brush control may remove legacy plants. For the ecological forester, managing biological legacies and regenerating a new cohort of trees may require more creative approaches to site preparation and competition control than are sometimes practiced. One innovative example of doing this comes from the Chippewa National Forest, Minnesota, USA. In appropriate forests, the national forest uses a disk harrowing attachment on a skidder to create furrows of exposed mineral soil, while at the same time severing near surface root systems of aggressive woody shrub competitors (Figure 7.14). This approach has the advantages of providing establishment sites for natural or planted regeneration and effectively reducing woody competitors. Moreover, application of disk harrowing in the stand can be finely controlled by the operator to avoid large dead wood and other legacy structures and plant populations.

Figure 7.14 *Skidder mounted disk harrow (a) and resulting site preparation (b) on the Chippewa National Forest, Minnesota, USA. Photo credit: Doug Kastendick.*

of this comes back around to the concept of legacy management. For most foresters, advance regeneration of trees has not typically been thought of as a biological legacy, but operationally this is exactly what it is and how it should be treated—that is, a feature of the old forest that is lifeboated into the new forest to maintain continuity across forest generations. As such, most foresters have ample experience with designing and implementing harvests in ways that protect advance regeneration if it exists in their forest. For the ecological forester, lifeboating advance regeneration may be broadened to include a wider suite of species beyond those having commercial value, including those with high habitat value or those that may provide future options for management under changing climate conditions (see Chapter 16).

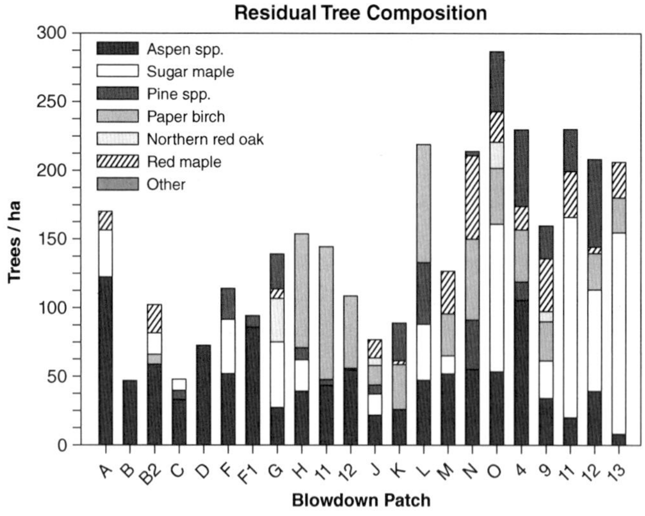

Figure 7.16 *Residual tree densities after a patchy windstorm in a mixed-wood forest, Minnesota, USA.*

How Much to Retain?

The second question that the ecological forester must address is how many legacies to retain. A glib answer to this question is that some is better than none and more is better than some (Figure 7.15). This answer actually reflects the traditional tendency to omit legacies from regeneration harvests, so in contrast, some is better than none. Again, natural disturbance is a useful guide for determining retention amounts. In our Minnesota, USA mixed-wood example, the number of trees surviving wind disturbance varied from high to low among disturbed patches (Figure 7.16), i.e., it was naturally variable. However in patches with high retention, most of the trees were from the small end of the diameter distribution (see Figure 7.9); these are less of a concern for emulation as legacies during a regeneration harvest. Rather, in this example, it is the residual aspen and pine that were larger, and generally occurred at low density, that would be the appropriate guide for retention. See *Part IV: Ecological Silvicultural Systems for Major Archetypes* for more examples of how retention levels were determined and implemented as part of ecological silvicultural systems.

The complement to residual live trees are the deadwood structures generated by natural disturbance. With heavy disturbances, there will be a lot of these within the disturbed area (Figure 7.17) and obviously the ecological forester would not consider using this type of structure as a model for deadwood retention and, regardless, there would not be this much deadwood in a healthy forest being harvested anyway. Again, a focus on retaining some larger downed logs and snags, if they exist, should guide retention. If they are not present (a not uncommon circumstance in many managed forests) then the ecological forester might

Figure 7.15 *(a) A single Norway spruce is retained in this harvest in Sweden. (b) Aggregates of Douglas-fir were retained in this harvest in Washington, USA. Photo credit (a): Shawn Fraver.*

Figure 7.17 *Deadwood after large blowdown, Havešová, Slovakia. Photo credit: Miroslav Svoboda.*

consider deliberate creation of deadwood legacies during a harvest (Box 7.3).

Retention with Selection Systems

Answering the question of how much to retain is important to consider when using selection systems as well. Amount of retention with individual tree selection takes on special meaning, as it largely relates to what is prescribed for the uncut matrix; we discuss this in more detail in the next section.

Determining how much to retain with group selection is more straightforward, as natural disturbances that are small patch- or multi-tree gap-scaled can serve as guides. As we have learned previously (Chapter 4), natural disturbances that are at a gap-scale (e.g., small patch blowdown, patchy fire) often leave some large trees alive (although

Box 7.3 **Creating deadwood legacies during a retention harvest**

What if there are no appropriate deadwood legacies to retain at harvest? That is, the stand has few large snags or downed logs considered desirable for ecological functions? This situation might readily occur in plantations or even natural forests that have been regularly thinned, such that there are few larger trees left to die. In these cases, the deliberate creation of deadwood at the time of harvest might be an option. Approaches can include **girdling** or high stumping of trees, including **cull trees** if they are large enough, at the time of the harvest. This is a straightforward way to enhance the levels of snag legacies (Fassnacht et al. 2015; D'Amato et al. 2015). In fact, this practice is used successfully by many different organizations. For example, National Wildlife Refuges in Michigan, USA use girdling and high-stumping to augment the large snag pool in harvests conducted to increase early succession habitat, and forest ownerships in Sweden employ high-stumping for snag creation (Figure 7.18). See Box 13.3 in Chapter 13 for further examples of how to integrate the recruitment and maintenance of deadwood legacies into ecological silviculture prescriptions.

Enhancing the amount of large deadwood on the ground during harvest can also be done by deliberately leaving some cut trees lying where they are, as well as leaving large branches of harvested trees. Applied less frequently is an approach that emulates the soil pit and mound structure resulting from trees blown down by deliberate pushing over of large trees with harvesting machinery (Figure 7.19).

While generally considered in the context of even-aged regeneration systems, creation of deadwood is relevant to selection systems as well, even single-tree selection. For example, some percentage of trees in either single-tree or group selection harvests can be high-stumped, or their large tops left in place, or some trees can be dropped and left in place, and some can be pushed over to create pits and mounds.

Figure 7.18 *Created snags in managed forests: (a) whole-bole girdling of jack pine to induce delayed mortality in a recently harvested forest in Michigan, USA; (b) high-stumping of jack pine in the same forest; and (c) creation of tall stumps for snags from harvested trees in Sweden. Photo credits: (a, b) Greg Corace (c) Matts Lindbladh.*

Figure 7.18 *Continued*

Figure 7.19 *Creating downed logs and soil pits and mounds, emulative of windthrow disturbance, by pushing over trees using harvesting machinery: a) a pitch pine pushed over during a harvest in 2007; b) the same pitch pine, now well-decayed, eleven years after the harvest. Photo credit: Bob Williams.*

Figure 7.20 *Group harvest opening with live tree legacies. White Mountain National Forest, New Hampshire, USA.*

Figure 7.21 *Large gap disturbance from blowdown, Havešová, Slovakia. Photo credit: Miroslav Svoboda.*

they may die over time with exposure or from secondary agents). Regardless, leaving at least a few live tree legacies in group selection openings provides a level of continuity to the pre-disturbance forest (Figure 7.20), with levels of within-gap retention dictated by regeneration ecology of species being favored (e.g., less retention when regenerating shade-intolerant species) and overall structural and compositional goals. For example, group selection prescriptions in mixed-conifer types in Sierra Nevada national forests in California, USA, have included structural retention in the harvest openings by requiring that all trees >76 cm diameter at breast height (dbh) be retained in the gaps; the intention is to retain these structures in perpetuity (Verner et al. 1992).

Gaps created by natural disturbance will also contain ample amounts of dead wood (Figure 7.21). While leaving the levels of deadwood characteristic of natural gap disturbance during a harvest is not realistic, leaving (or creating; see Box 7.3) some large deadwood in a harvest gap is in line with emulation of this dynamic and with application of the principle of continuity at the gap-scale.

How and Where to Retain Legacies?

The third question that the ecological forester must answer about legacy retention is where and how to retain legacies—that is, what is the spatial pattern of retention across the harvest unit? The question of spatial pattern is most readily considered in the context of even-aged and two-aged regeneration systems, where there is ample opportunity to vary the pattern of retention between dispersed or aggregated (i.e., into patches) across a harvest unit. Consideration of spatial pattern of retention within an even-aged harvest is important because some objectives are best met with dispersed retention and others with aggregated retention (Aubry et al. 2004; Franklin et al. 1997; Franklin et al. 2018) (Table 7.2). For example, aggregated retention is an obvious way to provide protection to understory plants and the forest floor during harvesting, whereas these populations and conditions may be more vulnerable to harvest disturbance with dispersed retention. Different patterns of retention also may have profoundly different effects on the growth and productivity of regenerating trees (Palik et al. 2003; Montgomery et al. 2013).

Dispersing large (old) trees and other structures individually across a harvest unit is generally a straightforward implementation; but aggregating trees in patches will take some sleuthing. With aggregated retention of trees, a range of diameters will likely be represented in the patch, especially more small versus large trees. With aggregated retention, the ecological forester should look to place aggregates in areas that include some large, old trees and potentially large snags or logs (Figure 7.22).

As the name suggests, one common goal of variable retention harvest systems is incorporating spatial heterogeneity of retention within a harvest unit, ranging from dispersed to aggregated retention within the same harvest unit (Figure 7.23). In fact, depending on the scale of observation, this kind of variability in post-natural disturbance structure is likely characteristic of most forest ecosystems. Thus, capturing this variation within a harvest unit would in fact be emulative of structural conditions resulting from many natural disturbances.

The question of where to retain legacies spatially also should be considered with selection-based regeneration systems, although the answer may be less obvious. Recall from Chapter 3 that selection systems remove individuals or groups of mature trees to initiate a new cohort. It follows from this that with selection, mature trees and other structures will be retained within the matrix of the forest,

Table 7.2 *A comparison of ecological and operational aspects of dispersed and aggregated structural retention. Adapted from Franklin et al. 1997, 2007, 2018.*

Characteristic or Objective	**Spatial Pattern**	
	Dispersed	*Aggregated*
Microclimate modification	Generalized over harvest	Localized within harvest
Retention of diverse tree sizes, species, and conditions	Low probability	High probability
Retention of large trees	More emphasis	Less emphasis
Retention of multiple vegetation layers	Low probability	High probability
Retention of snags	Difficult	Readily accomplished
Disturbance to forest floor	High probability	Low probability
Retention of structurally intact forest habitat patches	Not possible	Possible
Sources of woody debris	Distributed	Clumped
Distributed sources of tree energy for belowground processes	Yes	No
Windthrow hazard of residual trees	Average wind firmness greater, but trees isolated	Average wind firmness less, but trees have mutual support
Residual tree damage	High probability	Low probability
Tree form and geometry	Uniform	Variable
Distribution of fine fuels	Uniform	Variable
Impacts on regeneration growth	Generalized over harvest area	Concentrated in aggregates; minimal between aggregates

Figure 7.22 *Aggregation of trees in legacy patch. Patches may be selected to retain examples of the dominant forest condition (a), or to retain (protect) unique features, such as this fire-scarred red pine (with a woodpecker excavation) found in the center of the legacy patch (b).*

Figure 7.23 *Variable retention harvest unit in Washington, USA, displaying variation from dispersed to aggregated retention.*

but generally not (intentionally at least) within openings created for regeneration. The ecological forester, when addressing the question of where to retain legacies at harvest in selection systems, needs to consider both the matrix and the openings, particularly larger gaps, created in the matrix.

Retention in the Matrix

Maintaining the forest matrix in a mature condition is one of the most important environmental benefits of uneven-aged silviculture. However, for the ecological forester, an added consideration for the matrix is the explicit incorporation of retention by maintaining large (old) trees indefinitely as legacies (i.e., to live out their lifespan) and, as a consequence, provide a future source of large snags and downed logs. Examples of such approaches have been applied in several regions (Mitchell et al. 2000; USDA Forest Service 2004; Keeton 2005). Typically, they use retention by designation in the matrix and avoiding removing these structures in future harvests. Mixed-conifer forests of western North America are an example. Large old trees and their derivative structures (snags and downed boles) are constant and important structural elements of such forests (Covington and Moore 1994; Noel et al. 1998). A management goal for some ownerships is to maintain a population of large old trees in the matrix of these forests (e.g., Johnson et al. 2003).

Spatial Pattern in Group Openings

An important consideration with group selection systems is determining where to retain legacies in the openings created by harvest. Prescriptions based on group selection can readily incorporate retention of trees, snags, and downed boles, particularly in larger gaps (Figure 7.24); moreover, there may be opportunity in these openings to vary the spatial pattern of where these structures are retained. For example, retained trees can be centralized in the opening or left near the edge (Figure 7.25). Just as with the spatial pattern of retention in larger harvests, the location of retained trees within group selection openings may influence a variety of environmental and habitat characteristics at the gap scale (refer to Table 7.2).

How to Account for Retention over Time

The final question regarding retention is a temporal one. Specifically, how best to integrate future silvicultural activities with the legacies retained at harvest. Retained trees and other structures are dynamic elements of a stand. Future silvicultural treatments need to account for these changes in order to sustain the functionality of legacies. Moreover, understanding how legacies change and influence stands over time may provide opportunities to address additional goals of ecological silviculture.

Retained trees may survive for decades or even well over a century in many forests (Jönsson et al. 2007). Still, these trees will eventually die; silvicultural prescriptions should include provisions for adapting to changes in structural and functional conditions as this occurs. For instance, the "footprint" of a retained tree might increase due to windthrow, or as a result of increased safety concerns after a live tree transitions to a snag (Figure 7.26). Prescriptions should anticipate these dynamics by allowing for a buffer

Figure 7.24 *In this example from the Chippewa National Forest, Minnesota, USA, structural retention in a large opening included a legacy northern red oak, a red pine, and a pine snag.*

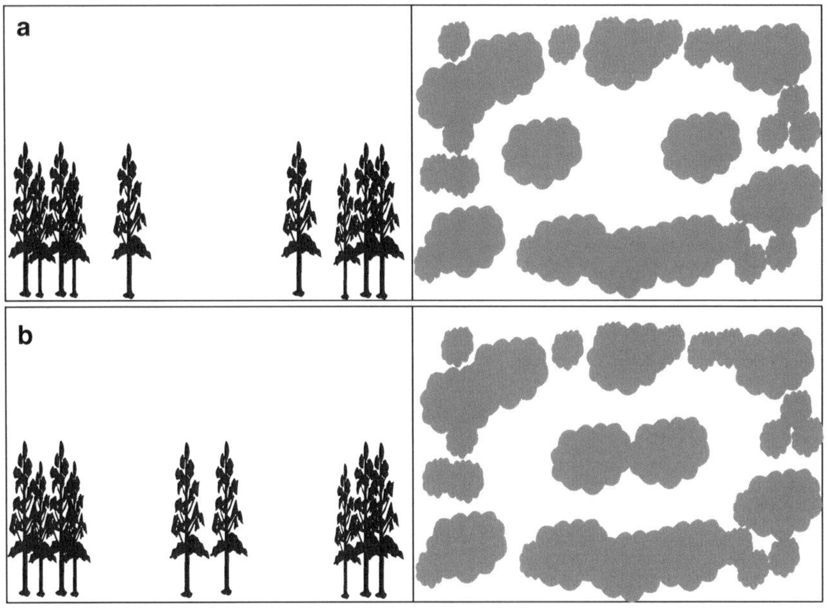

Figure 7.25 *Spatial variation in legacy retention within large group openings. (a) Retaining trees near to the gap edge will result in less alteration of microclimate and resource availability at gap center and has the effect of softening the edges of the opening. (b) Centralized retention may ameliorate microclimate in the gap center, but also potentially inhibit growth of less shade tolerant tree regeneration.*

around these dynamic structural legacies within which no machine traffic is allowed. Sustaining the functionality of legacies over time may rely on careful documentation of legacy location, particularly of the largest trees, snags, and downed logs in the stand, and this should be accompanied by clear descriptions of the long-term objectives for retention to ensure future harvests do not remove retained trees or heavily impact areas near aggregates or retained deadwood structures (Figure 7.27).

Changes in stands over time directly resulting from legacies left at harvest may present unique opportunities for a forester to further other goals of ecological silviculture, e.g., restoring native tree species diversity. For example (and as discussed previously), retained trees should be

Figure 7.26 *This retained live tree was snapped in a windstorm, increasing its ecological footprint in the stand; the resultant footprint should be buffered in future harvest entries, so as to sustain the functioning of both the snag and downed wood legacies.*

Figure 7.27 *Documenting locations and long-term objectives for retained trees is important for ensuring future silvicultural activities sustain the function of legacies. (a) Use of tree marking paint to designate retained trees with "L" for legacy in a northern hardwood forest in western Massachusetts. (b) Signs designating leave trees on the borders of aggregated retention patches in a Douglas-fir forest in western Washington.*

selected to represent the diversity of canopy trees present in a pre-harvest stand. Initially, this retention provides a diversity of mature tree lifeforms, habitat elements, and nutrient inputs in the regenerating forest. However, over time these trees may play an important role as a source of seed, as the stand progresses into the *Mature Forest Stage*. As such, areas around these retained trees may contain higher levels of advance regeneration of less common species than other areas of the stand. This may provide opportunities for the application of localized harvesting, e. g., individual- or group-selection, based on these locations, so as to increase the representation of these species in the *Old Forest Stage*. There are many other examples of how retention can be linked with concurrent or future silvicultural activities to further the long-term objectives of management (Box 7.4).

Box 7.4 **Linking retention with other silvicultural activities**

Although many of the decisions regarding legacy retention occur at harvest, it is important to take a longer-term view of how retention interacts with other silvicultural activities to address long-term goals for management. For example, retention may be tied to tree regeneration goals by planting shade-tolerant species near retained trees, given the lower light levels in these areas and, thus, the greater long-term survivability of shade-tolerant trees in this setting (Palik et al. 2003). Moreover, shade-tolerant tree species might be planted within aggregates themselves to promote the development of multi-storied conditions. As another example, the cover of shrub and forb species, important for many birds and small mammals, may be expanded laterally by initially limiting the application of release treatments in areas adjacent to aggregated retention patches and then later using precommercial thinning adjacent to aggregates to encourage the development of understory vegetation (Figure 7.28). As a last example, in fire-dependent forests characterized by woodland structural conditions, underburning or mechanical treatments might be applied as a follow-up to retention harvesting to restore historical woodland densities and open understory conditions (Figure 7.29).

Figure 7.28 *Functioning of lateral cover of shrubs present in retained aggregates can be enhanced by limiting the level of release treatments applied in areas immediately adjacent to the aggregate patches, such as in this regenerating jack pine forest in Minnesota, USA.*

Figure 7.29 *An example from a pitch-pine-scrub oak woodland (New Jersey, USA) of combining dispersed retention with additional silvicultural treatments to achieve long-term objectives for a stand. (a) The forest prior to treatment that included retention harvesting and whole-tree chipping. (b) The same stand eleven years after treatment. Photo credit: Bob Williams.*

Landscape Considerations

A final consideration to be aware of relates to how retention in stands manifests at landscape scales. In practice, the amount of retention within a harvest unit is a factor that may influence determination of harvest unit size. As foresters know, harvest unit size is often an important consideration, especially when even-aged approaches are the primary regeneration system being used. Specifically, large clearcuts can elicit strong public objection, often based on appearance, with maximum allowable clearcut sizes being stipulated by forest practices acts in some countries and regions. Size of a clearcut is also important in terms of impacts on various ecological characteristics (Franklin et al. 2018). Ecological and visual impacts of large clearcuts can be mitigated to varying degrees with even moderately low levels of retention (e.g., 10 to 15 percent). By doing this at the scale of the landscape, clearcut size becomes increasingly less important as retention level increases, and can become of little consequence in forest landscapes harvested with high amounts of retention (Figure 7.30). We will discuss the landscape context of ecological silviculture in more detail in Chapter 9.

Figure 7.30 *Harvest unit patch size loses much of its ecological and social significance as an increasing percentage of the forest is retained. With higher levels of retention, both the visual and ecological impacts of large harvest units are reduced, as in these examples from the western USA.*

What to Remember About Managing for Continuity with Legacy Retention

A key point to remember from this chapter is that manipulation of biological legacies during regeneration harvesting is the primary tool the ecological forester uses to ensure continuity between generations in managed forests. This is an important point to be aware of and to actively consider when writing prescriptions, because classical approaches to regeneration harvesting have not explicitly considered continuity through legacy management. In other words, if you are following normal regeneration practices you most likely are not managing for continuity.

Another point to remember is that legacy management is applicable and appropriate, regardless of the type of regeneration system used (i.e., even-aged, two-aged, uneven-aged regeneration systems), although the specifics of retention will differ with the scale of the regenerating disturbance. One thing that is common (among disturbance scales) when managing for continuity is that the ecological forester needs to answer four fundamental questions about retention when writing prescriptions: what to retain (e.g., large tree, particular species, condition of legacies), how much to retain (some is better than none), what is the spatial pattern of retention (dispersed, aggregated), and how to integrate legacy retention with other sivicultural activities to meet long-term management goals for a stand.

The ideas presented in this chapter are best thought of as general guidelines that are applicable to any forest and regeneration approach. Based on these guidelines, an important question when judging the ecological success of retention is whether the retained species, structures, and patterns represent the range of sizes and physical conditions, and spatial heterogeneity of legacies, that typically exist in the stand following a natural disturbance. If the answer to this overarching question is no, then the success of the retention harvest from an ecological standpoint may be in doubt and the fundamental questions (particularly the first three) should be addressed again. In addition, retention should be viewed with the same long-term perspectives that are used for encouraging the growth, maturation, and regeneration of living crop trees with future harvest entries and silvicultural treatments designed to sustain a diversity of structural legacies and enhance their functioning over time (addressing the fourth fundamental question). In later chapters we will return to the concept of continuity with legacy retention, specifically addressing how to incorporate legacy management into long-term silvicultural prescriptions for the archetypes presented in Chapter 6.

CHAPTER 8

Managing for Complexity, Diversity, and Heterogeneity in Established Stands

As you have learned, after a severe (tree regenerating) disturbance, a forest stand or patch may have substantial structural complexity resulting from the legacies of the previous forest that survived, or were created by, the disturbance. This legacy influence is the basis for the first foundational principle of ecological silviculture—managing for *continuity* across forest generations. But there is more to the complexity story than biological legacies left after disturbance. As a forest develops over time, it undergoes processes that result in the development of additional aspects of complexity, including heterogeneity of structural conditions across the stand and, often, greater species diversity. These are processes that are largely related to tree growth and mortality that may begin in the Young Forest Stage and proceed through the Mature Forest Stage, ultimately resulting in the kind of complex forest structure unique to old-growth forests.

Understanding how to manipulate and incorporate complexity and diversity of species into intermediate treatments in established stands is the basis for the second foundational principle of ecological silviculture, namely managing for *complexity* of structure and for the desired *diversity* of species. Moreover, the development of complexity/diversity is intimately related to the third foundational principle, i.e., consideration of *appropriate timing* of silvicultural interventions. Appropriately timed interventions facilitate the development of complexity and species diversity.

As you read further, it will be apparent that using silviculture to increase structural complexity in established but developing stands is contrary to classical approaches to intermediate treatments, which often deliberately or inadvertently maintain stands in simplified and homogeneous structural states. Moreover, forests managed for timber are often harvested before significant structural complexity or species diversity has had a chance to develop. Learning why and how to implement intermediate treatments differently from classical approaches is the topic of this chapter. Specifically, we will review the idea of complexity of structure and species diversity in the context of forest maturation. Then we will consider some of the silvicultural tools that can be used to facilitate the development of complexity, heterogeneity, and diversity in managed forests. Finally, we will consider the importance of timing silvicultural interventions to ecological cues. As in Chapter 7, which addressed continuity through retention harvesting, our discussion of complexity/diversity and timing of intermediate treatments will be a general overview, with later chapters addressing these principles as part of silvicultural prescriptions for the disturbance archetypes presented in Chapter 6.

Revisiting Complexity/Diversity in Maturing Stands

Many of the attributes of structural complexity that emerge as a stand matures are associated with the development of individual trees, including their size and form, their branch and crown conditions, and their bark and root characteristics. For example, as trees age and get larger, they develop more complex crown structures, including larger, often horizontal branches (Figure 8.1a). They may have more structurally intricate bark characteristics, including furrowed, flaking, or peeling bark (Figure 8.1b). Regardless of their size, as trees age, they may accumulate the physical effects of injury and disease, such as cavities and fissures (Figure 8.1c). As you have learned, the structural features of large or old trees have important ecological functions, particularly related to provision of unique habitat for other species of plants and animals (Figure 8.2; Chapter 5).

Accumulation of tree-derived structures from mortality is another important category of structural attributes to consider in maturing stands. As non-competitive tree mortality becomes increasingly important in older stands or patches, the abundance of large snags, dead boles, and branches on the ground increases (Figure 8.3). Moreover, this deadwood decays over time, such that a maturing forest will have an increasing amount of decayed deadwood from trees that survived an initiating disturbance or established

Figure 8.1 *Unique structural features associated with mature and old trees such as (a) the complex branch system on this 300-year-old red pine, (b) peeling bark on this paper birch, and (c) a decay cavity at the base of this old pine.*

after the disturbance (Figure 8.4), rather than from the legacy deadwood from the previous forest, which in fact may be largely gone in the late Mature Forest Stage (Spies et al. 1988). Also, as derivatives of windthrown trees, root-wads and associated tree-fall pits and mounds accumulate in wind-prone forests as they mature (Figure 8.5).

It is actually how attributes accumulate into a population of structures (e.g., trees, snags, decayed logs) that conveys complexity to a stand. For instance, you might assess the level of complexity of a forest by quantifying the number of large diameter trees, but also the diameter distribution of all trees. A stand with a greater diversity of diameters, inclusive of large trees, would be considered more complex than a stand with a narrow range of diameters, even if all of these trees were large (McElhinny et al. 2005). The same kind of assessments can be made for other characteristics, such as tree ages or the proportion of snags or deadwood in different size or decay classes.

A particularly important aspect of complexity in maturing stands relates to how trees and tree-derived structures are arrayed spatially (horizontally) in a stand. This is the aspect of complexity often referred to as spatial heterogeneity

Figure 8.2 *Old tree habitat in the form of (a) a cavity used by this red squirrel and (b) large horizontal branches used by this black bear cub. Photo credits: (a) Igor Drobyshev, (b) Greg Corace.*

Figure 8.3 *Deadwood attributes derived from large, old trees include (a) large snags, (b) downed trees, and (c) large branch systems.*

Figure 8.4 *Decaying deadwood, including (a) snags and (b) logs, both derived from trees that either survived a regeneration disturbance or established after the disturbance. Photo credit: Chris Gronewold.*

Figure 8.5 *Root-wads in old-growth forests from (a) recent blowdown in an old-growth black ash forest, Minnesota, USA and (b) an old blowdown in an old-growth eastern white pine-eastern hemlock forest, New Hampshire, USA.*

of structure; a structurally heterogeneous stand is one with a nonuniform distribution of various aspects of structure within the stand. Recall that this heterogeneity often results from an equally uneven distribution of non-competitive tree mortality (Chapter 5), which determines the spatial distribution of deadwood, tree crowns, and canopy gaps, as well as the distribution of light and soil resources, which in turn influences the ability of different tree species to establish and grow.

Another important feature of the maturing stand is an increase in plant species diversity, including trees, relative to the Young Forest Stage (Chapter 5). This is important ecologically because different tree species provide different types of resources (e.g., fruits or seeds, roosting structures) for other organisms. This increase in species diversity can often be related to the increasing heterogeneity of canopy conditions in the stand as a result of small scale disturbances (Halpern and Spies 1995). Shade-tolerant species

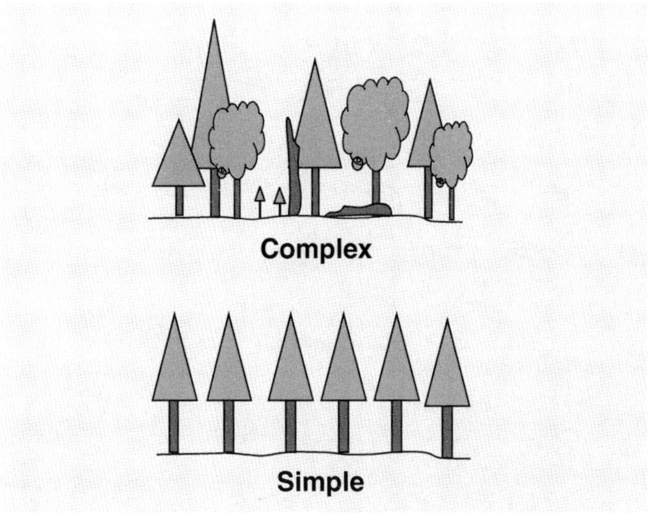

Figure 8.6 *Comparison of two hypothetical forests, one that is structurally complex versus one that is structurally simple.*

may establish in the understory of maturing stands, but it often takes small-scale canopy disturbance to increase resource availability locally and facilitate the growth of these species into larger sizes.

A simple illustration can help show how the various characteristics of structure convey complexity and diversity to a forest. Consider a hypothetical forest in both complex and simplified structural states (Figure 8.6; see also Box 8.1). An obvious difference in this comparison is that the complex stand has a wider range of tree sizes (and potentially ages), including some large trees, compared to the simplified stand, which has a much narrower range of sizes, no exceptionally large trees, and potentially no old trees. The complex stand has two tree species in abundance, versus the simplified stand composed of only one species. The complex stand also contains some larger snags and dead trees on the ground, versus the simplified stand that lacks large deadwood. Finally, the distribution of trees and crowns, or conversely, of canopy gaps, is spatially variable in the complex stand, leading to heterogeneity of structure and resources within the area of consideration. This is in contrast to the simplified stand where these structural conditions are arrayed in homogeneous distributions.

Armed with this review of forest complexity, we now turn our attention to thinking about how silviculture can be used to direct the development of complexity and diversity in established forests (Principle 2). This will be accomplished through use of intermediate treatments, particularly thinning and deliberate decadence creation.

The Complexity of Thinning: Timber Production versus Ecological Motivations

Silvicultural thinning, or the reduction of tree densities, has long had as a primary goal, and as typically applied, the improvement of growth, health, and quality of the remaining trees. Thinning treatments are often modeled on natural decline and mortality of trees during stand development. Used for this purpose, thinning may have a basic goal of removing trees that are likely to die in the near future from resource competition, thereby directing resources to and concentrating growth on remaining crop trees. In the case of commercial thinning, the harvested trees are of a size to be economically valuable; hence the operation is designed to capture mortality of usable wood before death actually occurs. Used in this way, thinning bypasses the process of competitive mortality and, in so doing, can be a valuable tool to hasten the development of large trees, a goal that is important for both classical and ecological silviculture.

Box 8.1 Comparing complexity among different forests

There are several things to remember about structural complexity when comparing different forest ecosystems or different conditions of the same type of ecosystem. First, consider that not all natural forest ecosystems are equally complex. For example, by most measures, an old-growth longleaf pine woodland pales in structural complexity to an old-growth northern hardwood forest (Figure 8.7a, b), but this is ok, as they are completely different types of ecosystems. To compare them directly is akin to comparing apples to oranges and concluding that they can't both be fruit since only one is colored orange. Still, the longleaf pine woodland is measurably more complex and heterogeneous than a loblolly pine plantation that might be growing on a longleaf pine site (Figure 8.7c). This is the second thing to remember: complexity comparisons are best made between a natural ecosystem and something that might replace it in a managed setting. Our interest is in the degree of disparity between the managed and natural condition for a particular type of ecosystem, not between different types of natural ecosystems. Finally, remember that the baseline for comparison of structural complexity in managed stands should be *healthy*, natural ecosystems, or the best available examples of the latter. For example, a long fire-excluded longleaf pine woodland may be quite complex compared to the loblolly pine plantation (Figure 8.7d), with the former having more tree species, a greater diversity of tree sizes, and potentially greater horizontal complexity, but this is not the appropriate comparison to make, since the fire-excluded woodland is not healthy from the standpoint of how it will function in terms of fire behavior, provision of habitat, and other ecosystem services. Similar comparisons of functional condition hold true for many other forest types.

Figure 8.7 *Structural comparisons of different forest types: (a) old-growth longleaf pine woodland, (b) old-growth northern hardwood forest, (c) loblolly pine plantation, and (d) fire-excluded longleaf pine woodland. Photo credits: (c) Erich G. Vallery, USDA Forest Service - SRS-4552, Bugwood.org, (d) William D. Boyer, USDA Forest Service, Bugwood.org.*

Figure 8.7 *Continued*

While competitive mortality as a natural process does tend to result in more homogeneous distributions of trees, silvicultural thinning carries this homogenizing process to the extreme. Thinning, especially pre-commercial thinning, is often distributed across a stand specifically to create a uniform distribution of crop trees (O'Hara and Oliver 1999), efficiently distributing access to resources to those individuals that will eventually be harvested. Thinning is also used to improve and standardize tree quality and form and to remove commercially undesirable species that established through natural regeneration. (Smith et al. 1996). Poor quality trees, as measured from an economic standpoint—e.g., those beginning to develop cavities or decay pockets—may be preferentially removed (Graves et al. 2000), in addition to removing lower value species. The long-term goal is to retain only healthy, ostensibly genetically superior trees of commercially desirable species.

In comparison, thinning in ecological silviculture invariably has more diverse goals than improving tree quality and stimulating growth of dominant and co-dominant trees, although this can certainly be an objective. Additional goals may include stimulating development of trees in subordinate positions, including seedlings or saplings or those at intermediate positions in the canopy (Bailey and Tappeiner 1998). Enhancing tree species diversity, as well as herb and shrub diversity, may be an important objective of thinning for ecological silviculture. Achieving such goals may involve removal or modification of dominant trees or "dominant thinning, crown thinning, thinning from above," in contrast to focusing only on small and intermediate tree removal ("thinning from below"), which is often a primary focus in timber-focused approaches (Box 8.2).

Another important distinction between classical thinning and thinning in ecological silviculture is that the goal

Box 8.2 **Thinning methods**

Thinning methods are defined based on how individual trees are selected for removal. **Geometric thinning** (syn. **row** or **mechanical thinning**) removes rows of trees without regard for crown position or quality to achieve a desired spacing or residual stocking. This approach is especially common in plantation silviculture and is often a first thinning entry to establish access for subsequent thinning treatments (Figure 8.8a). The other primary thinning methods are based on the **crown classes** of trees being removed and of those being favored. **Low thinning** (syn. thinning from below) focuses on favoring the growth of trees in the dominant and codominant canopy classes by removing trees in lower canopy classes (intermediate, overtopped, and occasionally some codominant trees; Figure 8.8b). This approach has the least impact on residual tree growth given it generally removes weaker competitors and tends to maintain fairly closed conditions in the upper canopy layer. In contrast, **crown thinning** (syn. thinning from above) has a great impact on residual tree growth, as this method favors the best quality dominant and codominant trees by removing other codominants and dominants (Figure 8.8c). **Selection thinning** (syn. dominant thinning) specifically targets the development of trees in lower crown classes by removing poorly formed dominant trees, whereas **free thinning** (syn. crop-tree release) applies a combination of the aforementioned methods throughout a stand to favor the best crop trees, regardless of crown position.

Figure 8.8 *(a) Row thinning in a loblolly pine plantation in central Florida, USA, (b) low thinning in a ponderosa pine forest in northwestern New Mexico, USA, and (c) crown thinning in red oak-northern hardwood forest in central Minnesota, USA.*

The focus of these approaches is often on improving crop-tree quality and increasing growth. An ecological forester may deviate from this construct to ensure removals are applied in a way that maintains less common species and a diversity of crown classes and tree forms to increase structural complexity and compositional tree conditions. Thinning methods (like crown thinning and free thinning) that have significant impacts on residual tree growth can also represent an important tool in the toolbox of the ecological forester. In particular, these methods can be applied to reduce the time needed to develop large-tree structures in forests in the Young Forest Stage (Singer and Lorimer 1997) and to accelerate the development of lower canopy layers by creating small canopy gaps through the removal of trees in the upper canopy (Figure 8.9). An important distinction with application of thinning approaches in an ecological silviculture context is that they should be applied to create heterogeneity of structure across a stand (e.g., see **Variable Density Thinning**), rather than the more typical uniformity of outcomes that result from a timber-focused application of thinning methods.

Figure 8.9 *Northern hardwood forest in north-central Vermont, USA two years after crown thinning. Note well-developed understory layer and scattered large diameter overstory trees favored by crown thinning treatment.*

of the latter is often to create structural and compositional heterogeneity throughout the stand, rather than to create spatially uniform stands of crop trees. Moreover, thinning in ecological silviculture invariably involves retaining a diverse mix of tree conditions in the stand, including non-commercial species and trees with little current or future economic value, such as understory maples in pine plantations (Figure 8.10a) or trees with poor form and defects, such as cavities (Figure 8.10b). For the ecological forester, the contributions these "nonstandard" trees make to biodiversity and wildlife habitat outweigh their limited economic value (Carey 1995). Post-thinning stocking levels should account for growing space occupied by these important ecological components.

Figure 8.10 *(a) Understory sugar maple in a red pine plantation. (b) Cavity in an oak used by pine marten on the Huron-Manistee National Forest, Michigan, USA.*

As mentioned already, the ecological forester may well be interested in using thinning to stimulate growth and hasten the development of large trees and the motivation may be the same as in timber-focused forestry: enhancing yield or growing high quality timber. In fact, doing so may be necessary to finance ecological forestry operations. Added to this, the ecological forester will also use thinning to grow large trees for their ecological benefits, such as habitat for nesting and roosting, for their influence on microenvironment, and for their contribution of food resources and fine fuels. The contribution of individual trees to these functions increases with size. Moreover, large trees are important for their derivatives: large snags, deadwood on the ground, tip-up mounds, and root wads. To reiterate, thinning to facilitate the development of large trees can also be important for meeting a broader range of ecological objectives (cf. Singer and Lorimer 1997).

Small-Scale Canopy Disturbance

As a forest develops, competitive-based mortality is augmented by small-scale canopy disturbances from wind, lightning, insects, disease, or fire (Figure 8.11). Small-scale canopy disturbances are a fundamental feature of natural stand development and contribute greatly to the development of spatial heterogeneity of structure, as well as an increase in species diversity (Chapter 5). Classical applications of thinning in older stands, even when using thinning of larger trees (i.e., crown or dominant thinning), homogenizes the structure and composition of a stand largely because the approach does not include an analog for gap-creating disturbances and focuses on maintaining full site occupancy across a given area by uniformly applying recommended residual stocking levels. In fact, timber management often has sought to avoid agent-based mortality, especially when it kills dominant trees and is spatially aggregated, i.e., creates canopy gaps. Since such mortality has typically not been viewed as a part of acceptable development processes in forests managed for timber, applications of silvicultural thinning to deliberately create structural heterogeneity have been limited. More recently, there is recognition that thinning analogs to natural tree mortality should include not only emulation of competition-based mortality, but also non-competitive, small-scale disturbance or gap-forming events (Franklin et al. 2007). As discussed next, integrating competitive and non-competitive mortality into thinning is precisely what an approach referred to as *variable density thinning* accomplishes.

Variable Density Thinning (with contributions by Derek Churchill)

Variable density thinning (*VDT*) as a silvicultural approach emerged out of a need to move structurally simple forests toward greater structural heterogeneity (Carey 1995). In fact, an objective for using *VDT* is to jumpstart the development of structural complexity and spatial heterogeneity in structurally simple stands and shorten the time required to reach levels of complexity and species diversity characteristic of mature forest conditions. *VDT* is perhaps best used in this context and these settings, that is, structurally simple forests (often plantations) where there is a desire to hasten the development of structural complexity, to increase species diversity, and instill heterogeneity of these features in a stand. In reality, many forests managed for timber are in a simplified state relative to their natural counterparts at a similar age, so there are ample opportunities to apply *VDT* in managed landscapes.

In concept, *VDT* approaches are based on the complex and variable structure that results from small-scale canopy disturbances and competition-based mortality in maturing stands (Chapter 5). In practice, *VDT* prescriptions are implemented by varying residual tree density across space within the stand, including bookends of un-thinned areas and openings, with intermediate (and variable) levels

Figure 8.11 *Small-scale gap forming disturbances including (a) root-rot gap in a red pine stand and (b) a wind-snapped sugar maple in a hardwood forest.*

Table 8.1 *Variable density thinning patch types*

Patch Type	Function	Size Range/Density
Skips-Unthinned	- Protect biological hotspots - Allow competitive mortality - Provide dark, moist habitats - Break up visuals	0.05–0.3+ ha
Gaps-Openings	- Harvest overstory - Regenerate new species - Promote understory- midstory development - Contain root rot pockets	0.04 to 0.4+ ha, larger for shade intolerant species and forests with greater canopy tree heights, 0–12 retention trees per ha in gaps
Heavy thin-Dispersed retention	- Grow big trees with large crowns - Stimulate epicormic branching - Promote understory- midstory - Regenerate new cohort	Single tree or cluster release Patch: 0.2 and > ha 10–20 percent relative density
Moderate thinning	- Increase/maintain crown development and growth of existing cohort - Reduce competitive stress, increase resistance to drought, insects, etc. - Stimulate understory development	- 25–45 percent relative density - Generally from below, but can be crown or dominant thinning to promote desired species/cohorts

of thinning and residual density in between (Franklin and Lindenmayer 2002; Harrington et al. 2005). Developing targets for spatial patterns of thinning/residual density can come from studies of old-growth or mature forest reference systems (Carey 1995; Larson and Churchill 2008) or from practical guidelines based on experience. The result of properly implemented *VDT* is much greater spatial variability in tree densities than occurs with classical thinning prescriptions and, consequently, greater heterogeneity of structure and environmental conditions within a stand.

The structural conditions created by *VDT* can be thought of in terms of three primary patch types, including unthinned areas (often termed skips), openings or gaps, and thinned areas. A fourth patch type may be added that consists of heavy thinning—essentially, gaps with retention trees, although ecological foresters should already be applying the principle of continuity, i.e., legacy retention, in gaps cut as part of VDT. The different patch types result in different functional contributions in the stand (Table 8.1). These include development of large trees (just like standard thinning), opportunities for establishment of new regeneration or release of existing woody and herbaceous species (not typically a goal of standard thinning), and creation of spatially variable and contrasting light, microclimatic, and habitat conditions (Table 8.1). Use of *VDT* alone may be insufficient to establish plant species of interest, in which case *VDT* may need to be combined with other regeneration and intermediate treatments aimed at manipulating composition (Box 8.3).

There are a variety of fairly simple approaches that have been used to implement *VDT* operationally (Box 8.4), including **skips and gaps** *VDT* (Carey 2003; Harrington et al. 2005) and localized release (O'Hara et al. 2012). For all approaches, the objective of marking is essentially to vary canopy patch types across the thinning unit. The amount of each canopy patch type and the size of the different patches will depend on the forest type, treatment objectives, and current conditions. In general, the thinned area comprises

Box 8.3 **Manipulating composition in established stands**

Successional processes in developing stands often result in increased tree and plant species diversity in the Mature and Old Forest Stages compared to the Young Forest Stage. This change reflects an increase in the richness of niches in the maturing forest as it gains a diversity of individual structures, as complex spatial arrangements of these structures develop, and as spatial heterogeneity of canopy, deadwood, and soil conditions increase. Essentially, niche richness reflects a diversity of conditions available to sustain high species and functional diversity of plants and other organisms.

For tree species, this compositional change may involve the establishment and increasing abundance (if previously established, i.e., advance regeneration) of shade-tolerant tree species. Moreover, increasing frequency of small-scale canopy disturbances in the Mature and Old Forest Stages may allow early or mid-successional species to maintain or reestablish a presence in the forest.

The compositional enrichments in developing stands are dependent on adequate seed sources being available in or in the vicinity of the forest as it develops (Kern et al. 2017). Such seed sources may be lacking, particularly if additional species are not commercially desirable and management actions have been aimed at eliminating them (Palik and Engstrom 1999). In such cases, deliberate actions to enhance tree diversity may be needed and should be included as part of the intermediate treatment toolbox.

Underplanting or direct seeding are two potential ways to add desired species back into depleted stands. Both are likely to be more effective when timed to other

Figure 8.12 *Variable density thinning in a sixty-year-old red pine plantation with eastern white and jack pines establishing from broadcast seeding in a harvest gap, along with volunteer hardwoods and balsam fir. Red Lake Wildlife Management Area, Minnesota, USA.*

Figure 8.13 *Disk harrowing in the foreground of a gap in a VDT red pine stand (background of gap received no site preparation). The harrow has the dual effects of exposing mineral soil and severing root systems of highly competitive beaked hazel. Chippewa National Forest, Minnesota, USA. Photo credit: Doug Kastendick.*

interventions that open growing space and reduce resource competition. For example, planting or seeding in stands that have undergone variable density thinning can potentially target mid- to intolerant species in gaps (Figure 8.12). With classical silvicultural thinking, regeneration would not normally be an objective associated with thinning; but recall a goal of *VDT* is to hasten the development of complexity/heterogeneity of the stand. As such, enrichment of species diversity as part of a *VDT* prescription may be a goal of the ecological forester. Planting or seeding may also be added to group selection harvesting if adequate and desired regeneration is not secured naturally.

Providing appropriate seedbeds for desired species, as well as controlling competing vegetation, may also be needed as part of a *VDT* prescription. Again, these activities are typically associated with regeneration harvesting, but are also in line with the broader goals of *VDT*. Treatments might include prescribed fire in appropriate forest types or chemical and mechanical site preparation (Figure 8.13).

These treatments may be needed to increase survival of planted seedlings, but also to provide improved conditions for natural regeneration. An added benefit of providing some open canopy conditions with *VDT* is that habitat for other (non-tree) plant species may be enhanced, including herbaceous and shrub species associated with the more variable canopy environment of stands in the Mature and Old Forest Stages.

Box 8.4 **Variable density thinning in practice**

An easily conceived approach to implementing *VDT* is to envision a grid pattern in the stand that is to be thinned (Figure 8.14a). Patch types are then assigned to each cell of the grid. Patch (grid cell) size should be defined by characteristics of the forest type (e.g., sizes of small natural gaps). In application, a wide range of sizes have been used for both gaps (e.g., 0.04 to 0.40 ha) and skips (e.g., 0.05 to 0.30 ha) depending on forest type and objectives. In the example (Figure 8.14b; Color Plate 8.1), 60 percent of the stand is thinned and 20 percent of the stand is in skips and gaps, respectively. The thinned matrix could receive the same level of basal area reduction throughout, or alternatively, residual density could vary from patch to patch within some predetermined range, for instance by using a table of values determined randomly prior to marking.

A variation on this approach to implementing *VDT* that is used in the Pacific Northwest, USA is to first apply a basal area or spacing based thinning prescription throughout the stand, including explicit instructions to vary density targets within some range. The second step is to then add skips, gaps, and heavy thinning areas of various shapes and sizes to create high contrast patches (Figure 8.15; Color Plate 8.2).

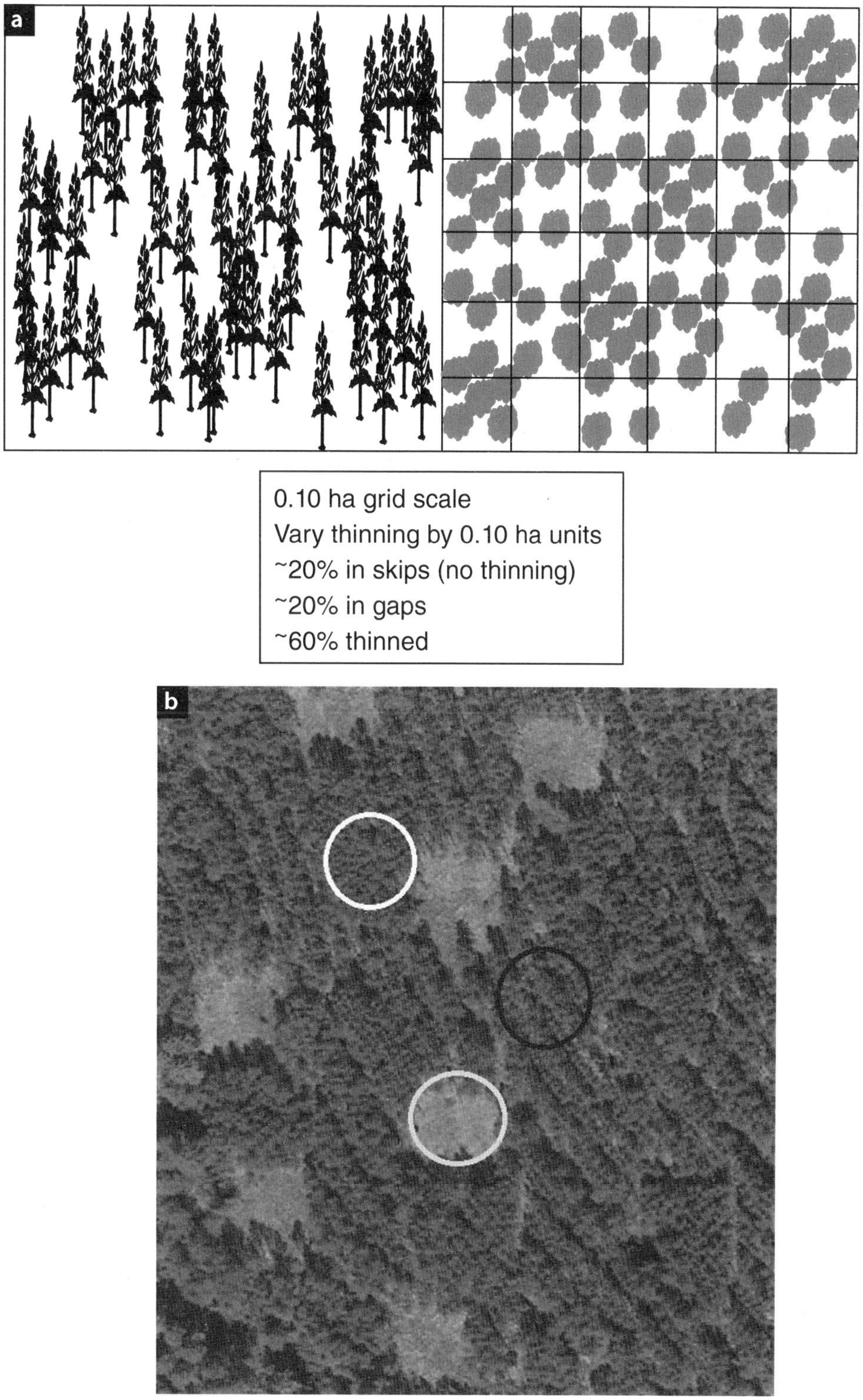

Figure 8.14 *(a) Grid approach for implementing variable density thinning. (b) Aerial view of VDT application in a forty-nine-year-old red pine plantation on the Chippewa National Forest in Minnesota, USA. Note that the white circle outlines a skip, the off-white circle a gap, and the black circle a row thinned area of the stand.*

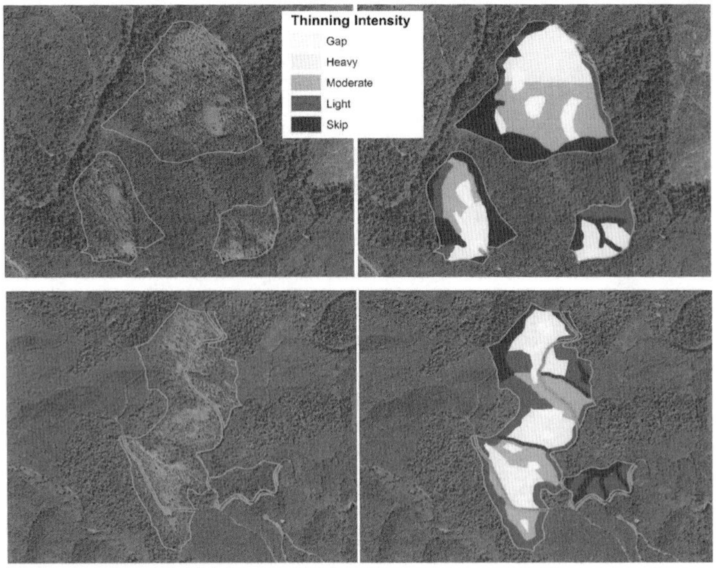

Figure 8.15 *Aerial images of variable density thinning in 40 to 50 year old Douglas-fir-western hemlock stands, illustrating skips, gaps, and varying levels of thinning. In the examples, skips are associated with edges and located along streams; more recent applications leave skips in the unit proper. Photo credit: Abraham Wheeler.*

Figure 8.16 *Deadwood from trees that predated the establishment of a white spruce plantation on the Superior National Forest, Minnesota, USA. These features are logical places to locate skips when marking for variable density thinning.*

Regardless of the approach, *VDT* prescriptions may be marked on the ground or through use of GPS enabled field tablets to identify different patch types. When marking, the forester looks for opportunities to tie patch types to existing features of the stand. For instance, gaps might be placed in areas that already have a natural opening or advance regeneration of a desired species. Similarly, skips might be assigned to protect sensitive features, such as a large snag or other deadwood (Figure 8.16), a small wetland, or a patch of desirable understory vegetation.

the majority of the treatment and is the background treatment or "matrix," while skips and gaps generally make up the balance of the treatment area. Determination of the amount and size of different canopy types should be based on consideration of the desired functions (Table 8.1), rather than cookie cutter guidelines.

A formalized approach to *VDT* is based on determining structural patterns in reference stands to determine thinning patterns in managed stands (Churchill et al. 2013). The approach is based on viewing structure in terms of individuals, clumps, and openings (i.e., *ICO*). The pattern of tree distribution is first identified in reference stands using a spatial clump identification algorithm (Plotkin et al. 2002). The reference information is then incorporated into prescriptions and tree marking for restoration of simplified forests. In practice, the *ICO* approach has proven a reliable method for providing predictably heterogeneous outcomes of thinning based on natural reference models appropriate to the forest type. It is especially well suited to frequent fire forests, but is being used in other forest types to manage for spatial patterns found in natural forests (Box 8.5).

For all *VDT* approaches, it is important to remember that the goal is creation of distinctly heterogeneous conditions across the stand and that this outcome will be influenced by patch sizes. Very small patch sizes may not result in distinctly different resource and habitat conditions across the stand. Similarly, large patches can result in gaps that approach the size of small clearcuts within the stand, essentially creating a new stand that is distinct from the original.

Another consideration is the goal of future entries into *VDT* stands. The next entry might focus on the thinned matrix, either prescribing additional thinning or creating additional gaps to stimulate understory and regeneration development. Alternatively, the next entry might expand existing gaps to release established regeneration. While a subsequent thinning of skips may be deemed appropriate, generally these areas are set aside indefinitely to allow successional dynamics to proceed unaltered.

Box 8.5 ***ICO* Thinning (*Contributed by Derek Churchill*)**

The *ICO* method is a stand-level tool to create highly variable, fine scale mosaic patterns commonly found in frequent fire forests (Larson and Churchill 2012) and other forest types (Larson and Churchill 2008). It is based on conceptualizing and quantifying forest structure in terms of widely spaced individual trees, tree clumps of different sizes, and openings. *ICO* prescriptions are guided by targets for these three elements that originate from reference stands. Clumps are defined by a maximum distance between trees, based on the average distance at which mature and old trees have interlocking crowns. Instead of marking for a range of basal areas, marking crews identify and track the number of individual trees and clumps they retain, while incorporating other leave tree criteria (Figure 8.17a). Small to medium openings, e.g. less than 0.13 ha, are automatically generated in the process of retaining clumps and the target density, as long as marking crews do not fill all growing spacing with trees.

As individual trees and tree clumps are readily visualized in the field, *ICO* can be a more intuitive and efficient approach to achieving heterogeneity than marking for variable density using basal area or spacing (Churchill et al. 2013). Another major difference between *ICO* prescriptions and spacing or basal area prescription is that there is not a per acre average density target on every acre. Instead, the goal is to retain a specified number of individual trees and clumps across the entire unit. This provides flexibility to vary density and pattern across a unit and tailor it to site conditions.

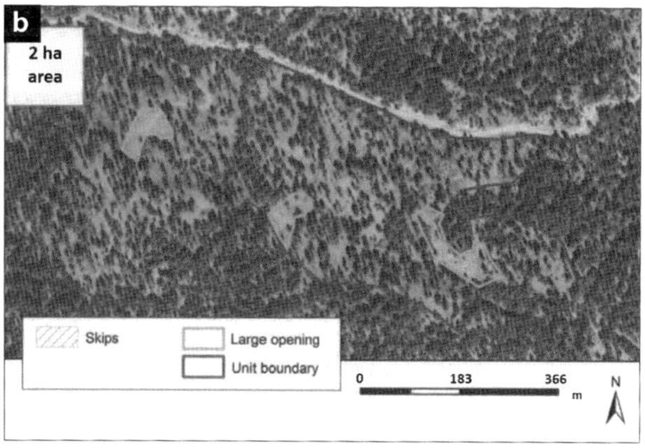

Figure 8.17 *(a) A crew member marking a four-tree clump for retention in an ICO marked ponderosa pine stand in eastern Washington, USA. (b) Aerial view of an ICO thinned ponderosa pine and Douglas-fir stand on the Okanogan-Wenatchee National Forest, Washington, USA. Note how the unit boundaries were used to create an additional "finger" skip that extends into the stand. A 2 ha square area is shown to provide a reference for patch sizes within the unit. Photo credit: Derek Churchill.*

In practice, *ICO* can be combined with skips and gaps VDT (Figure 8.17b; Color Plate 8.3). In the example, the treatment was designed to restore a pattern of individual trees, clumps, and openings consistent with reference conditions. Large openings were marked. Smaller openings occurred as a result of leaving clumps. Large skips (0.2 to 1.2 ha) were included to protect a riparian area, large Douglas-fir with dwarf mistletoe, concentrations of downed logs, and multistory patches. *ICO* results in fine scale variability in the thinned portion of a unit. By adding large openings (gaps) and untreated patches (skips) that are larger than the largest clumps, even greater contrasts in density, microclimate, and stand dynamics are created across a unit.

Modifying Selection Systems to Enhance Spatial Heterogeneity

In Chapter 7, we discussed modification of selection regeneration systems to be more inclusive of legacy retention (large live trees, logs, and snags), both within group openings and associated with single-tree selection. While this does address the need to enhance structural complexity of a stand by incorporating the principle of continuity, it does not necessarily follow that heterogeneity of structure in managed stands (*Principle 2: enhancing complexity/diversity*) will fully develop from legacy retention alone. Deliberate actions may be required to enhance heterogeneity of structure in established stands, as well as to allow large old trees to develop; neither of these conditions readily follow from formalized approaches to implementing selection systems (Box 8.6).

Another consideration for selection systems is that prescribed openings often have little resemblance to the size

Box 8.6 **The BDq method in selection systems**

Selection systems, particularly single-tree selection, are often implemented quantitatively using the BDq method or other stocking control approaches (O'Hara 2001). The structural outcomes of these approaches differ in several ways from the outcomes of single-tree or gap-based natural canopy disturbances. With the BDq method, a target residual basal area (B) and maximum diameter (D) are specified a priori, while the distribution of trees across diameter classes is determined by the diminution quotient (q), a value that reflects the ratio of the number of trees in diameter class a, to the number in diameter class a+*1* (Smith et al. 1997). With this approach, the resultant diameter distribution is a negative exponential, often referred to as a reverse-J distribution; multiple q-factors for different size classes has also been employed to manage for rotated sigmoid and other distribution forms (Leak 2002).

Specification of a target residual basal area is not necessarily in conflict with a natural disturbance-based approach to uneven-aged management; it is reasonable to expect a forester to select this target based on an understanding of stand responses to disturbance, the silvics of component species, and the length of the cutting cycle. However, many natural canopy disturbances result in substantial spatial variation in stand structure including basal area and canopy openness. In contrast, selection systems based on the BDq method may deliberately or inadvertently minimize spatial variation in structure in an attempt to reach a constant target basal area throughout the stand.

Both the selection of maximum tree diameters and *q* values have the potential to move stands structurally in directions having little or no natural analog. The theoretical goal of the *q* quotient, which ranges from >1 to 2, is to create "balanced," all-aged stands that will sustain timber yield—that is, stands with uniform ratios between successive diameter classes across the full range of diameters. In reality, there are few examples of natural forests that exhibit such balanced distributions (Matthews 1989). In fact, much of the theoretical basis for this approach was based on documentation of reverse-J shaped diameter distributions when stand data was aggregated over landscape scales (Meyer 1952), highlighting the lack of ecological justification for managing for such a distribution form at localized scales.

In practice, *q* values are selected based on the silvics of the tree species being managed and the proportion of the stand a forester wishes to have allocated to different size classes of trees. When high *q* values are selected, a great proportion of the stand is occupied by smaller (saplings and poles) trees, relative to low *q* values. High *q* values also select against larger diameter trees. These larger trees, often the oldest in the stand, are the hallmark of multi-cohort, old-growth forests and should be retained in some reasonable abundance if managing for ecological objectives is a priority. The use of distribution forms that have multiple *q* factors for different size classes, including lower *q* values for larger diameter trees, is one suggested solution to this issue (O'Hara 1998).

It is not likely that this dilemma can be overcome through careful choice of a single *q* value, as neither field research nor computer simulation has found a value that is relevant to both economic and ecological objectives (Hann and Bare 1979; O'Hara 2001). More importantly, it has been shown that stable multi-cohort tree populations can be achieved by diameter distributions other than a negative exponential, including rotated sigmoid forms that allocate a greater proportion of growing space to medium and large diameter trees (Goff and West 1975; Lorimer and Frelich 1984; Leak 1996; Goodburn and Lorimer 1999; James et al. 2004). Simulations also suggest that a negative exponential distribution may not maximize yields or optimize economic returns (Adams and Ek 1974; Bare and Opalach 1988; Kaya and Buongiorno 1989; Erickson et al. 1990).

distribution of gap-based natural disturbances. The latter generally exhibit a range in sizes, typified by a distribution heavily skewed toward smaller openings, but with inclusion of a few larger openings (e.g., Rebertus et al. 1997; Hanson and Lorimer 2007). In contrast, selection systems generally have minimal variation in opening sizes. Moreover, with group selection, openings may be consistently larger than the natural multi-tree gap disturbance regimes they are purported to emulate. For example, group selection in mixed-conifer types on National Forests in the Sierra Nevada Mountains, USA involve 1- to 2-ha openings. A gap of this size is many times larger than the size of the natural gaps (~ 0.08- to 0.10-ha) that are typically found in these forests (Knight 1997). At the opposite end of the spectrum, application of individual tree selection by definition avoids inclusion of larger multi-tree openings. However, even in forest characterized by small-gap disturbance regimes, some larger openings may be important for maintaining a less-shade-tolerant tree component in a forest that is otherwise dominated by shade-tolerant species (Hanson and Lorimer 2007; Kern et al. 2013).

Factors influencing foresters to deviate from a skewed distribution of opening size and instead select a narrow size range, or larger than natural opening sizes with group selection, include concerns about growth rates of reproduction, cost and operational difficulties associated with harvesting in small areas, ease of applying area regulation approaches with larger gaps, and generation of greater revenues during a single entry.

Regardless, the end result is that stands managed with selection can be homogenized spatially either by placing overly large group openings systematically across a stand or by use of consistently small single-tree openings throughout a stand. Structural homogenization can have implications for tree species composition in managed forests. For example, as historically applied, single-tree selection systems for northern hardwood ecosystems in the North American Great Lakes region effectively drive overstory composition to sugar maple dominance (Figure 8.18; Kern et al. 2013), whereas unmanaged mature and old-growth stands having a range of canopy opening sizes often support four or five species in abundance and ten or more species in total (Curtis 1959).

Modifying selection systems to avoid this homogenization involve use of opening sizes that vary spatially to more closely match the size distribution of gaps created by natural disturbances, rather than using a standardized opening size and placement. Varying harvest opening size to match the variability in opening sizes observed in natural stands is likely to result in a distribution skewed toward smaller average opening sizes, potentially including single tree openings, thus the distinction between group selection and individual tree selections as separate regeneration systems becomes less apparent and the use of systems that combine both elements (i.e., patch selection, Nyland 2007) becomes more appropriate for achieving ecological objectives. By varying opening size across a stand during an entry, the homogenization effect of a single opening size used throughout is avoided, enhancing the heterogeneity of canopy structural conditions within a stand if for no other reason than openings range in size from single-tree gaps to larger groups. It also follows with selection systems that the variation in opening size is inclusive of no opening at all within some areas of a stand, i.e., *skips*, much like with variable density thinning. The distinction from classical application of selection is that with an ecological approach, these skips may be set aside in perpetuity as areas where natural stand dynamics are allowed to proceed without intervention.

Figure 8.18 *Northern hardwood stand in northern Michigan, USA dominated by sugar maple as a result of long-standing single-tree selection that creates a narrow range of small opening sizes that favor this shade-tolerant species.*

The shape and orientation of gaps or openings is another important consideration that influences ecological functions and aesthetics. Circular and compact shapes maximize the area of gaps that are farther away from the canopy edge, thus increasing light availability and reducing seed dispersal from the adjacent canopy. More linear, sinuous openings can act as barriers to crown fire spread, impede insect dispersal, and may result in less windthrow in the adjacent canopy if oriented against the prevailing winds. Gaps oriented in an east-west direction will also receive higher rates of sunlight versus north-south oriented gaps, especially at higher latitudes. Varying gap shape and orientation, in line with the local disturbance regime, will result in greater spatial heterogeneity and complexity in forest development over time.

Operationally, implementation of a modified approach to selection as described above can still use a BDq (Box 8.6) or other stocking control approach but with modification. Typically with these approaches, there is an attempt to achieve the desired residual basal area uniformly across the stand. This is usually achieved by marking trees for selection within subareas of the stand and attempting to meet the target basal area in that subarea. The result is that at the whole-stand scale, the deviation of residual basal area from the average (target) is minimal. This is because the same opening size is being used throughout the stand. A simple alternative to this approach that will enhance heterogeneity of canopy structure is to vary the target residual basal area among subareas within a wider range around the target, again by using a wider range of opening sizes, from single tree openings to larger gaps, inclusive of permanent skips within some subareas (Box 8.7).

Decadence Creation

Until recently, deliberately killing or wounding of healthy trees, or even retaining defective trees of any species, was anathema to foresters given the prevailing objective of timber management. To maintain or even create trees that did not fit this objective (a focus on healthy crop trees and species) subtracted from the economic value of the forest and ran contrary to goals of stand improvement though removal

Box 8.7 **Managing for structural heterogeneity in northern hardwood ecosystems: Aitkin County (Minnesota, USA) Land Department example**

Mesic northern hardwood forests of the western Great Lakes region, USA have natural disturbance regimes characterized by frequent small to infrequent meso-scale wind events that remove individual trees, small groups of several trees, and patches of up 0.5 ha. The resultant opening size distribution spans a range of sizes, including many smaller openings (i.e., left skewed) and occasional larger openings.

The silvicultural approaches used in this forest type are often disconnected from this natural disturbance regime and resultant distribution of openings sizes. Historically, single-tree selection (and diameter limit cutting) have been used on many ownerships, resulting in a narrow range of small gap sizes, effectively driving composition to tolerant tree species, largely sugar maple.

The Aitkin County Land Department (ACLD, Minnesota, USA) was an early adopter of an approach to northern hardwood silviculture that runs counter to the classical single-tree selection approach. A management goal for ACLD includes production of quality hardwood sawtimber from a variety of species that span a range of shade tolerance, including sugar maple, northern red oak, basswood, paper birch, trembling aspen, yellow birch, black ash, and ironwood. As such, prescribing a range of opening sizes, inclusive of some larger openings, is important to the success of meeting this goal.

Specific silvicultural actions used by ACLD reflect a consideration for development of heterogeneity in opening sizes. Essentially, this entails using a spatially variable approach to selection that includes single tree removal combined with cutting small to mid-size openings, as well as leaving some uncut patches (skips) (Figure 8.19; Color Plate 8.4). This approach creates structural heterogeneity over space, fosters the development of multiple-age cohorts, and allows diverse tree species that vary in shade tolerance to establish.

Figure 8.19 *Northern hardwood silviculture as practiced by the Aitkin County Land Department (Aitkin County Minnesota, USA) to promote structural heterogeneity and tree species diversity. (a) Larger gap to promote establishment of mid-tolerant tree species. (b) Aerial view of managed stand (center of photo) that shows the wide range of opening sizes used. Photo credit (a): Mark Jacobs.*

Figure 8.20 *A tree marked for removal due to a defect in its bole.*

of defective, poor quality trees (Figure 8.20). However, all this was before the ecological value of decadent trees and large deadwood was fully understood by ecologists. Now foresters generally appreciate the value of large snags and logs on the ground as habitat for a variety of organisms, sinks and sources of nutrients and moisture, and geomorphic modifiers in riparian areas and streams. Decadent, declining, and injured trees with obvious defects are now recognized as the near-term sources of deadwood, the trees most likely to provide cavities, and an important food resource for a variety of organisms (Figure 8.21).

The other part of this story is that managed forest ecosystems are often lacking in large deadwood structures. In younger forests, this largely comes about because legacy structures were not retained during regeneration harvests (Chapter 7) and as just mentioned, dead and declining trees are often removed when the opportunity presents itself. The consequence of this is that many managed forests can use an infusion of snags and logs, or their precursors, declining trees.

This is where decadence creation comes into play. The technique involves deliberate killing of living trees or injuring them to induce decline, with the goal of creating snags or logs on the ground or setting in motion the process that will ultimately lead to deadwood. Felling trees to create large logs and branches on the ground is relatively straightforward (Figure 8.22). Creating snags can be more complicated. Often snags are created by girdling, which involves removal of the bark, phloem, and cambium in a band around the circumference of the tree. Girdled trees die because their roots are deprived of carbohydrates produced through photosynthesis; without a healthy root system, the tree no longer absorbs water and soil nutrients. Often trees are girdled near their base, which is done relatively easily from the ground (Figure 8.23a, b).

Figure 8.21 *Decadent and declining trees. (a) Tree with internal decay as evidenced by shelf fungus. (b) Sugar maple with an internal injury marked for retention in this stand. (c) Sugar maple with numerous cavities from woodpecker excavations. Photo credit (b, c): Chris Gronewold.*

Figure 8.21 *Continued*

Alternatively, trees may be girdled higher up on the bole, often just below the main crown (Figure 8.23c). Low girdling tends to hasten death because there are few stored carbohydrates available to the root systems. In contrast, high girdling can delay death because carbohydrates stored in the tree boles are available to the roots and there may be some live branches below the girdle that continue to produce some carbohydrates through photosynthesis. Until supplies are depleted, the tree can remain living.

Another approach for creating snags involves injuring trees in a way that induces decline and ultimately death, analogous to what might happen after a lighting strike. One technique used to do this involves placing an explosive fuse within the crown of the tree above the lowest live branches. The resultant detonation will remove most, but not all, of the live crown, resulting in a delayed death of the tree over several years (Figure 8.23d). Shorter snags can also be created by "high stumping" trees with a mechanized harvester or processor. While these snags are typically 3–6 m tall and will not have all of the functions of a full snag, they are an operationally efficient and inexpensive approach to increase snag and future downed wood levels during a treatment.

How Much Deadwood Is Enough?

How much deadwood or declining trees should an ecological forester consider creating? As with determining the number of structural legacies for retention, including deadwood legacies (Chapter 7), natural systems can provide a guide (e.g., Fassnacht et al. 2014). Remembering that a goal of intermediate treatments in ecological forestry is to hasten the development of the complex structure found in older natural stands, it is helpful to look at deadwood characteristics in mature and old-growth forests for guidelines on deadwood targets. For example, consider the number

Figure 8.22 *A red pine deliberately felled to create deadwood on the ground, Minnesota, USA.*

Figure 8.23 *Low girdling of (a) a black ash using a drawknife and (b) a sugar maple using chainsaw cuts. (c) High girdling of a red pine at the base of the crown. (d) Red pine snag created by removing the majority of the crown through use of detonation cord. Photo credits: (b) Karin Fassnacht, (c) Doug Kastendick.*

Table 8.2 *Snags in old-growth red pine forests in Minnesota, USA.*

Old-Growth Forest	Snags/ha > 13 cm and < 43 cm diameter	Snags/ha ≥ 43 cm diameter
Itasca State Park	44	10
Sunken Lake Natural Area	138	2
Lac La Croix	32	24
Scenic State Park	18	6
Ramshead Lake	56	22
Voyageurs Island	72	54
Pine Point Research Natural Area	68	20
Mean of red pine plantations	5	1

of snags found in old-growth red pine forests in northern Minnesota, USA (Table 8.2). While variable among sites, the numbers are generally much higher than those found in stands managed for timber, often plantations, for both smaller and large snags (Fraver and Palik 2012).

This is just an example of the kind of information that can be used to determine target numbers for deadwood creation. It is unrealistic to think that foresters will attempt to create snag numbers such as these in all stands. Moreover, deadwood target numbers should be ecosystem specific; number and sizes are relative to the ecosystem of interest. The size and age of created deadwood should also be considered, which may necessitate multiple entries into a stand to create deadwood from larger diameter trees as they age and to add new deadwood into the system as previously created deadwood decays.

There are, of course, safety concerns associated with creating snags, but to the extent possible, removal of snags for safety concerns and machinery trafficking around snags or downed logs should be avoided. A simple way to address both safety concerns and avoidance of trafficking damage is to target snag creation in the skips used in both *VDT* installations and modified selection approaches discussed above. Also, tradeoffs with economic objectives can be minimized by concentrating decadence creation on cull individuals where possible to increase both the long-term ecological and economic value of a given stand. Moreover, increasing use of mechanized harvesting equipment with reinforced cabs has greatly reduced safety risks associated with retaining snags. It has thus become operationally much easier in recent years to retain and create snags.

The Complexity of Time: Silvicultural Interventions at Ecologically Appropriate Time Intervals

As you learned in Chapter 5, it takes time for complex structure to develop in forest ecosystems. While silvicultural tools such as thinning, especially *VDT*, and decadence creation can accelerate the process, there may be no substitute for allowing forests to proceed through a developmental sequence that allows significant structural complexity and heterogeneity to accumulate. The importance of time relates to the third foundational principle of ecological silviculture (Chapter 3), specifically considering the importance of applying *silvicultural interventions at ecologically appropriate time intervals.* Recall from this earlier discussion that the use of ecologically based time intervals is a consideration for ecological silviculture as it applies to the timeframes necessary for recovery of forest ecosystem structure and function following a major treatment (i.e., a regeneration harvest), as well as the time interval that should be allowed between intermediate treatments, such as *VDT.*

First let us consider the factors determining time intervals between regeneration harvests, termed **rotation** age when using even-aged systems or **cutting cycle** with uneven-aged systems. Stand rotation lengths in North America have often been based on financial criteria related to maximizing discounted present net value (Davis et al. 2001) or on biological standards related to annual timber production (Newman 1988), based on culmination of mean annual growth increment. Similarly, cutting cycle lengths may also be based on factors such as optimizing financial returns or maximizing production over a given interval, with values generally ranging from 10–25 years depending on the forest type (Nyland 2007). Overall, the rotation ages suggested for forests managed ecologically will be well beyond those developed based on financial criteria (Curtis 1997), largely because the interest rates used in economic optimization tend to favor short-term financial returns (Bettinger et al. 2009). For ecological forestry, the problem with basing rotation age on culmination of mean annual increment is that in unthinned stands, time to growth culmination can be quite short relative to tree lifespans. For instance, volume growth of site index 52 m Douglas-fir culminates at around sixty-five years in unthinned stands (McArdle et al. 1949), despite the life span of this species regularly exceeding 300 years (e.g., Tappeiner et al. 1997).

Determining time intervals between regeneration harvests is often more complex in ecological forestry because considerations are not limited to wood production or financial returns. Rather, it is the time needed to develop desired structures or structural conditions that will be a large factor in determining the length of the recovery period between regeneration harvests (Box 8.8). Basing the timing of regeneration harvests on appropriate ecological recovery periods is potentially a contentious issue because, as pointed out above, these periods will almost always be much longer than rotations based on economic factors, may involve lower levels of removals during any given entry in the case of selection systems, and are likely going to be longer than those based on growth criteria as well.

In setting intervals between regeneration harvests, foresters have to consider the time needed for specific

Box 8.8 **Contrast between Recovery Periods and Rotation Length**

In timber-focused forestry, stands are invariably terminated before significant structural complexity and often tree species diversity has developed, even though substantial biomass may have accumulated (Figure 8.24). Consequently, forests managed for timber typically lack trees of very large diameter (relative to the potential for the species in question), have small amounts of large dead wood accumulation, particularly well-decayed dead wood, and are lacking in trees with unique structures (e.g., cavities, large limbs, heartwood, and brooms). Moreover, they are often low in canopy tree species diversity because the time between harvests is insufficient for tolerant species to establish or to advance into intermediate and co-dominant canopy positions. This problem is compounded when stands are deliberately or inadvertently managed for discontinuity during establishment, e.g., if no legacies are retained or only a single species is planted or encouraged. An example of this is seen in Great Lakes aspen forests. Most of these forests originated in the early part of the twentieth century after widespread logging followed by severe slash fires (Graham et al. 1963). Many of the stands that originated from this disturbance history are devoid of any significant biological legacies and are spatially homogeneous in structure (Figure 8.25a). Moreover, management prescription calls for clearcut harvesting of stands at 60 to 80 years of age—before significant structural complexity and compositional richness has had time to develop. In fact, Old Forest Stage aspen stands historically have been (and continue to be) referred to as *overmature*, suggesting that the stand should be liquidated, when the condition is actually healthy from an ecological perspective (Figure 8.25b).

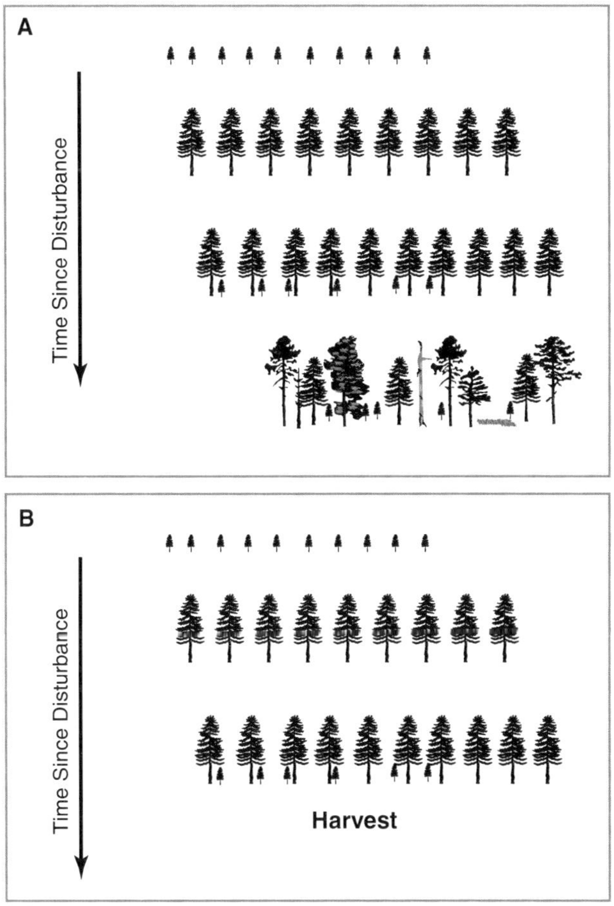

Figure 8.24 *Conceptual representation of the importance of timing for the development of structural complexity. (a) A young post-disturbance forest begins with simplified structure and composition (i.e., no continuity from the previous stand) and over time develops significant structural complexity due to tree growth and decline and small-scale canopy disturbance. (b) In many classical forestry approaches, the forest is harvested before significant compositional diversity and structural complexity has had time to develop.*

Figure 8.25 *a) A regenerating aspen stand in northern Minnesota, USA. The stand is devoid of biological legacies from the previous stand and this forest type is often harvested before significant structural complexity has had enough time to develop. b) An Old Forest Stage aspen stand in northern Minnesota, USA.*

structural elements and functional attributes to recover to a desired level, as well as the time period over which it is expected to be at that desired level. For example, a structural goal might include development of a population of larger diameter trees or large well-decayed snags. Another example might be the re-establishment of late successional tree species. Not only are long time intervals required to develop such structures or forest conditions but, once created, that structure or condition may need to be maintained for a significant time period before the next harvest.

It may be possible in some cases to integrate ecological and growth metrics for determining the timing of intermediate treatments and, ultimately, regeneration harvests. For instance, research has shown that the culmination of growth increments in some forests can be delayed for extended periods of time through periodic thinning (Curtis 1995; Williamson 1982; D'Amato et al. 2010), essentially pushing the timing of a regeneration harvest (in forest managed using even-aged regeneration systems) to older tree ages, allowing more complexity of structure to develop, along with large diameter, high-quality logs. In fact, thinning treatments are increasingly being applied to older forest stands in attempts to increase residual tree vigor and to promote and maintain structural characteristics of old-growth stands in forests managed for wood (e.g., Latham and Tappeiner 2002; Fajardo et al. 2007; Kolb et al. 2007). A consistent result across many of these studies (Latham and Tappeiner 2002; Bebber et al. 2004; Powers et al. 2009) has been the positive growth response of older trees to thinning (Box 8.9).

Box 8.9 **Extended recovery silviculture for red pine**

The Red Pine Level of Growing Stock Study on the USDA Forest Service's Cutfoot Experimental Forest in northern Minnesota, USA was established in 1949 to examine growth and yield of red pine thinned to and maintained across basal areas ranging from 13.8 to 32 m^2/ha (60 to 140 ft^2/ac) (Figure 8.26). A unique aspect of this study is that the red pine were eighty-five years old in 1949, with the various treatments thinned six or seven times between 1949 and 2009, offering an unprecedented opportunity to examine growth and structural development in old, maturing, but managed stands (D'Amato et al. 2010).

One important result from this study is that mean annual volume growth shows extended periods of continued growth due to periodic increases after the thinning began

Figure 8.26 *Red pine stands managed at (a) low residual basal area (13.8 m^2/ha) and (b) high basal area (32 m^2/ha) for extended periods of time. Trees were 85 years old at study establishment in 1949 and 145 years old when these pictures were taken in 2009. By 2009, stands (a) and (b) had been thinned six and seven times, respectively.*

at age 85. The stand-level growth increases, particularly within the higher stocking-level treatments, lend support to the notion that older forest stands are responsive to thinning treatments, despite peaks in stand- and tree-level productivity that typically occur at younger ages (Ryan et al. 1997; Smith and Long 2001).

Notably, the findings of this study lend support to the notion that the use of repeated thinning treatments can maintain greater levels of volume production beyond historically-applied rotation ages (i.e., 60–100 years; Buckman 2006; Gilmore and Palik 2006). The results also suggest that the application of thinning treatments after culmination of MAI can be used to maintain relatively high levels of stand production and the development of high quality trees for timber, while also achieving ecological objectives (e.g., development of tree diameters >60 cm) that require recovery lengths beyond those suggested by peak MAI (D'Amato et al. 2010). Similar findings and conclusions have been demonstrated in North America for coastal Douglas-fir (Curtis 1995) and more widely in European forests (Assman 1970; Pretzsch 2005). Collectively, these findings highlight the potential for using repeated thinning and extended recovery periods beyond typical rotation ages to simultaneously meet ecological and economic goals within forests composed of long-lived tree species.

It should be noted that allowing appropriate recovery periods between regeneration harvests for significant structural complexity to develop is not going to be an option for all managed forests. This is because the time interval between a regeneration harvest and development into the Old Forest Stage (Chapter 5) is much longer than may be practical in a managed setting. This is particularly the case for forest types that require long time periods to develop complex structure following stand-replacement disturbance events, such as the Douglas-fir-western hemlock forests described in Chapter 6. Few foresters are likely to have the time and space to consider managing forests on rotations of the more than 250 years that would be required to develop into the Old Forest Stage following a clearcut harvest in the Douglas-fir region. This takes us full circle to the importance of managing for continuity with legacy retention (Chapter 7) in these types of forests, as these enable the ecological forester to retain and create structural conditions on a shortened time scale, rather than having to allow them to develop afresh following each harvest, as would be necessary under strict even-aged silvicultural systems. In effect, retention treatments can partially substitute for time.

Recall from our discussion earlier in this chapter that thinning, particularly variable density thinning (*VDT*), is a tool to hasten the development of structural complexity in established strands. So, like retention at harvest, *VDT* can also partially substitute for time. Historically, the timing of thinning in timber-focused forestry is based on relationships between tree size and density using tools such as stocking diagrams (Box 8.10). The timing of thinning in stands managed ecologically is still a concern, but intervals between thinning are likely to be more varied, and likely longer, than in timber-focused forestry, and not so easily guided by available stocking guides. This is because of the greater diversity of structural elements, such as trees, and processes that are management foci in ecological silviculture, many of which do not lend themselves to classic crop-based thinning regimes that might be triggered based on basal area or relative density targets, without consideration for the presence of structural targets. In effect, maximal intervals between thinning may need to be considered, as opposed to the minimal interval that typically dominate the scheduling of thinning in timber-focused forestry.

Box 8.10 **Determining timing and stocking levels for thinning**

The timing and level of thinning applied to even-aged forest stands are largely guided by stand management diagrams based on the generally predictable relationship between average tree size and density in **self-thinning** populations of trees (i.e., **size-density relationships**). These diagrams are used by foresters to approximate the **stocking levels** that maintain full site occupancy, but are not too dense to create competitive conditions that would stagnate growth or induce the self-thinning mortality commonly observed later in the Young Forest Stage. The maximum stocking level on these diagrams either reflects the maximum size-density relationship based on the **-3/2 self-thinning relationship** for a species (i.e., Density Management Diagrams) or an average maximum density derived from **normal stands** above which growth declines below desired levels of increment (i.e., the **A-line** on **stocking charts**). Minimum stocking reflects the lower limit of full site occupancy by the tree population of interest (**B-line** on stocking charts). A limitation of these tools as it relates to ecological silviculture is that they have been largely derived from fully-stocked stands of a single-species, with a primary emphasis on generating stand-wide conditions that maintain optimal levels of growth. While this is a useful approach in areas where maximizing stand-level wood production is the primary objective, it generally results in low complexity of structural conditions, including homogeneity in live-tree density and spacing, canopy cover, and tree size across a management unit (Figure 8.27). These stocking level concepts can still have merit in relation to managing for complexity; however, to do so, they should be applied and varied at the sub-stand level to create high levels of within-stand heterogeneity in stocking conditions, including creating gaps or maintaining high density regions that would be deemed under- or overstocked based on stocking guides (see *Variable Density Thinning*).

Figure 8.27 *White pine forest uniformly thinned from below to lower level of fully stocked conditions based on white pine stocking chart (B-line) in the Adirondack Region of New York, USA.*

What to Remember about Managing for Complexity/Diversity in Established Stands

The ecological forester needs to understand the concept of complexity/diversity in developing forests, including its causes and the consequences of lacking these conditions. Armed with that knowledge, they can then begin to incorporate *Principle 2* into their prescriptions by devising actions that sustain or restore structural complexity, species diversity, and heterogeneity, or in fact, jumpstart the development of these conditions faster than what might occur naturally. At the very least, they need to recognize the importance of determining the timing of silvicultural interventions (*Principle 3*) to coincide with structural development triggers, rather than strictly basing these on growth or economic factors, as in timber-focused forestry.

Some specific points to remember relative to applying these principles to management:

1. The ecological forester recognizes that structural complexity and species diversity is highly variable among different types of *natural* forest ecosystems. Not every type of forest is equally complex or diverse. In fact some forests, e.g., frequent fire forests, may be fairly simple in structure. However, compared to their managed counterparts, most forests tend towards structural complexity, greater tree species diversity, and more spatial heterogeneity of structure.
2. Thinning is an important intermediate treatment for both classical and ecological silviculture, with a similar goal for both of developing large trees. However, thinning in ecological silviculture inevitably incorporates much broader objectives than increasing tree size and quality; in fact, enhancing the habitat value of large trees may be the primary driver for thinning.
3. The ecological forester often will devise thinning approaches that provide for spatially varying outcomes (heterogeneity) across a stand, e.g., variable density thinning. This same argument applies to the use of selection in uneven-aged regeneration systems. Varying spatial pattern of opening sizes is important for avoiding homogenization of stands in structure and composition.
4. The ecological forester will devise thinning treatments for the benefit of a diverse suite of species, including but not limited to crop tree species. Thinning to favor non-dominant and non-commercial species may be an important goal.
5. The ecological forester may need to take deliberate action to augment deadwood features in

managed stands. This might include felling large (potential valuable) trees to add deadwood to the forest floor or creating large snags through girdling or some other means.

6. Trees with defects, cavities, etc., while not economically valued, are generally not removed by the ecological forester, as they have immediate habitat value and are a likely near-term source of large deadwood.
7. Finally, time is an important ally of the ecological forester, specifically by incorporating appropriate time periods between interventions to allow stands to develop complex and heterogeneous structure. This consideration requires the use of ecological metrics to determine rotation ages (better thought of as recovery periods) and cutting cycles, rather than strict adherence to growth or financial metrics.

CHAPTER 9

The Context of Silviculture at Landscape Scales

It is only at larger spatial scales—which we refer to as the landscape scale—that the full expression of forest ecosystems, including their richness and functionality, can be fully appreciated. In fact, there is an entire subset of ecology—landscape ecology—that focuses on the effects of ecological processes at large spatial scales that range from hundreds to millions of hectares.

However, as you know, silviculture is practiced at the scale of individual stands. As such, a landscape perspective may not always be emphasized when developing silvicultural prescriptions. Despite this, foresters recognize that natural landscapes incorporate a large variety of patches (see definition below) that differ in composition, structure, and level of development, as well as patches of non-forested ecosystems, such as meadows, lakes, and streams (Figure 9.1). It is certainly possible to consider the population of patches in isolation, for example, by tallying characteristics of each. But it is only possible to fully appreciate the full variety and interaction of patches in a landscape by stepping back and going up in scale of observation and thinking.

Building on this perspective, it is also possible to take an inventory of managed stands in a landscape and consider the management histories and silvicultural treatments that have occurred or are planned for each. But again, this is insufficient to fully understand the interactions of these stands at a larger spatial scale and how silvicultural actions work to sustain or degrade the fundamental integrity of the landscape from an ecological perspective. Taking a landscape-scale view of managed stands is needed to understand how they interact as a collection of patches and as sources of emergent phenomena, such as productivity, movement of water, provision of large wood and sediments to aquatic ecosystems, and connectivity among patches as it affects movement of biota and materials.

The converse of the positive properties that emerge at the landscape-scale is the potential for stand-scale actions to create dysfunctional landscape conditions through, for example, fragmentation of forest into small and isolated stands or from cumulative effects that result from excessive levels of particular activities. Concern over the potential for dysfunctional landscapes resulting from forest management emerged in earnest in the 1980s and ultimately resulted in greater attention on the landscape perspective, and sometimes even legal requirements to consider it. For instance, the National Environmental Policy Act (NEPA) created a legal requirement on US federal forests to analyze ecological effects at large spatial scales, particularly those related to the collective impacts of management activities (Franklin et al. 2018).

Taking a larger-scale view of silviculture for sustaining the function of landscapes is the subject of this chapter. Specifically, we address the fourth foundational principle of ecological silviculture (Chapter 3): *planning and implementing silvicultural actions in the context of how these actions influence the structure and function of landscapes.* After reading this chapter, you will have an appreciation of some of the major organizing principles of landscape ecology and how they apply to ecological silviculture. You will have a better recognition of special places in the landscape, including both places of conservation and cultural significance, and how to plan silvicultural actions with their sustainability in mind. Finally, you will be introduced to a management construct called Shades of Green, which gives a perspective on how to incorporate the first three

Figure 9.1 *A forest landscape in the New Jersey, USA Pine Barrens. Note the variety of patches including burned (and burning) and unburned forest, forest of differing composition (different canopy shades), and a river in the background. Photo credit: Bob Williams.*

principles of ecological silviculture (continuity, complexity/diversity, time) in varying ways to satisfy diverse objectives across the landscape. This concept and the chapter at large draw on material from Chapter 5 of Franklin et al. 2018.

Landscape Organization and Silviculture

Landscapes can be viewed using several organizing constructs, including networks, gradients, and patchworks. We discuss networks and gradients only briefly here (see Franklin et al 2018 for a thorough discussion), but go into more detail on patchworks since in managed landscapes it is the accumulated activities in stands (i.e., silviculture) that create or modify patchwork structure. We also discuss several phenomena that emerge as a result of landscape structure and that have relevance to silviculture, including edges and edge effects and landscape connectivity.

Networks and Gradients

Networks offer a way of thinking about several important landscape features that are linear and result in high connectivity across the landscape. The most important natural network in forest landscapes are drainage systems, such as riverine watersheds (Figure 9.2) or extensive wetland complexes. Drainage networks have a high level of continuity, with the dominance of flows following the energy gradient upslope to downslope, for instance from intermittent headwater channels, to first order streams, to high order rivers. In riverine and other drainage networks, the structure and function of adjacent forest, particularly in the upper reaches (smaller order streams), are directly related to the health and functionality of aquatic ecosystems. We talk about the importance of this connection (via riparian areas) in detail in Chapter 15.

The most important human-created network in forest landscapes directly relevant to silviculture are road networks, e.g., skids trails and haul roads. Contrary to the dendritic (linear) nature of drainage networks, road networks are rectilinear. This can lead to intersections of roads and drainages that create major management challenges such as altered drainage patterns, barriers to movement of aquatic organisms, and destabilization of hill slopes and stream channels. Foresters need to be alert to the potential impacts of roads on natural flows of materials (e.g., water and sediment) and movement of organisms in the aquatic network and therefore have increasingly modified best management practices to sustain these functions through structures such as bottomless culverts.

Figure 9.2 *Suiattle River Watershed, Glacier Peak Wilderness, Washington, USA.*

Environmental gradients are another way of visualizing the organization of landscapes, particularly when large and environmentally varied landscapes are being considered. Environmental gradients are particularly evident in mountainous regions (altitudinal gradients), but they are typically present in various forms in all types of landscapes. For instance, climatic gradients involve systematic changes in precipitation, temperature, and wind along elevational, topographic, edaphic, latitudinal, or longitudinal continua. Soil gradients are also characteristic of many regions, often resulting from gradients in key soil-forming processes, such as in deposition of wind-carried (aeolian) dust or volcanic ejecta (tephra). It is important to keep gradients in mind when planning silvicultural prescriptions across larger project areas, as there may be need to make adjustments to objectives and approaches along the gradient.

Patchworks

Patchworks are the most relevant way to visualize the structure of landscapes in the context of silviculture, as it is the natural patchwork (type and array of forest conditions) and constructed patchwork (managed stands) that are the operating grounds of silviculture. Put simply, a landscape patchwork is a collection of patches of different types (Figure 9.3). *Patches* themselves are distinguishable areas that differ from other patches in size, shape, and condition, including vegetation or ecosystem type, such as forests, meadows, bare rock, and lakes. A goal in visualizing landscapes as patchworks is to identify areas that differ in conditions and underlying characteristics, facilitating efforts to assess and plan projects. Hence, landscape patchworks are conceptually and practically important to foresters planning silvicultural actions.

Each landscape has a dominant patch type, called the *matrix*. Generally, the matrix is the most extensive or most connected patch type in the landscape, e.g., a matrix of mature forest or of young forest. Another generic patch type often recognized in landscape ecology is the *corridor* (see discussion of connectivity below), which is usually a long and narrow patch connecting two patches of similar type.

Figure 9.3 *Landscapes can be viewed as collections of patches of contrasting conditions, as in this landscape of forest and pasture in the Czech Republic. However, it is important to recognize that there is no single natural classification of patches useful to all resource managers or stakeholders; patches in a landscape are defined in the context of specific applications or management concerns.*

Patches have attributes of size, content, persistence, and heterogeneity. Patch content will largely determine the level of contrast with adjacent patches and, as explained later, influences the level of connectivity in the landscape. Persistence relates to whether the patches are essentially permanent because of their physical nature (e.g., lakes and rock outcrops), semi-permanent because of management intent (e.g., reserved forest areas), or transient (harvested areas). There will generally be some minimal size of patch that is recognized in most landscape analyses and this is often based on what is practical. Many patches in forest landscapes will also exhibit internal heterogeneity (or patchiness) at spatial scales smaller than the minimum patch size that has been selected for that landscape; examples could be clumps of larger and older trees within an otherwise uniform younger stand (e.g., structural legacy retention) or small openings lacking tree canopy cover (e.g., gaps in a stand that had variable density thinning).

Timber-focused and ecological foresters will likely choose different criteria for recognizing patches in the landscape. The idealized vision of a forest landscape in timber-focused forestry is likely one composed of multiple age classes of even-aged stands (usually plantations) (Figure 9.4). In such a landscape, the patches themselves (e.g., sizes and shapes) reflect past decisions that resulted in their delineation, primarily based on site, economic, and operational considerations, such as soil productivity, ease of access, and available logging methods at the time. The patches are considered permanent in the sense that the intent is to grow repeated crops of trees within approximately the same patch boundaries, with the rotation age rising and falling as the desired rate of return, levels of investment in early stand treatments (release, thinning, and fertilization), logging and wood processing technology, and wood

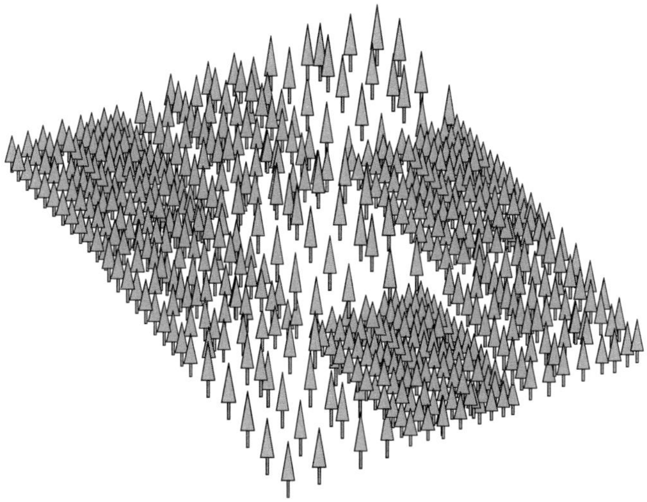

Figure 9.4 *A conceptual representation of a timber-focused forest landscape consisting of multiple age classes of even-aged plantations.*

product price forecasts change. Internally, patch content is structurally and compositionally homogeneous and keyed to the types and spacing of trees that will produce the desired product and return on investment.

In contrast, the ecological forester's ideal landscape is usually more complex than that of the timber-focused forester (Figure 9.5). The sizes of patches are likely to be larger (by coalescing smaller stands into larger ones) and the shapes are likely to be more irregular, reflecting boundaries that follow natural features such as changes in composition, soil, or underlying geomorphology. Within-stand structure is likely to be complex and heterogeneous, with variation in composition, tree sizes, and density.

For the ecological forester, natural landscapes and landscape processes provide a baseline for determining

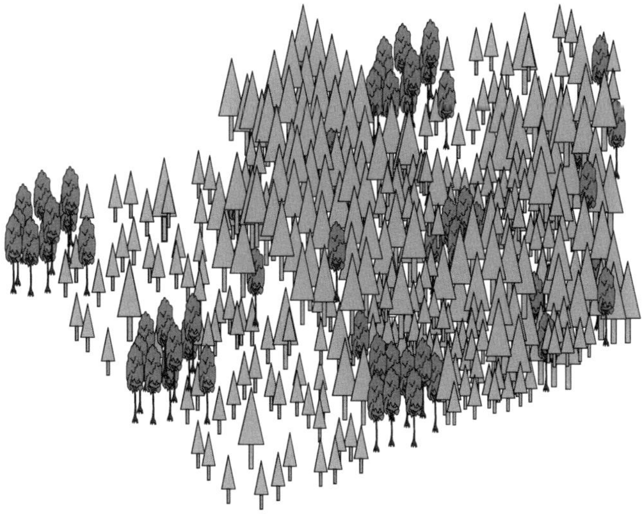

Figure 9.5 *A conceptual representation of an ecological forester's ideal landscape, consisting of stands of differing forest development stages (age), which better reflect underlying geomorphic and disturbance templates that determine size and shape, and greater compositional, tree size (age), and structural variation.*

Box 9.1 **Managed Forest Patchworks**

Resolving issues related to patch size and distribution in real landscapes depends upon many factors, including the management goals for the landscape and the type(s) of forest ecosystems being managed, including their natural disturbance regimes. Here we looks at these issues using two of our disturbance archetypes (Chapter 6), including forests characterized by infrequent severe disturbances (archetype 1) and forests characterized by chronic low severity disturbances (archetype 2).

Archetype 1 Landscapes

Archetype 1 forests are characterized by infrequent, severe disturbance, with wildfire being the most common disturbance agent. Douglas-fir-western hemlock forests, jack pine forests, and lodgepole pine forests are North American examples of this archetype. Landscapes dominated by archetype 1 forests will generally have well-defined and often large natural patches. Recall also that significant amounts of structural legacies (within-patch ecological content) are characteristic of disturbances for these forests.

An ecological forester taking a landscape view of archetype 1 forests needs to determine the appropriate sizes of managed stands (patches), as well as the content or condition of those stands following harvest, particularly the amount of structural legacies. While the natural patchwork generated by disturbance provides some guidance for this, recall that natural fire disturbances in archetype 1 landscapes may be very large, involving hundreds to many thousands of hectares. Managing for stands at this same scale as disturbance patches will rarely (or maybe never) be an option.

Given the challenges of generating the large patches historically created by natural disturbances, alternative guiding features are the habitat requirements of the native biota. This would include determining stand sizes and the types, amounts, and spatial patterns of legacy retention in archetype 1 landscapes that encompass the needs of wildlife species in these systems, including those requiring habitat conditions that immediately follow large, high severity fires. In fact, this is likely to provide a more realistic guide than the scale and condition of the historical disturbance patchwork.

Consideration of harvest impacts in small watersheds also provides an important reference point in selection of stand sizes, at least in some (mountainous) regions. Small stand sizes that result in chronic perturbation of watersheds by repeated entries during a harvest cycle should be avoided (Reeves et al. 1995; Reeves and Spies 2017). Instead, harvest unit patch sizes that result in a single entry during a rotation are preferable because they allow the harvested watersheds to undergo extended periods of recovery between harvests (Figure 9.6). Nonetheless, existing restrictions on maximum harvest unit patch sizes associated with state-level policies and third-party certification may limit the ability to increase harvest unit patch sizes to these scales.

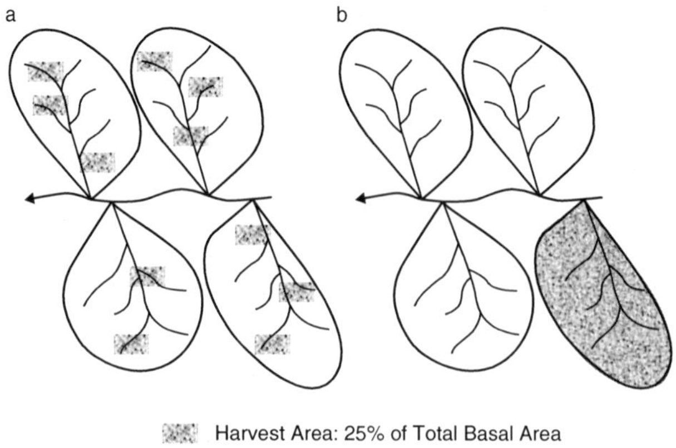

Figure 9.6 *Alternative approaches to distributing harvests over a series of headwater watersheds. In the staggered-setting scenario (a), small harvest units result in chronic perturbation of watersheds; they are repeatedly subject to harvesting during a cutting cycle or rotation. In the alternative scenario (b), a single large harvest unit (equaling the basal removed in a) results in a single entry during a cutting cycle allowing the watershed to undergo a long recovery period between disturbances (redrawn from Reeves and Spies 2017).*

One way to delineate stands in archetype 1 forests using a natural template is to recall that disturbances, while large and severe, are often spatially heterogeneous across a landscape. Consequently, the effects of disturbance, including tree mortality and the amount of structural legacies, are equally variable within the disturbed patch. This variation offers insight into a stand delineation approach that is still loosely based on the template of natural disturbance, whereby individual stands are delineated within much larger patches based on emulation of variation in disturbance severity and resultant legacies. With such an approach, stands (subpatches) would include a range of live tree and deadwood legacies. These stands would not all be treated simultaneously (i.e., not a temporal emulation of the natural disturbance), but such an approach would allow for the entire array of forest developmental stages (Preforest through Old Forest Stages) to be represented in the landscape at any one time. Public lands, such as US federal forests, are those that are most likely to incorporate this approach successfully, given their spatial extent, although other ownerships, including reservations of indigenous peoples and properties managed by conservation organizations (e.g., The Nature Conservancy) may also successfully use this approach.

Stand (patch) sizes based on this approach may still be larger than those used historically in timber-focused landscapes and may thus not be widely accepted by the public or conservation organizations. However, concerns over large harvest units in archetype 1 landscapes have largely been a consequence of the wide use of clearcutting historically, in which there was no significant structural retention. Harvest practices that include the retention of meaningful numbers of trees (see Chapter 7) are one important key to acceptance of larger harvest units by both conservation biologists and the public. As you will learn later, retention is also important because of its positive effect on connectivity in the harvested landscape.

Archetype 2 Landscapes

Archetype 2 forests are those subject to frequent low severity disturbance, generally fire. Examples from North America include longleaf pine and shortleaf pine woodlands in the southeastern coastal plain and ponderosa forests in the western United States and Canada. In these landscapes, distinct patchworks at larger scales generally are not evident because the stand-replacement wildfires that create patchworks, such as in Archetype 1, rarely occurred. Instead, natural frequent-fire landscapes are mosaics of fine-scale, low-contrast patches, that include representation of Preforest, Young, Mature, and Old Forest Stages at this fine scale (Chapter 6). Where broad landscapes of this archetype existed, the forest mosaic was continuous over large areas, undergoing localized changes in the composition and structure of the low-contrast, fine-scale mosaic in response to disturbance, forest development, and environmental gradients.

Most frequent-fire forest landscapes have undergone changes in structure during the last 150 years, primarily as a result of European settlement and associated land use (see Chapter 12). In the southeastern US, millions of hectares of forests dominated by longleaf and shortleaf pines have been converted to pine plantations. Native forests that remain are often overly dense with ingrowth of hardwood tree species. In western North America, disruptions of the frequent-fire forests began with grazing in the nineteenth century by millions of head of domestic livestock followed by timber harvest (including conversion to plantations during the latter half of the twentieth century), and fire exclusion (see Chapter 12). Remaining natural ponderosa pine and mixed-conifer forests are often dense and fuel-loaded. Wildfires occurring in these forests are now high severity and cause complete mortality, including historically fire-resistant old pines.

Ironically, as a result of the historical management of these types of forests described above, the major management need is not one of creating a more ecologically appropriate patchwork of stands, but rather erasing the patchwork altogether and replacing it with a continuous forest mosaic that has a structure highly resistant to wildfire and in which management activities on the ground occur at finer spatial scales. Thus, emulation of a natural landscape structure would involve applying fine-scale silvicultural treatments, including single-tree and small group harvesting, within the landscape, without demarcation of traditionally defined stands. Reintroduction of low-severity surface fires, or emulation of this disturbance, would be needed to restore/maintain the open, woodland structure of many forests in this archetype.

However, a challenge in some frequent-fire landscapes is—as a part of restoration—to provide for localized areas of denser forest conditions required by certain native biota. Examples in western North American include goshawks and some owls, including the threatened northern spotted owl. These birds and some of their prey require denser forests as nesting and roosting habitat as well as for foraging habitat. Hence, the landscape context for these forests needs to include management for these denser forest conditions, even while working to restore the larger landscape to a fire resistant woodland state. Retaining some portions of the frequent-fire landscape as denser forest also will likely be important on ownerships where continued timber harvest is intended. Embedding localized areas of dense, closed canopy forest with a matrix of fire-resistant woodland is one approach that can work in both cases.

appropriate patch sizes, shapes, content, and heterogeneity, but this will be tempered by technical, economic, and social considerations (Box 9.1). As in a timber-focused landscape, some patches will be permanent (e.g., lakes and meadows) or semi-permanent (e.g., riparian buffers), but more are likely to be transient because a patchwork based on a natural template will undergo segmentation and recombination into new patchworks over time. This will be especially common in landscapes that started with a timber focus but are now managed following a natural model. In these landscapes, the historical patch sizes and boundaries are often inappropriate for achieving ecological objectives. For instance, historically defined stands are likely to combine and coalesce into larger units of management that better reflect an underlying geomorphic template or the extent of natural disturbances, such as fire in frequently burned forests. Finally, patches created by ecological silviculture will generally incorporate much greater internal spatial heterogeneity (i.e., the patches will not be structurally or compositionally uniform) than in the timber-focused landscape.

Edge and Edge Effects

The concept of edges is related to the preceding discussion of patchworks in that edges are the boundaries between patches (including natural patches and managed stands). The importance of edges between stands will be evident to any forester who has had to deal with blowdown along recently exposed boundaries between cut and uncut forest.

Blowdown and other processes that occur along an edge are part of a suite of edge-related phenomena known as *edge effects*, which are the influences of adjacent patches on each other with respect to some parameter, process, or organism. Edges themselves may also provide a unique habitat condition in the landscape. The expression of edge effects

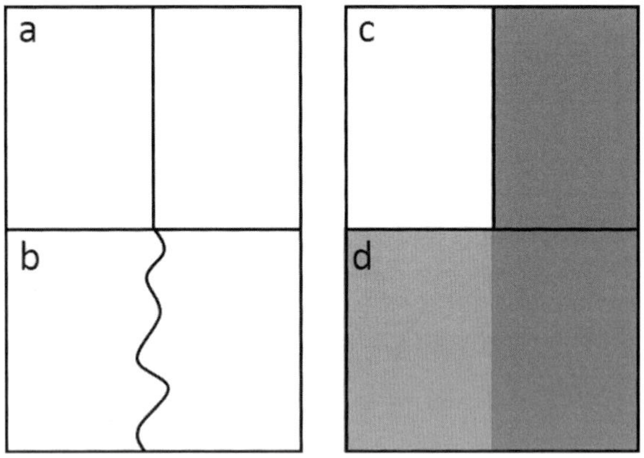

Figure 9.7 *Conceptual representation of different edge shape and contrast; (a) linear edge, (b) convoluted edge, (c) high contrast edge, (d) low contrast edge.*

(e.g., the depth of penetration of influences from one patch into an adjacent one) and characteristics of edge as habitat will be influenced by the physical attributes of the edge, such as shape and contrast (Figure 9.7). For example, a convoluted edge shape may offer greater resistance to blowdown compared to a linear edge. Similarly, blowdown may be lower along an edge between two patches of low contrast (e.g., old forest and a recent harvest with structural retention), compared to an edge between two patches of high contrast (e.g., old forest and a clearcut).

Edges can influence ecosystem processes. Productivity will be higher along a forest edge adjacent to an opening due to availability of additional light and potentially other resources. Tree mortality and reproduction of plant species is commonly affected by edges (e.g., Chen, Franklin, and Spies 1992), either positively or negatively depending on the species. Edges can also have profound influences on disturbances by either bounding or limiting their spread (Figure 9.8) or by producing conditions that are highly susceptible to disturbances, such as windthrow (e.g., see Franklin and Forman 1987).

Figure 9.8 *An edge between pine forest (left) and Atlantic white cedar forest (right) in the New Jersey (USA) Pine Barrens. In the photo, a recent fire stopped at the edge of the less flammable cedar forest. Photo credit: Bob Williams.*

Edges have both positive and negative impacts on the abundance of biota. Wildlife ecologists often focus on edges as distinct and desirable habitat for many game species, such as deer and elk. Their concept of edge is usually broad, including large areas of adjacent patches utilized by species that feed in openings and use the forest side of edges for hiding and thermal cover (Hunter and Schmiegelow 2011). However, some biota react negatively in varying degrees to edges and may require the conditions provided in the interior of patches—i.e., portions of patches free of significant influences from adjacent contrasting patches. Most often the patch interior-dependent species that have been recognized in the literature are forest species. The factors responsible for such interior forest dependence can include the need for specific microclimatic conditions (e.g., cooler, moister), a reduced potential for direct predation or nest parasitism, and an increased availability of prey species. Despite the emphasis on biota dependent on interior forest conditions, it is important to note that some of the organisms that utilize nonforested openings (e.g., meadows or Preforest Stages) are similarly sensitive to the influences of forests on the adjacent non-forest habitat.

Edge physical structure is fairly straightforward to define and measure, but it is the lateral extent of edge effects that is often of interest. The lateral extent of edge effects from one forest patch (or stand) into another can be difficult to determine, as it depends on the parameter being considered. For example, the depth of edge is shallow for soil temperature and radiation and deep for relative humidity and wind (Figure 9.9). Many variables can influence the depth of microclimatic edge, including season, time of day, weather

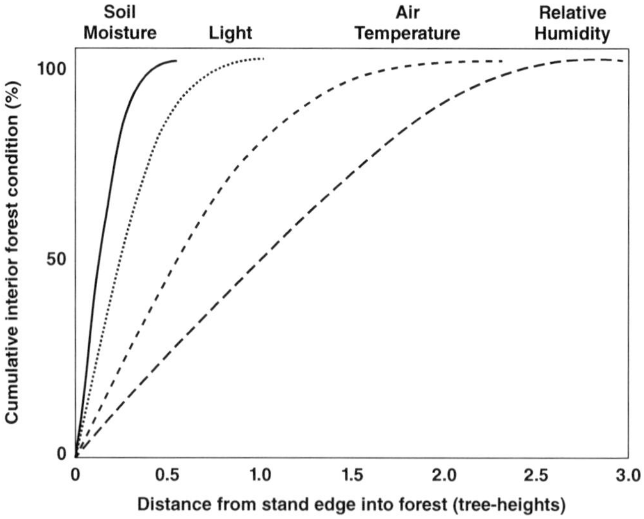

Figure 9.9 *The extent of edge influence (depth into the patch or stand) will differ with the parameter, organism, or process of interest. Illustrated here is a generalized representation of the extent of edge influence on different microclimatic variables within an old-growth forest patch adjacent to a clearcut. (Redrawn from Franklin et al. 2018).*

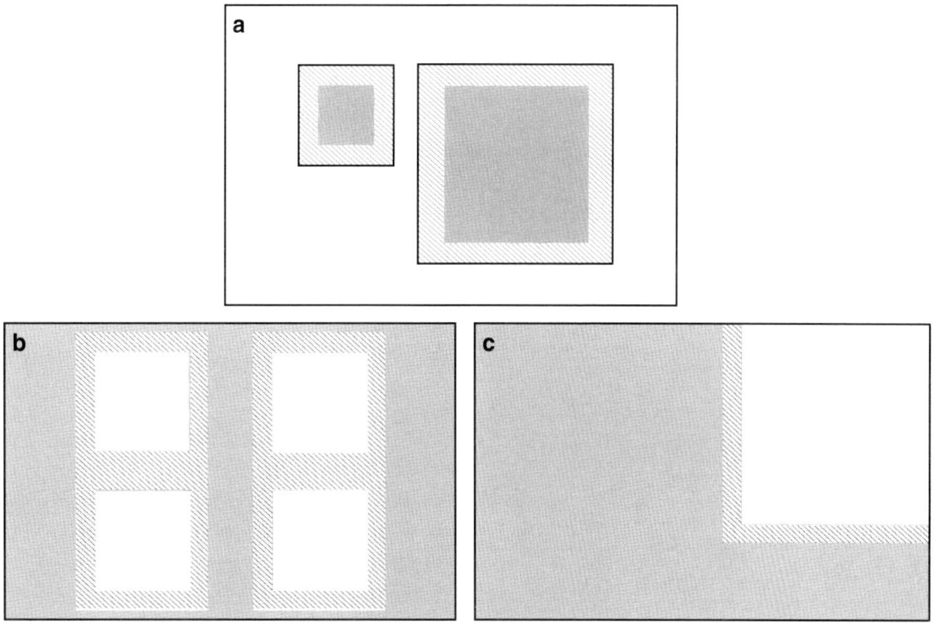

Figure 9.10 *Conceptual representation of edge area in relation to patch size and its influence on amount of edge in a landscape. a) The two patches (e.g., mature forest) set in contrasting patch conditions (e.g., recent clearcut) have the same edge width (light gray stippled area), but a greater proportion of the small patch is edge-influenced than the large patch. A consequence of the edge area to patch size relationship is that a landscape containing several small, dispersed harvests (b) will contain more edge area than a landscape with a single harvest of equal size (c).*

conditions, and slope angle and exposure (e.g., north or south). Extreme conditions—the worst case scenarios—are often emphasized in evaluating edge influences because it is often extreme conditions (such as those that occur on very hot and dry days) that control species distributions or processes, such as fire behavior.

As with microclimate, an array of factors will influence the expression of edge for other parameters of interest, including the distribution of different organisms such as invasive plant abundance, and ecological processes such as soil nitrogen mineralization. Thus, there is no single measure of edge effects applicable to all parameters of interest.

There are several points to remember about edges when considering the landscape context of silviculture. Generally, lower amounts of human-created edge are desirable, due to the often negative influences of this created edge on interior forest conditions. How can this be achieved? It is controlled by patch size distribution (see discussion on patchworks above). Smaller patches have higher edge to area ratios than larger patches; thus, a landscape that contains many small, dispersed, uncut and recently harvested units will contain more edge and more edge-influenced forest than a landscape with the same amount of uncut and harvested area concentrated in fewer, larger stands (Figure 9.10). When possible, the ecological forester works to reduce edge and edge effects by increasing patch (stand) size, which, as discussed above, may often be an outcome of moving toward use of natural models. Another point to remember is that edge effects are most extreme when the contrast between patches is high, such as between a recent clearcut with no retention and a mature forest (Figure 9.10). Contrast can be lowered by reducing the disparity between the adjacent patches, for example, by retaining a moderate to high level of live tree legacies in the harvest unit. As you have learned, such structural legacies are characteristics of many types of natural disturbance, so if you are applying natural models of silviculture in the landscape, you most likely are also reducing the amount of high contrast edge.

Connectivity

Connectivity (among patches) is an important attribute of landscapes, particularly as it affects the ease or difficulty for biotic movement. Landscapes that consist of highly contrasting patchworks are often viewed as having low connectivity. For example, a landscape composed of a mixture of clearcut and mature forest patches (Figure 9.11) is one in which mature forest biota may have difficulty dispersing. This may be because of a higher risk of predation in the clearcuts or because of unfavorable environmental conditions (e.g., too hot or too dry). Connectivity in landscapes can vary dramatically depending upon the organisms or materials of interest, which somewhat limits the value of generalities, but the concept of connectivity remains important to consider in the context of silviculture as it can be influenced by the structure

Figure 9.11 *A patchwork landscape of clearcut and closed forest patches. Photo credit: USDA Forest Service Southern Research Station, USDA Forest Service, SRS, Bugwood.org.*

of forest in stands across the landscape (see section on Shades of Green below).

Conservation biologists have sometimes emphasized the use of habitat corridors to provide for connectivity across large landscapes. While corridors can be important to specific biota, scientists and managers involved with the management of large areas generally recognize their limitations. In reality, the fundamental issue of landscape connectivity may be addressed in ways that reflect the cumulative actions of silviculture among stands. For example, during development of the Northwest Forest Plan (USDA Forest Service 2004), there was concern about successful dispersal of northern spotted owl (a subspecies of spotted owl listed as threatened in the US) between large forest reserves. Young owls had been observed to disperse randomly rather than use forest corridors. To improve the probability of successful dispersal, it was recommended that the matrix of federal forest between designated Habitat Conservation Areas be managed so that 50 percent of every quarter township (a quarter township is approximately 18.6 km^2) be forested, with an average tree diameter of 28 cm (11 inches) and at least 40 percent canopy closure, the so called 50-11-40 rule (Thomas et al. 1990).

This example illustrates how real world solutions to connectivity often involve altering the dominant conditions in a landscape through cumulative stand-scale actions (the result of silviculture). For instance, a landscape dominated by clearcuts is likely to be a hostile environment for mature forest biota, but if this landscape contains some protective cover in the form of islands of trees, logs, and other vegetative cover in recent harvest units (i.e., continuity of structure through retention), it may provide a safer environment for facilitating movement of mature forest species (Figure 9.12).

Areas of Special Concern

All managed forest landscapes will contain localized areas of exceptional conservation and cultural significance. The ecological forester needs to know the location of such areas and their specific characteristics (including why they need special attention), so that appropriate plans for their management can be developed and inappropriate activities avoided. Certainly, it is common practice as part of project planning to assess stands for internal features of conservation and cultural significance. Such features typically include species and areas of concern, e.g., threatened and endangered species; smaller aquatic wetlands and aquatic habitats, such as vernal pools; unique non-forest habitat, including rock outcrops and caves; first (Native) inhabitant use areas, such as camps or burial sites; and early settlement sites, such as logging camps or homesteads.

Within-stand features are important for the ecological forester to recognize and sustain; however, our concern in this section is the larger features in landscapes that may span several to many stands or landscape patches. These will be areas of broader conservation or environmental concern, including those that carry out critical functions in the landscape, such as regulation of water quality and quantity and provision of critical habitat. Areas of cultural concern at larger spatial scales again are likely associated with activities of Native or First Nation peoples, such as large collecting areas for food, medicines, or building materials or for vision quests.

Areas of Conservation Concern

In forest landscapes, areas of conservation concern with the greatest relevance to silviculture are often mid-spatial scale in size such that they occur adjacent to stands undergoing management as opposed to being located in them. They also may be part of larger stand complexes, but they are not landscapes unto themselves, such as large conservation reserves. Such areas will generally involve management strategies different from those used in the surrounding forest, but they should not necessarily be viewed as no management zones. In fact, such areas may actually require focused active management, particularly where important structures or processes need to be restored or maintained. Some examples of these areas follow.

Aquatic and wetland ecosystems

Aquatic and wetland ecosystems are the most extensive category of landscape elements that require special management consideration. They are present in almost all forest landscapes and have high ecological significance in terms of distinctive biota and important functions. An array of

Figure 9.12 *Clearcuts (a) can be impediments to the movement of many organisms when they lack protective cover in the early years following harvest. Harvested areas with significant retention (b) are more conducive to the movements of many animals from stand to stand because significant hiding cover is always present. Photo credits: (a)* Dark Clouds over Finnish Forest or Beauty of Forest industry? *by* talaakso *licensed under CC BY 2.0 (https://creativecommons.org/licenses/by/2.0/), (b) Doug Maguire, Oregon State University, Bugwood.org.*

aquatic ecosystems span multiple stands (that is, are landscape-scale features) including intermittent and permanent streams, rivers, lakes, and larger wetlands. The functional relationship of aquatic and wetland ecosystems to surrounding forests is covered extensively in Chapter 15. Suffice it to say here that aquatic and wetland ecosystems are a management concern because of their widespread occurrence, the role they play in movement of water and materials across a landscape, and the habitat they provide for biota including fish, amphibians, invertebrates, vascular plants, algae, fungi, and an array of microbial organisms (e.g., Jones et al. 2018). Moreover, there is great potential for forest management to either sustain, degrade, or improve the functionality and habitat value of aquatic and wetland ecosystems through silvicultural actions in adjacent forests (i.e., riparian forest). As you will learn in Chapter 15, application of the first three principles of ecological silviculture (i.e., continuity, complexity/diversity, and timing) is directly relevant to efforts to sustain or restore the functional connection between forests and aquatic ecosystems. Applying these principles in the bulk of the forest landscape addresses Principle 4: context.

Contrasting terrestrial habitat

Forest landscapes typically incorporate a variety of largely terrestrial non-forest patches that contrast with dominant forested portions of the landscape. These include distinctive rock and soil formations, as well as vegetated areas dominated by quasi-permanent herbaceous and shrub vegetation, or greatly contrasting tree vegetation. Such areas provide habitats that are important for elements of the regional biota.

Figure 9.13 *A sandstone outcrop in mixed pine forest in Rocky Arbor State Park, Wisconsin, USA. Photo credit: Yinan Chen. This file has been released explicitly into the public domain by its author, using the Creative Commons Public Domain Dedication.*

Figure 9.14 *Alvar vegetation on limestone bedrock in Västergötland, Sweden. Photo credit: Jgrahn. Used under the Creative Commons Attribution-Share Alike 3.0 Unported license.*

Distinctive rock features, such as outcrops and rockslides, are easily identified in the field (Figure 9.13). These provide habitat for vertebrate and invertebrate animals and may act as escape sites for some species, such as mountain goats and sheep. Additionally, these sites may provide habitat for populations of uncommon plant species and support distinctive vegetative communities.

Areas of unusual sediment accumulations or near surface bedrock are often sites that host distinctive biota. These areas include serpentine soils, limestone bedrock outcrops, and sand barrens. Serpentine areas, which develop soils with unusual chemical properties, are one of the best known examples of unique geological substrate (Kruckeberg 1967). A very distinctive biota is typically present on serpentine soils that includes trees and other plant species that can tolerate their unique chemical properties. Dry grasslands with scattered trees that form on thin soil over limestone bedrock include *alvars and cedar glades* (Figure 9.14). Finally, *sand barrens* occur on deep deposits of coarse sand and support distinctive ecosystems and biota that contrast with surrounding areas, such as scrub oak barrens on sand ridges surrounded by longleaf pine in the southeastern coastal plain of the US (Figure 9.15).

Meadows, shrub fields, and other natural non-forested openings are a diverse but easily recognized category of unique habitat in forest landscapes. It is important to recognize that these are distinct from Preforest Stage ecosystems, which are also dominated by vegetation other than trees (Chapter 5), because the former are permanent/semipermanent features. However, just as in the case of the Preforest Stage, permanent non-forest openings contribute uniquely to a forested landscape because of the contrast that they provide in terms of structure, function, and composition. Such non-tree dominated ecosystems invariably support species that are not found in the surrounding forest. They may also provide habitat for species, such as some ungulates, which utilize forested areas for cover and meadows and other open areas for foraging.

The charge to the ecological forester is to recognize the array of these features in the landscape and work to ensure their long-term integrity as open habitat. Appropriate management of these specialized habitats involves i) protecting them from direct damage (e.g., avoid using them as landings and for equipment trafficking) and ii) avoiding changes in cover (e.g., avoid converting them to forest), but also ensuring their integrity by providing appropriate buffers to reduce the potential for edge effects (e.g., excessive blowdown from adjacent forest edge into the feature).

Biological hot spots

The category of biological hot spots refers to habitats in forested landscapes that have unusual significance to the life histories of specific animal species, but that are not obvious to the untrained eye, as is often the case with the

Figure 9.15 *Scrub oak woodland located on a deposit of coarse sand, Florida, USA. Photo credit: William D. Boyer, USDA Forest Service, Bugwood.org.*

features described in the previous section. Two examples from western North America illustrate how diverse such hot spots can be: calving areas for elk and spawning habitat for anadromous fish.

Elk—Pregnant Roosevelt elk seek secluded areas for calving. They look for areas where they feel safe that are also sites with good forage and water availability. Suitable calving sites may be limited in area and not evident to a casual observer. Identifying and protecting such areas may be important as part of silvicultural planning on larger ownerships. Similar types of terrestrial hotspots include foraging areas that may provide rare food resources and overwintering habitats for large populations of animals, such as winter yarding sites for deer (Lindenmayer and Franklin 2002).

Fish—River drainages can be highly variable in their capabilities as spawning habitat for anadromous fish. Some reaches have optimal conditions for spawning and early survival of fry, while others are completely lacking in such habitat. For example, high quality spawning and rearing habitat for Pacific salmon in many streams and rivers of coastal northwestern North America are associated with debris jams formed by episodic debris flows from their tributaries (Benda et al. 2004; Reeves et al. 2003). The spatially and temporally variable input of large wood, organic material, and sediments from debris flows creates a mosaic of complex in-stream and off-channel habitats needed by salmon. In such river systems it is important to manage headwater forests for complex structural conditions, including large trees and logs, to ensure that when headwater areas fail, the resulting debris flows incorporate significant quantities of these structures (Figure 9.16). This example illustrates how maintenance of spawning habitat may involve silvicultural activities at locations well removed from the habitat itself.

Figure 9.16 *To provide for spawning habitat for fish, headwater drainages should be managed to ensure that periodic landslides and debris torrents contain the large quantity of large logs needed to sustain debris damns in the main stem, as in this example from Okanagan, British Columbia, CA. Photo credit: Leonora King used through the CC0 1.0 Universal Public Domain Dedication.*

Corridors and remnant patches

Corridors—Previously we discussed how connectivity in a landscape may best be addressed by increasing the landscape's overall connectivity through structural retention within patches or stands rather than corridors, which may have only limited effectiveness. Nevertheless, there are sometimes areas that have been used for daily or seasonal movements of some biota that need to be recognized and managed as corridors to accommodate their continued functioning (Lindenmayer and Franklin 2002). These corridors will often be related to specific topographic features, such as valleys that connect major habitat segments or prominent passes on key travel routes. Generally, the ecological forester will want to maintain known corridors intact for their travel function.

Remnant patches—Landscapes often include stands of old or unusual trees or uncut primeval forest. These have many important conservation functions, as well as important social significance. Old-growth remnants provide unique habitat for certain species and they may be sources of genotypes for species that otherwise no longer exist in the local landscape. Old trees and forests have particularly high value to people as preferred sites for recreation, reflection, and spiritual recharging. While there may be limited silvicultural activities occurring in these forests, it is important to know where they are in the landscape and to protect the integrity of their shared boundaries with managed stands.

Areas of Cultural Importance

Managed forest landscapes often contain sites of cultural and social significance. These can be of either historical or modern significance or both. They can be sites that have recreational, inspirational, religious, or patriotic (e.g., battlefields) significance. They can also be sites that did or do provide culturally important products, such as foods, medicines, or traditional construction materials. Identifying and protecting such sites can be important for gaining or maintaining the social license to practice forestry.

Cultural sites of concern are most often associated with indigenous peoples, such as Native American (US) or First Nation (Canada) bands and tribes. These sites include areas used for spiritual quests or other religious activities, for gathering plants used for food, medicine, making baskets, and clothing, and for acquiring materials for the construction of houses, poles, and canoes. On public ownerships in particular, these areas may be protected from active

Figure 9.17 *Huckleberry fields are fire-maintained, shrub-dominated ecosystems that are important cultural food source areas, particularly for the Native American and First Nation peoples of North America. Indian Heaven Wilderness, Gifford Pinchot National forest, Washington, USA. Photo credit: USDA Forest Service Pacific Northwest Region. Use licensed under CC PDM 1.0.*

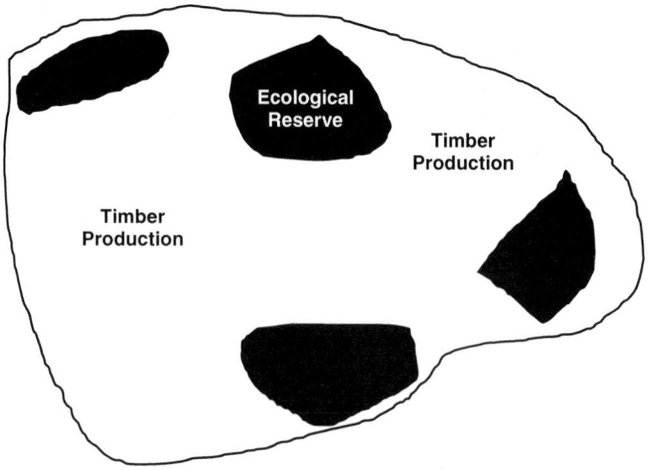

Figure 9.18 *A black-and-white forest landscape in which ecological reserves (shown here as black) are embedded in a matrix managed primarily for timber production (shown here in white) and without direct provisions for sustaining biological diversity. In such a landscape, the ecological reserves are intended to provide all habitat needs for forest biodiversity.*

management. However, maintaining some culturally important areas may actually require active management. For instance, huckleberry gathering areas in forested regions of North America are persistent early successional communities that were created and sustained on forested sites by fires of both natural and anthropogenic origin (Figure 9.17). Periodic disturbances, such as fire or tree removal, are required to prevent forests from dominating such sites and dramatically reducing huckleberry production.

Shades of Green Landscapes

The final context theme we address concerns the allocation of landscapes into different broad use categories (see Chapter 1). A common view of landscape allocation, shared by some foresters and conservation biologists, is that forest landscapes should be allocated into two broad categories, production forests and ecological reserves, with the latter being the locales for protection of biological diversity. Conversely, on the lands allocated to wood production, presumably management is free to proceed without significant constraints for conservation of biological diversity directly, although there may be legal constraints and voluntary guidelines designed to sustain ecosystem services that must be addressed in the production forests (Chapter 1).

Such a view of the landscape is effectively black and white, in that forests are either habitat for biodiversity or they are not (Figure 9.18; Color Plate 9.1a). This landscape perspective encourages management decisions that will actually tend to create a black-and-white reality, since there is no expectation in such a landscape vision that timber production areas would consider practices or constraints to sustain elements of biological diversity. A real world example of such a landscape allocation approach is found in New Zealand, where government owned native forests are protected from logging and maintained as ecological reserves, while timber management occurs strictly in plantations (Bennett 2015).

Neither foresters nor conservation biologists are well served by a black and white landscape designed to exclude consideration of biological diversity in forests actively managed for timber. From the forest management side of the equation, it is important to recognize that ecosystem production, including production of timber, is sustained by biota in the forest. The forests' biota are, in fact, the creative agents that convert resources of light, water, and nutrients into energy-rich carbon compounds, some of which are used in developing the physical structure of the forest. They are also the agents that break down dead organic materials, thereby freeing nutrients. Hence, the biotic richness of a forest is the basis of production, as well as all other ecosystem functions. In the long run, it is almost certainly going to be counterproductive for timber management to ignore the importance of sustaining the full array of biota found in the local forest ecosystem.

On the other side of the biodiversity coin, conservation biologists should seek to conserve biological diversity over much of the landscape, preserved and otherwise. This is important for maintaining the functionality of these lands as well as for sustaining species. Sustaining a high level of ecosystems functionality will not be accomplished with a network of isolated ecological reserves within a matrix of production forests if biodiversity is ignored in those production forests. Most reserves will be of such a size that they depend upon conditions in the surrounding landscape matrix to achieve conservation goals (e.g., Prugh et al. 2008). Moreover, there will never be reserves sufficient in extent and representativeness to achieve the goal of preserving most species (Lindenmayer and Franklin 2002).

This conundrum suggests the need for a different approach to landscape allocation than a black and white model, one in which conservation of biological diversity and ecological function is pursued concurrently with timber production, including on the same stands in the landscape (Bennett 2015). There have been several different models developed that speak to this perspective, notably the triad model (Seymour and Hunter 1992), which proposes allocating forest use into categories of intensive production, ecological reserves, and multiple-use/extensive forestry. In practice, forests managed to produce timber often constitute the bulk of the landscape and are the place where ecological forestry should be the main paradigm. The triad model is certainly an improvement over a black and white model, but only to a point, when it is interpreted to translate into a black, white, and green landscape, where everything in green is ecological forestry without recognizing a need for variation and flexibility of application among ownerships.

An alternative perspective that incorporates this flexibility is one that acknowledges the need for silviculture in the matrix to vary in the specifics depending on ownership (Palik and D'Amato 2017)—that is, a *Shades of Green* landscape (Franklin and Swanson 2008). In the Shades of Green model, conservation of biological diversity is pursued concurrently with timber production in most stands in the landscape, but to varying degrees, hence the differing Shades of Green among patches (Figure 9.19; Color Plate 9.1b). Franklin et al. (2018) describe it as seeking to create landscapes that are *Shades of Green*, in which all patches in the landscape contribute to the goal of sustaining biological diversity, even immediately following timber harvests. Most patches (stands) in such a landscape have significant *ecological content* by incorporating the foundational principles of ecological silviculture in some way. Patches that lack significant ecological content (white) are a minority in such a landscape, but are not unimportant, particularly if they take on added importance as places where intensive timber production is pursued.

Green patches in such a landscape will include mesoscale features of special ecological significance (see Section 2 of this chapter). Aquatic ecosystems are likely to be one of the more common and extensive of these features. Green patches will also reflect retention of legacy structures and organisms in harvested portions of the landscapes. Finally, a Shades of Green landscape will include a diversity of ecosystem types and conditions (developmental stages), needed to maintain the complete array of habitats and functions.

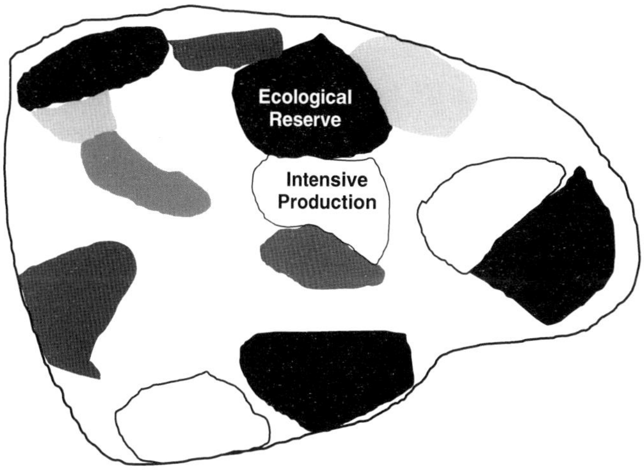

Figure 9.19 *A Shades of Green landscape (depicted as shades of gray in this figure, but see Color Plate 9.1) consists of patches (stands) that are managed to sustain elements of habitat and ecological functionality. This landscape can be viewed as a patchwork of shades of gray (or shades of green in real-world landscapes). Patches in the landscape retain ecological content (e.g., organisms and legacies), contributing to goals of maintaining ecological functions and biodiversity broadly, but to varying degrees. For instance, the ecological contribution of the lightest gray (i.e., light green) patches (matirx) may come primarily from retention of live trees, logs, and snags, whereas dark gray (i.e., dark green) patches may include full application of all our principles of ecological silviculture.*

Landscape allocation models, such as the Shades of Green model, are better suited to creating conditions that sustain biological diversity and ecosystem function broadly across the landscape, compared to a black and white model, and are probably more reflective of real-world influences of policy and politics on application of ecological silviculture than are the triad model and those like it.

What to Remember about Landscape Context

The ecological forester needs to be aware of landscape context (Principle 4) when developing silvicultural prescriptions, including how stand-scale actions are influenced by and impact landscape structure and function, the importance of special places in the landscape, and the need for variation in silvicultural approach depending on ownerships and objectives, even within the sideboards of ecological silviculture. This can be a daunting task for small ownerships (how do you keep tabs on what the neighbors are doing?), but it is not a foreign concept within some larger forest management organizations. For instance, on US National (federal) Forests, silvicultural prescriptions are developed in consultation with a large planning team charged with considering landscape goals and objectives for a project area (regional landscape), as well as forest-wide plans. What these and similar organizations may not be doing explicitly is incorporating all three foundational principles (continuity, complexity/diversity, and timing) into their prescriptions in ways that deliberately reflect consideration of landscape context and goals. For example, many organizations now embrace, in some fashion, retention of legacy trees during a harvest (i.e., implementing continuity), but they may not array this retention in space and time in ways that also foster greater connectivity of habitat across the landscape. Many organizations have spatial data layers of ecosystems based on underlying soil and geomorphic templates (recognizing the natural landscape patchwork), but they still delineate stands based on historical boundaries that were established based on other, non-ecological, criteria. Finally, some organizations do not fully consider the potential for ecological silviculture to contribute to conservation goals and timber production in the landscape, and instead view their landscape as black (ecological reserves) and white (timber production). Moreover, organizations that do add multi-value forest management into their landscape allocation may do so in a way that reflects a strict triad (black, white, green) model, rather than the Shades of Green model we discuss.

Our intent for this chapter has been to present some basic landscape concepts and ideas in a way that provides the tools to better consider the landscape context of your silviculture as you move it toward explicit incorporation of ecological principles. Ownership patterns and associated political and economic constraints can be quite complex in any given landscape, making it difficult for a single forester to ensure prescriptions sustain functions in a landscape context. Nonetheless, the intent of this chapter is to provide a framework for considering the impacts of stand-level silviculture on these broader processes and the ways in which they may be accommodated by working within and across ownership boundaries.

PART IV

Ecological Silvicultural Systems for Major Archetypes

CHAPTER 10

An Introduction to Ecological Silvicultural Systems

Silvicultural systems are the foundation of forest stand management and are typically thought of as the sequence of silvicultural treatments that occur over the life of a forest to meet a particular set of objectives (Nyland 2007). As mentioned in Chapter 3, these systems are typically named for the regeneration method used, given the strong influence of these methods on forest conditions and associated silvicultural treatments. Also, as historically applied, the emphasis of a silvicultural system is on establishing, tending, and harvesting a forest crop, with treatments focused primarily on affecting the composition, growth, quality, and health of desired tree species and crop trees of those species (Matthews 1989).

Given the long-term nature of forestry and the unexpected ecological or socioeconomic changes that can occur over the life of a stand or a cohort of trees, silvicultural systems are intended to be flexible and able to evolve in response to unforeseen changes or management outcomes (Smith 1962). This flexibility has provided opportunities in recent decades to integrate ecological objectives into classical silvicultural systems (Gustafsson et al. 2012); however, our belief is that a comprehensive ecological silviculture approach requires a reframing of the silvicultural systems concept into one that is centered on an understanding of natural disturbance regimes for a particular ecosystem (archetype models) and that deliberately incorporates the foundational principles described earlier in this text (*continuity, complexity/diversity, timing, context*).

With this reframing in mind, a general definition of *ecological silvicultural systems* is the long-term sequence of treatments adaptively applied for restoring, maintaining, and enhancing compositional diversity, ecological complexity, and heterogeneity, and informed by an understanding of natural disturbance dynamics and processes, i.e., our archetype models.

Recall from Chapter 6 that the four natural disturbance archetypes we discussed (*infrequent severe disturbance, frequent low-intensity disturbance, gap disturbance, mixed-severity disturbance*) serve as conceptual models of natural forest development against which timber-focused management models can be contrasted. The contrasts can be stark and have important functional consequences (Box 10.1). Keeping them in mind allows a forester to judge the efficacy of an ecological silviculture approach at meeting management objectives.

In the chapters that follow, you will note that we break from the traditional convention of naming silvicultural systems by their regeneration method (e.g., clearcut, shelterwood, selection) and instead name our systems after the prevailing natural disturbance regime as reflected in the archetype descriptions (Chapter 6). When combined with information about the specific ecosystem being managed, our silvicultural systems reflect the combination of disturbance and development model and forest type (e.g., *gap disturbance in northern hardwoods*).

In practice, the objectives addressed by an ecological silviculture approach are likely to be broader in scope than those associated with a timber-focused approach, and will include sustainability of a broad range of ecosystem services, such as ecosystem and watershed health, habitat and species diversity, and resilience in the face of major stressors like climate change. As such, and as you will see, ecological silvicultural systems will look different in construct than systems with a timber-focus.

One major difference relates to the timing of entries and activities. Silvicultural systems, as historically applied in even-aged contexts, generally have a linear progression beginning with stand establishment and culminating at final harvest. This linearity and finality leaves little opportunity for continuity through legacy management (Principle 1) and the timing of harvests and other treatments under this approach is generally guided by economic criteria or crop-tree development, as opposed to ecologically justified intervals (Principle 3). In contrast, our ecological silvicultural systems are organized around disturbance events and developmental conditions or stages, such that the application of treatments will be based on ecological cues that may occur over a loosely defined time frame and that do not necessarily follow in a linear progression.

For example, the conceptual beginning of an ecological silvicultural system is often an initial entry during the Young Forest or Mature Forest Stages, given a common

Box 10.1 **Natural versus Timber-Focused Models: Contrasts and Consequences**

For the ecological forester, it is important to judge the degree of discrepancy in structure and composition between forests managed using a timber-focused silvicultural system and the natural model for a particular forest type, as this reflects where along a continuum of trade-offs between ecological and timber-focused goals silviculture has or could take a forest (Palik and D'Amato 2017). While the details of these differences will vary with the forest, most of the contrasts will be similar regardless of the forest.

Tree species diversity

Most forest types contain at least a few tree species that either occur together or at various times during successional development. By contrast, in the timber-focused model for many forest types, only one or perhaps two species are favored and, in fact, other species may actively be discouraged at harvest or during site preparation or release treatments (Palik and Engstrom 1999). A potential functional consequence of reduced tree species diversity can be greater susceptibility to catastrophic mortality from insect pests or diseases, potential loss of unique habitat characteristics provided by diverse species, and limited recovery pathways following disturbance.

Stand age structure

Elsewhere we have discussed the oftentimes stark differences in age structure between the natural and timber-focused models (i.e., complex and diverse age structures in natural stands versus often narrowly even-aged in the timber-focused model). This may be particularly acute when additional tree species are discouraged (see above) since the contribution these species make to age complexity is lacking. One functional consequence of age simplicity is that even-aged stands, having little vertical complexity of tree heights, may be more susceptible to damage from high wind events (Hanewinkel et al. 2014; Mason 2002) and will lack the vertical structure required by many interior forest wildlife species for foraging and cover (Holmes and Pitt 2007).

Live and dead tree legacies

Given the predominant use of even-aged silvicultural systems in the timber-focused model, live tree legacies are generally limited or non-existent in most applications. Similarly, snags and large deadwood legacies are not typically retained in such a model, with future snags and cavity trees often actively discriminated against in timber stand improvement activities and selection harvests. Spatial patterns of legacies are not an issue of concern in the timber-focused model, because few of these structures exist. A functional consequence of the lack of live and dead tree legacies is that unique habitat features provided by these structures are limited or nonexistent. Moreover, there is limited diversity of resources and habitats, as generated by different spatial patterns of legacy retention, particularly of live trees, and downed woody material on the forest floor.

Stand density

Many forest types, when managed primarily for timber objectives, are maintained at higher tree density than is characteristic of the natural model. This can be true even in newly regenerating stands, where trees may establish over several years to decades after natural disturbance, compared to high density plantings in many stands managed for timber. Lack of spatial heterogeneity of density in the timber-focused model is also in contrast to the natural model for many forest types, particularly as stands or patches within stands mature and mortality of canopy trees from disturbance and disease and other stressors begins to occur. Functional consequences of high tree densities, as well as reduced heterogeneity of density, include loss of unique habitats associated with pre- and early Young Forest Stages and reduced diversity of resource and environmental conditions across stands over the course of development.

Maximum tree ages

Stands in the Old Forest Stage of most forest types will contain populations of old trees. Recently disturbed stands contain legacy trees from old cohorts. In contrast, stands managed for timber will contain few to no legacy trees from the previous stand. Moreover, rotation ages when using a timber-focused model generally do not exceed 150 years and are much shorter for most organizations. A functional consequence of this truncation in maximum tree ages is loss of unique structural habitat, such as large horizontal branches, furrowed or flaking bark, and cavities.

overarching goal of restoring or accelerating the development of complex conditions to forests simplified by past management or historical land use (Principle 2). This period of complexity enhancement is then followed by treatments designed to sustain, recruit, and enhance the functioning of legacies, while maintaining age structures consistent with those created by the prevailing natural disturbance regime. In contrast to timber focused approaches, there is no finality to an ecological silvicultural system and although trees are harvested based on merchantability standards, harvest timing and intensity is reflective of natural return intervals and disturbance severities.

It is important to note that even when pursuing a restoration objective with an ecological silvicultural system,

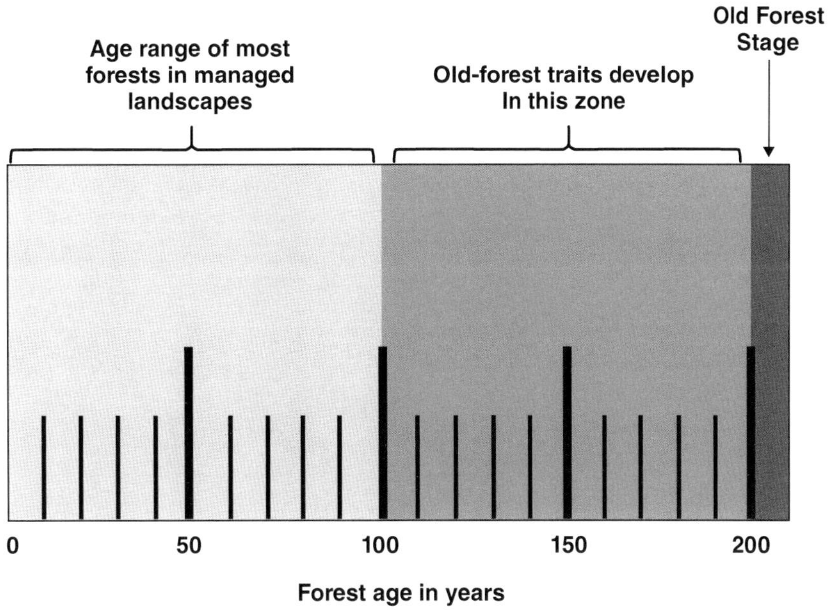

Figure 10.1 *Timeline of old-forest development. In most forests, it takes centuries for significant levels of structural complexity, compositional diversity, and heterogeneity to develop. Forests that are managed for timber will often be less than one hundred years old and have little old-forest characteristics. While passive management can lead to complexity, diversity, and heterogeneity, active management can greatly shorten the time needed to develop desired characteristics. (Adated from Hagan and Whitman 2004.)*

timber management can still be an important objective. In fact, securing revenue through timber harvesting is likely necessary to generate funding for other aspects of restoration. Once restored, or in forests that are already healthy and functional, ecological silviculture can be used to maintain desired characteristics, with timber production as one of the shared objectives. The point being that ecological silviculture is not simply an approach used to meet conservation objectives.

Ultimately, choice of treatments that constitute an ecological silvicultural system will vary among stands in a landscape and over time in accordance with different and changing objectives of management and past and current drivers of change. Thus, context (Principle 4) will play a central role in determining application of ecological silvicultural systems across a landscape and the resultant landscape patterns. In most regions, this landscape will likely include forests where time is the only driver of development to a desired level of complexity/diversity and heterogeneity, i.e., passively managed forests. While this condition has a place in the landscape (e.g., no-management reserves), many organizations do not have the luxury of waiting the often centuries-long time periods required for old-forest characteristics to develop, while still managing for timber (Figure 10.1). As you have learned, an ecological silviculture approach can greatly accelerate the development of complexity, diversity, and heterogeneity in forests managed for timber; thus, formulation of ecological silvicultural systems for different forest archetypes is an important next step in your development as an ecological forester.

A final point is that, as with classical systems, our ecological silvicultural systems are not meant as recipes pulled from a cookbook of prescriptions, but instead should embrace past guidance to approach these systems as working hypotheses for achieving long-term ecological silviculture objectives (Smith 1962). The following chapters present several examples of such systems applied to our specific archetypes and case study forest ecosystems, beginning with the *Infrequent severe* disturbance archetype in the next chapter.

CHAPTER 11

Silviculture for Archetype 1 Ecosystems: Forests Subject to Infrequent Severe Disturbances

This chapter focuses on ecologically based silviculture for Archetype 1 forests, i.e., forests subject to infrequent but severe disturbances, often wildfire. Forest ecosystems identified with this archetype are highly varied but share some common features. One or two tree species typically dominate from the Young Forest Stage (YFS) into the Old Forest Stage (OFS), but as the forest continues to develop, dominance may shift to other species as part of normal successional processes. The forests are often described as being even-aged, although on closer examination many actually contain multiple age classes of trees. Developmental sequences in these forests are often long because major disturbances occur at infrequent intervals, relative to the life-spans of the dominant tree species. Lacking a significant intermediate disturbance, developmental processes eventually result in structurally complex old forests (Figure 11.1).

This archetype encompasses a variety of forest ecosystems globally, including several types of moist temperate forests in western North America and cool to cold moist subalpine forests found in mountainous regions of both western and eastern North America. It is also found in boreal and subalpine forest regions throughout the northern temperate zones of the world (Table 11.1). Globally, Archetype 1 ecosystems are important sources of timber and when managed for wood production, the silvicultural model usually deviates considerably from a natural model of development. Our purpose in this chapter is to illustrate these differences and to outline a silvicultural approach based primarily upon processes and conditions that typify the natural developmental model for the archetype and that better incorporate our foundational principles.

To illustrate our silvicultural approach for this archetype, we will use the Douglas-fir- western hemlock (DF-WH) forest of western Washington and Oregon that has a disturbance regime best characterized by large, infrequent disturbances, including a significant high-severity component (Franklin et al. 2018). Because such disturbances burn over large and diverse areas under varying burning conditions for periods of weeks to months, the burns do typically display a variety of outcomes, including partially burned and unburned patches. We first review some important ecological features of the DF-WH forest in the context of the development model presented in Chapter 6. Next, we examine the history of silvicultural approaches that have been used for this ecosystem. As part of this discussion, we contrast the current approach to intensive timber production on industrial lands to a natural forest model. Next, we discuss opportunities for applying an ecological silvicultural approach to this forest type and, more broadly, the archetype, which is overtly based on a natural developmental model. Central to this discussion, we outline an ecological silviculture system for this forest. As you will see in this and subsequent silviculture chapters for the archetypes (i.e., Chapters 12–14), our system purposely includes application of the four foundational principles of ecological silviculture (i.e., **continuity, complexity/diversity, timing, context**) into prescriptions. We finish by highlighting some exemplary applications of ecological silviculture that incorporate a natural model approach.

The Douglas-Fir-Western Hemlock Ecosystem

Recall from Chapter 6 that the Douglas-fir-western hemlock forest is found in the Pacific northwestern US and coastal British Columbia, Canada, as well as in some moister, limited mid-elevation interior areas of northern Idaho and western Montana (US) and British Columbia (Canada). These forests are strongly dominated by conifers, including Douglas-fir and western hemlock, but also western white pine, western redcedar, grand fir, and Pacific silver fir. Forests adjacent to the Pacific Ocean may include Sitka spruce and, in northern California (US), coast redwood. Additional conifer species in the montane (mid-elevation) forests of the Cascade Range (US) include noble fir and mountain hemlock, while interior western variants can include western larch and even ponderosa pine. The conifer species found in this forest span a range of

Figure 11.1 *Examples of Archetype 1 Old Forest Stage ecosystems: a) jack pine in Michigan, USA, b) red spruce in Maine, USA, c) Sitka spruce-western hemlock in coastal western North American, and d) Pacific silver fir-hemlock forest in Washington USA. The term old forest is relative to the life span of the dominant trees; not all Archetype 1 forests will contain extremely long-lived trees, but all experience severe disturbance infrequently enough that old-forest characteristics develop. Photo credits: (a) Greg Corace, (b) Shawn Fraver.*

shade tolerances and collectively represent some of the longest-lived and largest statured tree species in North America (Table 11.2). Hardwood tree species are less abundant but still ecologically important, and include (depending on location) bigleaf maple, red alder, black cottonwood, bitter cherry, cascara, Pacific dogwood, golden chinkapin, Pacific madrone, Oregon white oak, tanoak, California black oak, and canyon live oak.

The disturbance regime of the DF-WH forest is characterized by large, infrequent (100 year to >400 year) wildfires that include significant areas of high or complete stand-replacement severity (Figure 11.2). The DF-WH forest can also experience severe wind events, landslides and avalanches, and volcanic eruptions. A multicentury-long developmental sequence is initiated following these disturbances (Figure 11.3). The developmental events and stages are described in detail in Chapter 6 (and in Franklin et al. 2018) and are only summarized here (Table 11.3).

Structural legacies of standing dead and downed trees are a particularly striking feature of this forest type following disturbances (Figure 11.4), which is a consequence of the massiveness of trees in the Mature Forest Stage (MFS) and OFS (Figure 11.5). However, the DF-WH forest is also distinguished by an extended Preforest Stage (PFS) that

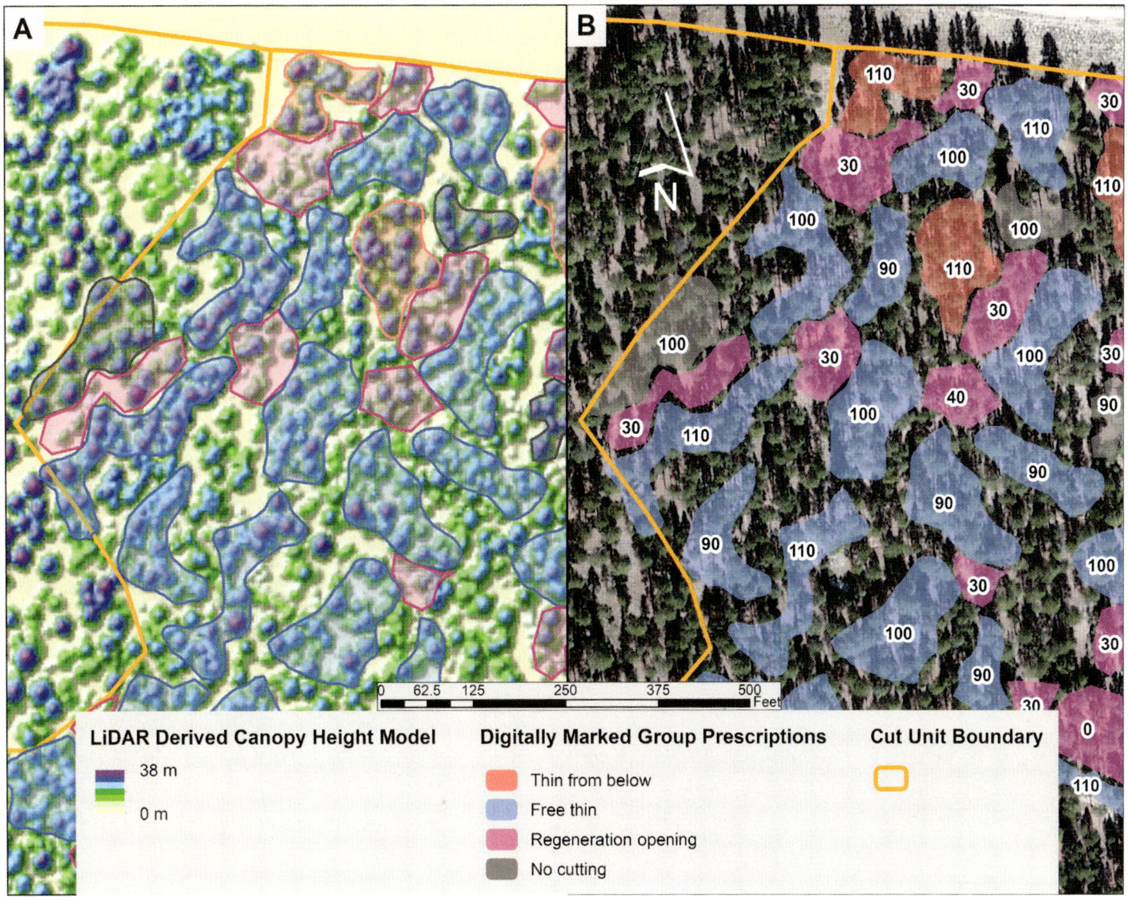

Color Plate 3.1. *Silviculture prescription for generating within-stand heterogeneity in ponderosa pine forests on the Coconino National Forest, AZ, USA. In the figure, fine-scale delineations represent differences in silvicultural prescriptions within larger stand boundaries, including different thinning intensities and regeneration openings.*

Color Plate 6.1. *a) Transect profile of an old-growth Douglas-fir–western hemlock forest, Cedar Flats Research Natural Area, Washington, USA. Illustration by Robert Van Pelt. b) Old-growth Douglas-fir-western hemlock forest.*

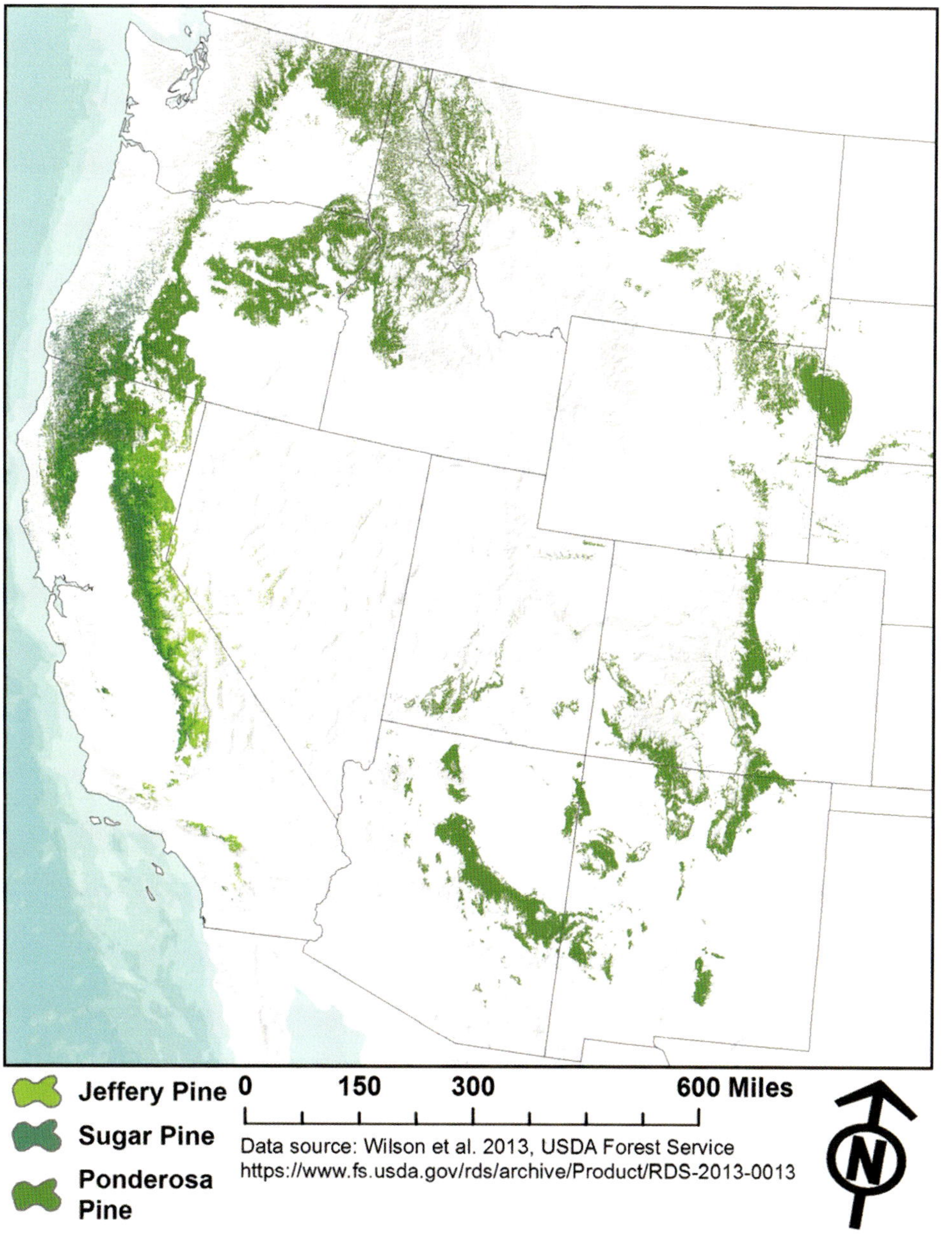

Color Plate 6.2. *Approximate distribution of historical frequent fire forests in the western US. The distribution of ponderosa pine, sugar pine, and Jeffery pine are used as surrogates for historical presence of a frequent fire regime. Other fire-resistant species, such as Douglas-fir or western larch, are also common. Note that these species can occur in moist, historically mixed-severity forest types as a minor component; these areas were removed from the map. Frequent fire forests in Mexico are not shown. Map courtesy of Kerry Kemp, the Nature Conservancy of Oregon. Data adapted from Wilson et al. (2013).*

Color Plate 6.3. *a) Example of a frequent fire forest with an active fire regime in the Eastern Cascades, Washington, USA. The low density, open canopy is dominated by large, fire resistant trees that are the structural backbone for this type of ecosystem. b) Profile of frequent-fire forest showing different canopy strata and patch types. Note the presence of regeneration thickets, old trees, and young to mature trees all within a small area (Illustration by Robert Van Pelt).*

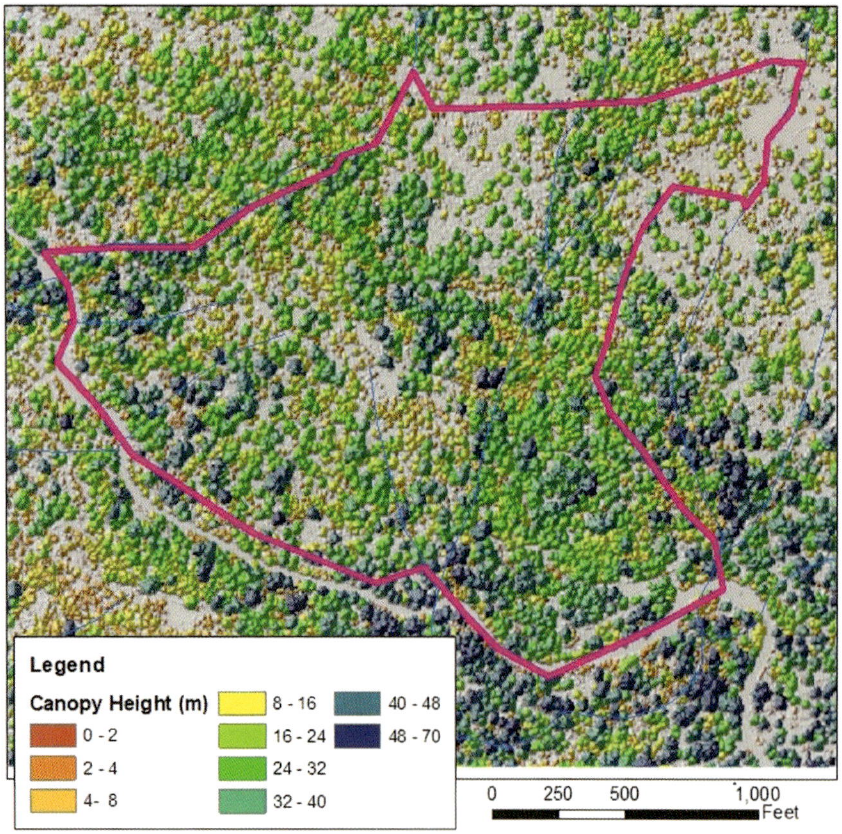

Color Plate 6.4. *Overhead LiDAR image of a frequent fire forest with an active fire regime in Yosemite National Park, USA. Note the small patch sizes of different structural conditions, ranging from clumps of old trees, to patches of mid-sized trees, to thickets of regeneration, to openings with few or no trees.*

Color Plate 6.5. *a) Profile of an old-growth northern hardwood forest with emergent eastern white pine, Porcupine Mountains, Michigan, USA. Illustration by Robert Van Pelt. b) Old Forest Stage northern hardwood forest in the Dukes Research Natural Area, Michigan USA. Note the range of tree sizes at a fine spatial grain.*

Color Plate 6.6. *a) Profile of an old-growth red pine forest. Illustration by Robert Van Pelt. b) Old Forest Stage red pine forest in Minnesota, USA. Photo credit: b) Shawn Fraver.*

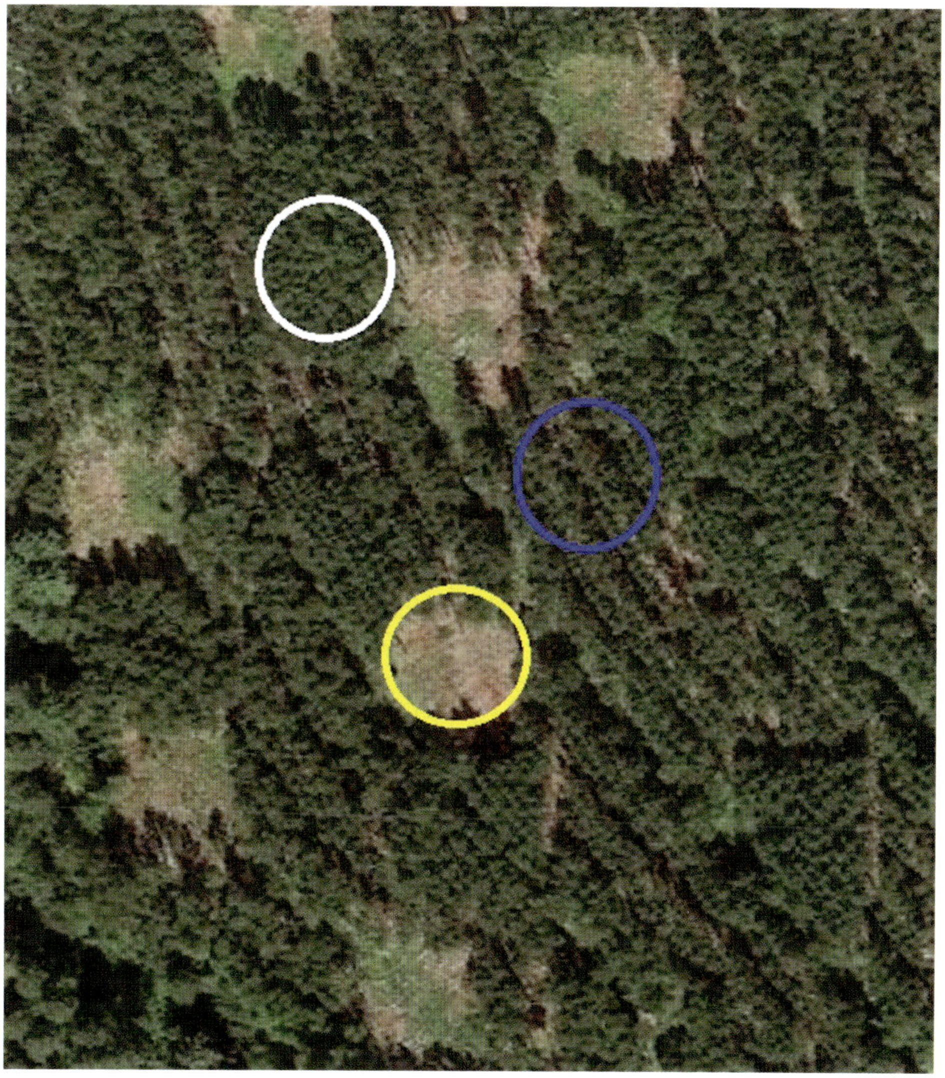

Color Plate 8.1. *Areal view of variable density thinning application in a 49 year-old red pine plantation on the Chippewa National Forest in Minnesota, USA. Note that the white circle outlines a skip, the yellow circle a gap, and the blue circle a row thinned area of the stand.*

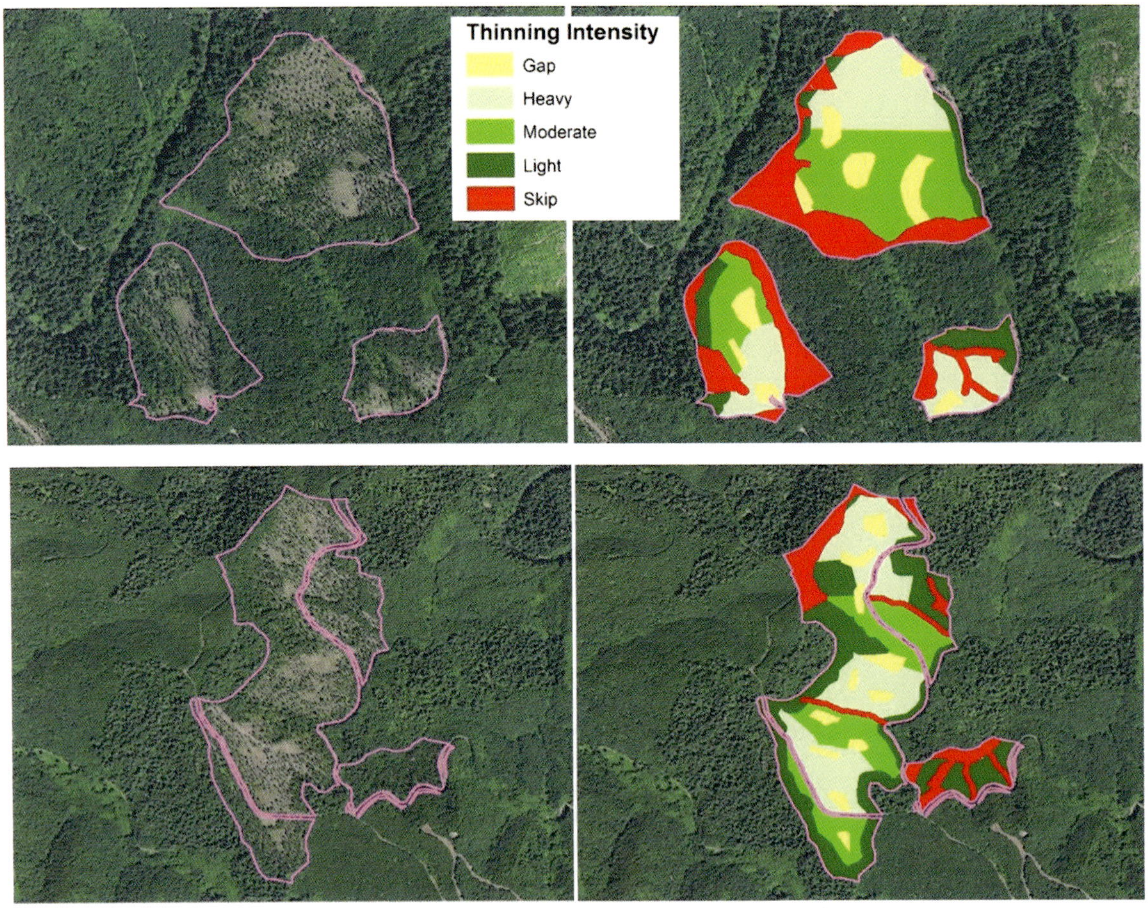

Color Plate 8.2. *Aerial images of variable density thinning in 40 to 50 year old Douglas-fir-western hemlock stands, illustrating skips, gaps, and varying levels of thinning. In the examples, skips are associated with edges and along streams; more recent applications leave skips in the unit proper. Photo credit: Abraham Wheeler.*

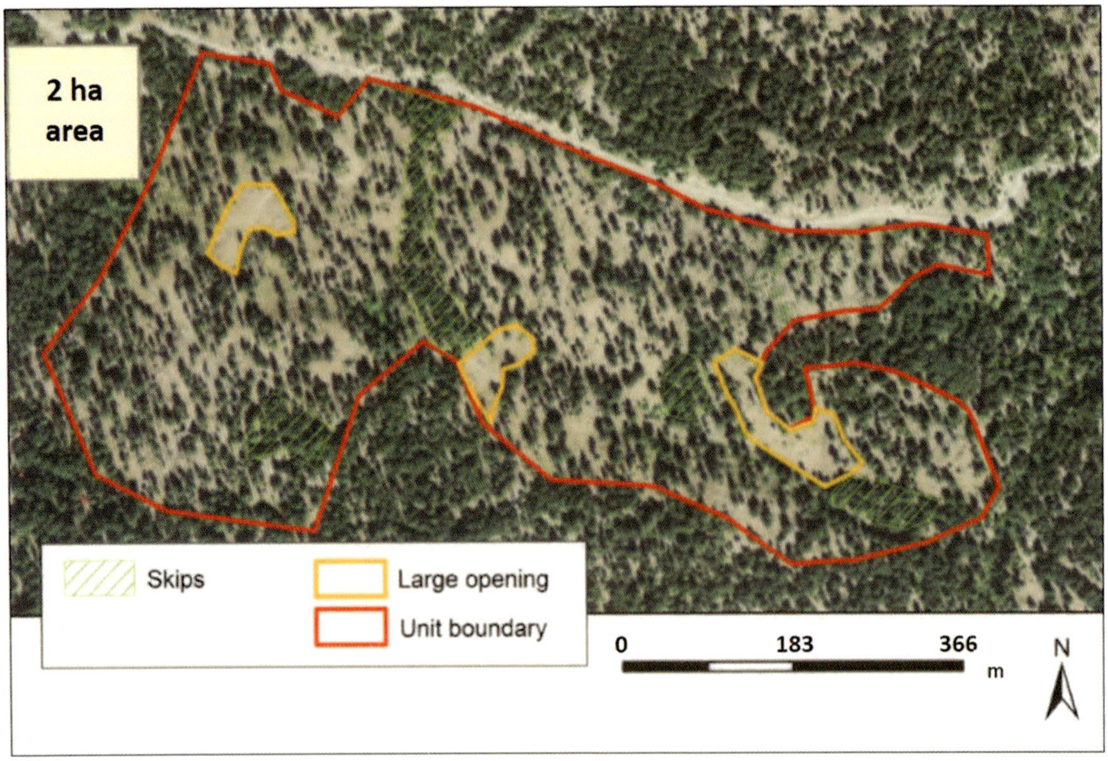

Color Plate 8.3. *Aerial view of an ICO thinned ponderosa pine and Douglas-fir stand on the Okanogan-Wenatchee National Forest, Washington, USA. Note how the unit boundaries were used to create an additional "finger" skip that extends into the stand. A 2-ha square area is shown to provide a reference for patch sizes within the unit.*

Color Plate 8.4. *Northern hardwood silviculture as practiced by the Aitkin County Land Department (Aitkin County Minnesota, USA) to promote structural heterogeneity and tree species diversity. a) Larger gap to promote establishment of mid-tolerant tree species. b) Aerial view of managed stand (center of photo) that shows the wide range of opening sizes used. Photo credit: a) Mark Jacobs.*

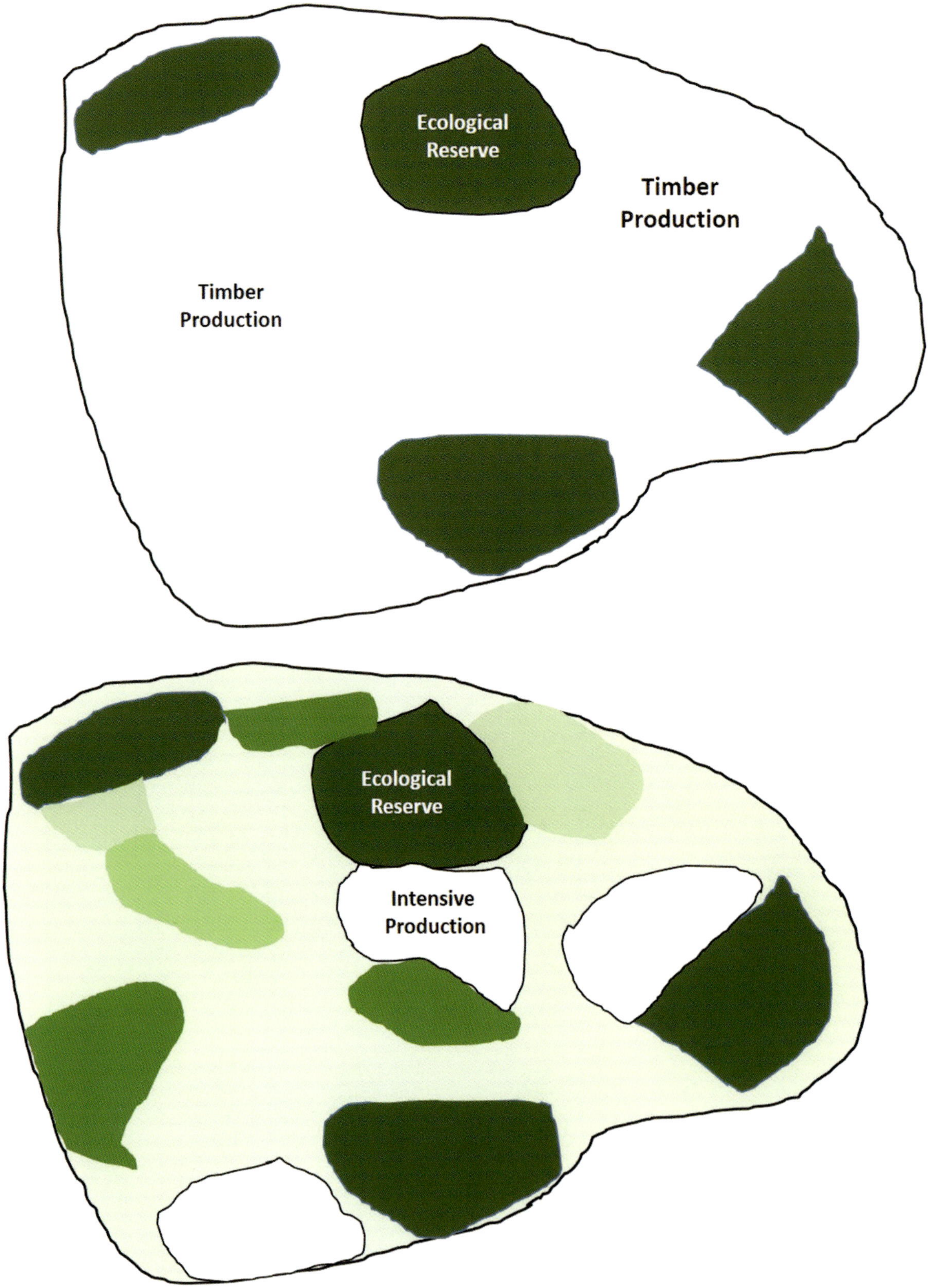

Color Plate 9.1. *a) A green-and-white forest landscape in which ecological reserves (shown here as dark green) are embedded in a matrix which is managed primarily for timber production (shown here in white) and without direct provisions for sustaining biological diversity. In such a landscape, the ecological reserves are intended to provide all habitat needs for forest biodiversity. b) A shades of green landscape consists of patches (stands) that are managed to sustain elements of habitat and ecological functionality. This landscape can be viewed as a patchwork of shades of green. Patches in the landscape retain ecological content (e.g., organisms and legacies), contributing to goals of maintaining ecological functions and biodiversity broadly, but to varying degrees. For instance, the ecological contribution of the light green patches may come primarily from retention of live trees, logs, and snags, whereas dark green patches may include full application of our principles of ecological silviculture.*

Color Plate 12.1. *a) Photo and stem map of a commercial, fuels reduction treatment in a Douglas-fir and western larch forest in northeast Washington, USA. Overstory trees are evenly spaced at wide spacing (14 m), all understory and midstory trees were removed by a hand crew (ladder fuel treatment), and most surface fuels (logging slash) were hand piled and burned. b) Photo and stem map of a commercial, dry forest restoration treatment in a Douglas-fir and ponderosa pine forest in central Washington, USA. Overstory trees were left in a pattern of widely spaced individual trees, clumps, and openings. Ladder fuels and surface fuels were treated with a prescribed fire after the commercial harvest.*

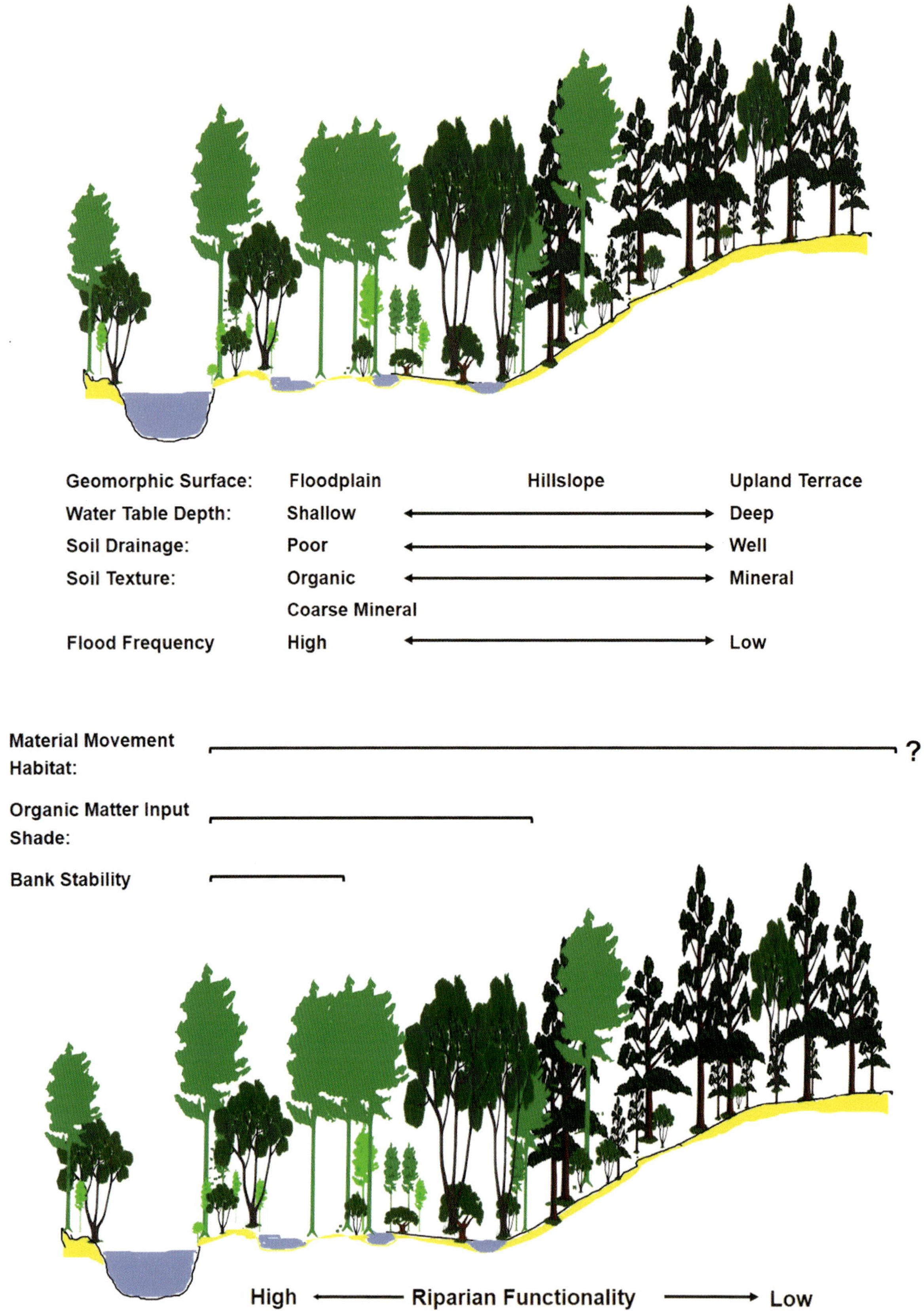

Color Plate 15.1. *a) Stylized example of the physical riparian ecotone. Note the changes in dominant tree species from flood-tolerant deciduous species nearer the stream to xerophytic coniferous species on the hillslope and terrace. b) Stylized example of the functional riparian ecotone. In the figure, material movement includes exchanges of water, sediment, and nutrients between the aquatic and terrestrial systems; organic matter input includes large dead wood, particulate, and dissolved organic matter.*

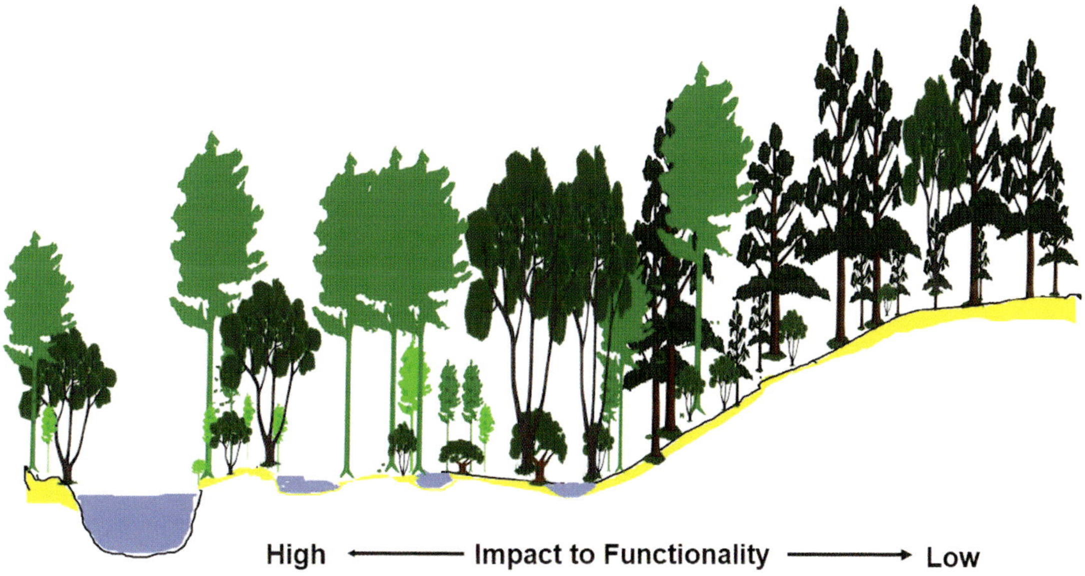

Color Plate 15.1. *c) Stylized depiction of an ecotone of impact (from management activities) to riparian functions.*

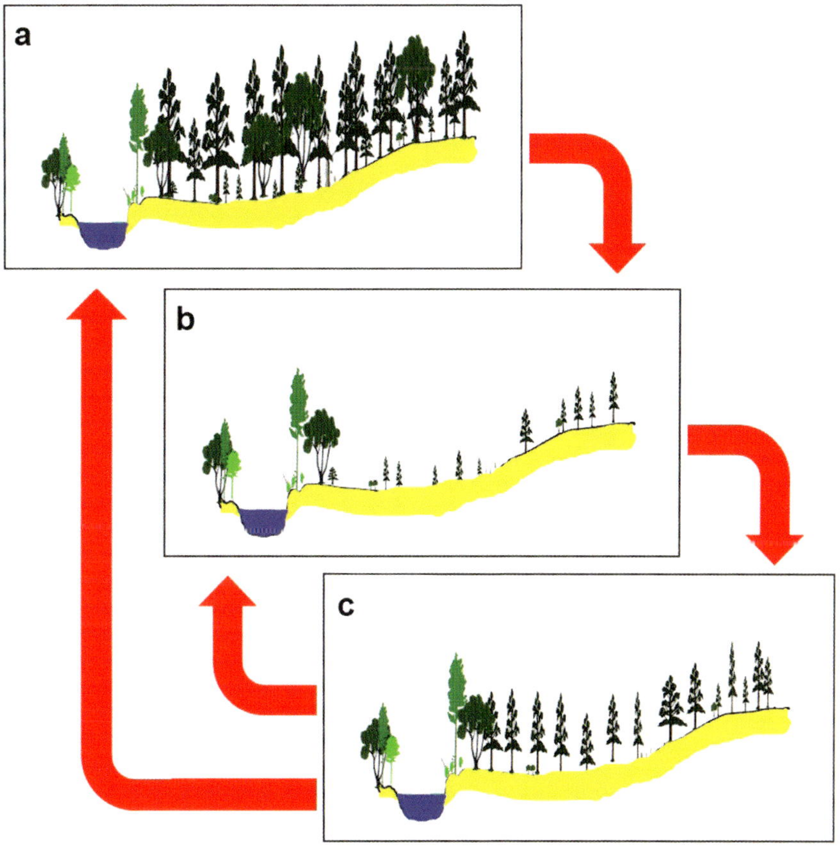

Color Plate 15.2. *Stylized depiction of multiple pathways of riparian forest development. a) The unharvested, mature or Old Forest Stage ecosystem containing large old trees that eventually will provide large dead wood to the forest and stream. b) The same forest after harvesting of the large, old trees. c) The forest after development into the Young Forest Stage. Timing harvesting on financial or productivity ques would likely cycle this forest between b) and c), whereas timing based on ecological development would allow the forest to cycle back to a) before another harvest is considered.*

Table 11.1 *Examples of Archetype 1 Ecosystems.*

Ecosystem	Distribution	Natural Disturbance	References
Moist temperate Pacific coastal conifer forest, including Douglas-fir-western hemlock	Pacific coastal northwestern US; coastal British Columbia, Canada	Fire and wind	Franklin and Dyrness 1988; Franklin, et al. 2002; Franklin et al. 2018; Chapter 6
Montane true fir- hemlock	Mid to high elevations of Pacific coastal northwestern US and British Columbia, Canada	Fire and wind	Franklin and Dyrness 1988
Subalpine forests characterized by species of true fir and spruce	Higher elevations in north temperate regions of world	Fire, wind, and insects	Barbour and Billings 2000
Lodgepole pine	Mid- to high-elevations in western North America	Fire and mountain pine beetle	Barbour and Billings 2000
Boreal forests of spruce and larch	Alaska, Canada, Russia	Fire	Barbour and Billings 2000
Jack pine	North central North America	Fire	Heinselman 1981

Table 11.2 *Selected silvical characteristics of conifer tree species commonly found in the Douglas-fir-western hemlock forest and its variants.*

Species	Shade tolerance	Maximum diameter (cm)	Maximum height (m)	Age potential (years)
Douglas-fir	Intolerant	150–300+	70–80+	1000+
Western hemlock	Very tolerant	90–120	65	500+
Western red cedar	Very tolerant	150–300	60+	1500+
Pacific silver fir	Very tolerant	90–110	55+	550+
Western white pine	Intermediate	110	65	450+
Grand fir	Tolerant	75–125	55+	250+
Sitka spruce	Tolerant	280–230	70–75	400+
Coast redwood	Tolerant	150–380	80–100	2000+
Noble fir	Intolerant	275	80	450+

Figure 11.2 *Large infrequent wildfires are the characteristic disturbances in the Douglas-fir-western hemlock forests of western North America and typically include large areas of stand-replacement mortality. Unburned and lightly and intermediately burned areas are also characteristic of these large burned landscapes. The photo shows the Warner Creek Burn of 1993 on the Willamette National Forest, Oregon, USA.*

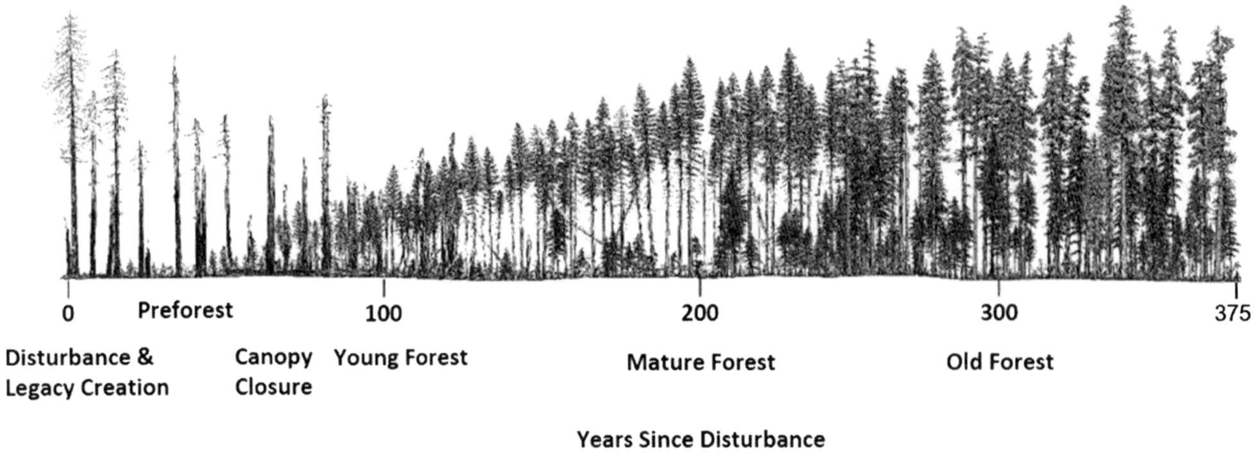

Figure 11.3 *Douglas-fir-western hemlock forests are characterized by multi-century developmental sequences, which are initiated by near complete stand-replacement disturbances. Illustration credit: Robert Van Pelt.*

can last four to six decades (Freund et al. 2014; Tepley et al. 2014). The PFS is probably longer and more ecologically important here than in any other temperate forest type. The PFS serves as an important driver of long-term structural and compositional complexity, including rapid growth of dominant trees during this period (Tappeiner et al. 1997).

Relevance of the Natural Model to Silvicultural Approaches

Will foresters attempt to emulate the natural developmental model and create all of these stages in their classic forms on all ownerships? Probably not, but it is important to understand the unique structural, compositional, and functional elements that each stage brings to the landscape. It is on large forested ownerships that are fully committed to ecological forestry, such as federal forests, where it is most feasible to work toward adequate representation of all of the developmental stages. However, as we will see later in this chapter, it is very possible to incorporate key structural and functional elements of the later stages (Mature and Old) into virtually any silvicultural prescription for any ownership.

The Evolution of Timber-Focused Silviculture in the Douglas-Fir-Western Hemlock Forest

Much of the early harvesting in Douglas-fir-western hemlock forests occurred on lands owned by forest industry and usually in high-volume, virgin, old-growth forests. Initially

Table 11.3 *Key characteristics of developmental stages in Douglas-fir-western hemlock ecosystems.*

Stage or Event	Attribute	Cause or Consequence
Disturbance and Legacy Creation	Canopy opened at whole-stand scales with extensive dead tree (snag and log) legacies	Results from near stand-replacing fire
Preforest Stage	Abundant legacies of dead trees in form of snags and downed logs; varying density and spatial pattern of residual live trees and regeneration; open environment free of tree canopy dominance	Diversity of habitat features available; lack of tree dominance allows diversity of plant life forms and associated fauna to flourish, resulting in proliferation of food webs
Young Forest Stage	Trees regain canopy dominance and grow rapidly	Strong competitive interactions among trees and with other plants; rapid accumulation of leaf area and biomass
Mature Forest Stage	Creation of canopy gaps as result of shift to agent-based tree mortality, increased decadence, and redevelopment of understory and midstory canopy layers	Redevelopment of understory communities, development of mid-story of shade-tolerant trees, significant dead wood (snag and log) additions, and other decadent features
Old Forest Stage	Structurally complex in terms of both individual structures and spatial heterogeneity, which includes canopy gaps, patches of reproduction and young trees, numerous large old trees and large accumulations of coarse woody material	Small-scale disturbances dominate; all developmental processes, including competition, are present; high levels of habitat diversity and of stability in ecological processes and conditions

Figure 11.4 *Wildfires in the DF-WH forest leave extensive biological legacies, such as large quantities of standing dead and downed trees, which are important structural elements of the subsequent Preforest and Young Forest Stages. Warner Creek Burn, Willamette National Forest, Oregon, USA.*

a cut-out and get-out mentality prevailed due to the abundance of the resource. Clearcutting was the preferred regeneration system, although abundant live trees (seed trees) were often left due to lack of markets and because they were believed to be necessary to provide a seed source. Natural regeneration of Douglas-fir was sporadic following burning of the associated logging slash, which promoted regeneration in some areas but eliminated both the live trees and regeneration in others. Over time, both Oregon and Washington (USA) instituted reforestation requirements that first included a focus on attempting to regenerate with seed trees and seeding, but ultimately have emphasized planting of commercial species with standards associated with a minimum density of free-to-grow seedlings 3–5 years after harvest. An outcome of these requirements has been a loss of the historic complexity in structural and compositional conditions that characterized the PFS and YFS in these forests, as most efforts now focus on rapidly establishing fully stocked stands of Douglas-fir and excluding fire. These efforts intensified as the supply of privately owned, virgin old-growth forest diminished following World War II. Much of what we know of today as high yield forestry for the DF-WH forest was pioneered at this time, ultimately moving silviculture to a highly agronomic approach, largely in plantations (Figure 11.6). At the time, this type of management was conducted largely by vertically integrated companies that were managing their forest lands to provide the raw material for their manufacturing facilities.

A major change in industrial forest land ownership occurred near the end of the twentieth century when these lands were spun off into **Real Estate Investment Trusts** (REITs) and **Timber Investment Management Organizations** (TIMOs) as a result of changes in tax laws (Mendell 2016). This resulted in intensified short-rotation, plantation-based forestry with management further constrained by a requirement for a high rate of return on capital (Franklin et al. 2018). With this history, it is not surprising that the silvicultural model currently used on industrial forest lands is very different from the natural model of the DF-WH ecosystem (Table 11.4) and incorporates little by way of the foundational principles of ecological silviculture into prescriptions. While many private land owners do not follow an industrial model in its entirety, considering it here does serve to highlight the potential for extreme deviation from the natural model (with a timber-focused approach) and some of the consequences of this deviation (Table 11.4).

The historical development of silviculture for DF-WH forest on federal (US) ownerships differs from that on industrial ownerships. On federal forests, the earliest approaches

Figure 11.5 *Douglas-fir-western hemlock forests develop a structurally complex Old Forest Stage during the multicentury periods that typically occur between high-severity disturbances, as illustrated in this old-growth forest in western Oregon, USA.*

Figure 11.6 *Landscape occupied by intensively managed Douglas-fir plantations in the Oregon Coast Range, USA. These are now generally owned or managed by Real Estate Investment Trusts (REITs) or Timber Investment Management Organizations (TIMOs), with the objective of high returns on capital and short rotations.*

Table 11.4 *Important differences in outcomes between the natural development model and the timber-focused (industrial) model for Douglas-fir-western hemlock forests.*

Feature	Natural Model	Timber-Focused Model	Functional Consequences[a]
Tree species diversity	Typically including up to five species, depending on site	Species other than Douglas-fir rarely planted, other species sometimes tolerated in low numbers	Reduced resource and habitat diversity, lower potential biodiversity; potentially higher susceptibility to insects/disease
Age structure	Ranges from broadly single-cohort to single cohort with low to moderate numbers of older trees	Narrowly even-aged single cohort	Potential for greater susceptibility to wind disturbance
Live tree legacies	Few to moderate numbers	None to few	Loss of unique habitat features and associated biodiversity
Snag and log legacies after a canopy disturbance	Abundant, containing a diversity of sizes including large structures	None to few and typically of small size	Loss of unique habitat features and associated biodiversity
Spatial pattern of legacies	Dispersed to aggregated legacies	Not applicable	Reduction in habitat diversity and associated biodiversity
Stand density	Diverse, including low density levels at all stages of development	Relatively high densities beginning at stand establishment	Loss of habitat associated with more open YFS
Spatial pattern of tree distribution	Heterogeneity in density and canopy cover	Homogeneity in density and canopy cover	Reduction in resource and habitat (including microclimatic) diversity and associated biodiversity
Maximum tree age	500 years and greater	50 to 80 year rotations	Loss of unique structural features and associated biodiversity
Developmental stages present on landscape	All stages present	Single stage (YFS) present	Major loss of functional capabilities, including habitat and associated biodiversity

[a]Examples of functional consequences of a timber model compared to the natural model.

to regeneration harvests generally focused on securing natural regeneration and involved retention of a few seed-trees at harvest to insure there would be a supply of seed for this regeneration. For a period of time, it was believed that regeneration occurred from seed stored in the forest floor, but this was disproved (Isaac 1943). Subsequently, five seed-trees per hectare was adopted as the minimum requirement (Isaac 1943). Clearcutting was ultimately adopted as the primary regeneration system (Isaac 1943), utilizing natural regeneration that resulted from seeds produced in adjacent uncut stands. In the late 1930s and early 1940s, selection management of Douglas-fir forests was proposed on federal lands but this was strongly opposed by forest researchers and never received serious consideration. Ultimately, even-aged regeneration systems prevailed based on the better growth rate of Douglas-fir under full sunlight and concern about partial cutting or selection ultimately converting stands to shade-tolerant species, such as western hemlock.

Harvesting expanded dramatically in DF-WH forests on federal lands following the end of World War II. At this time, the USDA Forest Service adopted the *staggered-setting* system of clearcutting (Figures 11.7, 11.8), in which clearcuts of 16 to 24 hectares were dispersed through areas of unlogged forest. Natural regeneration of the harvest units was expected from seed cast by the adjacent uncut forest following broadcast slash burning. The staggered-setting approach was believed to have several advantages over fewer, but larger, harvest units, including the potential to rapidly develop road systems in these previously inaccessible forest landscapes, the ability to limit development of extensive areas of logging slash (and associated wildfire risk), and the potential for natural regeneration.

Natural regeneration of trees using the staggered-setting clearcutting system did not meet expectations. By the end of the 1950s, enough experience had accumulated to make clear that prompt and adequate natural regeneration could not be depended upon. Consequently, a policy of artificial regeneration through planting of Douglas-fir was adopted. In the 1960s and 1970s, additional problems emerged with clearcutting, especially on more environmentally severe habitats, such as sites that were prone to severe frosts or very hot and dry or frost-prone sites, where plantations often

Figure 11.7 *Staggered-setting clearcutting regeneration system adopted by the USDA Forest Service in the Douglas-fir region forest following World War II. Among other advantages, this system allowed for rapid development of a permanent road network. Pictured is a pattern of dispersed patch clearcuts on the Gifford Pinchot National Forest, Washington, USA.*

Figure 11.8 *The staggered-setting clearcutting system ultimately resulted in the rapid fragmentation of the forest landscape, which, in turn, made the residual forest much more prone to disturbances, such as windthrow (Figure 4.16) and poor habitat for many old-growth related species.*

failed. In these locations, foresters began to use variants of shelterwood systems, leaving some mature Douglas-fir in harvest units, to help ameliorate challenging environmental conditions on these sites (Figure 11.9). This was encouraged by some experiments that demonstrated that isolated leave trees could survive on cutovers (e.g., Franklin 1963). Building on this experience, silviculturists also began to retain trees for wildlife and other ecological values on harvest units on federal forests in the late 1970s and 1980s (e.g., Buermeyer and Harrington 2002) and, as explained in Box 11.1, this ultimately gave rise to the concept and practice of variable retention harvesting (see Chapter 7).

Figure 11.9 *Application of a shelterwood system in a Douglas-fir-western hemlock forest on the Mount Hood National Forest, Oregon, USA. Shelterwoods were found to be necessary even when planting on environmentally harsh sites in the Douglas-fir region to assure prompt and adequate regeneration.*

Although generally not as extreme as the highly agricultural model used on private industrial lands, the traditional even-aged silvicultural approach used for decades on federal lands also failed in significant ways to incorporate the foundational principles of ecological silviculture and, consequently, sustain important forest functions and biota. Generally the goal was to terminate older stands and retain minimal levels of legacies, or none at all; that is, *no provision for continuity*. Thinning in existing stands focused on release of crop trees and generally involved removal of any hardwoods or conifer species considered to be undesirable due to their wood quality and/or growth rates; that is, *no provision for complexity*. Harvest rotations were scheduled on the basis of optimizing stand growth (i.e., culmination of mean annual increment) and few forests were to be regrown to MFS or OFS, nor was regrowth of old trees and related structures (large snags and logs) a goal; *timing of activities based on ecological development* was not a goal. Finally, silvicultural prescriptions were developed on a stand-by-stand basis, with little consideration of *a landscape context*. In the next section, we discuss the advent of approaches for silviculture in the Douglas-fir-western hemlock forest that are more overtly ecological and that better incorporate the foundational principles.

Box 11.1 The evolution of variable retention harvesting

The emergence of structural retention approaches, particularly variable retention harvesting (VRH) as it has come to be known, has been particularly notable in the Douglas-fir-western hemlock forest. This is, in part, because it contrasts so greatly with the deeply embedded practice of clearcutting with no consideration of biological legacies in this forest type. Some key events in the evolution of VRH follow.

Early retention approaches

The concern for providing for adequate quantities of wildlife structures in harvested areas, including live trees, snags, and logs, became evident as clearcutting of old-growth forests proceeded on federal lands in the 1970s. Thousands of hectares were being logged each year with little provision for adequate numbers of such retained structures, particularly of larger sizes. Various approaches were tried, including retention of small areas of snags and creation of snags from green trees left following the logging. None of the approaches appeared to adequately provide for a sufficient and continuing quantity of such trees.

Retention of live trees with current and potential wildlife value seemed to provide a solution. At the time, however, many silviculturists were critical of the concept, doubting both the necessity for providing such structures and believing that they would quickly blow down. Researchers with the USDA Forest Service began formally examining retention approaches on several experimental forests in the DF-WH region (Figure 11.10). Initially the level of retention tried was about 15% of the live basal area, with the leave trees dispersed relatively evenly over the harvest units. This level and pattern of retention was similar to what was being left initially in shelterwood harvests so it was known to be practicable from a logging standpoint but still sufficient to alter the microclimate and habitat conditions; unlike the shelterwood overstory, trees were to be retained through the next rotation. Various approaches for selecting leave trees were tried. In one early demonstration, two different marking strategies involved: i) primarily selecting trees that were strong dominants and would produce significant amounts of high-quality wood during the next rotation, and ii) primarily selecting trees that had high wildlife value, e.g., forked trees and trees with cavities, some of which would become snags and downed logs over time.

The initial focus was on dispersed retention in the belief that a pattern of well distributed trees was necessary to achieve ecological objectives (see Chapter 7). However, Plum Creek Timber Company became interested in the retention concept to improve both the visual and ecological content of their harvests. They experimented with various approaches and eventually concluded that aggregating retention in patches of various sizes and shapes could be very ecologically effective and superior from some management perspectives. For example, in the Cougar Ramp harvest unit (Figure 11.11), Plum Creek foresters were able to retain large soft snags in patches of live trees with minimal safety concerns and conduct logging and other future management activities on the harvested part of the unit with greater efficiency and lower costs.

Variable retention harvesting emerges

Several subsequent developments led to a more formal conceptualization of variable retention harvesting and a demonstration of its adaptability to a variety of forest conditions and management goals (Lindenmayer and Franklin 2002). In 1993, the provincial government of British Columbia (Canada) convened a scientific panel to review forest practices on northwestern Vancouver Island. Officially known as the Scientific Panel for Sustainable Forest Practices in Clayoquot Sound, the panel reviewed historical harvest practices in the region and recommended several major changes (Clayoquot Scientific Panel 1995). One recommended change was shifting from clearcutting to a retention harvesting system with a goal of at least 15% on each harvest unit. This panel coined the term ***variable retention*** to recognize that the specifics of the

Figure 11.10 *An early example of retention harvesting in Douglas-fir-western hemlock forests from the Willamette National Forest (Oregon, USA): a) dispersed 15 percent retention harvesting in a Mature Forest Stage stand in which the goal was to retain high-quality dominant Douglas-fir trees that would grow well during the next rotation and resist windthrow, b) the same stand twenty years later. Note several snags, which were created by artificially topping several trees at the time of the harvest.*

Figure 11.11 *Plum Creek Timber Company pioneered the concept of aggregated retention with the Cougar Ramp harvest unit, Washington Cascade Range, USA, in 1988: a) aerial view of this unit on which 15% of the forest was retained as unharvested blocks and strips, b) aggregated retention has proved to have many ecological and managerial benefits, such as the ability to retain a variety of tree sizes and conditions as well as soft snags, as evident in this photograph.*

amount and distribution of retention should be allowed to vary with management objectives and forest conditions. Use of the term variable retention harvesting has spread worldwide and generally is applied wherever retention approaches are used (Franklin et al. 1997; Gustafson et al. 2012).

A second major development occurred in British Columbia in1998, when MacMillan Bloedel made a corporate decision to transition from clearcutting to variable retention harvesting. This shift was to occur over five years, with a minimum target of 10% retention. In 2000, MacMillan Bloedel was acquired by Weyerhaeuser Company, who agreed to continue the transition to variable retention harvesting. In 2003, Weyerhaeuser reported that 100 percent of their harvests met the VRH definition and that the average retention level was 22 percent. The VRH program initiated by MacMillan Bloedel and completed by Weyerhaeuser demonstrated how a rich variety of retention approaches could be used under different conditions and management objectives (Figure 11.12).

Another major step forward in adoption of variable retention harvesting occurred in connection with the development of the Northwest Forest Plan for US federal lands within the range of the northern spotted owl. One provision of that plan required that any regeneration harvest had to include at least 15% retention. On national forest lands the retention was to take the form of both aggregated and dispersed retention. The Bureau of Land Management chose to do dispersed retention on the forest lands that it managed in western Oregon. With the advent of this plan and requirement, the era of clearcutting on federal forest lands in the DF-WH region was officially over.

Figure 11.12 *Four different examples of variable retention harvests carried out on MacMillan Bloedel Limited (subsequently Weyerhaeuser Company) forest lands on Vancouver Island, British Columbia, Canada.*

Figure 11.12 (*Continued*)

Ecological Silviculture for the Douglas-Fir-Western Hemlock Forest

We now turn our attention to overtly ecological approaches to silviculture in the DF-WH forest. Our premise is that many aspects of emulating the infrequent, severe disturbance developmental model, and incorporating the foundational principles, are applicable to DF-WH forests despite its long-protracted developmental sequence and its characteristically massive and structurally complex OFS. In fact, aspects of ecological silviculture have found their way into practice on many ownerships, including nonindustrial, privately owned forest, Native American and First Nation forests, community and municipal forests, and many state (US) and provincially managed crown forests (Canada). For example, much of the initial development and application of variable-retention harvesting (see Chapter 7) occurred on crown lands on Vancouver Island in British Columbia (Box 11.1).

Ecological silviculture generally has no place on DF-WH forests in industrial ownerships, which are now mostly institutional investors with high timber production goals and the constraint of a high return on capital. However, even these ownerships may have legal obligations or forest practice certification requirements (see Chapter 1) that may require them to consider ecologically-based management approaches in some form. For example, in the state of Washington, USA all private forest landowners have significant obligations associated with aquatic environments and fish, regardless of their financial goals.

Perhaps the most obvious potential application of ecological silviculture in the USA is on federal forests (USDA Forest Service and USDI Bureau of Land Management), which are tasked with meeting numerous environmental goals (Franklin et al. 2018). Although extensive areas of these forests are reserved from commercial timber harvest, there are large areas of younger forests that were planted following clearcutting beginning in the 1960s and are now prime candidates for silviculture based on an ecological approach (Franklin et al. 2018). Such an approach is what we turn to next.

An Ecological Silvicultural System for Douglas-Fir-Western Hemlock (Archetype 1) Ecosystems

Presented here is an ecological silvicultural system for the DF-WH forest based on a natural model of stand development and expressly incorporating the foundational principles we have discussed in this book. As explained in Chapter 10, we name our system after the prevailing natural disturbance regime for Archetype 1, that is, an *infrequent disturbance silvicultural system.*

For the DF-WH forest, the ecological silvicultural system can be used to address goals of providing both ecological and economic benefits through a series of activities that culminate with a variable retention harvest. However, elements of this silvicultural system are also of value where the goals are strictly ecological and there is not any intention of terminating the stand with a variable retention harvest. This might be the case when management is focused solely on restoration of natural forest conditions. Finally, silviculture may focus on restoration initially, say in older Douglas-fir plantations, but as part of an approach that also continues to manage for timber. With these different objectives in mind, the silvicultural system should not be viewed as a strict linear progression of actions beginning with a regeneration harvest, but rather as a family of potential actions, organized around developmental stages and applied according to the condition and developmental stage of the stand in question. Below we describe potential activities, including regeneration, tending, and harvesting treatments organized around developmental stages (Table 11.5). Beyond the first several decades, subsequent treatments could be viewed as regular entries into stands as part of a multi-aged group selection framework.

Table 11.5 *An ecological silvicultural system for Douglas-fir-western hemlock ecosystems.*

Developmental Stage/Event[a]	**Length**[b]	**Objectives/Consequence**	**Potential Silvicultural Activities**[c]
Disturbance and Legacy Creation	0	Provide for continuity of structure and function by leaving biological legacies; initiates a PFS	*Harvest*: variable retention harvest in MFS and treatment of slash, preferably by fire (either broadcast burn or piles)
			Deadwood creation: deliberate creation from live retention trees where large snags and logs are lacking (can be done with fire)
Preforest Stage	up to 60 years[d]	Provide for period of open, shrub, and herbaceous dominated habitat, as well as conditions for establishment of natural tree regeneration	*Invasive control*: reduce invasive plants that may develop
			Site preparation: to expose mineral soil/reduce duff to facilitate natural regeneration
Young Forest Stage (early)	20 years	Facilitate/establish new cohort of trees	*Regeneration:* supplement natural regeneration as needed
			Release of regeneration: as needed; avoid elimination of natural regeneration or understory and ground layer plant populations
Young Forest (after canopy closure)	60 years	Accelerate growth of individual trees; facilitate development of spatial heterogeneity in structure; create snags and logs as needed	*Thinning:* VDT of the main cohort; potential for multiple entries at intervals of 10–20 years
			Site preparation and competition control: in VDT gaps as needed.
			Enrichment planting or seeding: as part of VDT, as needed
			Decadence/deadwood creation: killing trees or inducing decadence in live trees as needed when deadwood is lacking
			Harvest: potential for VRH in some settings, resetting to *disturbance and legacy creation*
Mature Forest	70 years	Enhance complexity and spatial heterogeneity of structure; release and accelerate growth of desired trees; restore compositional diversity; create deadwood; potentially conduct a regeneration harvest	*Thinning*: VDT in the main cohort of the stand; potential for multiple entries and for incorporation of significant openings for establishment of a new mixed cohort of Douglas-fir and western hemlock
			Site preparation and competition control: in VDT gaps, as needed
			Enrichment planting or seeding: as part of VDT
			Decadence/deadwood creation: killing trees or inducing decadence in live trees when deadwood structures are lacking
			Harvest: potential for VRH in MFS stands, resetting to Disturbance and Legacy Creation
Old Forest Stage	150+ years	Minimal silvicultural needs	See text for explanation

[a]A regeneration disturbance initiates the developmental sequence or, for existing stands, the table is entered according to the developmental stage of the stand in question and actions are applied from that point onward.

[b]Reflects a general length of time that a particular stage might last, rather than the age of the stand after the initiating disturbance.

[c]Potential activities that might be pursued during the stage.

[d]This is the potential maximum length of time the PFS could last, based on the natural model. It is unlikely that most ownerships would purposely manage for this length of a PFS. Franklin and Johnson (2012) have recommended managing for a 30-year preforest period on federal lands in the DF-WH forest type.

Silviculture by Developmental Stage

Disturbance and Legacy Creation

In our DF-WH silvicultural system, the focus of a harvest is on live tree and dead wood retention and initiation of the PFS, both of which are defining characteristics of this forest. Variable retention harvesting emulates to varying degrees the patterns of tree death and live tree retention that occurs with severe, episodic natural disturbance. In practice, harvests are all heavy and will often include both dispersed and aggregated retention, since the two forms of retention achieve different purposes, but with highly variable levels of live tree retention depending upon the objectives of the landowner (Figure 11.13). The normal range of retention probabilities is probably 15 to 60 percent, assuming the goal is to provide areas that are suitable for development of early successional (PFS) biota and processes and for successful reproduction and early growth of the shade-intolerant Douglas-fir. Current recommendations on federal forest lands are for approximately one third of the harvest unit to be retained in moderate to large aggregates, which would include retention necessary to protect riparian habitat included within the harvest unit; this level is sufficient to sustain the majority of forest biota on the harvest unit. An additional 5 to 10% retention of individual and small clusters of trees is recommended on the harvested two thirds of the unit to provide for dispersed live trees, snags, and downed logs needed by early successional organisms. These levels of harvest severity provide environmental conditions that will both recruit multiple cohorts of the relatively shade-intolerant Douglas-fir, as well as maintain sufficient islands of closed forest to sustain most forest biota, including the shade-tolerant tree species such as western hemlock.

Retention of deadwood should be factored into the VRH as well. This includes existing large snags, large downed logs, and near-dead trees (for deadwood recruitment). Moreover, VRH offers opportunities to create deadwood structures from live trees, since machinery to fell or top trees is on site (Figure 11.14).

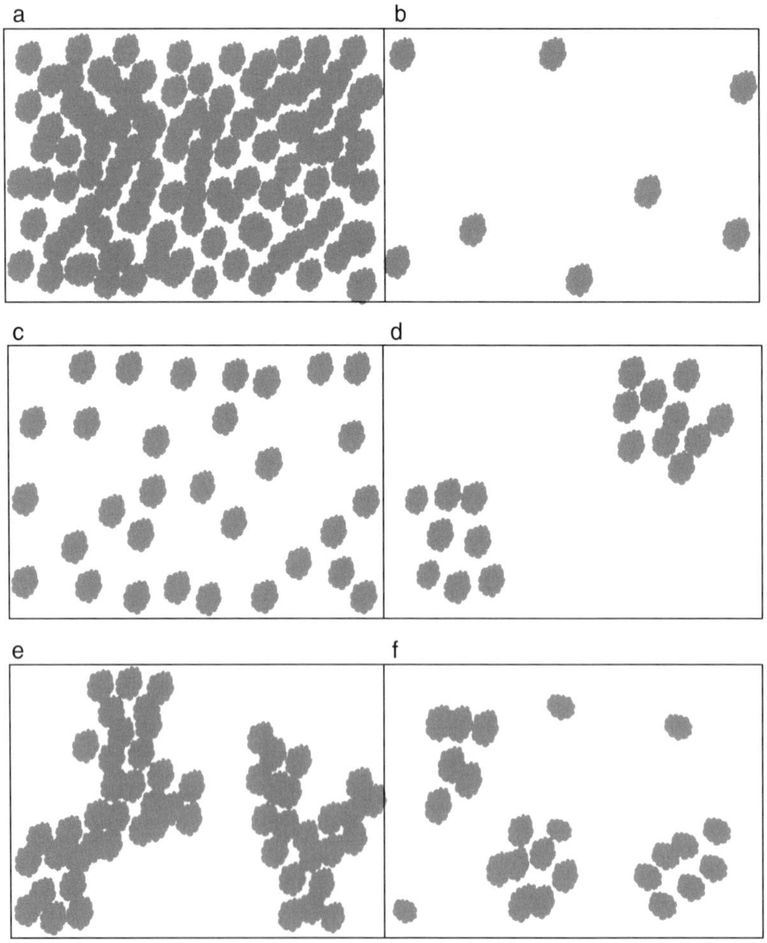

Figure 11.13 *Conceptual representation of various types of variable retention harvesting in Douglas-fir-western hemlock ecosystems, all of which reflect outcomes of heavy canopy removal emulative of severe infrequent disturbance. Starting with the uncut forest in the top left (a), retention may range from low to moderate levels of dispersed trees (b, c), to different levels and shapes of aggregated retention (d, e), to combined dispersed and aggregated retention (f).*

Figure 11.14 *Snags and downed logs are often deliberately created as a part of variable retention harvests in a Douglas-fir-western hemlock forest. Snags created through tree topping are visible in the harvested portion of a harvest unit with 40% aggregated retention. Paradise Hills DEMO site, Gifford Pinchot National Forest, Washington, USA.*

Preforest Stage

An extended Preforest Stage (PFS) is a distinctive feature of the DF-WH forest, and likely occurred in many settings after severe canopy disturbances. In a managed forest setting, the ecological forester can encourage a PFS by: i) allowing regeneration to develop naturally over several to many years; ii) delaying artificial regeneration for a number of years after harvest (Figure 11.15). Prescribed fire early in this stage is useful for managing landscape-level fuel continuity and resultant fire risk while also promoting spatial heterogeneity in growing space, plant communities, and regeneration within individual stands.

Eventually during the latter parts of a PFS, ensuring that tree regeneration is adequate will be an important consideration and if it is not, additional activities may be needed, including site preparation and competition control. The most common solution likely will be some tree planting, perhaps including elimination of competition at planting spots.

Young Forest Stage (early)

Silviculture in the early YFS will focus on development of the new cohort of trees. Assuming that adequate stocking has occurred through natural regeneration, there may be little need for further action to establish trees. Release of established regeneration, particularly by reduction of competing non-tree vegetation, may be needed, but this should be done so as to maintain a diversity of tree species.

Young Forest Stage (after canopy closure)

The primary emphasis in the later phases of the YFS will be introducing spatial complexity, advancing tree size, and encouraging understory redevelopment through the application of variable density thinning (Figure 11.16). The variable density thinning will result in areas that favor responses of understory and mid-story species, including hardwoods and shade-tolerant conifers. It should also result in retention of some areas of relatively high density, where environmental conditions will favor species that prefer heavily shaded environments. Although regeneration of Douglas-fir is not usually an objective, should it be pursued, then gaps will need to be 0.2 to 0.3 ha in size and may require some site preparation to control competition with the planted trees. Some part of the variable density thinning should include areas treated to maintain relative densities below 0.55 if the objective is to quickly grow some larger tree sizes (Tappeiner et al. 1997; Bailey and Tappeiner et al. 1998; Tappeiner et al. 2015). This could also be achieved through some form of crop tree thinning. Additional later thinning entries would again be variable density thinning

Figure 11.15 *Two views of well-developed Preforest Stage habitat in managed forest setting. (a) Blue Ridge, (b) Willamette National Forests, Oregon, USA.*

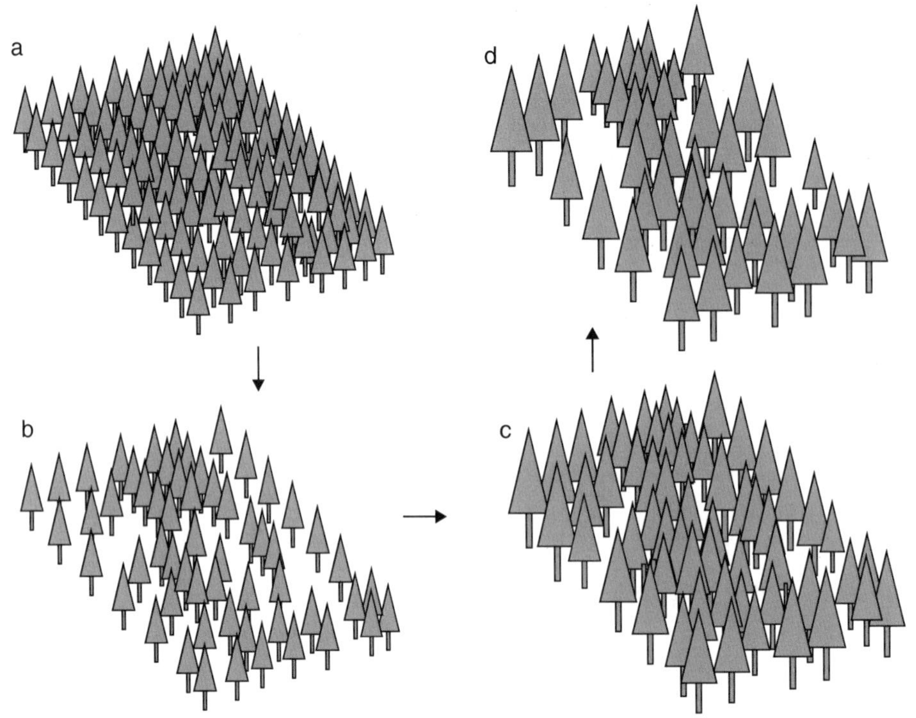

Figure 11.16 *Conceptual representation of variable density thinning (VDT) in a Young Forest Stage, structurally simple Douglas-fir stand. Starting from the unthinned stand (a), the initial application of VDT (b) creates some spatial heterogeneity of structure, releases some trees, and may create opportunities for regeneration in gaps. As this thinned stand further develops (c), a second VDT entry may be used to expand existing gaps and further thin the matrix (d). An additional objective in these thinnings will be to encourage development of plants in the understory and midstory positions.*

applied in a way that maintains or augments heterogeneity of density developed in the initial thinning (Figure 11.16) and could ultimately result in a multi-aged stand with regeneration in gaps. Release and acceleration of growth of crop trees will also be a goal of VDT (Box 11.2). The length of the YFS is long enough that two or three entries on a 10–20 year interval might be considered.

Growth rates of trees in the DF-WH forest are such that some individuals in the later part of the YFS may be large enough to contribute meaningful deadwood structures to the stand. As such, if deadwood is lacking, deliberate decadence or deadwood creation might be considered with a minimum goal of five large decadent trees or snags per hectare.

Finally, on ownerships where important financial constraints might prevent continuing young forests to later stages of development, stands in the later parts of the YFS can be subjected to a regeneration harvest that returns the system to the Disturbance and Legacy Creation event. This is possible because of the high rates of productivity and tree growth in Douglas-fir forests. If this occurs, it would take the form of a variable retention harvest (Box 11.3).

Mature Forest Stage

Stands in the MFS provide good opportunities for meeting objectives of ecological forestry with silviculture. If these forests have not been managed using an ecological approach, then their condition likely reflects development following a passive (no deliberate silvicultural activities) or a timber-focused model. In either case, the stands may be characterized by a simplified, homogeneous structure and be overly dominated by Douglas-fir. Nonetheless, past thinning treatments in stands managed with a timber-focused model may have advanced the development of understory vegetation, including advance regeneration of tolerant conifer species (Bailey and Tappeiner 1998). These structural and compositional conditions can be built upon and enhanced, depending on long-term objectives for a given stand.

There are two potential pathways for silviculture in such stands: i) VDT and other tending treatments to facilitate transition to a complex OFS or ii) VRH to reset the stand to a PFS (Figure 11.17). Prescriptions focused on transitioning toward the complex OFS may utilize patterns in understory vegetation and advance regeneration to guide application of VDT treatments.

Thinning in the MFS will be a VDT as outlined for the YFS. This may involve an initial thinning or it could build on an earlier conventional thinning, which might be a common history in mature plantations. An objective of thinning will include releasing trees with the greatest potential for long-term growth. The creation of canopy gaps should

Box 11.2 **Variable density thinning in Douglas-fir plantations.**

Ecological silviculture in YFS Douglas-fir plantations typically includes variable-density thinning (VDT), including skips-and-gaps VDT, but with more recent use of quantitative approaches, such as the ICO (Individuals-Clumps-Openings). Regardless of the approach, the objective is to create structurally heterogeneous stands, with this heterogeneity modeled on patterns observed in natural stands.

An important side-board for VDT prescriptions is the designated retention of existing old Douglas-fir trees, snags, and large deadwood (Muir et al. 2002) that may be found scattered within these plantations; that is, these should not be removed in a thinning. This provision may also include large stumps and hardwood tree species, particularly those of sufficient size to provide wildlife habitat.

The importance of unthinned areas (skips) is emphasized in these prescriptions, as well. Skips provide habitat conditions, i.e., denser, shadier, and moister conditions, in thinned stands that are not found with traditional area-wide thinning or in gaps. Skips provide refugia for a variety of organisms (e.g., some fungi, mosses, and amphibians) that are either intolerant of mechanical disturbance or more open environmental conditions. Skips may also provide hiding or resting cover for some species, such as ungulates. Some portions of pre-thinned stands are also likely to contain unique ecological conditions, such as springs, seeps, rock outcrops, concentrations of large standing or downed deadwood, or patches of hardwoods. These areas are also good candidates for skips in order to protect those features.

Increasing the rate of large tree development is also a goal in VDT in young stands. A broad knowledge base on thinning regimes for accelerating the growth of Douglas-fir has been developed over the last century, and this knowledge can also be readily transferred to achieving ecological objectives (Curtis et al. 1998). This includes applying crop-tree release in the matrix of VDT stands or across entire stands to maintain high rates of growth throughout and acceleration of development to the MFS (Curtis and Marshall 1993). Such density management treatments will also serve to stimulate responses in mid-story, understory, and ground layer plant populations (Bailey and Tappeiner 1998), providing future opportunities to enhance these elements and release new cohorts of trees during gap creation as part of VDT.

focus on areas with existing advance regeneration and a developed shrub layer to enhance the development of these features. Most likely, these areas will exist in the portions of the stand treated with thinning during the YFS. Deliberate decadence and deadwood creation might be pursued as a standalone treatment or during a thinning entry (or even during a VRH as described below) in the MFS, as stands in this stage will contain trees of a size suitable for creating deadwood structures.

Regeneration harvesting in the MFS should take the form of a variable retention harvest (VRH), as described under Disturbance and Legacy Creation. Depending upon

Box 11.3 **Variable retention harvesting in Young Forest Stage Douglas-fir plantations on industrially managed lands: adding some ecology to production forests.**

Douglas-fir stands and plantations in the later part of the YFS typically will have little structural complexity, including a lack of spatial heterogeneity. Intermediate treatments (e.g., thinnings) can help develop this complexity, particularly on public and other conservation-minded ownerships. In contrast, stands of this type in ownerships driven by financial considerations are candidates for clearcut regeneration harvesting. However, even in these instances, some elements of ecological silviculture may be appropriate or essential for meeting certain management objectives, such as obtaining third-party sustainability certification (e.g., Forest Stewardship Council) (see Chapter 1).

Variable retention harvesting can be a viable option in these cases. Application of VRH would include selecting areas for retention as aggregates that are associated with unique physical features of the stand, in particular, riparian habitat along streams and rivers, or other unique conditions in a stand, as well as some aggregates in the dominant forest condition. Dispersed retention of individual Douglas-fir may be included in the VRH prescription. Consideration also will be given to locating aggregates and individuals in ways that do not unnecessarily complicate logging operations.

Application of VRH in an industrial forest setting means that the majority of trees will be removed (thus lower levels of retention might be used than in other settings), so as to create an environment that is sufficiently large and open to provide suitable conditions for regeneration of trees, namely Douglas-fir. For example, the VRH prescription might include retention of 5% of trees in the form of small- to medium-sized (0.25 to 2.0 ha) aggregates, as well as individual or small clusters of trees. The retained trees would be maintained at least through the next rotation.

Is this an example of comprehensive ecological silviculture as we have presented it in this book? No, it is not, but it is an example of retention harvesting as we have defined it. Is this retention like a more conservation-oriented agency or organization might practice it? No, it is not, but we argue that this may be okay since in the context of industrial plantation management, what is being done is better than what might be done otherwise, i.e., no retention, from an ecological perspective (Palik and D'Amato 2017).

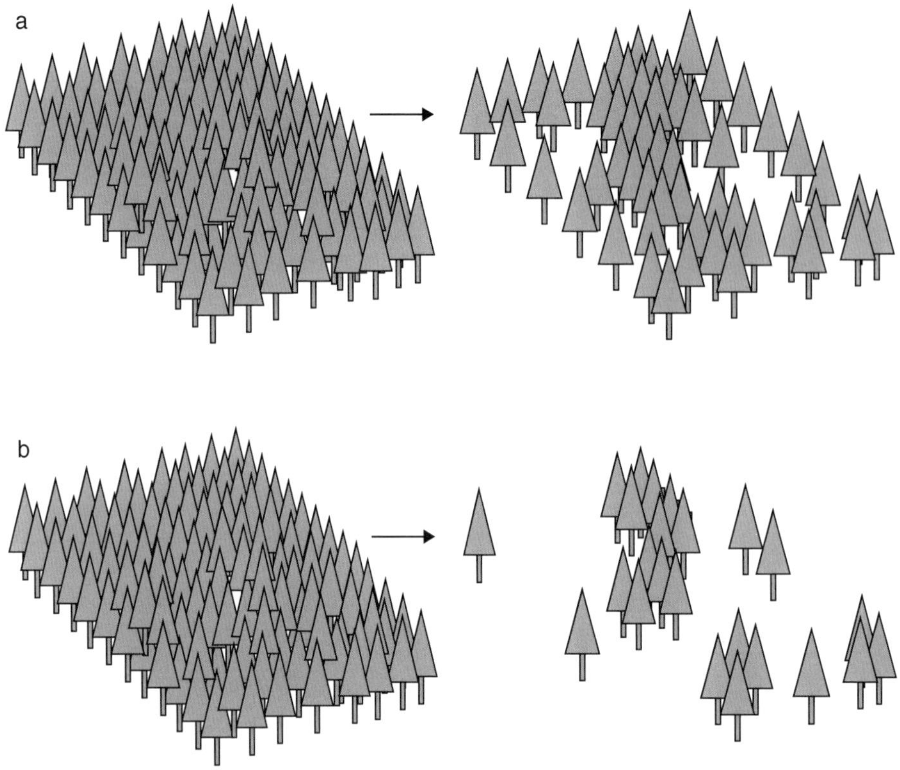

Figure 11.17 *Conceptual representation of two potential pathways for silvicultural treatment in Mature Forest Stage Douglas-fir-western hemlock forests. a) Variable density thinning may be used in structurally simple stands to hasten the development of complex structure. b) Mature Forest Stage stands may be harvested using a variable retention approach, resetting the stand to a post-disturbance, Preforest Stage, but enhancing structural complexity in the process.*

management objectives, the retention levels can vary widely in terms of both the amount and the pattern of retention, with resulting implications for various ecosystem functions, as described earlier (Table 7.2). In general, a dominance of aggregated retention is going to be best where there is concern for providing the open areas optimal for early successional or PFS species and for establishment and growth of shade-intolerant tree species, such as Douglas-fir. Significant amounts of dispersed retention (e.g., greater than 20%) can significantly impact the quality of early successional habitat and growth of shade-intolerant tree species. On the other hand, dispersed retention might be desired for such reasons as maintenance of root masses on unstable slopes, accumulation and retention of snow, or aesthetics. Such goals might be achieved with well distributed retention of 15 to 30 percent of dominant and co-dominant trees.

On ownerships where the goals during harvesting include both the creation of early successional habitat and lifeboating of forest-related species, the earlier prescription for harvesting on federal lands could be appropriate. This involves: (1) retention of about one third of the harvest area in larger aggregates (including those associated with riparian or other aquatic environments located within the harvest unit); and (2) harvest on about two thirds of the remainder with significant retention (5 to 10%) of single and small clusters of live trees, snags, and downed logs to provide the structural complexity that is needed as a part of early successional or PFS habitats. Such prescriptions should sustain most of the forest-related organisms and processes while providing good conditions of early successional biota and processes.

Most MFS stands in the contemporary landscape will not have been managed following an ecological approach, thus live tree legacies will be selected from the main (even-aged) cohort of trees, rather than from a population of older cohort trees in the stand. However, in stands that have been managed ecologically, such as those that have had a previous VRH, then there may be opportunity to select legacy trees from a population that is older (and larger) than the main cohort.

Old Forest Stage

Silviculture for timber management objectives is inappropriate in OFS DF-WH forests. While there may be a situation in which some type of silvicultural intercession could be useful, we can think of no general case in which an old-growth forest would be improved ecologically by a silvicultural treatment. Perhaps there could be a need for some activity, particularly in structurally simple stands or those that are depauperate in some particular native plant species. We assume that any activities would be explicitly related to specific conservation concerns or goals in the old forest in question.

Examples of Ecological Silviculture in Practice

Applications of the comprehensive silvicultural system described above (Table 11.5) can take many forms depending upon ownership, the nature of the stand, and on the management objectives. In this section, we highlight two examples of application by several different owners/stewards. The examples primarily focus on actions in discrete developmental stages, rather than application of our complete silvicultural system, which is often going to be the case.

Example 1: Accelerating Development of Structural Complexity in Simplified Stands

Many structurally simplified young Douglas-fir forests, largely plantations, were established during the mid- to late parts of the twentieth century on public and private lands with the intent that they would be managed primarily for timber production. Management objectives for many of these stands have shifted over time to include ecological objectives, such as protection of watersheds and provision of habitat for biological diversity. Accelerating the development of structurally complex and biologically rich conditions in younger plantations (addressing *Principle 2: complexity/diversity*) is now a major management goal on federal lands, other publicly owned forest, and some privately owned forests. For example, on federal ownerships in the US, extensive areas of Douglas-fir plantations were incorporated into Late Successional Reserves (LSR) established as part of the Northwest Forest Plan (Franklin et al. 2018). These plantations, many of which are in the YFS, are a primary focus for silviculture to accelerate development of structural complexity and species diversity (Figure 11.18).

The Washington State Department of Natural Resources also carries out significant amounts of silvicultural treatments that address *Principle 2: complexity/diversity* in established stands. Unlike the LSR stands discussed above,

Figure 11.18 *There are many Douglas-fir plantations representing simplified Young Forest Stage conditions within the boundaries of the Late Successional Reserves (LSRs) on federal lands in the Douglas-fir-western hemlock region, USA. Many of these are being treated to accelerate the development of structural complexity and to enhance their habitat quality and potential species diversity, as in the case of these plantations on Bureau of Land Management forest land in Oregon, USA. The photos illustrate the range of variation in intensity of thinning including: a) a small gap in front of a thinned matrix, b) uniform thinning of the matrix, c) retention of hardwoods and a large residual tree, and d) overall stand variability including gaps, a skip, and thinning of the matrix. Photo credit: Abraham Wheeler.*

the intent here is to eventually harvest these stands, but in the interim they are being treated to restore/enhance structurally complex conditions, including provision of nesting, roosting, foraging, and dispersal habitat for the northern spotted owl.

The core of prescriptions for ecological silviculture on these ownerships involves some form of variable-density thinning (VDT). As discussed above (Box 11.2), VDT is used to create structural richness and spatial heterogeneity and release crop trees (for both timber and ecological objectives), as well as other biotic elements, such as hardwood and other intermediate trees and understory plants.

Example 2: Legacy Retention in Maturing Stands

Increasingly, regeneration harvests in the DF-WH region incorporate structural retention (addressing *Principle 1—Continuity*), largely through use of **variable retention harvesting** (VRH), pursued to achieve multiple ecological and economic objectives. Examples of ownerships that regularly employ VRH include federal agencies in the US, the Washington (US) Department of Natural Resources (Figures 11.19, 11.20), and several industrial forest ownerships in the US and Canada (see Box 11.1).

Typically, VRH is applied to stands that were previously clearcut and planted with timber management objectives as the priority, but where contemporary objectives now include a much greater emphasis on sustainability of ecological services. This type of harvesting also may be applied to stands that originated from natural regeneration.

Prescriptions may include retention of at least 10 percent (and often considerably more) of tree-derived structures (snags and logs) in the form of small- to medium-sized (0.25 to 2.0 ha) aggregates and individual or small clusters of trees. Larger aggregates are typically associated with unique ecological features, including riparian areas, concentrations of deadwood, existing openings, and concentrations of hardwood species. Aggregates are also located in areas of

Figure 11.19 *Variable retention harvest in a fifty-year-old plantation on lands managed by USDI Bureau of Land Management in Douglas County, Oregon, USA: a) aerial view of the harvest unit, b) and c) ground views of conditions following harvesting. The more distant clearcut areas in (c) are industrially managed forest lands. Photo credit (a): Douglas Forest Protective Association.*

Figure 11.20 *The Washington (USA) Department of Natural Resources makes extensive use of variable-retention harvesting on the trust land that it is responsible for managing: a) aerial view of a landscape showing the complexity of the retention and of the harvested areas and b) a ground view of a DNR retention harvest showing both individual tree and aggregated retention. Photo credit: (a) Washington State Department of Natural Resources.*

the existing dominant forest condition to adequately provide that habitat in the regenerating stand. During the harvest, some of the retained trees may be converted into snags by topping or girdling.

VRH may include managing for a PFS. Various site preparation activities are often needed to better emulate the PFS, but also to facilitate natural regeneration of trees. This may include slash disposal, if possible using prescribed fire to stimulate the seedbanks of early successional herb and shrub species, seedbed preparation, and deadwood creation, all done in ways that lessen impacts to biological legacies. Regeneration of a new tree cohort and eventual reestablishment of a closed forest stand, while important, may be pursued less aggressively than a timber-focused model, by allowing for a PFS and for natural regeneration to develop relatively slowly.

The plan for tree regeneration takes advantage of on site seed sources for natural regeneration. Supplemental planting may be used for species or genotypes not well represented on the site, or when natural regeneration is below a minimum level of desirable stocking.

Eventually, stands treated with VRH will become candidates for variable density thinning, with the goal of enhancing species diversity and structural complexity, as well as developing merchantable trees for future harvests.

What to Remember about Silviculture for Archetype 1 Ecosystems

Globally, Archetype 1 ecosystems are often dominated by tree species with significant commercial value and have a long history of being managed for timber, often in plantations, as in our Douglas-fir-western hemlock example. When taken to the extreme, the timber-focused model for silviculture bears little resemblance to the natural model for most Archetype 1 ecosystems. When ecological (e.g., restoration) or both economic and ecological objectives are important, a silvicultural approach based on a natural model will be most relevant and, of course, will look very different than the timber-focused model. Some important considerations for the forester developing prescriptions based on the natural model include:

1. Legacy retention should be a central feature of harvesting, with retention inclusive of dispersed and aggregated retention, reflecting the often variable nature of natural disturbance severity in Archetype 1 ecosystems (*addressing Principle 1—continuity*).
2. An initial outcome of variable retention harvesting is initiation of a Preforest Stage of varying length, and immediate full site occupancy of regenerating trees in all settings is avoided (*addressing Principle 4—context*).
3. Natural tree regeneration is desirable. In fact, live seed tree retention and disturbance at the soil surface greatly facilitates natural regeneration (*addressing Principle 1: continuity*).
4. Silvicultural activities to encourage regeneration of all component tree species should be pursued, potentially including site preparation, competition control, and planting. However, stand-wide uniform application of site preparation and competition control should be avoided so as to avoid reductions in native plant species and creating uniform conditions across a stand (*addressing Principle 2: complexity/diversity*).

5. Thinning in Young and Mature Forest Stages can include crop-tree development, but this thinning will often be applied in a way that also creates heterogeneity of structure, inclusive of gaps and unthinned patches (*addressing Principle 2: complexity/diversity*).
6. Decadence and deadwood creation might be pursed in the Mature Forest Stage as these structural features are often lacking in managed forest landscapes (*addressing Principle 2: complexity/diversity*).
7. Substantially longer intervals between regeneration harvesting are likely to be the norm rather than the exception under a natural model (*addressing Principle 3: timing*), reflecting a desire to develop structures and conditions that depend on old trees.
8. Variability and heterogeneity at the landscape-scale, of harvesting and legacy creation, thinning, decadence creation, etc., are concepts to keep in mind when applying silviculture in Archetype 1 ecosystems. Put simply, if the goal is to promote native biodiversity and maintain more options for an uncertain future, not all stands should be treated alike (*addressing Principle 4: context*).

CHAPTER 12

Silviculture for Archetype 2 Ecosystems: Forest Characterized by Frequent Low-Severity Fire Disturbance

(Contributed by Derek Churchill)

In this chapter, we focus on silviculture of the Archetype 2 construct, that is, forests that are characterized by frequent, low severity disturbance, mainly wildfire. Recall from the description in Chapter 6 that in this archetype, low intensity (but often extensive) surface fires burn every 1–20 years and maintain a forest with a low density of larger old trees, an open canopy structure, regeneration often aggregated in patches, and a rich understory plant community of grasses and low shrubs (Figure 12.1). Developmental processes have evolved based on a feedback relationship with fire such that instead of progressing through a sequence of developmental stages after a major disturbance, as in Archetype 1, frequent fire forests persist as dynamic, fine scale, multi-age mosaics composed of all developmental stages over very long time periods and often over very large areas.

Tree species composition is dominated by fire adapted coniferous species, which are most typically pines, as well as a variety of oak species in many frequent-fire forests (Johnson et al. 2009). The pines, and to a lesser degree oaks, along with flammable grasses, produce pyrogenic litter that readily carries surface fire and creates a positive feedback to the fire regime. Succession to shade tolerant species over time is prevented by frequent fire. Similarly, the frequent surface fires regularly consume fine fuels, prune lower branches, and keep dead wood levels from accumulating. Frequent fire is thus a self-reinforcing disturbance that makes these ecosystems highly resistant to high severity fire as well as to periods of prolonged drought.

The frequent-fire archetype occurs in many different geographic regions of North and Central America, as well as Eurasia, spanning an immense range of climate and soil types (Table 12.1). It is important to note, however, that currently there are relatively few places where these forests have intact or restored frequent fire regimes. In forests that historically experienced frequent fire, the vast majority have undergone dramatic changes due to fire suppression, as well as timber harvesting, grazing, invasive species, and other anthropogenic disturbances (Covington and Moore 1994; Hessburg and Agee 2003; Addington et al. 2018).

Composition and structure of frequent fire forests can change rapidly if fires are suppressed and become less frequent. Major changes include succession to fire intolerant species, the eventual reduction or elimination of large, fire resistant trees, large increases in density and canopy cover, especially of small trees, altered understory plant communities, and elevated surface fuel loads (Figure 12.2a). In some cases, these changes have shifted the fire regime toward mixed and high severity fires and increased susceptibility to drought stress from a warming climate and associated large scale insect outbreaks (Miller and Safford 2012; Crockett and Westerling 2018; Stephens et al. 2018). Additionally,

Figure 12.1 *Example of a frequent fire forest with an active fire regime in a Jeffrey pine forest in the Sequoia-Kings Canyon National Park in central California, USA. The low density, open canopy is dominated by large, fire resistant trees that are the structural backbone for this type of ecosystem. Multiple age cohorts exist, including thickets of regeneration growing in openings. Photo credit: Marc Meyer.*

Table 12.1 *Examples of Archetype 2 Ecosystems.*

Ecosystem	Distribution	Primary Natural Disturbances	References
Ponderosa pine and dry mixed conifer	Throughout interior western USA	Low intensity fire, bark beetles	Chapter 6; Franklin et al. 2018
Longleaf pine	Southeastern coastal plain, USA	Low intensity fire, bark beetles	Mitchell et al. 2006; Kirkman and Jack 2017
Shortleaf pine	Ozark-Ouachita highlands, USA	Low intensity fire, bark beetles	Kabrick et al. 2007; Stambaugh et al. 2013
Neotropical pine forest	Caribbean islands and central American coastal plain	Hurricanes, low intensity fire, bark beetles	Myers and Rodríguez-Trejo 2009
Sierran mixed conifer and Jeffrey pine	California and northwestern Mexico	Low intensity fire, bark beetles	North et al. 2016
Oak and oak-pine forests of North America	Widespread	Low-moderate intensity fire	Sawyer and Keeler-Wolf 1995; Fule and Covington 1997; Johnson et al. 2009

timber-focused management has actively converted large areas of frequent-fire landscapes to plantations, such as in the southeastern US and the intermountain regions of North America, generally using an even-aged approach that includes fire suppression and results in very different structure and function than the natural model.

Fortunately, there are large-scale efforts underway in many places to restore frequent fire forests to approximations of their historic composition and structure, as we will describe later in this chapter (Figure 12.2b). This is being done to return them to a more fire resistant condition and increase their resilience to projected climatic changes (Reynolds et al. 2013; Hessburg et al. 2015), but is also being done within the context of managing for timber resources more sustainably. The silvicultural practices needed to do this present a unique and interesting set of challenges for the forester, chief among them being the need to often utilize a drip torch along with the chain saw.

The purpose of this chapter is to help you understand how to do this by providing models for silviculture, including restoration, that are emulative of the ecological construct for Archetype 2 ecosystems. For this, we will use our case study for the archetype: ponderosa pine and dry mixed-conifer forests of western North America, but with the ideas being broadly applicable to other Archetype 2 ecosystems (e.g., Box 12.1). We first review the important ecological

Figure 12.2 *Untreated (a) and restored (b) conditions in a ponderosa pine and Douglas-fir forest on the Confederated Salish and Kootenai Tribal Forest Lands in western Montana, USA. The high density of small and medium sized trees that filled in around the large trees after fire exclusion were removed with a noncommercial, variable density thin. Cut trees were hand piled and burned and then the whole unit was underburned with a prescribed fire.*

Box 12.1 **The ultimate frequent-fire ecosystem: longleaf pine (*Contributed by Brian Palik and Robert Mitchell*)**

The longleaf pine woodland ecosystems of the southeastern coastal plain of the US may be the ultimate example of a frequent-fire ecosystem—a bookend, so to speak—because of its nearly annual requirement for fire (Kirkman and Jack 2017; Franklin et al. 2018). It is also the best example of a frequent-fire forest ecosystem that is successfully managed for multiple values, including high quality timber, using a natural models approach to silviculture (Mitchell et al. 2006). These ecosystems are some of the most species rich in temperate North America, containing, for instance, nearly 25% of all plant species found in the USA and Canada (Kirkman and Jack 2017). They also contain large numbers of endemic plants and animals. Frequent surface fires, occurring at one to three year intervals, are the key to maintaining this level of biological diversity and the characteristic structure of longleaf pine forests (Figure 12.3), including low but spatially variable tree densities, patches of regeneration in larger openings (Figure 12.4), and a limited amount of hardwood trees.

While greatly reduced in area (95% reduction from historical range) the remaining intact forests are often managed for production of high-value timber products, including heart pine sawtimber and poles (McIntyre et al. 2008). The best examples of this management come from the Red Hills region of northern Florida and southwestern Georgia, USA, where a silvicultural approach termed the Stoddard-Neel system has been used for decades on large private ownerships maintained as bobwhite quail hunting preserves (Neel 2010; Way 2011). The Stoddard-Neel approach explicitly recognizes that no one attribute of the ecosystem should be favored at the expense of others. In longleaf pine ecosystems, these attributes include habitat for endemic and endangered species such as the red-cockaded woodpecker and the gopher tortoise, an aesthetically appealing and open park-like structure, species-rich native plant communities, and high quality timber, including heart-pine sawtimber and large poles.

Figure 12.3 *A longleaf pine stand that is subject to prescribed fire at one to three year intervals, illustrating the low tree densities, dominance of larger diameter trees, and the rich grass-dominated ground layer characteristic of these ecosystems. This stand at the Joseph E. Jones Ecological Research Center in southern Georgia, USA has also been harvested for timber using the Stoddard-Neel selection system (described in the text).*

Figure 12.4 *Longleaf pine regeneration is often concentrated in larger openings in the forest canopy, although it is also scattered throughout the forest matrix.*

Stoddard-Neel silviculture, while in concept an uneven-aged method based on single-tree selection, is in practice more akin to a variable-density thinning approach that results in spatially variable tree densities, openings of varied sizes, regeneration and release in these openings, and complex structural conditions, including canopy structure. One goal of the Stoddard-Neel approach is maintenance of a multi-cohort age structure at the stand scale, a characteristic of old-stage natural woodlands in this type. To accomplish this, tree selection criteria focus more on the trees left behind rather than those to be removed—i.e., a focus on legacy trees—so as to allow for development of large old trees mixed with patches of new regeneration. Trees with attributes of the sort timber-focused foresters might view as defects (trees with forks, damaged tops, basal fire scars, etc.) are often retained. Moreover, some trees with high commercial value are always retained if they are judged to have greater value for sustaining the long-term health of the ecosystem, such as providing wildlife habitat or a well-distributed source of pine needles for fueling fires (see next paragraph). Thus, selection of retained trees is based on ecological indicators, rather than some set of quantitative measures reflective of timber values, such as in the BDq approach (see Chapter 8). Moreover, some portion of dead and dying trees are explicitly retained during harvests because of their habitat value.

A key factor in selecting the longleaf pines that will be retained is how the trees contribute to the production and spatial distribution of needle fuels, which along with the bunchgrasses, are the primary fuel source for the frequent surface fires needed to maintain the longleaf pine ecosystem (Figure 12.5). Without these frequent fires, hardwoods can invade and gradually change the fundamental character of these pine ecosystems. If tree densities at the stand level are too low or if large canopy gaps develop, there may be inadequate needle fuels to carry surface fires, resulting in a consequent increase in hardwoods. Ultimately, some openings are needed to release longleaf pine regeneration. Hence, variation in tree density, from patches of high density to openings up to 0.10 to 0.25 ha in size, is a part of the desired stand structure.

The density variation has a temporal as well as a spatial component. Higher density patches of canopy longleaf pine (above 17m^2/ha) are needed to competitively inhibit hardwoods and to insure sufficiently high fire intensities to limit the survival of hardwood trees and shrubs. A reduction in density through harvesting (or small-scale natural disturbance) to a range of 7 to 17 m^2/ha allows for new seedling establishment but little height growth of grass-stage seedlings. Longleaf pine can remain in a grass stage that resembles a bunch grass and is highly resistant to fire for 10 to 15 years before height growth begins (Figure 12.6). A further reduction in tree density to below 7 m^2/ha is needed to initiate longleaf pine height growth. An important aspect of this dynamic is that larger openings with established and released longleaf pine regeneration need to be kept free of fire for several years—until seedlings are large enough in diameter and height to avoid mortality from fire. Eventually, however, these pines begin to contribute sufficient needle fuels to allow fire to reenter the former opening. This kind of temporal dynamic in tree mortality and the resultant variation in tree densities, needle fuels, and surface fires, is a characteristic outcome of lightning strikes, fire, bark beetles, and small-scale wind disturbance (Palik and Pederson 1996) and is exactly what the Stoddard-Neel approach to silviculture is designed to emulate.

Major efforts are currently underway to restore longleaf pine ecosystems throughout the southeastern US, including conversion of existing plantations of loblolly and slash pines to longleaf pine. One strategy involves gradual removal of the plantation overstory, using it as a source of fuels to allow periodic burning, with underplanting of longleaf pine seedlings (Kirkman et al. 2007; Kirkman and Jack 2017).

Figure 12.5 *Key to the longleaf pine ecosystem and the high fire frequency that it requires for its maintenance is the herbaceous understory dominated by bunchgrasses, such as wiregrass shown burning in this photograph. The combination of the wiregrass and the long needles of the pine make an excellent fuel for the frequent low severity fires. Photo credit: USFWS/Southeast, used with permission under the Creative Commons Attribution 2.0 Generic License. This image is the work of a US Fish and Wildlife Service employee, taken or made as part of that person's official duties. As a work of the US federal government, the image is in the public domain.*

Figure 12.6 *Longleaf pine spends an extended period (7 to 15 years) in a grass stage, during which it resembles a bunchgrass, grows little in height, and is highly resistant to fire.*

attributes of ponderosa pine and dry mixed-conifer forests, including disturbance, composition, and structure. We then review the history of forest management in these forests and contrast current timber-focused approaches with the natural model. The remainder of the chapter is focused on an ecologically based silvicultural system for ponderosa pine/dry mixed-conifer forests that includes principles, approaches, and an example of application.

Ponderosa Pine and Dry Mixed-Conifer Forests of Western North America

Recall from Chapter 6 that this forest type extends over most of the western intermountain region of North America from interior southern British Columbia, Canada south along the eastern slopes of the Cascade Range and the Rocky Mountains into northern Mexico (see Figure 6.10).

Ponderosa pine is the most common tree species, ranging from nearly pure forests on hot-dry sites to greater mixtures of ponderosa pine, Douglas-fir, and western larch on cooler-moister sites. Ponderosa pine and western larch are classified as shade intolerant, while Douglas-fir is considered moderately shade tolerant. Lodgepole pine can be a significant component on cold and dry sites and on the tephra deposition zone in central Oregon. Other species can include western white pine, sugar pine, Jeffrey pine, and incense-cedar.

The natural disturbance regime is characterized by surface fires every 5 to around 20 years (Heyerdahl et al. 2008; Reynolds et al. 2013; Addington et al. 2018). Fire return intervals fluctuated within this range based on variation in climate patterns (Heyerdahl et al. 2002). The structure of ponderosa pine and dry mixed-conifer forests with active fire regimes have a number of key characteristics. The forests are dominated by large, old trees that are viewed as the structural backbone of these ecosystems and that generally make up the majority of the basal area (Franklin et al. 2013, 2018; Hagmann et al. 2013). Old trees generally live 300–500 years, but can live up to 800–900 years (Figure 12.7). However, a full range of ages and tree diameters are generally present (Churchill et al. 2017). Tree populations are regulated by low levels of annual mortality across all age and size cohorts and relatively regular regeneration, which often occurs in dense thickets.

Recall that ponderosa pine/dry mixed-conifer forests do not generally progress through the disturbance/developmental stages as neatly as can the other archetypes, at least at the stand-scale (Chapter 5). Instead, most of the structural conditions and processes associated with the developmental stages are present within a fine-grained mosaic of 0.004 to 0.4+ ha patches characterized by a spatial pattern of widely spaced individual trees, clumps of 2 to more than 20 trees, and openings. New cohorts of trees generally recruit in dense thickets within openings. Fire, in combination with bark beetles, wind, and pathogens, thin out these clumps over time until they end up as small and sometimes large clumps of old trees. Large, isolated, individual trees also occur. New openings develop when groups or individual old trees succumb to insect mortality, crown torching, or other disturbances.

Figure 12.7 *A restored western larch frequent fire forest in Central Washington, USA. The old larch are 300–600 years old and are arranged in a mosaic pattern of clumps, widely spaced individuals, and extensive openings. Scattered regeneration is also present.*

Evolution of Timber-Focused Silviculture in Ponderosa Pine and Dry Mixed- Conifer Forests

Forest management in the ponderosa pine and dry mixed-conifer forest types has undergone numerous changes since Euro-American settlement of the western US. What began as large scale exploitation of old trees to fuel settlement of the West evolved into formal even and uneven-age silvicultural systems. In more recent years, the focus on US federal and some other lands has shifted to fire risk reduction and restoration of fire resistant conditions. In this section, we review the development of silviculture for this forest type and, ultimately, contrast the timber-focused approach with the natural model.

Early Exploitation, Grazing, and Fire Suppression (1860s ~ 1930s)

Beginning with westward expansion in the mid-1880s, ponderosa pine was extensively harvested to supply building materials for towns, railroads, mines, and fruit boxes (Graham and Jain 2005a). Flat terrain and easy access to millions of board feet (a board foot is a common measure of wood volume in the US; 1 board foot = 0.0024 m^3) of high quality, old trees was a major driver of rapid development in the West (Fiedler and Arno 2015). Early harvests were dominated by clearcutting of large tracks of land that accelerated with railroad logging in the early 1900s. With improved roads and the advent of truck transportation systems in the 1920s, selective harvesting of the larger, high quality trees became more prevalent.

Grazing, fire suppression, and cessation of Native American burning during this period combined to halt frequent fire in these forests (Hessburg and Agee 2003). Millions of sheep and cattle grazed the lush grasses of these forests in the late nineteenth and early twentieth century, which reduced fine fuel loads and the ability of surface fires to spread. Intensive efforts to suppress fire began after the massive 1910 fires in the states of Idaho and Montana, with fire suppression becoming the major objective of the US Forest Service. The change in fire regime coincided with a cool, wet climatic period that favored tree regeneration. The result was the beginning of densification of these forests and a shift to fire intolerant species on mixed conifer sites (Hessburg and Agee 2003).

Development of Silvicultural Systems for Ponderosa Pine (1940s to 1960s)

Formal uneven-aged or selection management began in response to major outbreaks of the western pine beetle and the associated mortality of high value mature and old ponderosa pine in the 1920s and 30s. These sanitation harvests utilized bark beetle mortality risk rating systems based on tree vigor in the 1930s (Dunning 1928; Keen 1943) (Figure 12.8). More formal uneven-aged systems based on these approaches eventually evolved that attempted to emulate the natural model to some degree (Meyer 1934; Pearson 1950). The Unit Area Control system developed for the Sierra mixed conifer forests provides a good example (Hallin 1954). Yet these silvicultural approaches steadily reduced old tree populations by removing low vigor and **overmature** trees in favor of younger, faster growing trees. Also, prescribed fire was not part of these management approaches, even though early foresters understood the critical role of fire in maintaining these ecosystems (Show and Kotok 1924; Weaver 1943; Cooper 1960).

Various forms of uneven-aged silviculture were widely applied to ponderosa and mixed-conifer forests on frequent-fire sites. These included everything from single-tree selection based on risk rating systems to much heavier harvests that left significant reserve stands. The heavier harvests were common in locations where the USDA Forest Service sought to support establishment of large mills (Langston 1995).

A major shift toward even-aged approaches began following World War II, with an increased emphasis on higher levels of wood production and greater management efficiencies (Youngblood 2005). Clearcutting, seed tree, and shelterwood methods were all adopted and used to different degrees across the west. Concerns about dwarf mistletoe also encouraged adoption of even-aged approaches. Overstory removal harvests were often employed on sites with significant development of advanced regeneration or younger tree cohorts. On moister sites that would support mixed conifer forests, faster growing, fire intolerant species often dominated advance regeneration, leading to major species conversion. While natural regeneration was the norm during this period, early efforts at planting began.

Figure 12.8 *Photo of sanitation cutting during a western pine beetle outbreak in a ponderosa pine forest in eastern Washington, USA. Harold Weaver, a leading early forest scientist who developed silvicultural approaches for ponderosa pine forests, is shown in the photo.*

Maturation of Timber-Focused Silviculture (1970s ~ Present Day)

Research on intensification of management in frequent-fire landscapes greatly accelerated in the last 50 years, along with its application to the management of ponderosa pine and mixed-conifer forests. Emphases included density management, growth and yield modeling, natural and artificial regeneration, site preparation, vegetation control, and harvesting methods. Synthesis publications on silvicultural systems were produced for every region in the western US by the mid-1980s (see Youngblood 2005).

Intensive, even-age plantation management approaches are currently the dominant management system on forestlands where timber production is the primary objective. Full fire suppression is a major element of this model, as prescribed fire is viewed as being an unacceptable risk to timber values. However, low to moderate productivity ponderosa pine and mixed-conifer forests generally are no longer owned and managed by organizations that seek high levels of return on capital, because of low returns and high fire risk. In addition, some of the private and Native American landowners still actively engaged in timber production do utilize uneven-aged approaches, such as Collins Pine, a family-owned timber company with holdings in Klamath Falls, Oregon, USA.

Many public, Native American, and private owners of ponderosa pine and dry mixed- conifer forests have shifted management of these forests toward such objectives as reducing the potential for high severity fires and restoration of other environmental values, as will be described later in this chapter.

Key Contrasts between the Natural and Timber-Focused Models

Perhaps more than in any of the 4 archetypes, management emphasizing timber production has created forests that are far out of alignment with the natural model for this forest. It has become abundantly clear after over a century of management that suppressing fire and growing dense stands to maximize timber production increases the risks of high severity fire and drought-related insect mortality. The timber-focused silvicultural model seeks to exclude fire and fully utilize growing space using an even-aged regeneration system. This approach is very different from the natural model in which periodic surface fire maintains a low density, multi-aged forest structure that rarely experiences a high-severity fire (Table 12.2).

The Shift to Fire Risk Reduction, Restoration, and Resilience

Widespread recognition of the dramatic changes in forest structure and composition due to fire suppression and past management emerged in the late 1980s and 1990s (Agee 1993; Covington and Moore 1994). The resulting increases in risk of high severity fire, coupled with expanding development in forested areas across the rural west, precipitated a societal shift in management objectives for these forests, particularly on public and Native American lands. Large fire years in the early 2000s created further pressure to address fire risk at a large scale (Finney et al. 2003; Westerling et al. 2006). As a result, reduction in fuels and high-severity fire risk supplanted timber production as a primary objective on public lands (HFRA 2003; Youngblood 2005). Outside of the Wildland-Urban Interface, this has further shifted from a focus on fuels and fire behavior to restoration of ecosystem functions in frequent-fire forests, including enhancing fire resistance.

Intensive timber production employing even-age management remains dominant on some private lands, particularly in the most productive areas of western pine forests, such as portions of the Sierra Nevada in California, USA and the northern Rockies. Objectives and silvicultural systems on state, Native American, and nonindustrial private ownerships vary considerably, with some focused on timber production, while others are developing hybrid approaches that integrate fire and drought resistance, with production of wood and generation of revenue.

Fuel Reduction

In order to meet the need to reduce fire risk, silvicultural guidelines for stand-level fuel reduction treatments emerged in the 1990's (Graham et al. 1999). These guidelines were mostly developed by fire ecologists, not silviculturists (Agee and Skinner 2005; Peterson et al. 2005; Johnson et al. 2007). Fuel reduction treatments are conceptually based on restoring the historical structure, composition, and fire regimes of these forests (Agee and Skinner 2005). In practice, however, the primary focus is on reducing fire using

Table 12.2 *Differences in outcomes between the natural model and a timber-focused silvicultural model for ponderosa pine and dry mixed conifer forests.*

Feature	Natural Model	Timber-Focused (even-aged) Model	Functional Consequences
Old trees	Dominate forest structure in terms of basal area and volume	Very few to no old trees	Lower resistance to crown fire and drought; few unique crown structures; lack of large snags and logs
Age and cohort structure	Multi-aged with 3+ age classes spanning 1–500 years	Single age class	Reduced genetic diversity; simplified structure; reduced management options
Canopy density	Low to moderate, 15–30% of maximum density	Moderate to high: 30–60% of maximum density	Higher susceptibility to crown fire, drought, bark beetles, and defoliating insects; reduced abundance and diversity of understory vegetation
Spatial pattern of tree distribution	Heterogeneous spatial patterns; mosaic of individual trees, clumps, and openings and associated developmental processes	Homogeneous spatial patterns; uniform spacing of trees; few or no openings	More uniform fire and insect disturbances; lack of openings results in reduced diversity in microclimate and biota, opportunities for regeneration, and accumulation and retention of snow
Species composition	Dominated by more fire-resistant species (ponderosa pine, western larch, and/or Douglas-fir); hardwood trees common in local areas	Dominated by single species	Reduced biotic diversity; greater vulnerability to insects and pathogens, especially in plantations of fire intolerant species such as grand or white fir
Surface fuels	Low fuel loadings and understory dominated by grasses and other herbaceous species	Moderate to high levels of fuel loadings	Higher potential fire intensity; greater risk of crown torching

thinning prescriptions comparable to those used in timber-focused thinning (Figure 12.9a, b; Color Plate 12.1a, b). These generally involve removing a majority of small and medium-sized trees to create a more open stand of larger, uniformly spaced trees. These treatments are intended to reduce potential for surface fires to grow into crown fires by removing ladder fuels (understory and mid-story tree layers) and reducing the potential for active crown fire spread by wide (9 m or more) tree spacing. The species preference is for fire-tolerant species, such as pine and larch. Prescribed fire or other mechanical methods are utilized to treat surface fuels (i.e. logging slash, downed wood, leaf litter, and tall shrubs) following the thinning.

Figure 12.9 *a) Photo of a commercial, fuel reduction treatment in a Douglas-fir and western larch forest in northeast Washington, USA. Overstory trees are evenly spaced at wide spacing (14 m), all understory and midstory trees were removed by a hand crew (ladder fuel treatment), and most surface fuels (logging slash) were hand piled and burned. b) Photo of a commercial, dry forest restoration treatment in a Douglas-fir and ponderosa pine forest in central Washington, USA. Overstory trees were left in a pattern of widely spaced individual trees, clumps, and openings. Ladder fuels and surface fuels were treated with a prescribed fire after the commercial harvest.*

Fuel treatments of this type have been conducted on millions of hectares across the interior west on federal, state, tribal, and private lands (Schoennagel and Nelson 2011). They have proven to be effective at shifting fire behavior from crown to low severity fires, even during extreme weather events (Fulé et al. 2012; Safford et al. 2012). Moreover, use of prescribed fire has slowly gained acceptance on some federal, state, and tribal land across western North America. However, smoke and safety concerns still limit its use. Unlike the pine forests of the southeastern US (Box 12.1), prescribed fire is relatively rare on private lands in the western US.

The focus on changing fire behavior and risk has its tradeoffs, however, as with any management approach driven by a single objective. The uniform spacing-based fuel reduction treatments simplify stands and create essentially widely space plantations. This eliminates habitat for wildlife species that require dense patches for cover (North et al. 2009). Spacing, combined with traditional concepts of tree vigor, often leads managers to thin clumps of old trees, which further reduces old tree populations. These plantation-like treatments also have aesthetic impacts for many stakeholders, who prefer more natural looking forests. From an ecological silviculture perspective, these fuel reduction approaches do not sufficiently incorporate the principle of complexity.

Restoration, Resilience, and Adaptive Capacity

Management objectives on federal and many other western forest lands in North America that were historically subject to frequent fire regimes currently have expanded to include broad-scale restoration of ecosystem functions and processes (Allen et al. 2002; Brown et al. 2004; North et al. 2009; Franklin and Johnson 2012). Activities are no longer just about reducing fuels and the potential for high-severity fire. Also, notably, in frequent-fire forests the objective is explicitly to restore resistance to disturbances, as well as resilience and adaptive capacity to a changing climate. Several important elements in these programs to restoring frequent-fire forests include: 1) retention of existing old trees and restoration of old tree populations; 2) restoration of hardwood species (e.g., quaking aspen) and communities in frequent-fire landscapes; and 3) restoring fine-scale, multiage, structural mosaics within the forest (Jain et al. 2012; Reynolds et al. 2013; Stine et al. 2014; Addington et al. 2018).

A fundamental goal of restoration is to shift forests from the high severity *boom and bust* pathway that they are currently in to a condition in which low-severity frequent fire will sustain a stable forest ecosystem (Figure 12.10). Use of historical reference conditions as a baseline for management targets is an important part of

Figure 12.10 *Schematic of forest development with and without frequent fire. F1 is a frequent, low severity fire pathway. F2 is a short rotation, high severity pathway. S1 shows historical conditions prior to fire exclusion. S2 shows typical current conditions after a century of fire suppression. If not restored, S2 is likely to transition to S3 and then onto an F2 pathway. With restoration (S2 to S4), a stand can be shifted onto the F1 pathway. Original figure concept from Andrew Larson.*

this approach (Churchill et al. 2017), but there is also the need to incorporate requirements associated with functional objectives, such as wildlife habitat for threatened species, as well as adjustments necessary as a result of climate change.

A broad consensus now exists that at the landscape level it is necessary to move from limited strategic fuel treatments to large-scale efforts to restore resistance and resilience at the scale of the entire landscape (Reynolds et al. 2013; Hessburg et al. 2015). These large scale efforts focus on realigning stand-to-landscape level structure, composition, and pattern with natural disturbance regimes, underlying biophysical conditions, and projected climatic changes.

It is also increasingly recognized that wildfire suppression policies need to allow more wildfires to burn, as mechanical treatments and prescribed fire will not be sufficient to restore and maintain western landscapes by themselves (North et al. 2015). Managed wildfire, or resource benefit fire, has been extensively utilized in remote designated wilderness and national park areas in the Rockies and Sierra Nevadas of the US for the last 30 to 40 years (Larson et al. 2013). It is slowly gaining greater acceptance and use on US national forests and tribal forests.

Restoring resilience has now become the primary objective on federal land in the western US, along with many other ownerships. Resilience is the broader concept defined as the ability of a system to persist through and recover from disturbance while maintaining its structure and function (Walker et al. 2004). In the case of frequent-fire landscapes, the goal includes resistance, which is the ability of an ecosystem to experience a disturbance without being significantly altered—which the natural frequent-fire forest was almost always capable of doing in the era of frequent low-severity fires.

However, increasing awareness of warming climate trends and corresponding increases in drought stress, insect outbreaks, and wildfires are prompting greater inclusion of adaptive capacity into the concept of resilience (Millar et al. 2007; Stephens et al. 2010; Schoennagel et al. 2017). Adaptive capacity includes the ability of a landscape to shift to more drought and fire tolerant species compositions, fuel structures, and landscape patterns that are aligned with future climate and fire regimes. This will include shifts to different forest types or non-forest vegetation in some places. A resilient landscape is also resistant to large-scale, high severity fires and drought induced mortality that can lead to rapid, destabilizing shifts in ecosystem types that make adaptation much more challenging.

An Ecological Silvicultural System for Ponderosa Pine and Dry Mixed Conifer Forest

Beyond Restoration: The Need for a Long-Term Silvicultural System

A silvicultural system based on a natural model is presented in this section that combines current restoration practices with a long term management strategy based on multi-age management. While current restoration practices are firmly rooted in ecological principles, restoration has not been framed as a formal silvicultural system in frequent fire forests (Sarr et al. 2004). Long-term management pathways and economic considerations for restoration also are not well developed. What comes next after completion of restoration of a frequent-fire forest is a question that has been raised frequently by managers and stakeholders.

Maintaining restored forests over time with periodic prescribed fire or managed wildfire certainly follows the natural model. Restoring a frequent fire regime will maintain fire resistant, open canopy conditions and drive self-sustaining forest developmental processes over time. A fire-only approach to maintenance is a logical option in places that lack economic goals, have challenging road access, or have productivity that is too low to support periodic forest harvesting.

Adoption of a fire-only management approach is not feasible on a large portion of the frequent-fire forest land in the western US. First, the amount of burning necessary would overwhelm adjacent communities and recreationists with smoke. Additionally, low risk burning windows are not long enough to burn the necessary number of hectares. Second, there will continue to be social demands for some level of timber production from forestland, including federal land, in order to help sustain local wood processing capacity, harvesting infrastructure, and rural economies (Franklin et al. 2018). Third, revenue will likely be needed to help defray the ongoing cost of stewarding US national forests as demands on federal budgets increase in coming decades. On most state, tribal, private, and other non-federal lands, revenue generation is a significant objective and need.

To address this reality, we lay out a multi-age silvicultural system that sustains a high level of resistance to high-severity fire, insects, and drought while also generating wood and revenue in perpetuity. The system includes restoration treatments for forests that have uncharacteristic density, composition, and fire risk and then adds a multi-age treatment for long term management of restored forests. This approach maintains the basic structure and function of the natural model, but trades some aspects of resistance and other ecological functions for timber production. This system provides significant flexibility for landowners to balance revenue needs with site-specific fire probabilities and the level of fire and drought risk they deem acceptable, in addition to other objectives.

Key Elements of an Ecological Silvicultural System

Key elements of an ecological silvicultural system are identified in Table 12.3. These are common to all treatment approaches utilized for forests in different structural conditions. Several of these elements are unique to frequent fire forests and some, in fact, are the exact opposite of approaches applicable to other archetypes. Collectively, they distinguish ecological silviculture in frequent fire forests from the other 3 archetypes. They are presented in a rough order of priority. We want to emphasize that not all elements have to be implemented to use this silvicultural system. Forest managers can develop a customized approach to their forest conditions and objectives by incorporating elements to different degrees.

1. Manage for tree species that are resistant to fire and drought. This is a foundational element in managing frequent fire forests. Many fire intolerant species grow faster, but are far less resistant to fire (and other disturbance agents). The use of the less tolerant species will greatly increase risk. While the fire-tolerant species are typically conifers (pines, larch, and Douglas-fir), species of oak are significant and fire-resistant components of many frequent-fire forest ecosystems.
2. Reduce and maintain surface (ground) fuels at low levels. High surface fuel loads increase flame lengths and soil heating, which can lead to excessive mortality of overstory trees. While downed wood is an important structural element of the frequent fire forest, large accumulations of downed wood are not desirable, nor are large patches of tall shrubs, unlike in the other archetypes. Leaving significant levels of logging slash to decompose in the woods should be avoided. Prescribed fire is the most effective tool to reduce surface fuels. Mechanical methods can be used to treat surface and ladder fuels where fire is not possible, although the effects on understory plant communities and soil processes will not be the same as with fire. Over time, managing for appropriate types and levels of fine fuels to carry low intensity surface fires is important. Small patches with accumulations of downed wood or tall shrubs can be retained and provide important habitat for some wildlife species.
3. Reduce ladder fuels and increase crown base height. Unlike the old growth stages of several of the other archetypes, multi-layered, vertically continuous

Table 12.3 *Key elements of ecological silvicultural systems for frequent fire forests and corresponding ecological effects. These elements apply to all structural conditions and prescriptions. However, not all guidelines have to be applied in all cases—foresters can incorporate elements to different degrees to create a customized approach for their particular situation.*

Element	Objective/Consequence
1. Manage for fire and drought tolerant species	Low levels of tree mortality when fires or droughts occur
2. Reduce surface fuels and maintain them at appropriate levels over time with prescribed fire or mechanical methods	Reduced flame lengths; avoid excessive soil heating and torching of individual trees; sustain spread of low intensity surface fire
3. Reduce ladder fuels and increase crown base height and maintain over time with prescribed fire or mechanical methods	Reduced potential for surface fires to transition to crown fires
4. Reduce and maintain low tree densities and canopy cover (i.e. low crown bulk density)	Create low potential for crown fire spread in the upper canopy; maintain tree vigor by reducing competition for water and nutrients
5. Promote large trees of fire-resistant species	Create greater resistance to fire with thicker bark and higher crown base heights; maintain overstory structure
6. Cultivate and maintain old trees	Same as large trees plus maintenance of genetic diversity and habitat functions
7. Restore and maintain a mosaic spatial pattern of patches of differing age and structural conditions	Reset fine- scale, gap-phase stand development processes; provide for habitat and other ecological functions associated with openings, clumps, and individual trees
8. Restore understory plant communities	Provide forage for wildlife species; suppression of excessive conifer regeneration; enhance nitrogen fixation; maintain fuel beds to sustain spread of low intensity surface fire
9. Promote broadleaf species in appropriate locations by removing competing conifers	Restore meadow habitats and stream flows that provide important habitat functions

canopies are not a desirable structural component in frequent fire forests. Both understory and midstory trees, as well as lower branches on overstory trees, can act as fuel ladders that facilitate transition of a surface fire into a crown fire. Minimizing vertically continuous canopies and increasing crown base heights (height to live crowns) in this way is important. Frequent fire forests generally contain multiple canopy layers or cohorts, but regeneration and mid-sized trees are rarely found growing underneath overstory trees. Prescribed fire and/or mechanical thinning treatments can be used to initially reduce and then prevent new development of fuel ladders over time. Some small patches with multiple canopy layers are ok, but these should be surrounded by open space to prevent crown fire spread if they do torch during a fire (Figure 12.11).

4. Maintain low tree densities. Managing for overall densities well below the carrying capacity of the site is necessary to keep risks of high severity fire and drought mortality low. This means maintaining stocking levels that are significantly lower than traditional density management zones. In general, we recommend 15–35% of maximum stocking across a stand, which roughly translates to 10–30% average canopy cover. Higher average stocking is possible, but fire and insect mortality risks will increase. Allowing stands to approach the zone of mortality in order to maximize growth is a high risk proposition. Openings should be common in a variety of shapes and sizes (Matonis and Binkley 2018). Conversely, dense patches and clumps within stands can have significantly higher density levels and experience insect or competitive mortality without threatening the whole stand.

5. Promote large trees of fire-resistant species. Larger trees are more fire resistant due to their thicker bark and greater separation of the canopy from the ground (i.e. higher crown base heights or height to live crown). Large trees provide for many other functions including seed sources, habitat, carbon storage, production of fine fuels, etc. The definition of *large* varies by region, but is generally greater than 50 cm in diameter in ponderosa pine and mixed conifer forests. In stands with few large trees, thinning from below and selecting trees with healthy crowns is recommended to accelerate the growth and development of large trees. Large trees generally comprise the majority of the basal area in natural frequent fire forests, but this does not have to be the case when timber production is a major objective.

Figure 12.11 *a) Effects of a prescribed fire used to treat excessive understory trees (ladder fuels) and surface fuels (logging slash and accumulated woody material) after a commercial restoration treatment in central Washington, USA. b) A prescribed fire conducted after a non-commercial small tree thin. Note the scorching of the lower crowns of the residual trees that will raise the crown base height, further reducing the ability of future fires to transition from a ground fire into a crown fire.*

6. Cultivate and maintain old trees. Old trees are the key to maintaining the continuity of forest structure and function over time. Most old trees are large, but some small old trees are often present in clumps. Old trees provide an expanded set of ecological functions over young, large trees. They develop even thicker bark, have complex crown architecture, and provide genetic diversity and higher habitat value. 150 years is often used as an approximate threshold for old trees (Van Pelt 2008). Retention during harvests and removal of most young trees within 2–3 crown widths are key strategies to sustain old trees, although some may need to be harvested to meet financial objectives. We reiterate that they should be primarily of fire resistant species.

Figure 12.12 *A treatment in a ponderosa pine forest in central Oregon, USA with openings and large clumps. Openings increase snow retention, alter fire and insect behavior, provide important foraging habitat for many wildlife species, and allow new cohorts to establish (see center of photo). Large clumps provide cover and nesting sites for many wildlife species. Note that a planned prescribed fire to treat surface fuels has not yet been implemented in the photo.*

7. Restore and maintain a multi-age, mosaic spatial pattern. The long term goal is to reset the fine scale, gap phase development of natural forest development by restoring characteristic patterns of individual trees, clumps, and openings (Figure 12.12). A wide range of age cohorts can be present across a stand. Clumps can consist of single age cohorts or can develop several age classes within them over time. Clumps with vertically continuous canopies should be rare, however. Skips, which are dense patches significantly bigger than the largest clumps, can be included to meet specific functional needs, but are not essential (Figure 12.13). Ensuring that openings are present in a variety of shapes and sizes is critical for many functions.
8. Restore and monitor understory plant communities. Herbaceous plants (mostly grasses) and shrubs play a key role in carrying low intensity surface fires. Surface fires in longleaf pine forests, for example, are dependent on wiregrass and the needles suspended in the leaves, allowing them to dry out quickly and carry fire. The understory is also the basis of the food web for most of the biota. Certain understory species play a key role in fixing and replenishing soil nitrogen lost during fires (Newland and DeLuca 2000). A healthy understory can also suppress tree regeneration, thus reducing the growth of ladder fuels. In many cases, restoration treatments focused on trees will also restore desired

Figure 12.13 *a) Provision for some small patches of dense trees within restored frequent fire forest ecosystems in order to provide habitat for species and processes that require a more heavily shaded microclimate. Such a patch is illustrated here, these are often described as skips in restoration projects—i.e., small portions of forests left untreated. b) Northern goshawk nest in dense patch in northeastern Oregon. Photo credit: Trent Seager.*

understory communities by opening the canopy in a heterogeneous manner. However, it is important to work with botanists or other knowledgeable people to understand what species and conditions are desirable and take appropriate actions as needed. These may include planting desired species, mechanically treating excessive shrubs, treating invasive species with herbicides, etc.

9. Promote broadleaf species in appropriate locations. Quaking aspen, cottonwoods, oaks, and other sprouting hardwood species provide important habitat and other functions. Removing conifers to promote these species is a common strategy in appropriate locations, such as riparian areas and wet meadows (Figure 12.14). Some conifers, especially old trees, can be left in patches of broadleaf species. In some frequent fire systems, oak species may make up a major portion of the overstory. In other systems, however, hardwoods can quickly take over a site and therefore must be managed carefully. This can occur, for example, in longleaf pine ecosystems

Figure 12.14 *Aspen and meadow restoration treatment in central Oregon. Removal of conifers that had encroached on this wet meadow allowed aspen to quickly reestablish. Stream flows and cover of meadow grasses also increased after the treatment. The site is now a habitat hotspot providing high quality forage and nesting habitat. Photo credit: Jerry Franklin.*

Silviculture by Forest Structural Condition

Current conditions in ponderosa pine and mixed-conifer forests vary widely based on management history, species, productivity, and other ecological factors. Tying silvicultural treatments to developmental stages is challenging for frequent-fire forests as they do not progress through developmental stages across whole stands. Hence, our ecological silvicultural system for frequent-fire forests is organized around common structural conditions. Foresters implementing this system will often start with stands that highly depart from the natural model where the first order of business is to reduce risks of mortality from fire and drought and shift composition and structure into a condition that can be managed more sustainably over the long term. In forests that are closer to the natural model, multi-age treatments that mimic the natural disturbance regime are utilized. In addition, high severity fires can occur in this forest type and such forests may require a different set of silvicultural treatments to move them toward the natural model.

Six common structural conditions that cover the spectrum of current conditions in frequent fire forests and their associated silvicultural treatments are described and illustrated below (Table 12.4) (Figure 12.15). While these six conditions certainly do not capture every condition that exists, an ecological forester can use them to tailor treatments to their specific situation. In many cases, stand condition may be a mix of two conditions. Remember that the key elements described in the preceding section (Table 12.3) apply to treatments for all of the conditions below.

1. Dense, Even-Aged Structural Condition

This condition is generally a product of clearcut harvests and subsequent development of a dense even-age forest either from natural regeneration or a mix of planting and natural regeneration. This type includes young blackbark ponderosa pine stands that were intensively logged in the early part of the twentieth century (Figure 12.16a), as well as more recent plantations. This condition can also originate from a high severity fire followed by dense natural regeneration. Low numbers of old or large legacy trees may be present (e.g. 5 to 12 trees per hectare), but are often completely absent. Some understory trees may exist as well. High density makes this type susceptible to drought stress, bark beetle attacks, and crown fires.

Prescriptions are primarily designed to reduce stand density, retain the larger trees, shift composition to fire tolerant species, and create a variable spatial pattern of tree density (Figure 12.16b). The long-term goal is to shift stands from a simplified, even-age structure to a multi-aged, spatially variable condition with an old tree component. In

Table 12.4 *An ecological silvicultural system for ponderosa pine and dry mixed conifer forests based on forest structural condition.*

Current Structural Condition	Description	Primary Objectives	Potential Silvicultural Activities
1. Dense, even-aged with sufficient fire tolerant component	Single cohort plantations or naturally regenerated forest; young to mature (15 ~ 120 years); may have a sparse lower cohort; density is above desired levels for fire resistance; composition is dominated by fire resistant species	Maintain or accelerate growth of trees; shift composition to fire tolerant species; create variable spatial (*ICO*) pattern; reinitiate clumped regeneration in openings; reduce surface fuels; restore understory plant communities	-Variable density thinning (VDT); commercial with larger diameter trees; noncommercial in units with small diameters -Single or multiple VDT entries to get to desired density levels -Prescribed fire or mechanical fuels treatment to reduce slash, thin excessive understory trees, and restore understory
2. Dense, multi-cohort with sufficient fire tolerant component	Multilayered forest; often dominated by high density of fire intolerant species, but sufficient density of desired species present; generally the result of natural regeneration	Shift composition to fire tolerant species; greatly reduce understory and mid-story layers (fuel ladders); maintain vigor of leave trees, especially old trees; reduce surface fuels; restore understory plant communities	-Non-commercial thinning of under/mid-story; retain clumps of fire tolerant species in openings -Species based, commercial thinning favoring fire tolerant species if needed -Prescribed fire or mechanical fuels treatment to reduce slash and other surface fuels, and restore understory
3. Dense, two-cohort with sufficient fire tolerant component	Overstory cohort of fire tolerant species with low to moderate density; dense under/mid-story layer; often the product of thinning, shelterwood, or variable retention harvest, but can result from high or mixed severity fire	Manage lower cohort to reduce density and enhance vigor of overstory; shift to fire tolerant species, reduce surface fuels; restore understory plant communities	-Noncommercial thinning of lower cohort -Prescribed fire or mechanical fuels treatment to reduce slash other surface fuels, and restore understory -Prescribed fire alone may be used when lower cohort is small

Current Structural Condition	Description	Primary Objectives	Potential Silvicultural Activities
4. Dense, lacking in desired species or forest health conditions	Overstory lacks sufficient density of fire tolerant species or major forest health issues are present; generally, multilayered forest dominated by fire intolerant species, but can be single cohort and/or dominated by fire tolerant species with major forest health issues	Retain existing healthy, fire tolerant trees; regenerate new cohort of fire tolerant species; reduce surface fuels; restore understory plant communities	-Variable retention harvest focused on retaining healthy, fire tolerant trees -Non-commercial removal of under/mid-story layers; retain fire tolerant species in openings -Prescribed fire or mechanical fuels treatment to reduce slash and other surface fuels, and restore understory -Plant fire tolerant species; rely on natural regeneration; or combination of both
5. Open, multi-cohort	Open canopy forest with range of tree sizes arranged in ICO spatial pattern; composition is primarily fire tolerant species; dominated by large and old trees, but this may vary depending on management history and objectives	Maintain and perpetuate open canopy, multi-cohort structure, and ICO spatial pattern; retain and/or promote development of large and old trees of fire tolerant species	-Multiage treatment every 10–30 years if landowner has economic objectives; follow with prescribed fire or mechanical fuels treatment -If landowner does not have economic objectives, can manage only with prescribed fire every 10–20 years, or mechanical fuels treatments
6. Post-high severity fire	70–100 percent of trees dead and still standing; shrub and herbaceous layer is generally abundant; natural regeneration is variable; much of the organic soil layer burned off, soils may be hydrophobic	Ensure establishment of new cohort of fire tolerant species; protect soils; prevent excessive accumulation of dead trees on the ground over time that can increase severity of re-burns	-Salvage harvest that retains live trees and a portion of dead trees depending on economic and ecological need -Plant fire tolerant species; rely on natural regeneration, or combination of both; consider stock from climate adapted seed zones

stands dominated by fire resistant species, an ICO (Individual Trees, Clumps, and Openings) approach to variable density thinning can often work well (see Chapter 8, Box 8.5). Species-based prescriptions can be more efficient in multi-species stands where simply retaining the fire-resistant species will achieve the desired density levels and spatial variability goals.

For units where commercial thinning is possible, average density targets for these treatments generally range from 50 to 150 trees per hectare or 9–18 m^2/ha of basal area (20–35 percent of max SDI or stocking), depending on several factors. Targets may be lower on very harsh sites or where there are lower levels of healthy trees of the desired species. In young plantations that are not of commercial size, a noncommercial variable density thinning can be used to reduce density all the way to target density. Alternatively, a two or three entry thinning approach can be used for both noncommercial and commercial thinning. This will reduce stand density and shift forest development toward a multi-age structure more gradually and produce revenue periodically over time.

2. Dense, Multi-Cohort Structural Condition with Acceptable Levels of Fire Tolerant Species

These forests result from infilling of young trees after fire exclusion combined with logging of some portion of the large and old fire resistant trees. These forests are most common on mixed-conifer sites where shade-tolerant species establish and grow rapidly in the understory following cessation of frequent fire. This condition also can occur on productive, pure ponderosa pine sites. The higher the productivity, the more rapidly trees fill in and grow into the midstory and ultimately the overstory. Some of the infill trees may be large and close in height to the legacy old trees. Crown fire risk in these stands is typically very high.

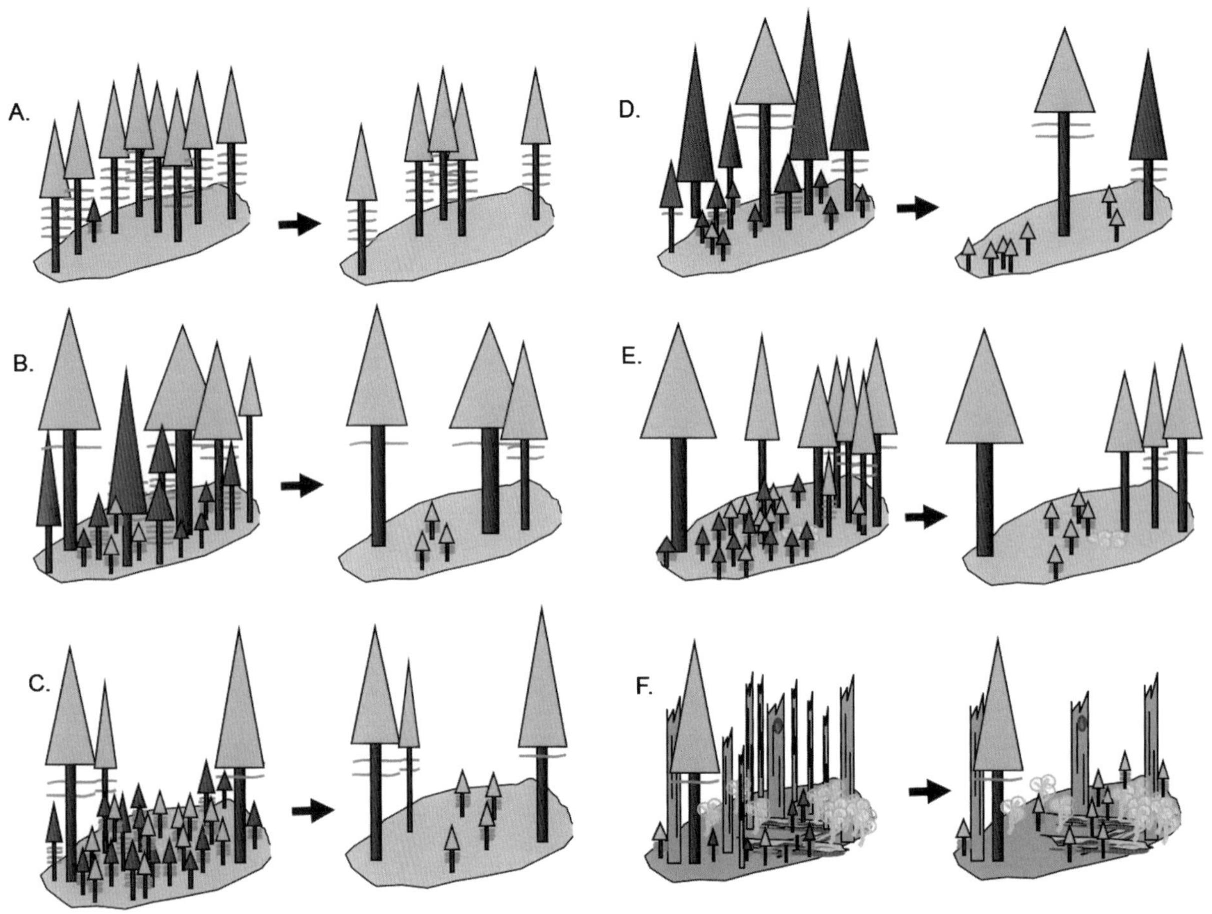

Figure 12.15 *Schematic of six common structural conditions and post-treatment conditions. Light gray trees are fire tolerant species and dark gray are fire intolerant. Panel A displays a dense, even-aged forest treated with a variable density thin. Panel B displays a dense, multi-cohort forest treated with a species-based thinning that removed fire intolerant species and retained most fire tolerant species. Panel C displays a dense, two-cohort forest where a small tree thin was used to reduce the density of the lower cohort and remove fire intolerant species. In panel D, a dense forest dominated by fire intolerant species is treated with a variable retention harvest that retains the healthy, fire tolerant trees and is subsequently planted with desired species. Panel E displays an open canopy, multi-cohort forest with fire tolerant species; a multi-age treatment thins throughout the diameter distribution and creates an opening to release low cohorts. Panel F displays a forest after a high severity fire where a salvage treatment removes a large portion of the small and medium sized small trees while retaining larger snags for future habitat; the forest is then planted with desired species.*

Figure 12.16 *a) Pre-treatment conditions in a dense, even-aged, naturally regenerated ponderosa pine forest in central Oregon, USA that originated after clearcutting in the early 1900s. b) Post-treatment photo at the same location as 12.16a. A commercial variable density thinning treatment was conducted that followed the principles in Table 12.3 and used an ICO prescription. Photo credit: Craig Bienz.*

Prescriptions in this type have two primary objectives. First is to decrease fire risk and drought stress for the existing large and old fire tolerant trees. This is accomplished by removing understory and midstory trees from within a distance of 1 to 2x the dripline (e.g., 6 to 12 m) around of large and old trees, thereby reducing ladder fuels and moisture competition. The second objective is to restore the desired species composition, density, and spatial arrangement of lower cohorts by thinning the small trees in the remainder of the stand. In stands that have a good range of tree sizes of fire-tolerant species, conditions resembling the natural model can be restored relatively quickly.

On mixed-conifer sites, most or all trees of fire-resistant species often need to be left to meet density targets. These prescriptions can often employ a very simple, species-based approach that can result in a good mix of individual trees, clumps, and openings. An example would be to remove all white fir less than 90 cm diameter and leave all ponderosa pine. Some fire-intolerant species may be retained depending on the site and management objectives.

In cases where densities of fire-tolerant species are high (e.g., on pure ponderosa pine sites), variable density thinning can be used. Prescriptions should first focus on retaining and releasing the existing larger and old trees, and then thinning younger cohorts to achieve desired density levels and a variable spatial pattern with multiple cohorts. A portion of the larger trees may be removed to meet economic objectives if necessary.

In units where a commercial sale is possible, the merchantable trees are typically removed, which is then followed up with a noncommercial treatment of the smaller diameter classes. These small-tree thinnings are often called ladder fuel treatments, as they focus on thinning the understory and midstory cohorts, which are ladder fuels for fires. Where markets exist for small diameter material, it may be possible to do the entire treatment commercially. In cases where the trees that need to be removed are not large enough or do not provide sufficient volume for a commercial sale, a noncommercial ladder fuel treatment is conducted.

A part of the silvicultural prescription should include prescribed fire or mechanical treatments of post-treatment fuels, particularly where logging slash and accumulated deadwood are abundant. If these surface fuels are left untreated, flame lengths of any future fires can be high, resulting in significant mortality of the retained trees.

3. Dense, Two-Cohort Condition That Includes an Adequate Fire-Tolerant Component

This structural condition is often the result of natural regeneration following a heavy thinning, shelterwood, or dispersed variable retention harvest. This condition can also occur after a mixed severity fire or insect disturbance that partially kills the overstory and is followed by natural regeneration. The low to moderate density overstory usually consists of desired fire-tolerant species, while the lower cohort is densely stocked and may include a significant shrub component. On mixed-conifer sites, the lower cohort is often dominated by fire-intolerant species. In the early years after the harvest or natural disturbance, the younger (reproduction) cohort does not pose a threat to the overstory in terms of ladder fuels, crown fire torching, moisture competition, and/or associated drought stress. As the trees grow, however, moderate and then high crown fire risk will develop. This can occurs after 10 to 15 years on more productive sites (Stephens et al. 2012), but after 20 or more years on less productive sites.

Treatments for this condition depend upon the size and density of the younger cohort. Prescribed fire may be sufficient if the lower cohort is small enough to be effectively thinned with a low intensity fire. If managing the lower cohort for future commercial entries is a high priority, prescribed fire may not be acceptable as it can produce highly variable outcomes. Also, as the lower cohort increases in size, a hotter prescribed fire will be necessary to kill these trees. At some point, the risk of killing a significant portion of the overstory becomes too high to use prescribed fire.

In cases where fire cannot be used, a mechanical, noncommercial thinning treatment can be used to thin the lower cohort to a low overall density (e.g., 25 to 125+ trees per ha) dominated by fire tolerant species. Spatially, the younger cohort should be removed from a distance of 1–2 times the dripline of any remnant overstory trees and retained elsewhere in the harvest unit in small to large clumps of 2 to 30+ trees, along with some widely spaced individual trees. Depending on the amount of slash left after the thinning treatment, a prescribed fire or pile and burn treatment will generally be necessary to reduce surface fuels.

4. Dense Stand Condition Lacking Healthy Trees Or Trees of Desired Species Or Both

This condition encompasses a range of stand structures where there are insufficient healthy trees of fire-tolerant species to retain in either a commercial or non-commercial thinning treatment. Thus regenerating a new cohort of fire-tolerant trees is necessary to achieve long-term objectives. This condition is most common on mixed-conifer sites where the fire-tolerant component was harvested, existing trees of any species have been heavily impacted by insect (e.g., defoliators, such as spruce budworm), and disease (e.g., heavy dwarf mistletoe infections). Another example would be plantations where inappropriate species or offsite seed stock were used, resulting in low vigor trees.

A regeneration treatment is generally necessary to address the primary need to establish a new cohort of trees. The amount and spatial location of retention will be a function of the density of existing fire tolerant, healthy trees. In some cases, treatments may have moderate to high retention levels (e.g., 30% of basal area) and will essentially be a variable retention harvest (see Chapter 7). In cases with very poor tree vigor or major insect and disease issues, retention may need to be low (e.g., 5%). In stands heavily infected with dwarf mistletoe, for example, high retention of infected overstory trees will lead to infection of future regenerations of that species (Figure 12.17a). These treatments may be commercially viable, or may not pay for themselves due to high levels of defect and/or partial mortality of the overstory. The cost of dealing with young, failing plantations can be quite high. As with most treatments in frequent fire forests, followup noncommercial thinning and prescribed fire or a mechanical fuels treatment will be needed to treat the remaining small diameter trees, logging slash, and other surface fuel levels that are often very high (Figure 12.17b).

In most cases, planting will be necessary to establish a new cohort of fire tolerant trees. On mixed conifer sites, early pre-commercial thinning will often be necessary to remove abundant and fast growing regeneration by fire intolerant species that can overwhelm the fire tolerant trees that were planted or established naturally. Prescribed fire can also be an effective tool for early density control to ensure another cycle of infilling with fire intolerant species does not begin. Finally, planting seed stock from a seed zone that is better adapted to projected future climates should be considered.

Density and spatial pattern of planting must also be determined. The timber-focused approach of high density, evenly spaced reforestation (e.g. 3.65 x 3.65 m planting) will set forests up for future drought stress and fire risk and require non-commercial thinning to reduce density. Thus planting at lower densities (e.g. 100 to 250 trees per ha) with high quality planting stock is recommended. Site productivity, natural regeneration potential, future fire and drought risk, and management objectives will need to be factored in to determine the exact planting density. Vegetation control may also be needed on sites where shrubs are likely to vigorously respond after fire. Planting is also an opportunity to establish a variable spatial pattern with clumps and individuals, while maintaining some openings. Planting at closer spacing in favorable planting spots and avoiding planting in unfavorable patches (shallow soils, etc) can achieve this. Remember that natural regeneration on such sites is typically spatially variable, a model that can be followed in artificial regeneration.

Figure 12.17 *a) A Douglas-fir forest heavily infected with dwarf mistletoe in north central Washington, USA. While mistletoe brooms provide high quality nesting habitat for a variety of birds and small mammals, a forest with this level of infection will not persist as the infected trees will eventually die. Younger Douglas-fir and future regenerations will become infected. The treatment for this condition is a regeneration harvest that retains any healthy trees. To balance the need for a new cohort and to maintain the functions of the larger, older infected trees, a portion can be retained in clumps and non-host species planted, such as ponderosa pine or western larch. b) A low density, dispersed retention harvest that removed most of the previous stand that was dominated by grand fir and Douglas-fir mostly infected with dwarf mistletoe. Healthy Douglas-fir and ponderosa pine were retained. A prescribed fire was conducted after the commercial harvest to kill remaining understory and midstory trees that were infected and had poor crowns, as well as to consume accumulated ground fuels and remaining logging slash. Ponderosa pine was then planted.*

5. Open, Multi-Cohort Stand Condition

This structural condition represents the target condition for frequent fire forests. Multiple cohorts exist across the stand in a range of clump sizes, but are generally spatially segregated so that clumps with multi-layered canopies are rare. Overstory density and canopy cover are low to moderate, and surface and ladder fuels are low. Large and old trees of fire-resistant species occupy a large proportion of the basal area, but this will vary depending on management history and economic objectives. Stands do not have to be dominated by large and/or old trees to be in this condition. What is most important is that they have the structure to support the stand development processes associated with the natural disturbance regime, primarily low severity fire.

Foresters have two options for this condition. The first is to do nothing as the forest is in a desired condition. Within 10-20 years, however, a maintenance treatment will be needed to deal with infill of trees and keep fire risk low. Fire, either prescribed fire or managed wildfire, can be used to maintain low levels of surface and ladder fuels and perpetuate natural development. Alternatively, mechanical fuel treatments can be used.

Box 12.2 **Multi-age management in frequent fire forests**

Multi-age silviculture that mimics a natural model for frequent fire forests can achieve goals of 1) restoring and maintaining resistance to high severity fire, drought, and insect attacks, and 2) generating wood and revenue in perpetuity. Although multi-age management has not been widely adopted, we anticipate that many ownerships that are currently conducting restoration treatments will adopt it once they complete the restoration phase. A general guide to this approach is provided by O'Hara (2014).

Multi-age management in frequent fire forests involves periodic entries using a flexible mix of group and individual tree selection, preferably using spatial patterns that mimic the clump based developmental and mortality patterns of natural frequent fire forests (Larson and Churchill 2012, Rodman et al. 2017). Commercial removal of trees, non-commercial thinning of smaller trees, and surface fuels treatments can all be accomplished during the same entry. At its core, the system seeks to harvest some portion of the periodic mortality of trees from the medium and large sizes classes that occur in a natural frequent fire forest. Due to the fact that snags and downed logs are consumed by fire relatively quickly in the natural model, loss of ecological function from removing wood is lower than in other archetypes. However, some provision does need to be made for providing sufficient levels of snags and down logs in order to provide for habitat needs of a broad array of biota.

A number of ecologically based, multiage approaches are currently used in ponderosa pine and mixed conifer forests (Emmingham et al. 2005; O'Hara 2005), as well as other frequent fire forests (Baker et al. 1996; Neel 2010). These include systems based on regulating the diameter distribution of a stand such as the BDq method (Fiedler et al. 1988; Bailey and Covington 2002), SDI allocation (Long 1995), and leaf area allocation (O'Hara et al. 2003). Other relevant systems include area control based systems, such as unit area control (Emmingham et al. 2005; Graham and Jain 2005b; Youtz et al. 2008) and group selection (York and Battles 2008).

While diameter based systems such as BDq can certainly achieve a range of ecological and economic objectives, we suggest that approaches based on area are a better fit to the natural model. It is also easier to maintain spatial segregation between different cohorts or canopy layers or, to put it simply, to avoid trees growing beneath other trees. Area based systems seek to maintain patches of different forest structure in a specified portion of the stand area. For example, 20–40% of the stand area can be in openings with clumps of regeneration, 10–20% in groups or patches of saplings, 10–30% in medium trees, and 20–40% large trees. Group or patch size is generally small (e.g., 0.02 to 0.2 ha). Youtz et al. (2008) provide an excellent description of this approach.

We emphasize, however, that multiage systems are very flexible in a number of areas (Emmingham et al. 2005; Graham and Jain 2005b). First, there needn't be an upper diameter limit above which all trees are harvested, as is the case with classic BDq. Second, prescriptions for the size of clumps, groups, and patches can be based on the number of trees instead of area (e.g., an ICO prescription, see Box 8.5). Visualizing and tallying numbers of trees in a group when marking can be easier than estimating area. Group, clump, and opening size and shape need not be rigid and can change over time. Third, groups or clumps do not have to be strictly even-aged as in classic group selection, although multistory clumps with significant ladder fuels generally should be rare. Finally, rigid regulation of structure is not a goal as cutting cycles, unit area proportions, and clump size distributions can vary from one entry to the next.

We suggest a cutting cycle of approximately 10–30 years. The length of the cutting cycle depends on the amount of growing stock following a harvest, the desired volume removal of the next entry, and site productivity or growth rate (O'Hara 2014). Road access, logging system, cash flow, and other operational considerations are also major factors in determining what is appropriate for a specific site. Higher productivity sites will generally be on the lower end of the range to avoid buildup of surface and ladder fuels that will elevate crown fire risk. Longer cycles may require an intermediate prescribed fire or other surface fuel treatment. This can be accomplished in a cost effective manner using large prescribed fires that treat many stands at once.

The second option is to conduct a multi-age treatment. These treatments consist of both thinning and large tree removal. Clumps of small and medium sized trees are thinned, leaving the larger and more vigorous trees. Over several entries, clumps are reduced to small to medium clumps of larger trees (~2-9 trees) or widely spaced individual trees. A portion of the larger size classes is harvested during each entry, generally by removing entire clumps of larger trees. This creates new openings where clumps of regeneration are actively promoted.

Restoring and maintaining an old tree population in perpetuity is possible and may be a desirable part of this management regime, such as on federal forest lands where ecological integrity is an over-riding objective. Elsewhere decisions on providing for old tree populations will depend upon the balance between ecological, financial, and other social goals. As old trees die, replacements can be selected from larger size classes.

Multi-age treatments are coupled with prescribed fire or other surface fuels treatments, as well as non-commercial, small tree thinning as needed to manage ladder fuels (e.g. seedling and saplings). Once a stand is in this condition, it can be managed in perpetuity using this same treatment approach (Box 12.2).

6. Post-High Severity Disturbance Condition

This condition applies to high severity wildfires or insect outbreaks (stand-replacement) disturbances, as well as to sites that were intensively harvested and then transitioned to a landowner utilizing ecological silvicultural principles (Figure 12.18). Although rare, high severity, stand replacing fires and insect outbreaks did occur historically in ponderosa pine and dry mixed conifer forests, as well as in other frequent-fire forest types. They are currently increasing in spatial extent (Harvey et al. 2016), a trend projected to accelerate with climate change if current forest structure is not restored. Post-disturbance management in frequent-fire landscapes subject to such events will therefore be an ongoing challenge facing ecological foresters. The question of whether to conduct salvage logging after a stand replacing fire (Figure 12.19) involves a complex set of tradeoffs (see Box 12.3).

With or without salvage, natural tree regeneration can be limited in large areas subject to high severity wildfire, particularly if there are few surviving trees to provide seed sources (Chambers et al. 2016) (Figure 12.18). Additionally, unfavorable environmental conditions in the years following the fire can lead to high mortality of any natural seedlings that do develop (Haffey et al. 2018; Stevens-Rumann et al. 2018). Such circumstances can result in frequent-fire landscapes shifting to woodlands, grasslands, or chaparral for long periods or even permanently (Allen 2007; Shive et al. 2018). Changes in vegetation of this type sometimes may be desirable from a landscape perspective as non-forest patches can impede future high severity fires. However, climate projections suggest that a significant portion of the forested land base in the western US will no longer support forest by the middle of the century and even less by the end of the century (Clark et al. 2016; Keane et al. 2018). If this circumstance becomes a reality, all landowners, including private and federal, will be affected. Major disturbances, such as wildfires and drought-induced insect epidemics, are projected to trigger these shifts (Johnstone et al. 2016; Littell et al. 2016).

Figure 12.18 *A large patch of high severity fire on the Colville Indian Reservation in north central Washington, USA in 2015. As very few trees survived the fire, this dry ponderosa pine forest is unlikely to regenerate naturally. Furthermore, the area is at the dry end of the range of ponderosa pine, and climate models predict that it will not support forest in the future. The decision of whether to plant or not is difficult.*

Figure 12.19 *An untreated (left) and salvaged (right) forest after a high severity fire in a dry mixed conifer forest in northeast Washington, USA. This unit is part of a study designed to demonstrate different levels of salvage and to test effects on fuel progression and the potential for reburns to kill future regeneration. A similar basal area of dead trees was retained in the treatment on the right as would have been left in a green (pre-fire) restoration treatment. This forest is located on the Colville National Forest as in part of the 2015 Stickpin fire.*

Box 12.3 **Salvage after high-severity wildfires**

Despite our best efforts to manage for low-severity fires in frequent-fire forests, severe wildfires driven by extreme weather will occur. We recommend that the issue of appropriate actions following stand-replacement disturbances, such as wildfire, be addressed as part of the overall management plan before such disturbances occur. Questions about salvage logging and its potential effects on ecological function are difficult to address and thus need time to work through.

In general, salvage is carried out primarily to recover economic values and makes little direct contribution to ecological recovery. The burned forest retains significant ecological values, including the structural and functional legacies of snags and down logs; many decades will pass before the forest begins to generate large woody debris again. The burned forest provides habitat for a variety of animal species. Salvage can eliminate much of this legacy, disrupt vegetative recovery, and cause damage to soils and waters (Lindenmayer et al. 2008; Peterson et al. 2009).

On the other hand, wildfires in densely stocked stands will typically generate substantially more dead woody fuels than would have existed historically following a stand-replacement fire event (Peterson et al. 2015). Such events were very uncommon in the frequent-fire forest landscapes and so would be a novel circumstance, one with little historical precedent. The fine fuels will disappear relatively quickly but some of the medium and larger fuels will persist for decades and can make it difficult to conduct subsequent prescribed burns or to suppress a subsequent wildfire. Hence, some removal of fire-generated fuels may be ecologically justified (Franklin and Agee 2003). There may also be economic or other social benefits from salvage.

If a decision is made to salvage dead wood we recommend that it be conducted so as to minimize impacts on ecological values. The same quantity and type of dead wood should be retained on site as would have existed there if a stand-replacement wildfire had occurred in a historic or restored stand—i.e., retaining the larger and more decay resistant trees (Figure 12.19). Material removed in the salvage should be small and medium-sized stems of species more prone to decay (e.g., grand/white fir). Salvage needs to occur promptly to capture such material, which emphasizes the importance of having a policy in place on post-disturbance activities in the management plan. Of course, any salvage operations need to minimize negative impacts on soil and aquatic values.

Finally, landscape-level perspectives are essential in salvage planning. Larger spatial-area considerations include provisions for habitat requirements for snag-dependent species, such as white-headed and black-backed woodpeckers and many fur-bearing mammals. These are likely to necessitate retaining significant unsalvaged areas. It is also critical to assess the distribution of various types of legacies, such as large snags and dense snag patches, across the burned landscape. Salvage can easily end up being focused on limited areas within the landscapes that have high levels of dead wood legacies, often because those end up being the locations that can carry the cost of salvage. This can result in the burned landscape being salvaged for big wood—in effect, a high-grading of the burned landscape of its best wood legacies, which absolutely should be avoided.

Hence, the first step following a high-severity fire is to decide whether restoring a forest ecosystem on the affected site makes sense given climate predictions and the objectives of the landowner. Climate models can be used to help inform this decision (Sanchez-Meador et al. 2017). If the answer is yes, then planting is often required to achieve sufficient density and species composition. Reforestation offers a unique opportunity to reset species composition to fire and drought tolerant species and introduce planting stock from different planting zones or even new species that are likely to be better adapted to the future climate. Climate projections for local areas can be helpful in making such decisions. Considerations for planting provided in the earlier discussion of structural condition 4 apply here as well.

Examples of Ecological Silviculture in Practice

There are many excellent examples of ecological silviculture in frequent fire forests in the western US. As discussed earlier in this chapter, large scale efforts are underway to reduce fire risk and restore conditions resembling the natural model on National Forests, state and federal wildlife areas, and private landowners with primarily ecological objectives. However, almost all these landowners are still working through the large backlog of acres that need restoration and thus have not shifted into managing and maintaining forests aligned with the natural model. Fortunately, there are a few examples of such management, including landowners with long-term economic objectives. Perhaps the best example of this (although outside of the western US) is the Stoddard-Neel system that has been used for decades in longleaf pine forests in the Red Hills region of northern Florida and southwestern Georgia, USA (Neel 2010; Way 2011), which is described in Box 12.1. In this section, we describe an additional organization from the western US, the Klamath Tribe and the Fremont-Winema National Forest, which uses a distinctly ecological approach for restoration.

Klamath Tribe and Fremont-Winema National Forest

The western half of the Fremont-Winema National Forest, Oregon, USA, which was formerly the heart of the Klamath Tribal Reservation, is currently being managed under a master-stewardship agreement between the Forest Service, Klamath Tribe, and other partners (Hatcher et al. 2017). Under this agreement, the Tribe comanages approximately 162,000 ha of national forest. To date, great progress has been made in achieving most of the goals laid out in the management plan the tribe developed for their former land (Johnson and Franklin 2008). These goals include landscape-scale restoration of frequent fire forests, meadows, and other habitats that approximate historical conditions, as well as enhancement of mule deer populations, protection of cultural sites, and fish habitat improvements. The tribal forestry department supports the Forest Service with NEPA planning, develops the operational prescriptions, and carries out the implementation of most of the projects. Tribal crews lay out and mark the timber sales. To date, approximately 4,000 ha have been treated or sold and over 60,000 ha are in the planning and layout stages. The long-term plan is to treat the majority of the former reservation over the next several decades. The Lomakatsi Restoration organization and the Nature Conservancy of Oregon are also formal partners and assist with implementation.

This area is located in south-central Oregon, USA, on the eastern side of the Cascade Mountains, east of Crater Lake National Park. Ponderosa pine and dry mixed-conifer forests dominate the landscape, although extensive lodgepole pine forest are present in cold air pockets and former meadows. Historically, this area was a classic frequent fire landscape with 10–12 year fire return intervals and extensive, low density forest dominated by large and old ponderosa pine (Hagmann et al. 2013). Much of this old tree population is intact due to successful litigation by the Tribe to stop regeneration harvesting by the Forest Service in the 1980s and early 1990s.

Figure 12.20 *a) Untreated mixed conifer forest in the Red Knight project area on the Fremont-Winema National Forest, Oregon, USA. White fir has encroached and grown up around the old ponderosa pines. b) Post thinning conditions in an adjacent unit. The Klamath Tribe designed this prescription that removes most of the white fir and retains ponderosa pine. Both commercial volume and a large amount of noncommercial sized trees and slash were removed. Note that the surface fuels have not been treated yet.*

The tribe elected for a restoration approach in their management plan that follows ecological forestry principles. Indeed, these principles were developed, refined, and operationalized through work for the Tribe by several authors of this book (Johnson and Franklin 2008; Franklin et al. 2013). Current conditions are dominated by dense, multistory structural conditions that have a sufficient ponderosa pine component (Table 12.4). Dense even-aged and two-aged forests are also common structural conditions. Silvicultural prescriptions include both variable density thinning and species based treatments that focus on removing white fir and lodgepole pine (Figure 12.20). Three key aspects of prescriptions include 1) retention and radial release of all old trees (over 150 years), 2) aggressive removal of white fir and lodgepole pine where they have encroached due to fire exclusion, 3) and formal use of the ICO method (Box 8.5) to restore fine-scale, spatially variable, multiage mosaic patterns. Aspen and meadow restoration (Figure 12.14) is another important treatment. Individual timber sales are large (typically 600 to 1200 ha) and seek to treat the majority of the acres within the area. Most treatments are commercially viable and are generating sufficient revenue to pay for follow-up ladder and surface fuel treatments. Long term multiage management (once restoration is complete) is not currently part of the plan, but is being discussed.

Three major challenges have emerged in carrying out this ambitious project. First, large-scale prescribed fire treatments necessary to keep up with the pace of mechanical treatments are limited by available burning windows, air quality regulations, and continuing social opposition among some tribal members and the larger public. Put simply, it is currently not feasible to burn the necessary acres to maintain this forest with frequent fire. The second challenge is the need for a well-trained tribal workforce to plan and implement this work. A major goal of the restoration effort is to provide employment to tribal members, yet turnover on seasonal work crews is high. Also, the tribe relies to a significant degree on consultants and partners (Lomakatsi) for technical expertise. The final challenge is maintaining social license for the program within the tribe, especially when treatments dramatically change forest conditions over thousands of hectares. Some tribal members remain opposed to logging based on their experience with past regeneration harvesting by the Forest Service. Others can be alarmed by the sudden change to customary locations and the impacts of harvesting. Tribal leaders and natural resource managers are engaged in an ongoing effort to listen to concerns, explain the rationale for treatments, and make adjustments to treatments where needed to maintain social support.

What to Remember about Silviculture for Archetype 2 Ecosystems

Frequent-fire forests have several unique attributes that make both their ecology and silviculture very distinct from the other three archetypes. First and foremost, the natural structure and composition of these forests is highly adapted to fire and, in turn, this requires frequent, low severity fire to be sustained over time. Second, the natural model in this archetype does not progress through a sequence of developmental stages after a major disturbance, but instead persists as dynamic, fine-scale, multiage mosaics of all four developmental stages within a stand. Fire suppression combined with management designed to maximize timber production have resulted in dense forests that are susceptible to high severity fires, drought stress, and related insect mortality.

Over the past twenty years, large scale efforts to reduce fire risk, restore ecosystem function, and enhance resilience in ponderosa pine and mixed conifer forests have resulted in silvicultural approaches based on the natural model. The key attributes of ecological silviculture for Archetype 2 will include:

1. Managing for an open canopy condition and density well below carrying capacity (20–35% of maximum stocking) is critical to maintaining low risk of crown fire and drought stress.
2. Species composition dominated by fire and drought tolerant species is a necessity.
3. Surface and ladder fuels (understory and midstory trees and lower crowns of large trees) need to be reduced and maintained at low levels to prevent surface fires from transitioning into a crown fire. Prescribed fire is often the best approach ecologically, but mechanical methods can also be used (or a combination of the two).
4. Large and old trees of fire tolerant species are the structural backbone of frequent fire forests and are the most resistant to fire. Maintaining and/or growing a cohort of large trees is thus a key strategy, although the amount maintained will depend on economic objectives.
5. Maintaining a fine scale mosaic of different patch types across stands is key to sustaining natural forest development. Patch types should include clumps of different sizes, thickets of regeneration, openings, widely spaced individual trees, as well as larger dense areas, wet meadows, and aspen groves where appropriate.
6. A large portion of current frequent fire forests are highly departed from the natural model due to fire suppression and past management. High density, multistoried canopies, and dominance by fire intolerant species make these forests highly susceptible to high severity fire.
7. Restoration treatments based on variable density thinning combined with prescribed fire are generally the first step to addressing these issues. If prescribed fire is not possible, mechanical methods to treat logging slash and other surface fuels can be used.
8. Once forests have been restored to conditions in line with the natural model (or close to it), multiage management can be conducted over time to maintain low fire risk and produce income and wood. A combination of individual tree and small group selection that mimics natural stand development, that ideally includes prescribed fire, is recommended. For forests with no long-term economic objectives, periodic prescribed or wildland fire can maintain natural developmental processes over time.

CHAPTER 13

Silviculture for Archetype 3 Ecosystems: Forests with Frequent Gap-Scale Disturbance Regimes

This chapter focuses on ecologically based silvicultural approaches for Archetype 3 (forests with frequent gap-scale disturbance), as discussed in Chapter 6. To illustrate our approach, we will use the northern hardwood-dominated forest found across large portions of northeastern North America as our case study, but the construct and ecological silviculture treatments we discuss for these forests will be applicable to other Archetype 3 ecosystems (Table 13.1). These approaches reflect the four principles of ecological silviculture presented in Chapter 3 and include: (1) gap-scale live and deadwood retention in multi-cohort management to encourage ***continuity***; (2) emulation of mesoscale wind disturbance to enhance compositional and structural ***complexity***; (3) using natural patterns of canopy disturbance to guide the ***timing*** and severity of regeneration harvests, and (4) use of a multi-scaled approach for considering regeneration harvests in a landscape ***context.***

Recall from Chapter 6 that northern hardwood forests are largely defined by the dominance of sugar maple with American beech also constituting an important canopy associate throughout much of the range of this forest type (Figure 13.1). The canopy layer of these communities can be comparatively species rich relative to the other archetypes with minor canopy species most frequently being yellow birch, eastern hemlock, white ash, and basswood. The low canopy bulk density (i.e., availability of canopy fuels) of these forests and the often humid climate regimes where they occur make severe fire an extremely rare component of northern hardwood systems, with disturbance regimes dominated instead by frequent gap- and meso-scale disturbances driven by wind, ice, insects, and disease (Frelich 2002). Remember from Chapter 6 that return intervals for stand-initiating disturbances are often two to three times the average life spans of canopy trees and age structures are correspondingly strongly uneven aged (Table 13.2). Although advance regeneration is the primary regeneration mechanism favored by frequent gap-scale regimes, the range of canopy openings and seedbed conditions historically created by wind in these forests allows for species with a wide range of silvical characteristics to occur in the canopy layer (Table 13.3).

Northern Hardwood Forests as a Natural Resource

Northern hardwood forests are an economically and culturally important part of forested regions of northeastern North America, with uses ranging from veneer and sawtimber to maple syrup production (Figure 13.2). These forests cover over 20 million hectares in the Lake States and northeastern US (Oswalt et al. 2014) and over 14.6 million hectares in eastern Canada (Bédard 2016) and are a central component of the forest-based industries in these regions. For example, average annual harvest levels from these forests from 2012 to 2016 was 16.7 million m^3 in New England and New York. During this same period, Quebec and Vermont produced 26 million gallons of maple syrup representing over 70 percent of global production of

Table 13.1 *Examples of Archetype 3 Ecosystems.*

Ecosystem	Distribution	Natural Disturbance	References
Northern hardwood forests	Northeastern North America	Wind, ice, insects, disease	Chapter 6
Lenga forests	Tierra del Fuego, Chile and Argentina	Wind, insects, fire, disease	Fajardo and de Graaf (2004)
European beech forests	Western Europe	Wind, insects, disease	Rugani et al. (2013)
Japanese beech forests	Japan	Wind, insects, disease	Yamamoto and Nishimura (1999)

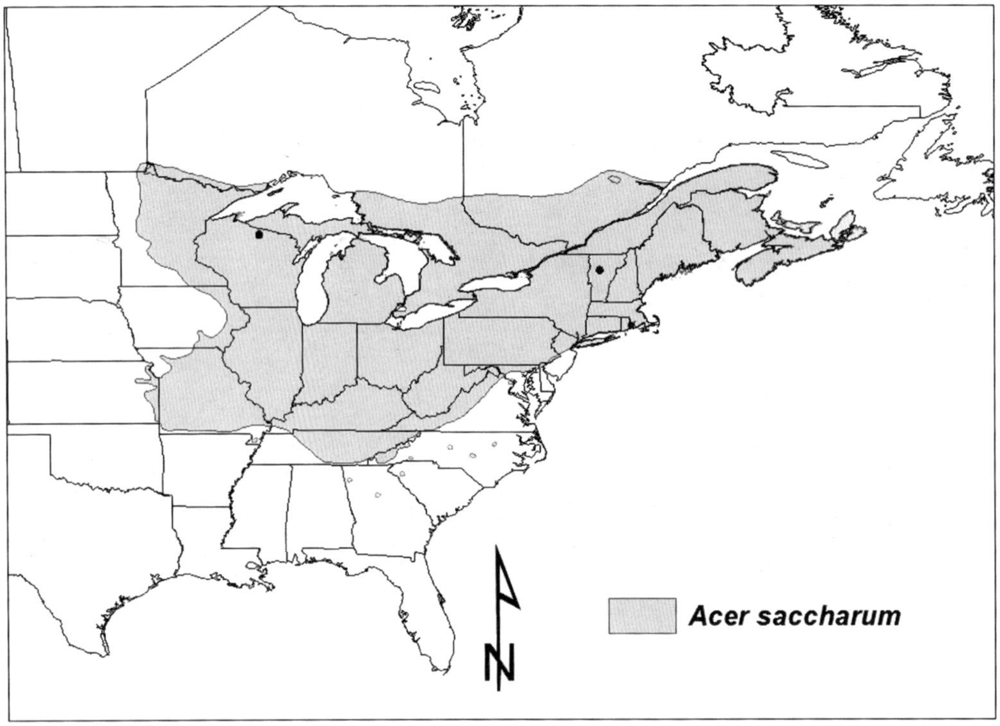

Figure 13.1 *Distribution of northern hardwood forests in northeastern North America, as defined by the range of sugar maple (*Acer saccharum*). Black dots on map indicate locations of examples of ecological silviculture in practice highlighted at the end of this chapter.*

Table 13.2 *Attributes of developmental stages in northern hardwood forest ecosystems.*

Stage or Event	Attribute	Cause or Consequence
Disturbance and Legacy Creation	Spatially heterogeneous live and dead tree legacies, including tip-up mound structures, seedling and sapling layer representing diversity of regeneration mechanisms; small (single-tree) to moderate openings in matrix of undisturbed forest	Results from recurring low severity disturbances and less frequent, mesoscale events, largely generated by wind and ice
Preforest	Generally absent	Rapid site capture after canopy disturbance through release of advance regeneration, vegetative response of broken but surviving trees, and establishment of seedlings from soil seedbanks or windblown seed
Young Forest	Generally a gap or patch-level condition that may last 50 to 80 years; stratified mixture early on with shorter-lived species, such as pin cherry, aspen, and paper birch forming upper canopy strata and with slower growing mid- and shade-tolerant species occupying lower canopy positions; species in lower canopy positions during this stage (yellow birch, sugar maple, beech) ultimately dominate all canopy layers in later stages; within patch legacies of older residual trees surviving wind disturbance present	Mixtures of multiple species representing range of regeneration mechanisms; differences in early height growth rate and longevity cause shift in dominance from early to later successional species as stage progresses
Mature Forest	Gap-scale disturbance from death of larger canopy trees increases heterogeneity of structure, including age-class diversity; large dead and decadent trees accumulate	Windthrow and breakage of canopy dominants allows for recruitment of cohorts of shade-tolerant species (sugar maple, American beech, eastern hemlock); tip-up mounds serve as important substrates for establishment of mid-tolerant species (yellow birch)
Old Forest	Dominated by shade-tolerant species (sugar maple, American beech, eastern hemlock) across a broad range of tree ages and canopy positions, with a lesser, but significant mid-tolerant yellow birch component; closed canopy, uneven-aged structures; diverse tree sizes, including very large trees, generally following descending monotonic or rotated sigmoid diameter distribution; large amounts of deadwood, including hollow snags and logs; spatial heterogeneity of structure; regeneration in patches	Long-term, frequent gap-disturbance interacts with dominance by shade tolerance to result in diversity of live-tree size and age structures; increasing levels of heartwood decay on large diameter canopy trees generates complex deadwood conditions; multiple tree-fall gaps and mesoscale wind events create heterogeneity and foster tree diversity

Table 13.3 *Select silvical characteristics of canopy tree species commonly found in northern hardwood forests in northeastern North America. (From Anderson et al. 2001 and Leak et al. 2014).*

Species	Shade tolerance	Preferred seed bed	Minimum gap size for canopy ascension (ha)	Early relative growth rate	Sprouting ability	Relative site requirement
Sugar maple	Tolerant	Litter	0.05	Slow to moderate	Moderate (stump)	High
American beech	Very tolerant	Litter	0.05	Slow	High (roots)	Low
Yellow birch	Intermediate	Humus mix	0.1	Moderate	Low (stump)	Medium to high
Basswood	Intermediate	Litter	0.1	Moderate	High (stump)	Medium to high
Paper birch	Intolerant	Humus mix	0.5	Fast	Moderate (stump)	Low
White ash	Intermediate	Litter	0.08	Moderate	Moderate (stump)	Very high
Red maple	Tolerant	Litter	0.08	Moderate	High (stump)	Low
Aspen (big-tooth and trembling)	Very intolerant	Moist mineral	> 1	Very fast	High (roots)	Low
N. red oak	Intermediate	Litter	0.2	Moderate	Moderate (stump)	Medium
Black cherry	Intolerant	Litter	0.5	Fast	Moderate (stump)	Low
Red spruce	Tolerant	Litter, moist mineral	0.05	Very slow	None	Low
Eastern hemlock	Very tolerant	Humus mix	0.05	Very slow	None	Low
Eastern white pine	Intermediate	Moist mineral	0.1	Moderate	None	Low

Figure 13.2 *a) Sawtimber and veneer sugar maple on a harvest landing in northern New Hampshire, USA. b) Tubing used for sap collection as part of a maple sugaring operation in central Vermont, USA. Photo credit: b) Mark Isselhardt.*

this valuable non-timber forest product. Output of maple syrup from these forests has increased by more than 200 percent since 1980 with this use expanding into many areas historically used for producing wood products (Farrell and Chabot 2012). Although this chapter focuses on the ecological silviculture of northern hardwood forests being managed largely for traditional forest products, the continued expansion of the syrup industry may generate a future need to balance the impacts of syrup production on forest ecosystem functioning and the ability to meet long-term silvicultural goals for an area, including maintenance of a diversity of canopy tree species and market options.

Evolution of Timber Production Silviculture for Northern Hardwood Forests

Northern hardwood forests are naturally mixed, in terms of overstory tree species composition and timber quality. Correspondingly, the earliest modern harvesting in northern hardwoods was primarily selective cutting that focused on removing either a preferred species (e.g., red spruce and eastern white pine in the eighteenth and nineteenth centuries, Kelty and D'Amato 2006) or the largest, high-quality individuals in a given area. This period of early selective cutting was followed by more intensive waves of harvesting in the late nineteenth and early twentieth centuries that, along with agricultural abandonment, led to large areas of single-cohort forests—a strong departure from the multi-cohort, old-growth conditions naturally characterizing these forests. In areas where selective cutting had been the primary harvesting approach, forests now often lack the large tree component or the tree quality necessary for long-term sustained management and in many cases have shifted toward strong dominance by shade-tolerant species, namely American beech (in the northeast USA) and sugar maple (in the Lake States, USA) in all canopy layers.

Formalized silvicultural approaches for managing northern hardwoods were developed in the 1920s and 1930s and focused on generating a sustained yield of commercial species from areas experiencing exploitive logging (Eyre and Zilgitt 1953; Gilbert and Jensen 1958). Given the primary commercial interests in sugar maple, a shade-tolerant species, early silvicultural guidelines emphasized uneven-aged approaches, particularly single-tree selection, to generate "balanced," uneven-aged structures for the sustainable management of these forests. In contrast to the other archetypes, these early approaches were justified based on natural tree size structures observed in old-growth forests dominated by shade-tolerant trees in Europe and eastern North America (Meyer 1952). An unfortunate misapplication of these natural benchmarks was the translation of size structures observed at broad spatial scales (reverse-J-shaped distributions) to management guidelines designed for stand-scale application. These management guides were intended to inform single-tree selection systems that maintained a descending monotonic target size structure based on regional empirical relationships (e.g., Arbogast structure; Arbogast 1957) or through application of BDq approaches (see Chapter 8). One ecological consequence of this management approach was the general loss of less tolerant species from these forests due to the creation of only small canopy gaps during single-tree removals (Webster and Lorimer 2005). In addition, although economically justified, marking guides gave priority to removal of "unacceptable growing stock," including declining and economically defective trees (i.e., trees with cavities or decay), resulting in a significant reduction in cavity trees, snags, and downed woody material and associated biota in these forests (Figure 13.3, Figure 13.4).

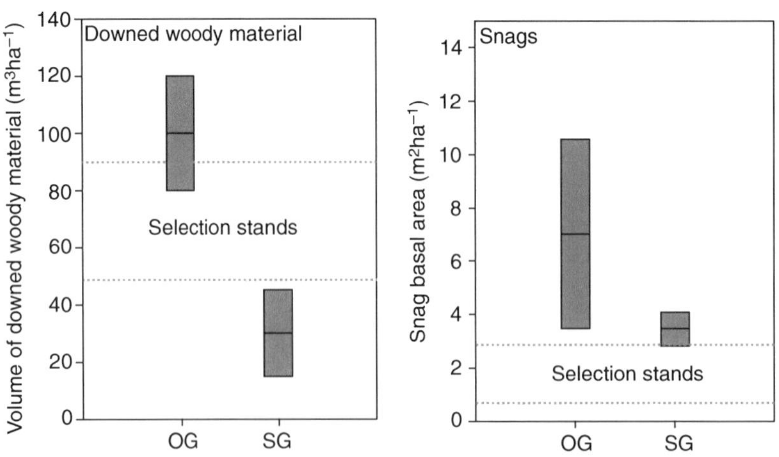

Figure 13.3 *Range of downed woody material volumes and snag basal areas found in old-growth (OG) and second-growth (SG) northern hardwood forests. Second-growth stands represent stand ages 65–100 years and no history of management. The range of values observed in stands managed with single-tree selection are also presented between dashed gray lines. Values for old-growth and second-growth are from McGee (2003); Gore and Patterson (1986); Tritton (1980); Goodburn and Lorimer (1998), and single-tree selection stands are from Goodburn and Lorimer (1998); Kenefic and Nyland (2007).*

Figure 13.4 *a) Old-growth northern hardwood forest and b) stand managed with single-tree selection for 30 years. Both stands are on the Mousseau Experimental Forest in Quebec, Canada.*

Concerns over the long-term compositional outcomes of single-tree selection, namely overwhelming dominance by American beech or sugar maple, has led to a shift in some regions toward the application of group-based selection methods (Leak and Filip 1977; Figure 13.5). In most cases, these systems are being applied to increase yellow birch abundance and typically rely on harvest gaps ranging from 0.1–0.13 ha (0.25–0.33 acre) in size (e.g., Kern et al. 2012), with application over time resembling area regulation approaches, but at the stand level. Absent from these approaches has been the recognition of the importance of incorporating retention of live and dead tree legacies into regeneration entries to meet ecological objectives, even if a greater diversity of tree species is now being supported (Kern et al. 2017). As such, the systems often generate a mosaic of different even-aged conditions at the stand scale. Moreover, the uniform application of opening sizes at regular intervals (i.e., fixed cutting cycles) has overlooked the wide range of canopy opening sizes and frequencies of gap and mesoscale disturbances historically characterizing these forests (Lorimer and White 2003; Table 13.4).

Based on the aforementioned approaches to timber-production silviculture in northern hardwood forests, there are numerous key differences in the structure, composition, and ecological function of areas managed with these management regimes relative to those observed in natural northern hardwood forests (Table 13.4). These include a lower diversity of tree species, canopy gap sizes, and live-tree structures, as well as a lower abundance of structural legacies, namely cavity trees, snags, and downed woody material. Moreover, historical disturbance regimes often generated uneven-aged forests that were dominated

Figure 13.5 *Group selection harvests in hemlock-northern hardwood forests at the a) Bartlett Experimental Forest, NH, USA and b) Argonne Experimental Forest, WI, USA.*

Table 13.4 *Key differences in outcomes resulting from a natural development model and the timber-focused management model for northern hardwood ecosystems.*

Feature	Natural Model	Management Model	Functional Consequences[a]
Tree species diversity	Typically including up to 10 species	Trees other than shade-tolerant, sugar maple, often discriminated against	Higher susceptibility to insects/disease; lower habitat quality due to lack of mast-producing species (American beech) and preferred nesting and foraging sites (yellow birch)
Forest age structure	Uneven-aged, but often dominated by two to three cohorts reflecting mesoscale disturbance events	Uneven-aged with balanced spatial distribution of cohorts	Lower overstory species diversity due to uniform gap size
Live tree legacies	Few to many within gaps	Few to none	Loss of unique habitat and seed sources
Tree derived legacies after a disturbance	Abundant; diverse sizes including large hollow snags/logs	Limited number, typically of small size	Loss of unique habitat
Spatial pattern of legacies	Dispersed to aggregated, reflecting single- and multiple-tree mortality	Not present	Reduction in resource and habitat diversity
Stand density	Relatively high at all stages of development	Comparatively low	Loss of deep shade microsites, lower midstory and deadwood carbon stocks
Spatial pattern of tree distribution	Heterogeneity in density and canopy cover	Homogeneity in density and canopy cover	Reduction in diversity of resource and microclimate conditions; loss of tree diversity
Maximum tree age	250–400 year tree life spans	Maximum diameters (tree-level rotations of 100–120 years)	Loss of unique structural features of old trees

[a]Examples of functional consequences of a management model compared to the natural model.

by two to three cohorts (Lorimer and White 2003), whereas most current management aims to create a balanced distribution of age classes in space and time. The consequence is much lower spatial variability in structural conditions and a lack of admixtures of old-tree legacies and the many functions they support within regenerating cohorts. The ecological silviculture system outlined later in this chapter integrates an understanding of the natural model for this forest with past silvicultural experience to maintain the key ecological functions and structures outlined in Table 13.4 within areas managed for timber production.

Current Challenges to Northern Hardwood Silviculture

Like many other forest types around the globe, the structure and functioning of contemporary northern hardwood forests has been negatively impacted in many portions of its range by invasive, non-native plants, insects, and pathogens and increases in herbivore densities. Of particular importance have been the impacts of beech bark disease on northern hardwood forests in northeastern North America and European earthworms in the upper Lake States, USA. Both of these stressors interact synergistically with elevated herbivory levels from eastern white-tailed deer, resulting in understory conditions that present significant barriers to regeneration of most canopy tree species, including dense American beech thickets, sedge mats, and dense fern cover (Figure 13.6). As such, prescriptions focused on achieving ecological silviculture objectives that are in concert with the structural and compositional complexity for the natural archetype now frequently include treatments specifically designed to address the challenges posed by these modern-day stresses. This underscores the importance of adapting and adjusting ecological silviculture approaches in response to both management outcomes and emerging stressors. This is something we will return to in Chapter 16.

Figure 13.6 *a) Dense American beech understory in northern hardwood forest affected by beech bark disease in northern New Hampshire, USA. b) Pennsylvania sedge mats in northern hardwood forest in north-central Wisconsin, USA impacted by European earthworms and high deer browsing pressure.*

An Ecological Silvicultural System for Northern Hardwood (Archetype 3) Ecosystems

The silvicultural system for Archetype 3 is the **gap-scale silvicultural system**, reflecting the pervasive influence frequent windthrow events have on the structural and compositional outcomes of northern hardwood forests at stand and landscape scales. Unlike many classical silvicultural systems in which treatments begin following a regeneration harvest and progress through stand development, both the current condition of northern hardwood forests and their prevailing ecological model require an alternative approach. In particular, much of the northern hardwood forest across northeastern North America is either in the late Young Forest Stage (YFS) or Mature Forest Stage (MFS). This current condition reflects the legacies of widespread land use, including agricultural abandonment, clearcut harvesting for industrial wood use (e.g., charcoal, pulp), high-grade harvesting for sawtimber in the nineteenth and early twentieth centuries, and more recent applications of selection silviculture (Kelty and D'Amato 2006; Figure 13.7). As such, an ecological silvicultural system in these forests will often begin midway through forest development with the application of silvicultural activities focused on accelerating structural and compositional conditions toward those found in the natural, prevailing condition in these forests, i.e., the Old Forest Stage (OFS) (Lorimer and White 2003). Remember from Chapter 6 that these conditions include gap-scale mosaics of the same developmental stages that often occur at much broader spatial extents for other archetypes.

Silviculture by Developmental Stage/Event

This section describes the silvicultural activities that constitute the gap-scale silvicultural system as applied to northern hardwood forests in northeastern North America (Table 13.5). As with any silvicultural system, application of this approach to other Archetype 3 forest types will require modifications to reflect the silvics of constituent species and local ecological and socioeconomic context. Organization of the system is by developmental stages so that decisions regarding what silvicultural activities to apply can be guided by dominant ecological processes and conditions in a given stage with acknowledgement of their linkage with structural and compositional outcomes in subsequent stages. As described above, the context used in the following discussion of the application of the gap-scale silvicultural system is in the prevailing stand conditions for northern hardwood forests across northeastern America, i.e., second-growth forests in the late YFS and MFS.

Disturbance and legacy creation

The processes of disturbance and legacy creation occur predominantly at the gap scale in northern hardwood forests with the general goal of creating structural legacies and stand conditions favorable for recruiting a new cohort of trees. Natural canopy disturbances are a key process for maintaining diversity in the tree community of these forests by releasing advance regeneration of shade-tolerant species, such as sugar maple, eastern hemlock, and American beech, while creating microsite conditions favorable for recruitment of small-seeded, mid-tolerant species, particularly yellow birch (Box 13.1). In addition, these canopy disturbances create and maintain uneven-aged forest conditions with a wide range of tree sizes and ages, including large canopy trees frequently exceeding 50 cm (19.7 in) in diameter (Goodburn and Lorimer 1999).

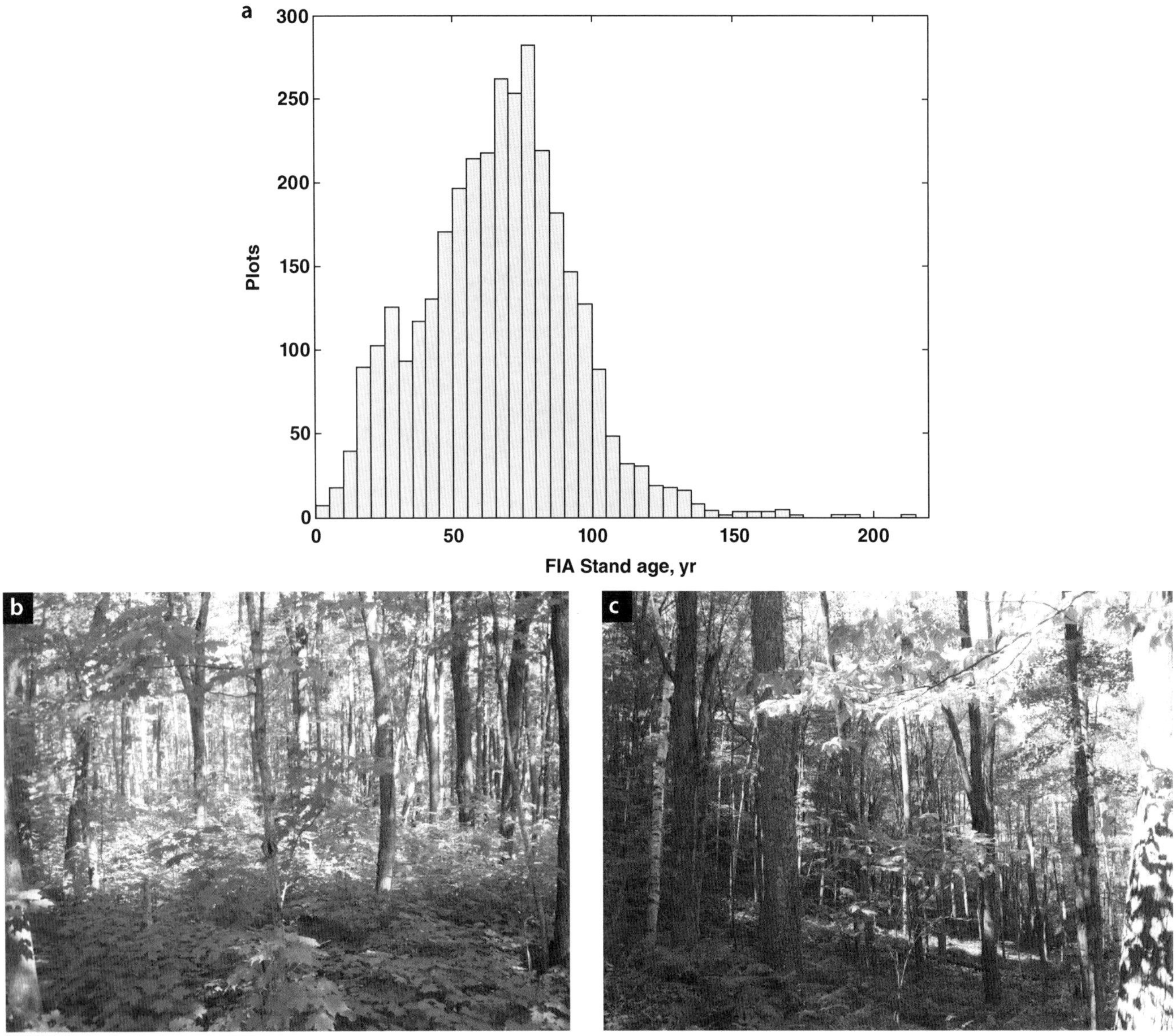

Figure 13.7 *a) Age-class distribution of northern hardwood forests across the northeastern US based on USDA Forest Service Forest Inventory and Analysis data. Note that many northern hardwood forests are in late Young or Mature Forest Stages, such as examples from northeastern Minnesota and Vermont, USA in b) and c), respectively. Correspondingly, most silvicultural systems are beginning at this stage of development.*

As mentioned above, initial entries for this silvicultural system are often occurring in structurally and compositionally homogenous stands in the late YFS and MFS. Correspondingly, a desired outcome for treatments applied during the disturbance and legacy creation stage is the introduction of structural and compositional complexity using regeneration methods—including variable density thinning, continuous cover irregular shelterwoods, or group selection approaches—that promote a range of spatial conditions and environments.

The regeneration method employed at this stage will largely be a function of current forest conditions in terms of dominant tree species and overall stand quality (i.e., stocking of acceptable growing stock from an economic perspective). For example, in cases where the forest understory is strongly dominated by shade-tolerant species, such as sugar maple or American beech, regeneration methods that emulate mesoscale wind events (e.g., group selection or continuous cover irregular shelterwood; Figure 13.10) might be applied to create the range of gap sizes that historically allowed for recruitment of less tolerant species. Similarly, these same methods might be more appropriate in forests with lower timber quality stemming from past exploitive harvests (Nyland 1992) as greater volumes are removed and more opportunities are afforded to shift growing space to individuals and species with long-term economic potential relative to other methods (cf. Strong et al. 1995).

Table 13.5 *An ecological silvicultural system for northern hardwood ecosystems.*

Developmental Stage/Event[a]	Objectives/Consequences	Timing (years)[b]	Silvicultural Activities
Disturbance and Legacy Creation[c]	Provide for continuity of structure and initiate Preforest Stage (PFS)	0	*Harvest*: in gaps as part of group selection, variable density thinning, or continuous cover irregular shelterwoods in late YFS, MFS, and OFS[d] *Thinning*: Crop tree release of dominant/codominant trees outside of gaps to accelerate development of large trees
Preforest	Short-lived. Release existing advance regeneration and create conditions for new seedlings to establish; set the stage for tree regeneration	0–5	*Site preparation*: removal of understory American beech where thickets have developed; mechanical, patch scarification to expose mineral soil seedbeds for yellow birch (consider mechanically pulling or pushing over trees with harvesting equipment to create tip-up mounds)
Young Forest (Initiation)	Facilitate/establish new cohort of trees at gap scale	Beginning between 5 and 10 years	*Release*: if needed, use to encourage development of yellow birch and sugar maple where American beech competition is an issue; use mechanical methods to avoid impacts to desired species
Young Forest (after canopy closure)	Accelerate growth of trees in cohort; initiate new cohort in other areas in MFS to create spatial heterogeneity in structural and compositional conditions	50–70 years after cohort establishment	*Thinning*: Crop tree release of dominant and co-dominant sugar maple and yellow birch; variable density thinning in spatially homogeneous stands (i.e., even-aged, second-growth stands)[d] *Harvest*: creation of new single-tree and multi-tree canopy gaps at 15–20 year intervals in stands with patchy distribution of structural conditions (apply site preparation treatments from *PFS* above in group openings)
Mature Forest	Enhance spatial heterogeneity; release and accelerate growth of desired trees; enhance structural complexity; initiate new cohorts	70–150	*Thinning*: Crop tree release of dominant/codominant sugar maple/yellow birch in cohorts initiated in earlier stage; group selection, variable density thinning, or continuous cover irregular shelterwood at stand-scale[d] if first entry *Decadence/deadwood creation*: in stands lacking these features; girdling of canopy trees as part of crop-tree release; felling of cull *Harvest*: large group or patch openings with legacy tree retention; can be part of variable density thinning or initial treatment in conversion to multi-cohort condition (apply site preparation treatments from PFS above in group openings)
Old Forest	Enhance structural complexity; restore/maintain uneven-aged structure; emulate natural dynamics	+150	*Decadence/deadwood creation*: in stands lacking these features (see above) *Thinning*: crop-tree release in cohorts *Harvest:* single-tree and multiple-tree gaps at 15–20 year intervals (apply site preparation from PFS above in group openings)

[a]A regeneration harvest initiates the developmental sequence; however, these stages are largely occurring at the gap or mesoscale in the northern hardwood archetype. Moreover, based on the general condition of the northern hardwood forest across much of North America, the first entry into these systems is typically occurring when the stand-level condition is either nearing the end of the YFS or in the MFS (Figure 13.7).

[b]Timing does not reflect strict year(s) of activity, but rather reflects how long a particular stage might last.

[c]In the northern hardwood archetype, the scale of the disturbance/legacy creation event is generally at the gap-scale as part of group or patch selection with legacy-retention harvests or as part of variable density thinning.

[d]In contrast to other forest archetypes, stand-replacing disturbance is quite rare in northern hardwood forests occurring at return intervals two to three times the average lifespan of canopy trees (200–250 years) and initiation of developmental stages is at the gap or mesoscale in a stand. Retention of live trees, as well as deliberate felling and retention of some canopy trees for deadwood within harvest gaps is a critical component of the Disturbance and Legacy Creation stage.

Box 13.1 **Microsite considerations during the disturbance and legacy creation stage to enhance compositional complexity**

Northern hardwood forests are unique relative to other temperate forests covered in this book in terms of the seedbed requirements for dominant tree species. In particular, many species, particularly those with a greater seed mass like sugar maple, do not require a mineral soil or downed woody material seedbed for establishment, and instead compete best in undisturbed leaf litter (Table 13.3). These species are also generally tolerant of shade and therefore can build up a large seedling bank in the absence of canopy disturbance. Nonetheless, two historically important canopy species, eastern hemlock and yellow birch, are quite small-seeded and rely on the exposed mineral soil substrates created by windthrow or the stabile, moist conditions provided by well-decomposed logs for establishment (Figure 13.8; Marx and Walters 2008). As such, ecological silvicultural approaches that solely emulate natural canopy disturbance and resultant canopy conditions through harvesting need to consider that microsites created by these events often serve to release existing advance regeneration at the exclusion of these small-seeded species. This is particularly an issue when harvests occur on snow-covered ground, minimizing soil scarification by harvesting equipment (Bolton and D'Amato 2011). To address this issue and maintain the compositional ***complexity*** of these forests, the gap-scale silvicultural system requires specific treatments to provide a diversity of seedbed conditions during the Disturbance and Legacy Creation stage. One approach is to create artificial tip-up mounds by winching down or pushing over trees with harvesting equipment (cf. Keeton 2006); however, this strategy likely is not operationally or economically feasible on most ownerships. At minimum, the silvicultural system should include the use of site preparation treatments or, where appropriate, snow-free logging operations to create exposed mineral soil seedbeds and also should retain legacies to serve as future downed log substrates. Logging during snow-covered conditions provides many benefits for sustaining long-term site productivity, including minimizing impacts to root systems of residual trees and reducing soil compaction and erosion. As such, relying on incidental scarification via harvesting equipment may not be possible for sites with limited operability during summer and fall months or where site-level guidelines require that harvests occur in snow-covered conditions. In these cases, follow-up site preparation treatments may be an added investment necessary to ensure suitable microsites are available for small-seeded species on a given site. Because regeneration harvests are generally occurring at the gap-scale, the use of site preparation equipment for these treatments, such as a salmon blade, that can be mounted on a small to medium-sized crawler tractor is most appropriate (Figure 13.9). In short, ecological silviculture that emulates natural canopy disturbances cannot overlook the microsite diversity created by these events and should incorporate the appropriate site preparation techniques for their creation in modern forests.

Figure 13.8 *a) Yellow birch establishing on exposed mineral soil microsite created by windthrow in northern New Hampshire, USA. b) Downed hemlock log with yellow birch and eastern hemlock seedlings in western Massachusetts, USA.*

Figure 13.9 *Use of a salmon blade within harvest gaps to reduce impacts of (a–b) dense sedge mats and (c–d) sugar maple advance regeneration on recruitment of yellow birch and eastern hemlock. Photo credit: Dakota Fassnacht*

It is important to note that while our emphasis in this and the other silviculture system chapters is primarily on ecological outcomes of silvicultural treatments, long-term entries as part of this and any other ecological silvicultural system also need to be operationally and financially viable. Therefore, initial entries as part of this system should consider how future harvests might be sustained based on quality and economic potential of trees left after treatments. This may be particularly important on smaller, private ownerships that often include long-term economic objectives along with non-timber objectives related to wildlife habitat and aesthetics.

Regardless of the regeneration methods used, there are several key structural and compositional outcomes that are the focus of initial entries in this ecological silvicultural system (Table 13.5). First is recruitment of a new age class of trees to begin transitioning toward the multiaged

Figure 13.10 *Northern hardwood forest three years after establishment cutting as part of continuous cover irregular shelterwood in southern Quebec, Canada.*

structures characteristic of the OFS of this archetype. Second is the advancement of the large-tree component of the stand through application of thinning treatments, specifically crop-tree release, to portions of the forest outside of regenerating areas (Box 13.2). Third is recruitment and retention of dead and live-tree legacies in harvest gaps and in matrix areas receiving thinning treatments through designation of leave trees and girdling and deliberate felling of (often) lower value overstory trees (Box 13.3). Finally, representation of intolerant and mid-tolerant canopy species should be increased through creation of suitable microsites and adequate light environments in areas where new age classes are being recruited (Box 13.1), as well as through retention of these species, where possible, to maintain seed sources on site. Achievement of this suite of conditions will provide an important initial template for subsequent entries focused on establishing and maintaining the mosaic of structural and compositional conditions characteristic of the OFS.

Preforest Stage

Remember from Chapter 6 that the PFS is of limited duration in Archetype 3 forests given the rapid capture of canopy gaps by advance regeneration and new seedlings. Silvicultural activities during this stage are largely focused on creating the range of resource (light, water, nutrients) and microsite conditions that historically allowed for the recruitment of a diversity of tree species following gap and meso-scale canopy disturbances (Box 13.1). The need for and types of treatments applied during this stage will depend primarily on the season of harvesting during the Disturbance and Legacy Creation-stage. On sites where harvesting occurred during the summer or fall, logging equipment often generates the incidental scarification required to create suitable microsites for new seedlings of light-seeded species. If harvests occurred on snow-covered ground, mechanical site preparation may be necessary to create mineral soil seedbeds.

Regardless of season of harvest, site preparation treatments to remove all or a component of existing American beech advance regeneration are often required during the PFS to allow for establishment of other species, particularly on sites with a history of beech bark disease and past partial harvesting (e.g., Figure 13.6). Similarly, in the Lake States, USA, removal of some existing sugar maple advance regeneration may be necessary on sites where partial harvesting has led to dominance of this species in the seedling and sapling layers. In both cases, these site preparation treatments are necessary to create the competition-free microsites afforded by tip-up mounds and well-decayed logs following natural disturbance in these forests.

Young Forest Stage (early)

The early YFS requires little silvicultural intervention in northern hardwood forests, as this is a period of rapid development of the cohort, including early stratification into

Box 13.2 **Accelerating large tree components through crop tree release**

Large living and dead trees are common attributes of the OFS, but are generally quite rare in contemporary, managed landscapes (Whitman and Hagan 2007). Although these structures may naturally develop in managed stands if given time, deliberate silvicultural treatments can also be used to advance development of these features as part of planned activities in an ecological silvicultural system. Pioneering work in ecological silviculture in northern Wisconsin demonstrated the utility of crop tree release in accelerating the large-tree component in second-growth northern hardwood forests (Singer and Lorimer 1997). This work found that, regardless of age, the duration required for canopy sugar maple to reach diameters typical of the OFS (>50 cm, 19.7 in) was reduced by 40–70 years compared to unmanaged areas when partial or full crop tree release was applied on a 10–15 year cutting cycle. This finding should not be surprising to any forester familiar with hardwood tree response to thinning treatments; however, it illustrates the utility of repurposing classical methods toward achievement of ecological objectives.

Accordingly, application of crop tree release as part of a gap-scale silvicultural system in the northern hardwood archetype borrows from our historical understanding of thinning and canopy tree growth in these ecosystems. As with past approaches, a portion of canopy trees are identified to serve as crop trees in areas not receiving regeneration harvest (i.e., in non-gap areas) during initial entries. Importantly, our use of the term "crop" differs from the classical designation, as the "crop" in our case includes trees designated for future commercial removals as well as permanent retention as legacies. For northern hardwood forests in the young forest stage, a general target density of crop trees should be 100–150 trees per hectare (40–60 trees per acre), with thinning treatments focused on releasing at least three sides of the canopy and reducing stand-wide stocking levels to 18.4-20.5 m^2/ha (80-90 ft^2/acre). The distribution of this stocking across a given stand may be quite heterogeneous given crop-tree release is often not uniformly applied across a stand (i.e., its application is dictated by where potential crop trees exist). This target density is based on a long-term goal of having 40–90 trees per hectare (16–36 trees per acre) > 50 cm (19.7 in) DBH, a structural hallmark of the OFS (Lorimer and Frelich 1997; McGee et al. 1999; Whitman and Hagan 2007), and acknowledges both natural mortality of these trees as well as the potential for future commercial removals over time.

Box 13.3 **"Morticulture" to maintain deadwood legacies and continuity in northern hardwood ecosystems.**

Silviculture has not typically focused on increasing the abundance of dead trees in the forest and, as a result, most comparisons of managed forests to their natural archetypes often highlight deadwood abundance as one of the greatest points of disparity (e.g., Figure 13.3; Table 13.4). In principle, managed forests will always have lower levels of standing and downed dead trees given the need to harvest trees to supply society's wood and fiber needs and pay for treatments. Moreover, a fundamental principle of classic silviculture is to redirect and capture wood that would otherwise be lost to competitively induced mortality to productive human use. Nevertheless, ecological foresters should pay as much attention to dead tree development as they do to live trees when developing prescriptions given the critically important ecological functions of these legacies. These functions, such as nurse logs, moisture retention, and nutrient cycling, also serve to enhance long-term economic productivity by influencing tree recruitment and productivity, adding further importance to managing for these forest elements. Deadwood is a dynamic component of forests and correspondingly, prescriptions incorporating these legacies should not only base target levels on an understanding of natural amounts for a given ecosystem, but must also consider the temporal and spatial dynamics of deadwood inputs over time.

As with silvicultural prescriptions for live tree components, the key details for implementing deadwood treatments should consider treatment severity, timing, and spatial configuration. For live-trees, harvest intensity is often based on the desired residual stocking in an area after a treatment; however, analogous stocking charts and density management diagrams do not exist for establishing post-harvest levels of dead tree stocking. Instead, the use of natural benchmarks derived from studies in old-growth ecosystems in the forest type of interest are one option for establishing natural benchmarks of deadwood abundance to guide "morticulture" decisions (sensu Harmon 2001). As mentioned above, we cannot retain all trees on a site given economic and operational constraints, so these natural benchmarks should be viewed as an upper bound of potential deadwood stocking. For example, numerous studies across old-growth northern hardwood forests have documented coarse woody material levels ranging from ~80-120 m^3/ha (1140-1715 ft^3/ac; Figure 13.3), with 30–40 percent of this volume composed of large logs > 40 cm DBH (15.7 in; Goodburn and Lorimer 1998; McGee et al. 1999) corresponding to average densities of 94–141 large logs/ha (38–58 large logs/acre). These values were used as a benchmark to guide ecological silvicultural treatments in northern Wisconsin, USA, where foresters deliberately felled or girdled stems to generate coarse wood volumes approaching 65 percent of natural levels as part of canopy treatments designed to generate multi-cohort stands (Figure 13.11). The percentage selected in this example (i.e., 65 percent) represented an upper threshold to what was economically feasible relative to harvest revenues generated. This value can be lower or higher for a given forest depending on economic objectives and stand value. One option for increasing economic feasibility is to focus deadwood management on diseased, dying, or cull individuals or species of lesser commercial value where possible.

The spatial configuration of deadwood treatments is also critical to consider and, as with many other ecological silvicultural treatments, will be based upon canopy disturbance patterns observed in a given archetype. For example, accumulations of deadwood are often spatially aggregated in northern hardwood forests and other Archetype 3 forests in the OFS, reflecting the legacies of windthrow and other gap-scale events (Woods 2004). As such, a proportion of trees typically removed when creating harvest gaps for regeneration (i.e., group openings in selection harvests or variable density thinnings) should be felled and retained at each entry to emulate these spatial patterns. As discussed above, the proportion can be based on natural deadwood volume benchmarks and a general volume conversion for "average sized" canopy trees (e.g., 0.85 m^3/tree or 30 ft^3/tree in second-growth northern hardwoods) can be determined prior to marking to accelerate decisions regarding how many overstory stems are needed to achieve deadwood targets at each entry. Of course, all deadwood pieces are not created equal from an ecological function standpoint and priority should be placed on recruiting deadwood from larger diameter trees (> 40 cm or 15.7 in DBH) given the disproportionate importance of large logs and snags in providing for habitat, carbon storage, and erosion control (McComb and Lindenmayer 1999). Unfortunately, these trees often represent the highest economic value in a stand, so marking rules will need to balance objectives related to maintaining large living and dead trees on site over time with economic returns from high value trees. One example of this balance was developed as part of the marking rules for ecological silviculture systems developed for second-growth northern hardwood forests in Wisconsin. In this approach, up to ten large trees (>40 cm or 15.7 in DBH) per hectare (four trees per acre) were designated for deadwood creation; however, if these trees made up more than 50 percent of all trees marked for harvest (i.e., deadwood creation plus those removed for wood products), marking would substitute up to thirteen smaller diameter trees (25-40 cm or 9.8-15.7 in DBH) per hectare (five trees per acre) for deadwood creation (Hanson et al. 2012).

Finally, the timing of deadwood treatments is one of the elements of "morticulture" often overlooked, as a past tendency has been to focus on restoring OFS deadwood conditions during an initial entry. Although such approaches may restore deadwood levels to those approaching old growth in the short term, they may not provide the wide variety in wood decay states and conditions that are a hallmark of most forests in the OFS, regardless of archetype. To address this concern, provisions for deadwood creation should be made at each harvest entry to create "pulses" of fresh standing and downed wood. In northern hardwood forests, this may occur every 10–25 years depending on the cutting cycles being

Figure 13.11 *Deliberate snag (a) and downed woody (b) material creation across second-growth northern hardwood forests in northern Wisconsin, USA. The goal of the treatments was to generate deadwood volumes that approached 65 percent of those found in old-growth stands in the region.*

used and would result in a mosaic of decay states across a given stand over time. In addition, retaining live tree legacies in harvest openings and across the stand to live out their natural lifespans will also provide delayed inputs to deadwood pools. An added emphasis on deadwood management will require some recalibration of timber-focused models of marking and harvesting (e.g., felling without bucking; Figure 13.12); however, in many cases deadwood management is consistent with timber-focused management goals, including improving timber quality through felling of culls and creation of suitable seedbeds for desired regeneration. As such, blending morticulture with silviculture in a given prescription should be something readily achieved by the ecological forester.

Figure 13.12 *a) Group selection opening with live-tree and deadwood retention in second-growth northern hardwoods in western Massachusetts, USA. Only cull trees were felled and left for downed woody material. Despite the ecological intent of deadwood retention in the prescription, the operator still bucked downed logs, highlighting the importance of communicating the silvicultural intent of treatments to operators. b) The same gap seven years later, showing the diversity of regenerating tree species, including yellow birch and pin cherry, and the advancement of wood decay.*

canopy layers reflecting the differences in height growth patterns and shade tolerance in the regenerating species. In the event that there are few regenerating yellow birch saplings relative to the shade-tolerant species released during the Disturbance and Legacy Creation stage, localized, mechanical release treatments can be applied around these individuals to insure their representation in the canopy. Nevertheless, the best course of action from a silvicultural and ecological perspective during this period is generally to do nothing, particularly if attention has been paid to creating suitable microsites during Disturbance and Legacy Creation and the PFS.

Young Forest Stage (after canopy closure at gap level)

Cohorts initiated during the Disturbance and Legacy Creation event will continue to develop as stratified mixtures as the YFS progresses, with shade-intolerant and mid-tolerant species, including paper and yellow birch and aspen (trembling or bigtooth), forming upper canopy strata and shade-tolerant species, including sugar maple, American beech, and to a lesser degree, eastern hemlock, dominating lower strata. Pin cherry, a noncommercial yet ecologically important species, also often dominates the upper stratum during the early part of the YFS, and typically naturally senesces 20-40 years after gap formation. Ecological foresters should resist the temptation to remove this non-merchantable species in the YFS, as the large volumes of nutrient rich foliage and soft mast provided by pin cherry early in forest development are critical for facilitating regeneration of species like sugar maple and supporting a diversity of wildlife species (Leak 2003).

The focus of silvicultural treatments in the developing YFS is primarily on accelerating large trees through thinning, namely crop tree release (Box 13.3). These treatments generally occur 50–70 years after the disturbance and legacy stage as trees in the regenerating cohort grow through the pole size class (13-28 cm; 5-11 in DBH). The intent is to create large, future legacy and crop trees at the cohort-scale; however, crop tree release may also be applied as a first entry across pole-sized, even-aged second-growth stands to create a distributed source of large canopy trees. In the latter case, these thinning treatments would be applied at the stand scale in conjunction with regeneration methods, such as group selection, variable density thinning, or continuous cover irregular shelterwoods, to transition to a spatially heterogeneous, multiaged condition (see *Disturbance and Legacy Creation Stage*, above).

Mature Forest Stage

As with many other archetypes, the MFS is the prevailing current condition for northern hardwood forests in most regions (e.g., Figure 13.7). Depending on past management history, this condition may include areas of advance regeneration of shade-tolerant species, reflecting past partial harvests and/or a high proportion of cull individuals in the canopy due to past exploitive harvesting. The focus of treatments during this stage will work with these conditions to further transition structure and composition toward those found in the OFS. Remember from the disturbance and legacy creation stage that the primary objectives of treatments during the MFS are recruitment of new age classes, continued advancement of the large-tree component, recruitment and retention of dead and live-tree legacies, and increased representation of intolerant and mid-tolerant species.

Variable density thinning is one approach to achieve the aforementioned objectives, particularly if there are > 100 canopy trees per hectare (40 per acre) or 40 percent of current stocking with future commercial potential (cf. Bédard et al. 2014). Ideally, this approach would entail creating canopy gaps ranging from 0.04–0.2 ha (0.1-0.5 acre) across 10–20 percent of the stand area, with gap placement favoring areas with a concentration of cull or existing advance regeneration (Figure 13.13). The aforementioned prescription represents an ideal case; however, the distribution and abundance of acceptable growing stock and advance regeneration is rarely uniform in northern hardwood forests in the MFS owing to past management and the influence of natural disturbances. To accommodate this variability and provide for future operability, a general rule of thumb when implementing variable density thinning in these forests is to restrict harvest gaps to no more than 30 percent of the harvest area and to maintain residual basal areas of 14–16 m^2/ha (61–70 ft^2/ac) at initial entries. Such targets have been effective in maintaining high levels of future economic potential, while also transforming stand structures from even-aged to two-aged conditions that include mid-tolerant species in the new cohort (Bédard et al. 2014). In harvest gaps, within-gap legacy creation through retention of live trees (Figure 13.14) and deliberate creation of deadwood is an important element to include as part of silvicultural activities (Box 13.3). As with the YFS, crop-tree release can be applied to the remainder of the stand to accelerate large-tree development, with those trees selected for release reflecting individuals and species with the greatest potential for long-term growth. Subsequent entries during this stage might occur on 15–20 year intervals and would concentrate on creating new canopy gaps in areas receiving thinning treatments in the previous entry. These gaps could either be placed in areas away from initial gaps to reflect the often isolated nature of multiple tree-fall gaps, or could be used to expand the extent of previously created gaps to reflect the larger coalescence of disturbed areas over time observed after meso-scale events (Raymond et al. 2009).

Figure 13.13 *Examples of live-tree retention at the gap-scale as part of northern hardwoods ecological silvicultural systems (a-c).*

Figure 13.14 *View of the thinned matrix areas from across harvest gaps in variable density thinnings in northern hardwoods in northern New Hampshire, USA.*

Old Forest Stage

Very few areas on the landscape currently reflect the OFS for the northern hardwood archetype and those that do are typically small areas in ecological reserves where silvicultural activities are excluded. In some cases, release treatments to reduce the impact of invasive plant species or protection measures to minimize browse impacts on regeneration might be applied in these places; however, manipulation of canopy trees as part of an ecological silvicultural system is unlikely. Instead, the opportunity to apply silvicultural treatments in this stage of development represents a long-term goal for areas that have returned to these conditions following time and the long-term application of the silvicultural system described across the previous developmental stages (Figure 13.15).

The management regime for areas that ultimately attain stand structural and compositional conditions analogous to those found in the OFS would largely reflect application over time of a mosaic of the treatments described for the previous developmental stages. A primary activity would entail regeneration harvests through a combination of single-tree and group selection openings that emulate the frequent, gap-scale regime characterizing this archetype and would contain the microsite and legacy elements described in previous stages. Excellent, classic guidelines exist for implementing these treatments in northern hardwood forests (Leak et al. 1987) with an important departure here being the permanent retention of some individuals in larger size classes at the group and stand-level to maintain large living and dead tree legacies over time. The intent of these harvests would be to maintain the multiage structure of the OFS through frequent (15–20 year) initiation of the PFS and YFS in space and time using harvest gaps that included larger openings (> 1 ha; 2.5 acre), reflecting mesoscale events. Retention of legacies in prior stages would preclude the necessity for application of crop tree release treatments for large tree development across the entire managed area, but these treatments could still be applied at the cohort-level to accelerate live-tree structural development.

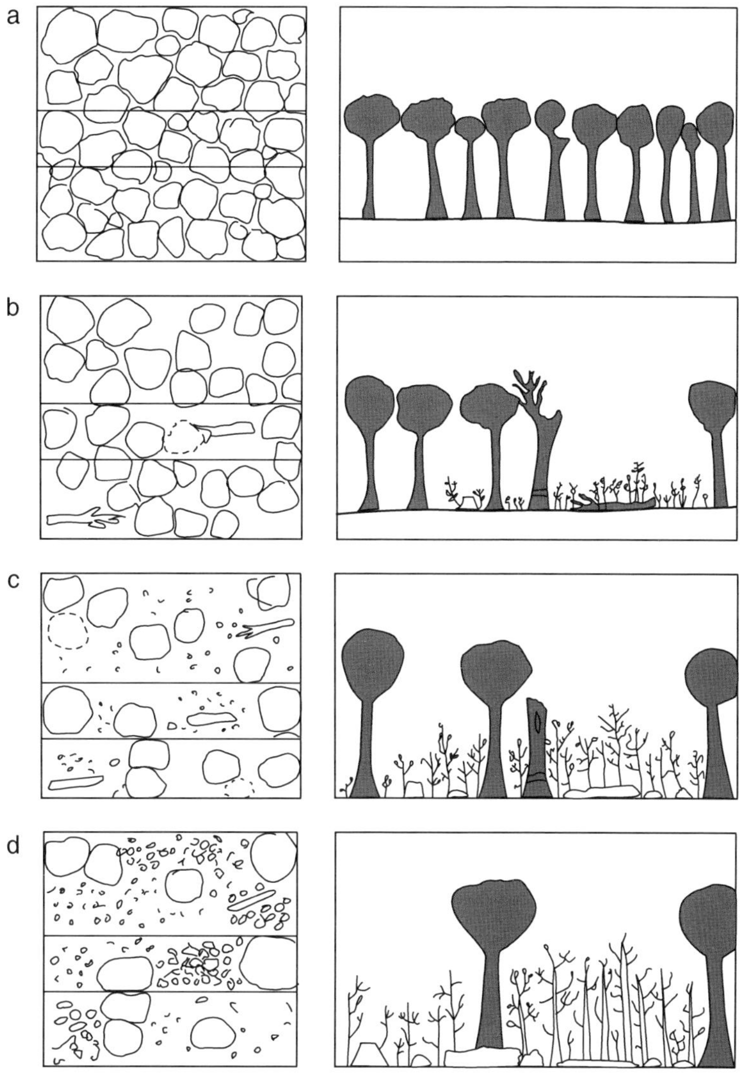

Figure 13.15 *Application of gap-scale silvicultural system to second-growth northern hardwood forest. Left panel represents overhead view with right panel profile based on transect demarcated through overhead view. a) Second-growth forest in early Mature Forest Stage prior to treatment. b) Disturbance and Legacy Creation event applied to Mature Forest Stage using variable density thinning (continuous cover irregular shelterwood) designed to recruit new cohort in multi-tree canopy gaps, accelerate large tree development through crop-tree release, and recruit live and dead tree legacies through designation of leave trees and creation of snag and downed wood structures. Combination of single-tree and group removals are applied c) 15 and d) 30 years after initial entry to accelerate development of compositional and structural conditions characteristic of Old Forest Stage. Note "removals" include snag creation at each entry (indicated by dashed lines in overhead view).*

Examples of Ecological Silviculture in Practice

Many ownerships rely primarily on the timber production model for northern hardwood forests (see *Evolution of timber production silviculture for northern hardwood forests*, above); however, there has been an increased integration of ecological principles into northern hardwood silviculture over the past several decades. This inclusion of ecological objectives has reflected both the broader appreciation for the value of ecological silviculture approaches across the global forestry community, as well as the economic compatibility of integrating many aspects of ecological silviculture with more classical models used for managing northern hardwood forests, including the use of multiage systems and management for mature structures. In addition, ownership patterns across the range of northern hardwood forests are quite complex, with many areas dominated by family forest ownerships. Correspondingly, the rising use of ecological silviculture in this forest ecosystem may reflect a greater compatibility of this approach with many landowner objectives relative to one simply focused on timber production (e.g., D'Amato and Catanzaro 2007). Although there are many agencies and individuals applying the principles of ecological silviculture in practice (Table 13.6), we chose to highlight two that reflect applications of these principles

Table 13.6 *Examples of organizations and agencies that recommend some aspect of ecological silviculture in their northern hardwood management guides.*

Entity	Ecological Objectives	Approach
Wisconsin Department of Natural Resources http://dnr.wi.gov/topic/forestry	Provision for biological legacies, wildlife habitat; restoration of old-growth forest conditions; production of diverse ecological benefits	Use variety of canopy gap sizes to promote range of tree species and develop vertical structure; retain >15 trees per hectare (seven per acre) to complete natural lifespan; consider increasing large tree component through crop-tree release and managing individual trees to diameters >61 cm (24 in)
Aitkin County Land Department, Minnesota https://www.co.aitkin.mn.us/departments/Land/landhome.html	Provide full range of habitat features in terms of forest age classes, species, and structures at the stand and landscape scale	Use of variable density thinning and group selection with crop tree release to perpetuate uneven-aged stands that include mid-tolerant species; retain snags, den trees, and mast trees; maintain multi-layered canopy and mature-forest conditions for wildlife
Michigan Department of Natural Resources http://www.michigan.gov/dnr/forestry	Increase structural and compositional diversity at the stand and landscape scale	Use group selection to improve vertical and horizontal structural diversity; retain mixture of overstory tree species to live out lifespan and contribute to woody material pools; create microsites for small-seeded species through retention of nurse logs or exposure of mineral soil; consider underplanting of underrepresented but historically important species (eastern hemlock)
Quebec Ministry of Natural Resources and Wildlife https://mffp.gouv.qc.ca/english/publications/forest/ecosystem-based-management.pdf	Maintenance of stand and landscape biodiversity	Silvicultural guidelines in general include employing knowledge of natural disturbance to determine the best silvicultural system with an emphasis on methods using partial harvesting and retention to maintain critical and sensitive habitat features.

on public and family forest lands: 1) a partnership of the Vermont Department of Forests, Parks and Recreation (VT FPR) and Audubon Vermont and, 2) the Wisconsin Department of Natural Resources (DNR).

Silviculture with Birds in Mind, Vermont Department of Forests, Parks and Recreation and Audubon Vermont

Forests cover almost 80 percent of the land base in the state of Vermont, USA, yet unlike many other regions of the US, the vast majority of these forests are owned by family forest owners (76 percent of total acreage; Foster et al. 2017). Engagement of these owners in active forest management and stewardship through landowner assistance programs has been quite modest in many regions, as these generally have a central emphasis on management plans for long-term timber production—an emphasis incongruous with most stated family forest owner objectives (Butler et al. 2014). Instead, surveys of these ownerships consistently point to wildlife habitat as a primary reason they own their land. Vermont had been successful in enrolling many landowners in their current use program, which does require periodic timber harvests. Yet foresters within the Vermont Department of Forests, Parks and Recreation (VT FPR) recognized an opportunity to further engage these enrolled landowners and broaden the appeal of active forest stewardship by focusing on this repeatedly stated, primary motivation for family forest ownership. To do this, VT FPR worked with Audubon Vermont to develop the program, *Foresters for the Birds* (FFB), which applies ecological and, in some cases, timber production silviculture to integrate songbird habitat management with timber management (Hagenbuch et al. 2011).

The focus of the FFB program is on using silviculture to create habitat for forest songbirds, represented by twelve priority bird species, all of which rely on the range of forest conditions contained in the OFS of northern hardwood ecosystems (Hagenbuch et al. 2011). Like most other regions dominated by northern hardwood forests, Vermont has a preponderance of second-growth forests having relatively little structural or compositional diversity. Accordingly, the FFB program presents a long-term vision and associated silvicultural strategies for family forest owners to increase the structural and compositional complexity of their forest land. To accomplish this, the FFB approach relies on several different options ranging from variable density thinning to irregular shelterwoods to increase the representation of key songbird habitat features, including complex horizontal and vertical structure, while also

Figure 13.16 *Northern hardwood forests managed as part of Audubon Vermont (USA) Foresters for the Birds program. Variable density thinning is designed to increase vertical and horizontal complexity in structure and composition over time to benefit several wood warbler species that require habitats found in the Old Forest Stage.*

improving timber quality for long-term economic objectives (Figure 13.16; Hagenbuch et al. 2011). For example, applications of variable density thinning recommended through this program typically combine group selection harvests, ranging in size from 0.004–0.04 ha (0.01–0.1 ac) across 5–15 percent of the stand, with crop tree release between groups on 74–173 trees per hectare (30–70 trees per acre). Attention is paid during gap placement so as to release advance regeneration of conifer species to increase this component for select bird species, such as black-throated green warblers. The use of girdling to release crop trees and increase deadwood components is also encouraged. Moreover, an emphasis is placed on viewing crop trees through the lens of bird habitat needs, including retaining species of value for foraging (yellow birch), individuals with large crown forms, and mast-producing species. Implementation of these approaches varies across ownerships; however, the consistency of these ecological approaches with landowner objectives has been clear with 261 family forest management plans now integrating these methods and over 100 foresters participating in the FFB program since its inception in 2011 (S. Hagenbuch, personal communication).

How well does the FFB approach address the four principles of ecological forestry?

The FFB approach, through its emphasis on creating deadwood and retaining live-tree legacies, including those serving critical habitat functions, demonstrates Principle 1, continuity. Principle 2—complexity, is at the core of the overarching goal of the FFB approach, as variable density thinning is used to create complex vertical and horizontal structural conditions while encouraging a diversity of canopy tree species through group openings and retention of less common overstory species. The long-term provision of bird habitat is a core goal of FFB and treatments are presented in the context of the duration of suitable habitat for different bird species following their application (e.g., 1–30 years following gap formation for eastern wood-pewee). Timing future treatments based on the duration of and need for an array of the habitat conditions at the stand and landscape scale over time is an excellent example of Principle 3, timing. Finally, although landowner outreach and engagement is occurring throughout the state, a primary focus of FFB is on landowners near larger, protected priority forest blocks in an effort to enhance the functioning of these areas in providing habitat for migratory bird species in the region, an illustration of Principle 4, context. Through engaging these owners, a broader goal is to incentivize forest ownership and protection to limit future fragmentation and development of these key forested areas.

Big Tree Silviculture, Wisconsin DNR

Northern hardwood forests constitute a large portion of the landscape of Wisconsin, USA (>1.5 million hectares; 3.8 million acres) and the Wisconsin Department of Natural Resources (DNR) is responsible for managing a portion of these forests on their own state forest lands, as well as providing broader silvicultural guidance for the management community in the state. These forests were primarily in the OFS at the time of European settlement (Frelich 2002), but were greatly transformed and eliminated by intensive harvesting and land clearing in the late nineteenth and early twentieth centuries, resulting in a current second-growth resource generally between 60–110 years old. In recognition of this historical change and growing awareness of the value of old-growth forests, the Wisconsin DNR, in collaboration with several forest scientists, embarked on a decade-long, comprehensive study of remnant old-growth areas to develop an understanding of the structure and functioning of these forests (Fassnacht et al. 2013). In addition, a primary focus of this research was to identify any differences that existed in structure and composition of these remnant, natural systems relative to those found in contemporary, second-growth forests being managed with a timber-focused model (e.g., Goodburn and Lorimer 1998, 1999). This work culminated in a broad initiative within the Wisconsin DNR to integrate management for old-growth structural and compositional conditions into statewide forest planning (WI DNR 2004 a, b).

The silvicultural approaches developed by the Wisconsin DNR to restore structural and compositional conditions found in the OFS are best exemplified by the operational demonstration work implemented as part of the Managed Old-Growth Silviculture Study (MOSS). This work packages a suite of classical and ecological silvicultural approaches to increase structural and compositional

Figure 13.17 *a) Aerial view of a portion of Northern Highlands American Legion State Forest in northern Wisconsin, USA being treated with continuous cover irregular shelterwood emulating mesoscale wind disturbance. Initial entries have created a range of canopy disturbance severities at the stand level reflecting historic damage from these events. Regeneration harvests focus primarily on four openings ranging in size from 0.4–1.2 ha that contain within-gap retention of 35–65 percent of the pre-harvest basal area. Areas surrounding openings are thinned with variable intensities to reflect variability in localized mortality. b) View of 1.2 ha area receiving regeneration harvest as part of meso-scale wind emulation five years after establishment cutting. a) Photo credit: Laura Reuling*

complexity across large forested areas and builds upon general guidance for "old-growth silviculture" developed by the Wisconsin DNR (WI DNR 2004 b). In contrast to a historical focus solely on sugar maple, regeneration harvests and associated site preparation treatments are applied to increase the representation of yellow birch and other historically important canopy species (e.g., Figure 13.9). This includes using continuous cover irregular shelterwoods to emulate the mesoscale wind disturbances historically critical for yellow birch establishment and the maintenance of structural and compositional complexity (Hanson and Lorimer 2007; Figure 13.17). In addition, downed dead trees and snags are deliberately created from cull and non-commercial canopy species at each entry to enhance the abundance of these legacies. Trees above certain diameter thresholds (e.g., >50 cm; 20 inches) or representing locally rare canopy species are permanently retained in matrix areas around harvest gaps to restore future large-tree components and preserve unique seed sources and ecological functions—a retention approach in stark contrast to historic marking guides for northern hardwood silviculture (Arbogast 1957).

How well does the Wisconsin DNR address the four principles of ecological forestry?

The Wisconsin DNR approach to managing for old-growth forest conditions clearly addresses Principle 1 (continuity) by creating coarse woody material and permanently retaining large diameter and rare canopy species. Applying regeneration methods that rely on a range of canopy opening sizes and spatial patterns of removal, like the meso-scale wind approach, embraces Principle 2 (complexity) by creating and maintaining a diversity of structural and compositional conditions spatially and temporally. The frequencies of future entries are reflective of their natural occurrence, with smaller gap creation occurring at greater frequencies than those harvests creating larger gap or stand-wide disturbances, thus addressing Principle 3 (timing). Finally, the influence of context (Principle 4) in guiding these silvicultural decisions is harder to assess; however, by applying a diversity of regeneration methods and retention strategies to address objectives associated with restoring OFS conditions, there is variation in approach at landscape scales.

What to Remember about Silviculture for Archetype 3 Ecosystems

Archetype 3 ecosystems are typically dominated by shade-tolerant species of high commercial value. As a result, historical management of these ecosystems has primarily used uneven-aged silvicultural systems to produce timber from desired species, with little consideration for the compositional and structural diversity characterizing these forests under a natural development model. Although classical uneven-aged management contains elements of the natural model (namely the promotion and maintenance of multiage structures), there are key ecological outcomes of the gap-scale silvicultural system that are quite different for the model that is only timber-focused. These include a higher diversity of canopy species, the presence of large live and dead tree legacies, and a greater range of canopy disturbance severities in space and time. The management of forests containing these features is particularly important in the context of management objectives related to the maintenance of long-term forest productivity—including for timber—and biodiversity conservation as well as resilience and adaptation to the many stressors now facing these ecosystems (see Chapter 16).

To achieve these and other ecological objectives, remember that application of the gap-scale silvicultural system to Archetype 3 ecosystems will include:

1. Initial treatments applied to create a spatially heterogeneous mosaic of structural and compositional conditions at stand and landscape levels that include gap-scale regeneration harvests to promote transition to multiage conditions and crop-tree release treatments to accelerate large live-tree development (*addressing Principle 2: complexity and Principle 4: context*).
2. Live-tree and deadwood legacy retention and creation as an integral component of all regeneration harvests, including within-gap retention of large canopy trees and felling and leaving of deadwood legacies in harvest gaps to emulate pulses of large woody material (*addressing Principle 1: continuity*).
3. Creating a range of canopy gap sizes through harvesting over time for initiating the Preforest Stage within stands and across the landscape in reflection of the historical size and frequency of gap- and meso-scale disturbances in these ecosystems (*addressing Principle 3: timing and Principle 4*).
4. Regeneration harvests that favor all component tree species and include specific site preparation treatments or seasonal harvesting considerations (snow-free harvests) to create competition-free, exposed mineral soil seedbeds for the recruitment of small-seeded species (e.g., yellow birch and eastern hemlock) in northern hardwood ecosystems (*addressing Principle 2: complexity*).
5. Designation and management of ecological and economic "crop-trees" for recruitment of large tree structures occurring in a range of spatial configurations across stands and managed areas (*addressing Principle 4*).
6. Long-term transition from initial treatments that focus on increasing spatial heterogeneity in live and dead tree structures to gap-scale harvests that maintain a within-stand mosaic of developmental stages and structures as development progresses (*addressing Principle 2*).

CHAPTER 14

Silviculture for Archetype 4 Ecosystems: Forests with Mixed-Severity Disturbances

In this final silviculture systems chapter, we focus on the Archetype 4 construct, that is, forests that are characterized by a mixed-severity disturbance regime: mixed in terms of severity and outcome among stands and over time and for some forests, mixed in terms of agent of disturbance. Recall from the description of Archetype 4 (Chapter 6) that these ecosystems are characterized by several features that in combination distinguish them from the other three archetypes (i.e., infrequent severe disturbance; frequent, low severity disturbance; moderate severity, gap-based disturbance). Specifically, canopy disturbances in Archetype 4 forests can range from infrequent, near stand-replacing events to heavy, but often less than stand-replacing events from fire, wind, root-rot fungi, or other agents. Moreover, these forest may also experience more frequent, but less severe disturbances that may not result in tree regeneration, typically surface fire. The canopy tree community can be dominated early by shade-intolerant species and this dominance may be maintained over time, even into the Old Forest Stage (OFS). Additional tree species are often included in the canopy, however, with these species spanning a range of shade tolerance and successional status. After more intense disturbance (either at the stand or within-stand—hereafter patch—scale), there may be an extended Preforest Stage (PFS), or this stage may be relatively short-lived, grading early into a Young Forest Stage (YFS). As a result of the variable canopy disturbance regime, the age structure of Archetype 4 forests can range from broadly **single cohort** to distinctively **two- or three-cohort**. Finally, Archetype 4 forests are characterized by high levels of structural complexity and spatial heterogeneity of structure and composition in the Mature Forest Stage (MFS) and OFS, as well as among stands across a landscape.

Globally, Archetype 4 ecosystems, when managed for timber, may often deviate considerably from the natural model. Consequently, there is ample opportunity to bring what you have learned about the principles of ecological silviculture to bear when conceptualizing an ecological silviculture system for forests in this archetype and for writing prescriptions that better reflect an ecological model. The purpose of this chapter is to help you do this by providing a model for silviculture that is more reflective of the ecological construct for Archetype 4 ecosystems than the typical (often timber-focused) model. For this, we will again turn to our case study for this archetype—red pine forests of the western Great Lakes region of North America—but we emphasize that the information presented should be broadly applicable to other Archetype 4 ecosystems (Table 14.1; Figure 6.38).

For the red pine ecosystem, we first briefly review environmental characteristics and ecological features in the context of the development model presented in Chapter 6. Next, we will examine the silvicultural approaches that are used for this ecosystem when timber production is a primary objective and contrast resultant features of the timber-focused management model to those resulting from natural disturbance and development. Then we will outline an ecological silvicultural system for this forest type (and archetype) that is more overtly based on a natural model, allowing better incorporation of the fundamental principles of ecological silviculture into prescriptions. We finish by highlighting several organizations that are already incorporating ecological principles to varying degrees into silviculture for the forest type.

The Great Lakes Red Pine Ecosystem

Recall from Chapter 6 that our red pine ecosystem case study occurs in the western Great Lakes region of the United States and southcentral Canada. Historically, this ecosystem occurred in mixed-species stands, but also occurs in stands dominated by red pine on sites having **xeric** to **dry-mesic** soil moisture regimes. Other species of varying abundance in this forest can include eastern white pine, jack pine, balsam fir, white spruce, trembling and bigtooth aspen, red maple, northern red and bur oak, and paper birch, species spanning a wide range of silvical characteristics (Table 14.2). These forests are structurally complex and heterogeneous in structure across most stages of forest development, especially in the OFS (Figure 14.1).

Table 14.1 *Examples of Archetype 4 Ecosystems.*

Ecosystem	Distribution	Natural Disturbance	References
Great Lakes red pine	Northern Minnesota, Wisconsin, Michigan, USA; Ontario, Canada	Fire, wind, root-rot fungi	Chapter 6
Scots pine	Northern Europe, Fennoscandia	Fire	Kuuluvainen 2009
Norway spruce	Central Europe, Northwest Russia	Wind, cold mortality? Bark beetles, wind	Trotsiuk et al. 2014 Khakimulina et al. 2016
Interior western moist mixed conifer	Northern and central Rocky Mountains, USA; Interior Rocky Mountains, Canada; Eastern Cascades, USA	Fire and to a lesser extent insects and pathogens	Agee 1993 Hessburg et al. 2016 Schoennagel et al. 2004
Table Mountain pine-pitch pine	Appalachian Mountains, USA	Surface fire, canopy disturbance (insect, wind, ice, drought)	Brose and Waldrop 2006
Douglas-fir-western hemlock	Dryer sites in western Cascade Range and coastal mountains in Washington and Oregon, USA	Fire, wind	Tepley et al. 2013

Table 14.2 *Select silvical characteristics of tree species often found in Great Lakes mixed pine (red pine) woodlands in North America. Modified from the Wisconsin DNR Silviculture Handbook (https://dnr.wi.gov/topic/ForestManagement/silviculture.html).*

Species	Shade tolerance	Preferred seed bed	Early relative growth rate	Sprouting ability	Maximum life span (years)
Paper birch	Intolerant	Moist mineral	Fast	Moderate (stump)	200
Red maple	Tolerant	Litter	Moderate	High (stump)	150
Bigtooth and trembling aspen	Very intolerant	Moist mineral	Very fast	High (roots)	100
N. red oak	Intermediate	Moist mineral with litter	Moderate	Moderate (stump)	400
Bur oak	Intermediate	Mineral	Slow	High (stump)	500
Eastern white pine	Intermediate	Moist mineral	Slow to Moderate	None	450
Red pine	Intolerant	Moist mineral	Slow	None	400
Jack pine	Very intolerant	Moist mineral	Slow to moderate	None	185
Balsam fir	Very tolerant	Moist mineral	Moderate	Limited (layering from prostrate stems)	200
White spruce	Tolerant	Moist mineral	Moderate	None	300

Figure 14.1 *Old Forest Stage red pine forests. a) Itasca State Park, Minnesota, USA, b and c) Seney National Wildlife Refuge, Michigan, USA. Note multiple tree species, large dead wood, diversity of tree sizes, and patchy canopy. Photo credits: (a) Shawn Fraver, (b, c) Greg Corace.*

The disturbance model for the red pine ecosystem is characterized by frequent **surface fire** and infrequent **crown fire**, along with wind and root-rot disturbances. Studies show that prior to European settlement, mean return intervals for surface fires ranged from 5 to 75 years, with shorter return intervals in higher use areas by indigenous people. These fires reduced hardwood competitors, thinned populations of pines, and maintained a woodland structure. Historically, mean fire return intervals for crown fires resulting in pine regeneration were on the order of 150–250 years. However, surface and crown fires were linked in that the former occasionally developed into crown fires that killed overstory trees. Importantly, fires that killed trees were often (but not always) patchy within stands, and thus they could be less than stand replacing and certainly varied in their severity across a landscape. The key developmental attributes for this forest result largely from the heterogeneous nature of the fire regime (Table 14.3). Additionally, large-gap or patch-scale tree mortality from wind and root-rot fungi may also be important drivers of heterogeneous stand structures and variation in composition, reinforcing the heterogeneous structure resulting from patchy fire (Figure 14.2).

Red Pine Forests as a Timber Resource

Red pine forests are an important timber resource in the western Great Lakes region of North America. The province of Ontario, Canada reported over 1 million ha (nearly 2.5 million ac) of red pine forest (reported with eastern white pine) in 2011 (Watkins 2011) and over 156,000 m^3 (>5.5 million ft^3) of red pine was harvested from 2012–2013 from these forests (https://www.ontario.ca/page/annual-report-forest-management-2012-2013). In the three US Lakes States of Minnesota, Wisconsin, and Michigan, the estimated area for red pine forests in 2016 was over 800,000 ha (>2,000,000 ac). Over 76 percent of this is plantation forest, most of which is on sites where red pine occurred

Table 14.3 *Key attributes of developmental stages in red pine ecosystems.*

Stage or Event	Attribute	Cause or Consequence
Disturbance and Legacy Creation	Spatially heterogeneous live and dead tree legacies ranging from dispersed to aggregated; matrix of open, disturbed forest with live trees individually or in patches or large openings in matrix of less disturbed forest	Results from near stand replacing to patchy disturbance mostly from fire, but also wind and root-rot fungi
Preforest Stage	Herbaceous and shrub dominated; may be extended and important or locally absent, lasting from < 1 to several decades	An early post-disturbance period dominated by herbs and shrubs due to slow establishment of pines; alternatively, prolific hardwood sprouting may bypass PFS in stands or patches within stands
Young Forest Stage	May last 40 to 70 years for new cohort pine; red pine, the MFS and OFS dominant species, may establish early, often with mixtures of other species	Densities of regenerating pine may be low initially, increasing slowly over decades; mixtures of multiple species are important
Mature Forest Stage	Red pine dominated, but with other tree species, including eastern white pine, sometimes abundant; shorter lived conifers (jack pine, balsam fir, white spruce) decline; natural thinning occurs in dense patches of pine; small-scale disturbance increases heterogeneity; larger dead and decadent trees accumulate	Patchy surface fires limit recruitment of sensitive species into the canopy; recruitment can occur in fire-skips; patchy fires creates openings for pine establishment; root-rot gaps are foci for hardwood establishment
Old Forest Stage	Dominated by red pine, with shared importance of eastern white pine; patchy canopy (woodland) structure; broadly even-aged or two-, three-cohorts; diverse tree sizes including large pines; some large individuals of other species; abundant large dead wood including snags; heterogeneity of structure; regeneration in patches	Long-lived pines accumulate complex structural attributes on individual trees; patchy surface and crown fires, wind, and root-rot gaps, create heterogeneity and foster tree diversity

Figure 14.2 *Large patch mortality of red pine from Armillaria root-rot fungi. Photo credit: Doug Kastendick.*

naturally. Annual harvested volume in 2016 of red pine from these forests was over 1.6 million m^3 (>57 million ft^3). Depending on the state examined, this level of harvest ranked third to fifth among all species in all forest types (USDA Forest Service Forest Inventory Data Online; http://apps.fs.fed.us/fia/fido/index.html). Timber product classes from red pine include **cabin logs**, **utility poles**, **sawtimber**, **pulpwood**, and **chips** (Figure 14.3).

Timber-Focused Silviculture for Red Pine Forests

You should get the sense from the discussion above that red pine forests are actively managed and an important resource for timber products. How they are managed relative to an ecological model of stand dynamics is where we will now turn our attention. Historically, for most organizations and ownerships, even-aged silvicultural systems have been the predominant management approach for red pine forests. This has been true in both natural origin and plantation forests (Figure 14.4).

Specifically, the predominant silvicultural system has included clearcutting, some type of site preparation, planting of largely red pine at high density, competition control, one or two commercial thinnings, and a regeneration harvest on a rotation of 50 to 150 years, depending on the organization or ownership. Seedtree and shelterwood systems, while not without precedent (Figure 14.5) (Eyre and Zehngraff 1948; Ricker and Zasada 1958), have largely been abandoned in timber-focused settings. One reason for this relates to concern over potential disease issues in some locales (Box 14.1), but more generally, abandonment of seedtree and shelterwood systems across

Figure 14.3 *Recently harvested red pines sorted to a) pulp logs (left) and saw logs (right) and b) utility poles and cabin logs. Photo credits: (a) Eli Sagor, (b) Doug Kastendick.*

Figure 14.4 *Natural (a) and plantation (b) origin red pine stands managed using even-age silvicultural systems. Photo credit (b): Doug Kastendick.*

Figure 14.5 *Red pine, eastern white pine, and jack pine regeneration established from the seed of trees left after an initial cutover of old-growth forest in 1915 in northern Minnesota, USA. Photo credit: Forest History Society.*

Box 14.1 **Red pine shoot-blight and multi-cohort stands.**

Understory red pines are particularly vulnerable to fungal pathogens that cause shoot blight, including *Sirococcus conigenus* and *Diplodia pinea*, with symptoms including browning of needles, death of branches, and mortality of regeneration (Ostry et al. 2012). Outbreaks of *Sirococcus* may be triggered by wet springs, shade, or high humidity (Ostry et al., 1990), while outbreaks of *Diplodia* occur in moisture-stressed trees or when wounded by hail or shoot insects (Nicholls and Ostry 1990). Both diseases are most severe on regeneration growing near infected overstory trees since the latter trees serve as the source of inoculating spores (Figure 14.6), although logging debris and cones can also be a source of *Diplodia* inoculum (Ostry et al. 2012).

Given that infected overstory trees are the primary source of inoculum for infecting regeneration, managing for more than one age cohort of red pine has been discouraged by many organizations. Simplifying age structures by managing for even-age stands can greatly reduce the potential for shoot blight diseases, but unfortunately, doing so runs contrary to the natural age structure of red pine forests, which include two- and three-cohort and broadly single-cohort stands, all of which put regenerating red pine in close proximity to overstory trees.

An approach to managing red pine stands according to a natural model when there is known or suspected shoot blight issues is to diversify composition of trees in ways that are fully consistent with the natural dynamic of the ecosystem and in line with our ecological forestry principles. For example, a regeneration harvest may employ retention of live red pine as legacies, with regeneration focused on other species native to the ecosystem, notably eastern white pine, jack pine, and oaks. A future regeneration harvest might retain mature eastern white pine and other species, while regenerating primarily red pine. This alternation of pine dominance breaks up the association of mature and regenerating red pine in time and reduces potential for mature trees to infect seedlings, but it still maintains a complex age structure and mixed-species composition in the stand, both characteristics of the natural ecosystem.

Incidence of shoot blight may be higher with lack of surface fires (the norm for contemporary examples of this ecosystem in the region), and thus reintroduction of surface fire could potentially reduce shoot blight occurrence. Doing so would be in line with emulation of natural disturbance dynamics.

Figure 14.6 *(a) Red pine regeneration growing in close proximity to older cohort of red pine. (b) Red pine seedlings infected with Diplodia shoot blight.*

the range of red pine is likely due to lack of patience with the often extended seedling recruitment phase that results from infrequent mast seed years (and that allow a PFS to occur). Thus, silvicultural systems that result in multi-aged structure are rare on the regional landscape and their omission is a great example of incongruity between a natural model of development and regeneration expectations in a timber-focused model, in terms of timing and levels of regeneration stocking at harvest. A general timeline of activities for the timber-focused silvicultural approach can look fairly similar among ownerships (Table 14.4).

Relationship of a Timber-Focused Model to an Ecological Model

In the past, it has been common to present the timber-focused silvicultural model for red pine (Table 14.4) as

Table 14.4 *Timeline of silvicultural activities in red pine forests managed with a timber-focused model.*

Year	Activity	Objective/Consequence
0	Clearcut harvest in summer, fall, or winter depending on the organization and objectives	Create high light conditions for seedlings; limited attention given to legacy retention beyond minimal recommendations in best management guidelines (see Chapter 3); leads to establishment of even-aged stands
1	Site preparation in summer after harvest, using herbicides or mechanical methods	Reduce competing vegetation; potential to inhibit already established regeneration and other vegetation; may reduce potential for mixed-species stands
2	Plant in spring; density recommendations range from 1000 to 2500 trees/ha (400 to 1000 trees/ac) with 1500 to 2000 (600 to 800) most typical	The goal is full site occupancy by red pine early in development; largely bypasses an extended PFS or YFS
2–4	Application of browsing control in fall in some locales as needed	Needed in many locations due to high white-tailed deer densities
4–5	Competition control using herbicides or mechanical release	Ensures site dominance of planted red pine; can reduce associated tree species, shrubs, and herbaceous species
30	First thinning	In plantations: row thinning or row thinning combined with low thinning between rows; In natural stands: strip thinning combined with low thinning in leave strips
45	Potential second thinning	Additional row thinning in plantations or free thinning around crop trees working off of existing harvested rows or strips (in natural stands)
60	Potential third thinning	"………."
50–150	Final harvest	Typically clearcutting, often with discouragement of retention of red pine due to disease and competition concerns

emulative of the natural disturbance and development model for the native forest type. This reflects the prevailing belief that stand-replacement fires and the ensuing even-age forest development best characterize the ecosystem. The misguided notion that clearcutting emulates stand-replacement disturbance is not unique to this ecosystem and there has been ample research and discussion over the years to dispel this idea for many different forest types (e.g., Hansen et al. 1991; Bergeron et al. 1999; Kuuluvainen and Siitonen 2013).

Yet the argument has not always been sufficiently tempered for the Great Lakes red pine ecosystem. While it is true that in some places, near-stand-replacing fires occurred and the resultant age structures can be described as broadly single-cohort, the simplified management model fails to capture the details of the natural model in many other ways. Moreover, given that the natural disturbance model is better characterized as spatially variable within and among stands (in severity) and that age-structures are much more complicated than narrowly even-aged, the timber-focused model falls far short of capturing the rich structural complexity and heterogeneity of this ecosystem, both within stands and across the landscape (Table 14.5).

An Ecological Silvicultural System for Red Pine (Archetype 4) Ecosystems

From the above discussion, it should be clear that a timber-focused silvicultural model for red pine falls short of emulating natural developmental dynamics of red pine ecosystems, nor does the approach specifically address the stand-scale principles of ecological silviculture (continuity, complexity/diversity, timing) in a meaningful way. Doing so is where we now turn our attention, specifically by framing an ecological silvicultural system for red pine ecosystems based on a natural model of development, one that expressly incorporates the foundational principles.

As discussed in Chapter 10, we name our system after the prevailing natural disturbance regime for Archetype 4, that is, a *mixed-severity silvicultural system*, reflecting the variable disturbance severity and resultant structural outcomes among stands. Below we describe the silvicultural activities associated with regeneration, tending, and harvesting for this system.

Silviculture by Developmental Stage/Event

Our ecological silvicultural system for red pine ecosystems (Table 14.6) is organized around major developmental

Table 14.5 *Key differences in outcomes resulting from the natural development model and the timber-focused model for red pine ecosystems.*

Feature	Natural Model	Timber-Focused Model	Functional Consequences[a]
Tree species diversity	May include up to 10–12 species in varying abundance	Trees other than red pine actively discouraged or at best tolerated in low numbers	Higher susceptibility to insects/disease in single-species stands; reduced habitat diversity
Pine age structure	Broadly single-cohort to two- or three-cohort	Narrowly even-aged	Greater susceptibility to wind disturbance due to loss of crown height diversity
Live tree legacies	Several to many	Few to none	Loss of unique habitat
Deadwood legacies	More abundant; diverse sizes including large structures	Limited number and typically of small size	"..........."
Spatial pattern of legacies	Dispersed to aggregated legacies	n/a	Reduction in resource and habitat diversity within a stand
Stand density	Relatively lower densities at all stages of development	Relatively higher densities initially and maintained over time	Loss of habitat associated with PFS; shortened transitions between early stages
Spatial pattern of tree distribution	Heterogeneity in density and canopy cover	Homogeneity in density and canopy cover	Reduction in resource and microclimate diversity
Maximum tree age	Potential for 200–300+ year tree life spans	50–150 year rotations	Loss of unique structural features of old trees

[a]Examples of functional consequences of a timber-focused model compared to a natural model.

Table 14.6 *An ecological silvicultural system for red pine ecosystems.*

Developmental Stage/Event[a]	Timing (years)[b]	Objectives/Consequence	Potential Silvicultural Activities[c]
Disturbance and Legacy Creation[d]	0	Provide for continuity of structure and function and initiate PFS	*Harvest*: variable retention harvest in MFS or OFS *Variable density thinning (VDT):* gaps cut for VDT may trigger a legacy creation event in the gap[e] and initiate a PFS in the gap *Deadwood creation:* deliberate retention and/or creation of down deadwood and snags as needed at harvest
Preforest	1 to 5+ years; could be 20+ years	Provide open, non-tree dominated habitat; ultimately sets the stage for tree regeneration	*Competition control*: reduce over-abundance of woody shrubs that develop due to fire exclusion *Site preparation*: to expose mineral soil/reduce duff to facilitate tree regeneration Prescribed fire, mechanical, or other means to achieve competition control and site preparation objectives
Young Forest (early)	5 to 30 years	Facilitate/establish new cohort of trees	*Regeneration*: assess existing regeneration, if needed plant, direct seed, or capture mast seed crop *Site preparation:* to facilitate regeneration from seed or planting *Browse control*: until threshold height is reached *Release of existing regeneration*: as needed; avoid elimination of natural regeneration or ground-layer plant populations

Developmental Stage/Event[a]	Timing (years)[b]	Objectives/Consequence	Potential Silvicultural Activities[c]
Young Forest (after canopy closure)	30 to 70 years	Maintain or accelerate growth of trees; initiate spatial heterogeneity in structural conditions	*Thinning*: VDT in even-aged stands or in younger cohort of 2+ cohort stands; potential for 2–3 thinnings on an approximately 15 year interval *VDT gaps:* site preparation, establish regeneration, competition control
Mature Forest	70 to 150	Enhance heterogeneity; accelerate growth of desired trees; enhance structural complexity; enrich tree composition	*Thinning:* VDT in even-aged stands or in younger cohort of 2+ cohort stands; potential for multiple entries *VDT gaps:* site preparation, establish regeneration, competition control, expand openings *Tree species enrichment:* planting/seeding as needed, often in VDT gaps *Decadence/deadwood creation*: including snags and logs; may include inducing decadence *Prescribed fire* or other means to reduce aggressive shrubs and fire-sensitive hardwoods *Harvest*: in older stands variable retention harvest (see above)
Old Forest	+150	Enhance structural complexity; restore or maintain woodland structure; enhance tree composition	*Decadence/deadwood creation*: (see above) *Thinning*: VDT in structurally homogenous, but old stands *VDT gaps:* site preparation, establish regeneration, competition control, expand openings *Tree species enrichment:* planting or seeding underrepresented species, often in VDT gaps *Prescribed fire* or other means to reduce aggressive shrubs and fire-sensitive hardwoods and to expose mineral soil seedbeds

[a]A regeneration harvest initiates the developmental sequence, or in the case of existing stands, the table is entered according to the developmental stage of the stand in question and actions applied from that point onward.

[b]Timing reflects a general length of time that a particular stage might last.

[c]Potential activities that might be pursued during some point in the stage

[d]Note that the term "legacy creation" is meant to include both retention of legacies from the pre-disturbance forest, as well as actual creation of new structures (largely deadwood) from residual live trees.

[e]VDT that includes larger gaps may initiate a disturbance/legacy event within a stand and return the user to time zero in the context of the gap, but not the whole stand, and in that gap, PFS and early YFS conditions and actions may apply.

stages (see Chapter 5). In most cases, these activities will be applied at the whole-stand scale. However, because these forests can develop as two- or even three-cohort stands, in some cases activities identified for a particular stage might be applied to patches that encompass cohorts within a stand. Within each developmental stage, there may be several actions listed. This is not to say that all of these actions will be pursued concurrently. Rather, these are the potential actions that might occur at some time during the stage.

In general, we apply the same framework to **natural origin** stands and **plantation**s, but as explained in the accompanying text, some of the specifics will differ between the two settings. However, despite their different origins, characteristics of plantations and natural origin stands that have been managed primarily for timber are similar; they both tend to be overly dense, limited in tree diversity, and structurally simplified. Consequently, a similar silvicultural approach can often be used in both settings.

Disturbance and Legacy Creation

In our red pine model, the focus of a harvest is on legacy creation (i.e., retention from the pre-disturbance stand or, in the case of deadwood, potential creation from live trees) and initiation of the PFS of varying length, rather than an exclusive focus on full site occupancy by tree regeneration early after the harvest. At scales of whole stands, variable retention harvesting is used to emulate to varying degrees the patterns of tree death and structural retention that occurs with natural (stand-scale) disturbance.

Harvests involve overstory removal with varying amounts and patterns of live tree retention (Figure 14.7). Retention

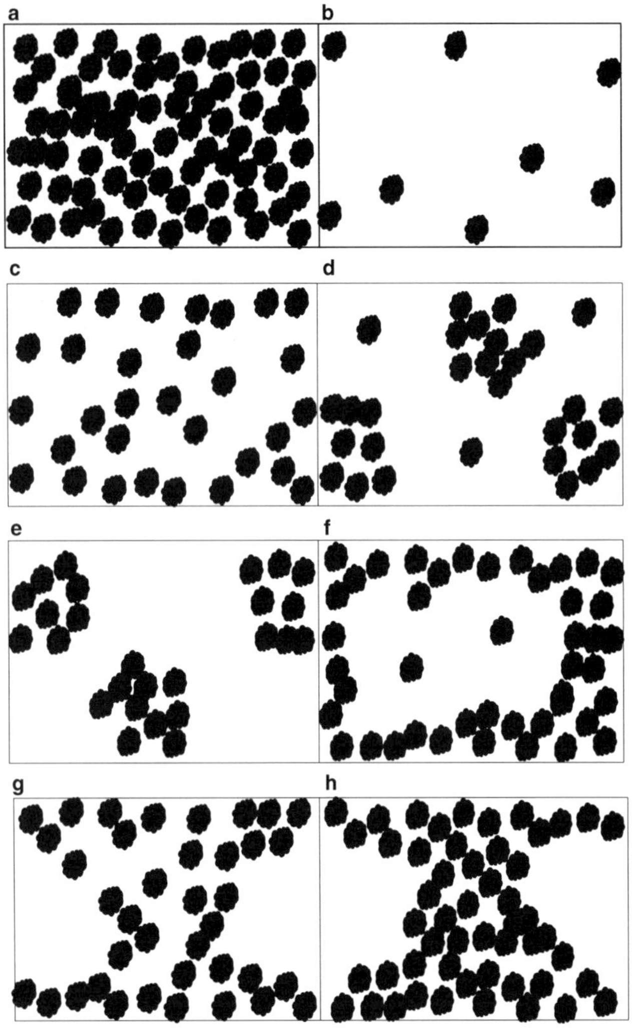

Figure 14.7 *Conceptual representation of various types of retention harvesting in red pine ecosystems, all of which reflect outcomes of canopy removal with retention. Starting with the uncut forest in the top left (a), retention may range from different levels of dispersed trees (b, c), to aggregate retention with or without some dispersed trees in the harvested matrix (d, e), to large patch openings with a few retained trees in the openings (f), to multiple smaller patch openings with or without thinning of the forest between the openings (g, h). The latter conditions (g, h) may also be reflective of structural conditions in larger gaps after variable density thinning.*

may range from a few dispersed individuals to higher levels of dispersed retention (Figure 14.7 b, c). Retained trees could be aggregated in the harvested matrix, with or without some dispersed trees between aggregates in the cut matrix (Figure 14.7 d, e). Large openings could be the dominant feature of the harvest, with retention occurring in, around, and between openings. For instance, a harvest may consist of one large central opening, surrounded by retained trees with some internal (to the opening) retention as well (Figure 14.7f). Alternatively, the pattern could consist of several smaller patch cuts with either thinned or unthinned forest connecting multiple patches (Figure 14.7 g, h). The possible patterns are numerous and all emulate to varying degrees the potential patterns of surviving trees after natural disturbance.

Variable retention harvesting as described above may resemble several classic silvicultural systems that retain trees at harvest, including continuous cover irregular shelterwood, shelterwood with reserves, and seed tree with reserves. In these variants of the corresponding even-aged systems, the residual trees are left for some period of time longer than is necessary to secure regeneration, i.e., they are left for reasons other than or in addition to securing regeneration. However, the motivation for leaving trees unharvested in timber-focused applications is likely not for ecological objectives. Rather, they may be left for a period of time needed to accumulate growth and value onto larger logs. Thus, there is a meaningful difference in the functional roles of residual trees in these classical systems compared with VRH.

Retention of other legacy structures, particularly deadwood, should be factored into regeneration harvesting as well, leaving, for example, large snags and nearly dead trees (for future deadwood recruitment) and avoiding damage to large downed wood. Harvesting may also offer opportunities to create deadwood structures, should these be lacking, since machinery to fell or top trees will be on site.

Some general quantitative guidelines for retention have been provided in Palik and D'Amato (2019), based on studies of old and mature stands. These include size recommendations for retained trees and snags, as well as densities of these structures. For instance, a general rule of thumb is that 75 percent of retention trees should come from the top 25 percent of the diameter distribution of the pre-harvest stand, favoring retention of larger trees. A variable retention harvest that is patchy and less than stand replacing might include at least 76 retained trees per ha (varying this number from fewer to greater depending on objectives). If the mean diameter of these trees is 40 cm, then this density corresponds to a retained basal area of approximately 9.6 m^2/ha. Historical photographs of old-growth stands after natural disturbance are supportive of this general density and basal area as a mean around which to vary retention amounts (Figures 6.33, 6.36). A reasonable rule of thumb for the addition of deadwood during the harvest is to create three to five logs on the ground, as well as three to five snags per hectare, that are 30 cm or greater in diameter, with the assumption that mortality of retained live trees will provide more of these structures over time.

Disturbance and legacy creation is also a feature of gaps cut as part of variable density thinning. If these gaps are large enough, they may display some features of the forest developmental sequence occurring at the stand-scale after disturbance, particularly legacy creation, a PFS, and the early YFS with establishment of new regeneration. Thus, there may be a fuzzy distinction between thinning and a regeneration harvest in terms of regeneration outcomes; the former may grade into the later depending on the specifics of the thinning (i.e., if it includes large gaps as part of VDT). Alternatively, in practice, openings generated by VDT may be small, such that early seral conditions are limited, providing only marginal PFS conditions and limited regeneration opportunities in a stand.

Natural origin versus plantation

Generally, there may be little distinction in application of a harvesting and legacy creation in natural origin stands and plantations; VRH can be used in either setting. This is because most stands that have been managed for timber, whether natural origin or a plantation, will lack significant structural complexity (particularly multiage structure and deadwood) and thus will benefit similarly from legacy creation as part of a harvest. What is likely to differ between stand types is the nature of trees that are available as legacy structures. Plantations will generally be overwhelmingly dominated by red pine having minimal size variation. Thus, retention trees will generally be limited to these individuals. Conversely, in natural origin stands that have reached the MFS or OFS, potential legacy trees will likely include: 1) a wider size range of red pines; 2) larger individuals of other tree species; and 3) trees that are decadent or dead. Thus, the options for legacies, as either live trees or dead or dying trees, tend to be greater in natural origin stands. The forester should capitalize on this opportunity to diversify structure and composition as part of the disturbance/legacy creation event.

Preforest Stage

A PFS is a rare thing in the timber-focused landscape given organizational objectives to achieve full site occupancy by crop trees quickly after harvest. The reality is that a PFS (i.e., a condition where herbaceous and shrub vegetation dominate the site and make important contributions to biodiversity), likely occurred in some locations after canopy disturbances (Figure 14.8). In some instances, however, the PFS may be short-lived or even largely nonexistent locally when, for example, advance regeneration of shade tolerant species is released by harvest or sprouting hardwoods (particularly aspen species), developing dense populations soon after harvest. Note that both of these examples may be more of an artifact of long-term fire exclusion, rather than a regularly occurring feature in the natural development model.

In a managed forest setting, the ecological forester can encourage a PFS by not rushing to establish tree regeneration immediately after harvest. While many organizations have requirements to achieve full stocking of regeneration within some specified time period after a harvest (e.g., five years), a PFS (albeit a short one) may still be possible with some advance planning, e.g., by delaying artificial regeneration for several years after the harvest. When a time sensitive regeneration requirement is not an issue, relying more on natural regeneration can lead to an extended PFS. For most red pine forests, natural regeneration of pines and other species may accumulate slowly, potentially extending to decades (see Figure 6.32).

Several other activities might need to occur during the PFS, notably site preparation and competition control. Some site preparation that exposed **mineral soil** and reduced **duff** layers may have occurred as a consequence of harvesting, particularly if this occurred during a snow-free period.

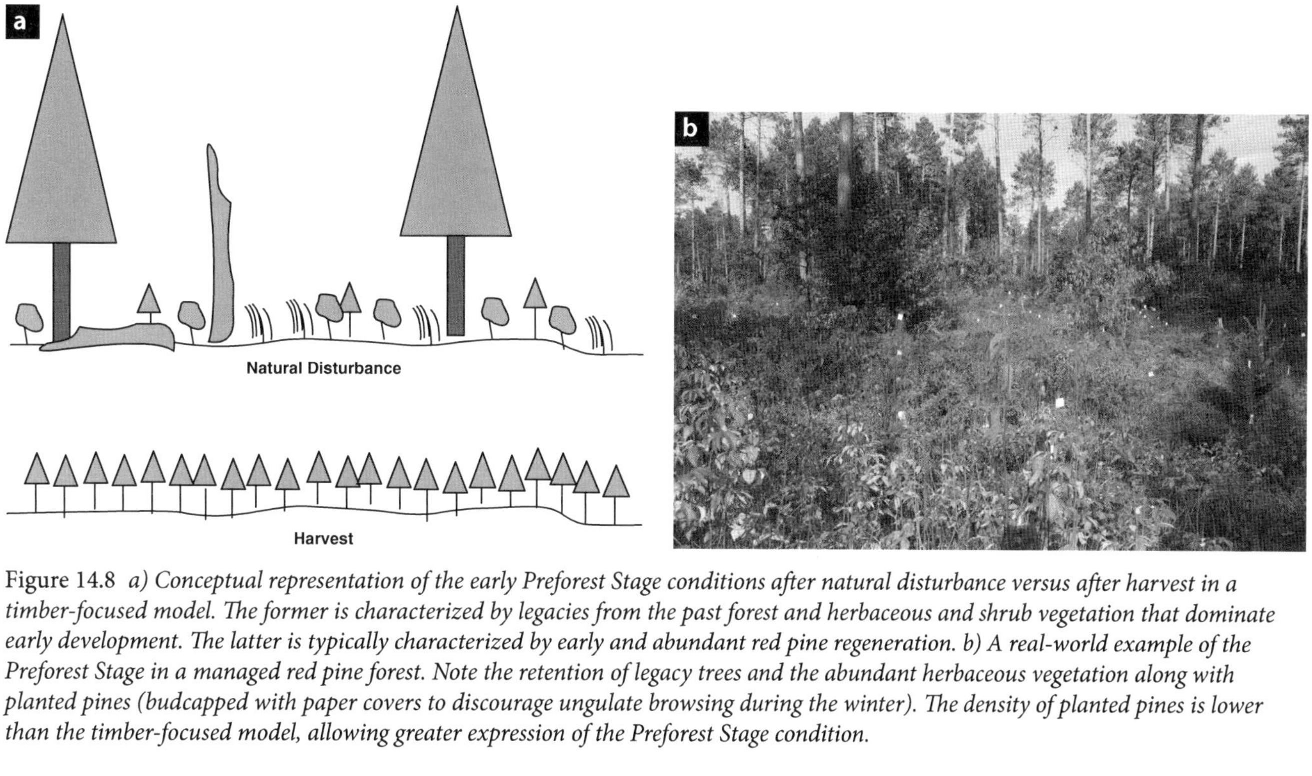

Figure 14.8 *a) Conceptual representation of the early Preforest Stage conditions after natural disturbance versus after harvest in a timber-focused model. The former is characterized by legacies from the past forest and herbaceous and shrub vegetation that dominate early development. The latter is typically characterized by early and abundant red pine regeneration. b) A real-world example of the Preforest Stage in a managed red pine forest. Note the retention of legacy trees and the abundant herbaceous vegetation along with planted pines (budcapped with paper covers to discourage ungulate browsing during the winter). The density of planted pines is lower than the timber-focused model, allowing greater expression of the Preforest Stage condition.*

Figure 14.9 *Pines established from seed in a large gap after a site preparation burn. Red Lake Wildlife Management Area, Minnesota, USA.*

Figure 14.10 *Beaked hazel proliferation after harvest in a red pine stand. Photo credit: Eric Zenner.*

However, additional site preparation might be needed at some point during the PFS to increase the likelihood of establishment from seed, particularly of pines. Ideally, this would involve prescribed fire, but in reality, mechanical methods will likely be used. Timing site preparation to precede a **mast seed crop** may increase the probability of successful regeneration from seed; this is rarely done in the contemporary managed forest landscape, but has been demonstrated to be successful in some settings (Figure 14.9) and was used successfully in the past (Figure 14.5). With site preparation, it is advisable to avoid area-wide application of any method so as to reduce the potential to inhibit established regeneration of trees and other vegetation and to minimize disruption of large wood on the ground.

Competition control may also be warranted during the PFS, particularly if invasive shrub species, including hazels and raspberries, become abundant in the stand after harvesting. These native shrubs may proliferate in openings due to fire exclusion, potentially inhibiting tree regeneration, but are also likely to outcompete herbaceous plants and other less abundant shrubs (Figure 14.10).

Natural origin versus plantation

A PFS will generally look different between natural origin stands and plantations that have been harvested. The latter are often lacking in diversity and abundance of native herbs and shrubs that would typically increase in abundance after a harvest in a natural stand. Other than red pines, there might also be a lack of seed sources for other tree species in plantations, thus limiting the potential for natural regeneration of other species. On the other hand, invasive shrubs may be equally prolific in both settings, especially if soil disturbance is significant. Thus competition control may be needed in either stand type. Moreover, site preparation, to prepare for seed establishment, would be equally important in both stand types.

Young Forest Stage (early)

Silviculture in the YFS will focus on establishment of a new cohort of trees. This should involve an assessment of natural regeneration (or artifical regeneration if added in the latter part of a PFS) that has accumulated during the PFS. If this is adequate there may be little need for further action to establish trees. If regeneration is inadequate, actions may be taken to facilitate or supplement it. This might include additional (or even initial) site preparation, as described above, to facilitate regeneration from seed, either from on-site sources or through **broadcast seeding**. For many ownerships, regeneration will include planting of pines, including red pine, but also jack and eastern white pines, but in an ecological system, planting densities will be at the lower range of those typically recommended in timber-focused settings, which can be 1,000 to 2,500 trees/ha (400 to 1,000 trees/ac). For the ecological forester, silviculture will be aimed at regenerating mixed-species woodlands, so an important consideration for regeneration is that artificial regeneration should supplement, not supplant, natural regeneration, particularly of non-pine species, which may be resprouting (hardwoods) or establishing from seed in some abundance (Figure 14.11).

Early release of established regeneration, particularly reduction in invasive shrubs, may be needed, but this should be done in a way to avoid reducing associated tree species in favor of red pine exclusively. Browse control on pines, particularly that caused by white-tailed deer, is often needed until seedlings reach a height that puts their upper apical buds above the reach of browsers.

Natural origin versus plantation

Silvicultural actions in the YFS are likely to be distinctly different in detail between natural-origin stands and plantations. In general, the latter will lack a significant population

Figure 14.11 *A red pine stand several years after a variable retention harvest with planted pines (bud capped) and natural regeneration of hardwood species.*

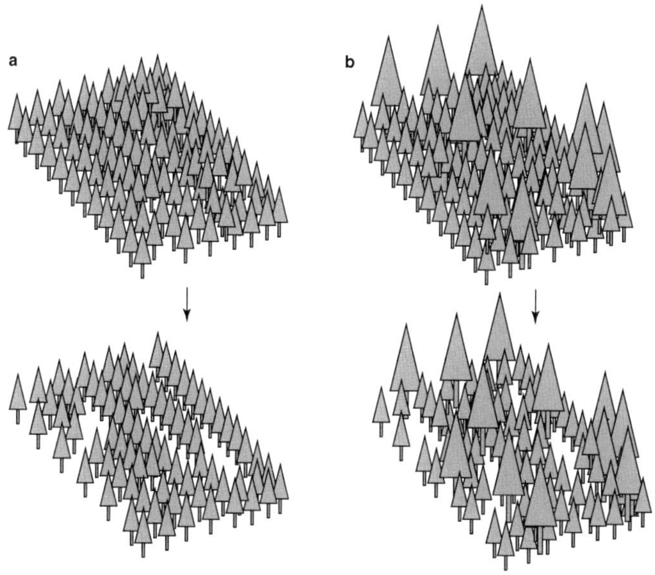

Figure 14.12 *Conceptual representation of thinning in the Young Forest Stage. a) Variable density thinning (VDT) in an even-aged stand. b) VDT of the younger cohort in stands with two cohorts.*

of natural origin regeneration of the mix of tree species characteristic of the ecosystem. Moreover, seed sources for these species will generally be absent in any abundance in the stands themselves. As such, regeneration activities in plantations are likely to be heavily focused on artificial regeneration, but if possible this should include a variety of tree species. This may include planting conifer species in addition to red pine, but also hardwood species. Alternatively, direct seeding might be used for all species, particularly non-pines, something most organizations do not currently pursue as part of a timber-focused management program.

Young Forest Stage (after canopy closure)

Depending on seedling growth rates, there may be a continued need for competition control around tree regeneration in the later part of the YFS. However, the primary emphasis in this stage will be on thinning established cohorts. Thinning should take the form of variable density thinning in an even-aged stand or within a younger cohort of trees in stands that already contain two or more cohorts (Figure 14.12). If thinned early, this might be a precommercial thinning, but eventually would be applied to trees large enough to be commercially valuable. Additional (later) thinning entries would again be a variable density thinning applied in a way that maintains or augments heterogeneity of density developed in the initial thinning. The time span of the YFS is long enough that two or three entries on a 10–15 year interval might be considered. If VDT includes cutting of larger openings, then there is a potential need for site preparation, regeneration (seeding, planting), and completion control in the openings, as described above.

Natural origin versus plantation

Application of variable density thinning in the YFS may be different in the two stand types, primarily because the spatial pattern of trees in natural stands will be less uniform than in plantations, so it often might be easier to move stands toward greater heterogeneity of structure. Secondly, natural stands are likely to contain populations of tree species other than red pine that may benefit from removing pines around them. In contrast, early thinning in plantations in the YFS will likely involve red pine exclusively. In either case, the common goals of thinning are to release growing space for trees, whether for crop trees or habitat trees, accelerate the development of structural heterogeneity, and potentially establish a new cohort of trees in openings as part of variable density thinning.

Mature Forest Stage

Stands in the MFS are fairly abundant in the region and thus present ample opportunities to meet objectives of ecological forestry with silviculture. If these forests have not been managed using an ecological approach, as described up to this point, then their condition will likely reflect development with fire exclusion or management following the timber-focused model; —e.g., high stem densities, high abundance of red pine, narrowly even-aged, abundant invasive woody shrubs, and fire-intolerant trees in the understory. In these stands, there are two pathways for silviculture; 1) harvesting/legacy creation leading to a PFS, or 2) tending treatments to further transition stands to the OFS. In actuality, these two pathways are not mutually exclusive in that tending/transition treatments may include variable density thinning that could create some larger PFS openings.

Thinning in the MFS may continue with the approaches outlined for the developing YFS. This could include an initial variable density thinning, if not done earlier. Alternatively, it could build on an earlier VDT, including thinning

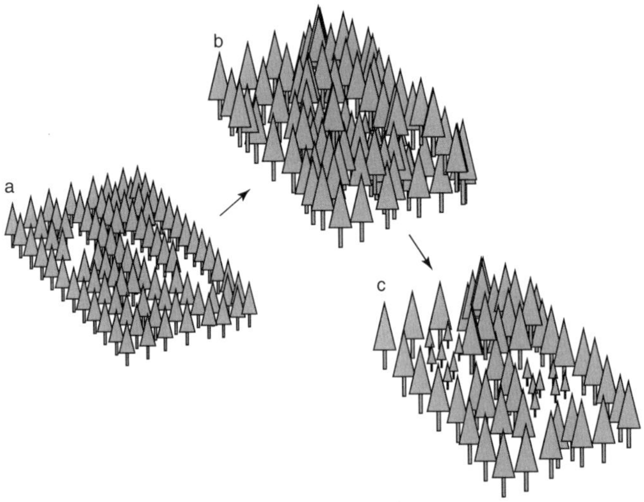

Figure 14.13 *Conceptual representation of thinning in the Mature Forest Stage. a) Stand after an initial variable density thinning in the Young Forest Stage. b) This same stand after maturation into the Mature Forest Stage. c) Thinning that builds on the earlier variable density thinning to expand gaps, release younger cohort regeneration in these gaps, and further thin the mature matrix.*

the matrix, expansion of existing gaps, or even addition of new gaps (Figures 14.13, 14.14). In the MFS, an objective of thinning is likely to include releasing trees with the greatest potential for long-term growth. As in the YFS, larger gaps created through VDT might warrant site preparation, regeneration, and completion control. Enrichment planting or seeding may be needed to establish underrepresented tree species. This may focus on VDT gaps, or potentially the matrix for more shade-tolerant species. Prescribed fire, herbicides, or mechanical treatments might be used to reduce abundance of aggressive shrubs and fire-sensitive hardwoods to enhance opportunities for new pine establishment and/or growth of advance regeneration.

Figure 14.14 *A plantation origin Mature Forest Stage red pine stand after a second entry that expanded existing gaps that were cut earlier as part of a variable density thinning, so as to release established regeneration. Photo credit: Doug Kastendick.*

Superficially, it may seem that VDT is just another name for group selection (i.e., cutting larger openings to secure regeneration, while tended the existing cohort of trees between the openings). A distinction is that group selection (and other selection methods) seeks to achieve a balanced allocation of growing space to each cohort of trees created over time, so as to ultimately have a stable age distribution and sustained flow of wood. This is not the intent of VDT—i.e., balanced allocation of growth space to cohorts and stable age distribution—since it likely would be applied only once or twice before a true regeneration harvest is prescribed. Moreover, group selection as typically applied will not necessarily result in spatial heterogeneity of structure, which as you have learned, is a key outcome of variable density thinning.

When regeneration harvesting is pursued in the MFS, it will involve a variable retention harvest (VRH), as described above (see Disturbance and Legacy Creation). Legacy creation as part of a VRH focuses on trees, deadwood structures, and conditions that take extended time periods to develop. Live tree retention may range from low to moderate density dispersed retention, aggregate retention (including unharvested leave patches), or large gap/intact matrix retention (Figure 14.7). While a PFS may precede development of regeneration, ultimately a VRH sets the stage for establishment of tree cohorts.

In the contemporary forest landscape, deadwood structures are generally in short supply in stands that have been managed for timber. As such, deliberate deadwood and **decadence** creation might be pursued as a standalone treatment or during either a thinning entry or during a VRH in the MFS. Stands in the MFS will contain trees of a size and age suitable for decadence and deadwood creation, including those not considered economically valuable (i.e., cull), yet of high ecological importance as deadwood legacies.

Natural-origin versus plantation

Dead and decadent trees are likely to be naturally more abundant in natural-origin stands versus plantations; so deadwood/decadence creation may be a lower need in the former. As discussed already, natural origin stands are likely to contain some measure of trees from species other than red pine; these might be favored for release in thinning or retention in harvesting. These trees are likely limited in plantations. Plantations also are likely to have less advance regeneration that can be released with thinning or harvesting due to the generally high overstory stocking in these forests.

Old Forest Stage

It is unlikely that silviculture for timber objectives would be pursued in red pine stands that have already transitioned into the OFS, so harvesting, if conducted at all in OFS stands, would likely be pursued for reasons other than timber extraction and would likely be part of a variable density thinning rather than a variable retention harvest. This is not to say that stands managed for timber will not contain populations of old trees and their derivative structures; if managed ecologically, stands managed for timber should eventually have these features (as a result of actions taken in earlier forest stages). The distinction is that red pine stands that have transitioned into the OFS are rare and likely to be managed for services other than timber, whereas stands managed ecologically and inclusive of timber objectives will benefit from having structural features characteristic of OFS stands.

It is likely that stands in the OFS may still be in need of some degree of restoration of structure and composition due to long-standing fire exclusion and other factors. Thus, if the management objective is to restore these ecosystems to a condition more reflective of the natural model, then various silvicultural activities may be appropriate. In these cases, silviculture is a substitute for time, an example being deliberate decadence creation, where snags or downed dead wood are lacking. While these features will accumulate with time, deliberate decadence creation may be used to add these features on a faster timescale. In this case, candidate large pines might be girdled to enhance snag habitat. Deliberate decadence creation in old-growth forest is not widely practiced, and may even be considered contrary to old-growth forest conservation, but doing so is consistent with an objective of enhancing complexity/diversity (Principle 2) in established stands as part of an ecological silviculture approach.

Another role for silviculture in the OFS, when fire exclusion has been longstanding, is dealing with the often overly dense hardwood understory, including the populations of invasive shrubs that develop in such stands (Figure 14.15a). Moreover, long fire-excluded forests may have overly high levels of fuels, leading to increased risk of catastrophic wildfire. Prescribed fire or mechanical removal for both competition control and fuels reduction (and potentially for some ownerships herbicides to reduce competitors) have potential roles in restoration given these conditions (Figure 14.15b).

OFS stands may benefit from variable density thinning to reduce tree densities, which better reflects a woodland structure and expedites the development of structural heterogeneity. This may initiate a PFS in large openings and ultimately can lead to establishment of a new cohort of trees. Gaps created through VDT may benefit from site preparation to increase opportunities for natural regeneration from seed, particularly in fire excluded forests, which are the norm rather than the exception for this ecosystem. Artificial regeneration and competition control in gaps may also be warranted. As in the MFS, regeneration activities in VDT gaps may focus on establishing underrepresented tree species.

Natural-origin versus plantation

Most natural origin stands that have reached an OFS are likely to be displaying some old-growth characteristics.

Figure 14.15 *(a) Well-developed hardwoods in a fire-excluded old red pine stand in Minnesota, USA. (b) Intermediate treatments, such as prescribed fire, may be used to restore a condition more reflective of the natural model including open canopy and understory conditions and reduced fuels. Photo credit: (b) Jack McGowan-Stinski.*

Moreover, there is likely to be limited public support for doing more than allowing time to pass in natural OFS stands. A notable exception is that most old stands will have had fire-exclusion for decades and, consequently, will have an overabundance of fire sensitive plants, including trees and shrubs. Reducing these populations with prescribed fire or mechanical methods may be warranted. Plantations on the other hand, even those that have reached an OFS, are prime templates to pursue restoration through silviculture and there is likely to be less public opposition to pursuing management in plantations. Such stands can start to resemble natural-origin stands in some characteristics, e.g., tree size and size variation (Figure 14.16). However, even these stands would still benefit from some silvicultural intervention. This may include reduction of overabundant fire-sensitive plants, enhancement of tree species diversity, increased spatial heterogeneity of structure, and creation of deadwood. Thus, as in MFS stands, silvicultural activities in old plantations may include variable density thinning, targeted thinning around larger individuals of other species that may have recruited into an old plantation, deadwood or decadence creation, direct seeding or planting to add missing tree species, and site preparation in openings to expose mineral soil for seed establishment.

Figure 14.16 *This red pine plantation in northern Minnesota, USA dates to 1900. After decades of growth and mortality, the planting rows are difficult to find and are obscured by ingrowth of fire-sensitive vegetation. The 2017 photograph shows that some trees have grown to large size and that substantial size variation has developed over 117 years, both characteristics associated with old natural forests.*

Examples of Ecological Silviculture in Practice

There are many organizations and ownerships that use the timber-focused model (or a variant thereof) outlined in Table 14.4 as their primary silvicultural approach. However, some of these organizations may also be incorporating ecological principles into their management to varying degrees, even when timber production is the primary objective. In fact, beginning in earnest in the early twenty-first century, several public organizations began to recommend silvicultural guidelines for red pine that were more inclusive of ecological objectives and practices (Table 14.7). These approaches often attempt to incorporate one or several of the principles of ecological silviculture, but they are generally not packaged into

Table 14.7 *Examples of organizations and agencies that recommend some aspect of ecological silviculture in their red pine management guides.*

Entity	Ecological Objectives	Approach
Wisconsin Department of Natural Resources http://dnr.wi.gov/topic/forestry	Provision of wildlife habitat; production of diverse ecological benefits	Retain some mature red pines at harvest if shoot blight is not a concern; retain snags that are not a safety concern; consider rotations of 120–200 years; retain wildlife trees during thinning; consider variable density thinning to diversify composition and structure
Michigan Department of Natural Resources http://www.michigan.gov/dnr/forestry	Restoration of native forest communities to benefit wildlife and species of concern; sustain native biodiversity	Employ a natural management strategy consisting of shelterwood or seed tree with retention of a mixture of trees; harvest in snow-free period to scarify soil, creating seedbeds for oak and pine; inter-plant red pine among other species, not necessarily in rows
Ontario Ministry of Natural Resources and Forestry https://www.ontario.ca/document/tree-marking	Maintenance of stand and landscape biodiversity	Silvicultural guidelines include employing knowledge of natural disturbance to determine the best silvicultural system and protection of habitat features, including stick nests, cavity trees, interior forest, mast trees, super canopy trees, and rare vascular plant populations.

comprehensive silvicultural systems (as in Table 14.6). There are a few organizations that do package their silvicultural approaches into more comprehensive ecological systems. We highlight two such examples below: Seney National Wildlife Refuge and the Chippewa National Forest.

Seney National Wildlife Refuge, Michigan, USA

The Seney National Wildlife Refuge (SNWR) located in the eastern Upper Peninsula of Michigan, USA is part of the United States Fish and Wildlife Service's National Wildlife Refuge System. Based on policies and directives such as the 1997 *Refuge Improvement Act* and the more recent Biological Integrity, Diversity, and Environmental Health policy, silviculture is used at the SNWR to achieve objectives related to restoring and sustaining habitat, particularly in fire-dependent ecosystems that have a long history of fire exclusion (Corace et al. 2009). Historically, mixed-species ecosystems dominated by red and eastern white pines occurred on most upland sites found on glacial outwash and sand ridges at SNWR (Burger and Kotar 2003). This ecosystem is broadly characteristic of much of the eastern Upper Peninsula of Michigan, USA, where mixed-species pine forests comprised approximately 38 percent of the pre-European settlement forest (Zhang et al. 2000). As of 2008, less than 1 percent of the primary red pine and eastern white pine forests of the region remained and most of these have been structurally and compositionally altered due to past management, including fire exclusion (Drobyshev et al. 2008a, b).

Fire history research at SNWR has shown that frequent, non-stand-replacing fires were common (Drobyshev et al. 2008a). These low-severity surface fires usually created small gaps, or resulted in minimal mortality to overstory trees, while maintaining a relatively open understory (Figure 14.17). The disturbance regime resulted in mixed-species pine stands with complex age structures (Drobyshev et al. 2008b). In contrast, contemporary second-growth pine forests have increased dominance of fire-intolerant species (e.g., trembling aspen and red maple), are less structurally complex, and have less red pine and eastern white pine regeneration (Drobyshev et al. 2008b).

Figure 14.17 *Surface fire in a Mature Forest Stage red pine stand, Seney National Wildlife Refuge, Michigan, USA. Photo credit: Greg Corace.*

A focus of pine forest management at SNWR is on restoring composition and structure in simplified, second-growth stands (Corace et al. 2009), with a goal of creating characteristics found in old-growth stands that have experienced frequent, low-severity surface fires. Such management not only meets ecosystem restoration objectives, but also provides breeding habitat for a diverse assemblage of neotropical migratory bird species (Corace et al. 2014). To meet these objectives, SNWR managers use a silvicultural approach designed to create more open, patchy canopy structure, reduce the dominance of jack pine, and increase establishment and growth of red pine and eastern white pines. The approach involves patch cutting of primarily jack pine greater than 15 cm (6 in) diameter to create variable-sized openings (2–5 ha) for red pine and eastern white pine establishment and reserving larger red and eastern white pine trees in these harvest areas (Figure 14.18). Additionally, where dense red pine saplings exist, they are thinned to increase growth rates of crop trees. Mechanical **scarification** of the soil to enhance establishment of light-seeded red and eastern white pine is accomplished with tracked harvesters during harvest operations. Moreover, efforts are made to retain all existing large snags, and new snags may be created where needed by girdling or topping larger jack pine. When possible, these mechanical treatments are followed with prescribed fire after fuel loadings have been decreased through harvesting, thus achieving both restoration and hazardous fuels reduction objectives. Monitoring of pine regeneration in these treatments verifies that eastern white pine regeneration responds favorably to the mechanical treatments alone, while red pine regeneration requires the follow-up prescribed fire treatment (Nyamai 2013).

How well does the SNWR approach address the four principles of ecological forestry?

The SNWR approach clearly addresses Principle 1—continuity through retention of trees and snags at harvest. To the extent that thinning is used for ecological purposes of developing large pines for their unique structural features, as well as creating heterogeneous stand structural conditions, the SNWR approach is also addressing Principle 2—enhancing complexity and diversity in established stands. Given that the SNWR pine ecosystems are managed for non-game wildlife habitat (Corace et al. 2014), most red and white pines will be allowed to live out

Figure 14.18 *(a) A stand of mixed-species pine with high density of jack pine and small red pine before treatment. (b) The same stand after removal of most of the jack pine and thinning of red pine, with retention of larger pines. Photo credits: Greg Corace.*

their natural life spans, thus timing management based on ecological queues—Principle 3—is also addressed. A key question for the SNWR is the degree to which landscape variation in forest condition will be promoted by the management approach. For instance, what is the proportion of different developmental stages across the landscape and are all stages represented to the extent desired, including a PFS in at least some locations? Also, does the approach generate levels of heterogeneity in composition and structure at landscape scales in patterns similar to that resulting from natural disturbance?

Chippewa National Forest, Minnesota, USA

Red pine-dominated forests of the USDA Forest Service's Chippewa National Forest (CNF), located in north central Minnesota, USA, are managed for a variety of ecosystem services including wildlife habitat, biodiversity, recreation, and timber. Much of the CNF red pine forests historically had a woodland structure, with an open and patchy canopy, and contained a variety of tree species in addition to red pine, including eastern white pine, jack pine, several oaks, paper birch, white spruce, balsam fir, and aspen (MN DNR 2003). These ecosystems, which occur on sandy soils, formed on **outwash** plains and **ice contact** features in the formerly glaciated landscape and were historically common in the broader region (Frelich 1995).

Fire disturbance reconstructions on the CNF demonstrates that the pre-European settlement landscape was characterized by frequent surface fires and infrequent crown fires (Guyette et al. 2015). Surface fires occasionally became crown fires, resulting in varying levels of tree mortality, while maintaining an open patchy canopy and relatively open understory. The disturbance regime resulted in mixed-species pine stands with complex age structures (Fraver and Palik 2012). In contrast, contemporary second-growth red pine forests are typically narrowly even-aged and overly dense, with increasing sub-canopy dominance of fire-intolerant hardwood (e.g., trembling aspen, paper birch, and red maple), and have limited natural regeneration of red pine or other conifer species.

Historically, silvicultural approaches for red pine forests on the CNF were focused on plantation establishment and even-age regeneration systems similar to the typical sequence of events with the timber-focused model (Table 14.4). A notable distinction from the general approach that was used widely on the CNF involved the windrowing of **logging slash** into distinct lanes to facilitate planting of pines between these lanes. As we discuss further below, this practice has important legacy consequences for ecological silviculture. Silviculture in natural origin red pine stands also followed the timber-focused model, with one or two thinnings of the largely even-aged red pine cohort followed by clearcutting and replanting.

Silviculture for red pine on the CNF has evolved considerably over the last decade and is now more cognizant of natural models and gives greater attention to the foundational principles. These changes largely focus on regeneration harvests, but also include some aspects of management in established stands. Characteristically, regeneration harvests in natural fire origin stands generally include substantial live tree retention, including red pine when shoot blight fungus is not a concern (see Box 14.1) or other species when it is a concern (Figure 14.19). Live tree retention may be dispersed or aggregated (Figure 14.20), or be implemented using a large gap/intact matrix approach (Palik et al. 2014). Rotation lengths between regeneration harvests are now prescribed to be at least 150 years, with some trees left to live out their natural life spans. Typically, multiple species of pine are planted after regeneration harvests and

Figure 14.19 *Retention harvests in red pine forests on the Chippewa National Forest, USA. (a) Eastern white pine is retained when shoot blight is evident in mature red pine. (b) Red pine can be retained when shoot blight is of less concern.*

natural regeneration of other native species is encouraged. A deviation from a natural models approach for regeneration is that planting densities of pine are typically much higher (up to 2,000 trees/ha) than what would have occurred naturally (Figure 14.21) due to concerns about achieving full site occupancy by five years after harvest as required by regulation.

Thinning in MFS natural origin stands (Figure 14.22) tends to reduce stocking closer to the B-line on the regional red pine stocking curve (see Figure 16.4), which is consistent with a historical woodland structure, but this may still not create much spatial heterogeneity in densities. In some cases, variable density thinning (Chapter 8) has been used to facilitate development of more heterogeneity of structure (Figure 14.23).

Variable density thinning using a skips and gaps approach has been used in plantation origin red pine stands to jump-start development of structural heterogeneity (Figure 14.24) and to diversify tree species composition. Achieving the later objective capitalizes on the common practice in the 1970s of windrowing logging slash into 5–8 m wide corridors, with 12–15 planting rows between. These rows are now often storehouses of non-pine plant diversity, especially deciduous tree and shrub species (Figure 14.25), adding to woody species diversity in the stands.

How well does the CNF approach address the four principles of ecological forestry?

The CNF approach to red pine regeneration harvesting clearly addresses Principle 1—continuity through retention of trees at harvest. Thinning to lower stocking in late YFS and MFS stands helps to hasten the development of large pines, both for their timber value and for their unique structural features and results in greater diversity of subcanopy tree species, thus addressing Principle 2—enhancing complexity/diversity in established stands. Moreover, where variable density thinning is used in established plantations

Figure 14.20 *Retention of pines in the Chippewa National Forest, USA, including individual trees (left background) and small aggregates (center).*

Figure 14.21 *Planted red pine regeneration in recent retention harvest on the Chippewa National Forest, USA. Note paper caps over the terminal buds of smaller pines used to discourage browsing by white-tailed deer.*

to enhance structural heterogeneity, Principle 2 is again being addressed. Use of extended rotations (150+ years) between regeneration harvests, much longer than is typical on timber-focused lands, and allowing some trees to live out their full life span, addresses Principle 3—timing silvicultural intervention based on ecological indicators. Principle 4 (context) is a bit more difficult to assess. There is variation in how retention harvesting is implemented across the landscape, e.g., favoring red pine or eastern white pine retention depending on the incidence of shoot blight pathogens and other factors, which addresses the context principle. As with many organizations, there is a general lack of appreciation/understanding for the roles of a PFS or an extended YFS for biodiversity, hence across a landscape, stands after harvest tend to be similarly dense with pine regeneration early in development.

What to Remember about Silviculture for Archetype 4 Ecosystems

Globally, Archetype 4 ecosystems are often dominated by longer-lived tree species with high commercial value. Consequently, there may be a long history of managing for timber in these ecosystems. Further, the silvicultural approaches used historically have been structured to optimize timber production, largely using an agricultural model. This model bears only limited resemblance, in terms of action and outcome, to the natural disturbance and development dynamics of these ecosystems. When a broader suite of objectives rises in importance, inclusive of ecosystem and watershed health, biodiversity, and social acceptability, the model for silviculture of Archetype 4 ecosystems will look very different from the timber-focused model, reflecting its strong relationship to the natural disturbance and developmental model for the archetype.

With this background in mind, the important attributes of ecological silviculture for Archetype 4 will include:

1. Legacy retention will be a central feature of harvesting, with this retention being variable in spatial patterns among stands in a landscape to reflect the variable nature of natural disturbance in these ecosystems (*addressing Principle 1—continuity and Principle 4—context*).
2. An initial outcome of variable retention harvesting is initiation of a Preforest Stage of varying length in at least some locations, rather than immediate full site occupancy of regenerating trees in all settings (*addressing Principles 1 and 4*).
3. When tree regeneration is pursued, densities across stands, or in cohorts within stands, may be lower than when using a timber-focused model,

Figure 14.22 *Standard thinning of mature natural origin red pine on the Chippewa National Forest, USA.*

Figure 14.23 *Variable density thinning in natural origin red pine on the Chippewa National Forest, USA. Note standard thinning in the foreground, a gap in the far center background, and skips (unthinned patches) in both the left and right background.*

Figure 14.24 *Variable density thinning in plantation red pine on the Chippewa National Forest, including (a) standard row thinning and (b) skips (left) and gaps (foreground).*

Figure 14.25 *An old logging slash windrow in a red pine plantation overgrown with deciduous tree and shrub species.*

reflecting the woodland structure of at least some of these ecosystems (e.g., our red pine case study).

4. Silvicultural activities to encourage regeneration of all tree species found in the ecosystem should be considered, potentially including site preparation, competition control, direct seeding and planting. However, stand-wide application of site preparation and competition control should be avoided to better maintain native plant species and to create more heterogeneous conditions across a stand (*addressing Principle 2—complexity/diversity*).
5. Thinning in Young Forest Stage and Mature Forest Stage stands can have an objective of crop-tree development, but this thinning will often be applied in a way that also creates or maintains heterogeneity of structure in a stand, releases wildlife habitat trees, and may include gaps and unthinned patches as part of variable density thinning (*addressing Principle 2*).
6. Decadence and deadwood creation might be pursed in Mature Forest Stage and even Old Forest Stage stands, as these structural features are often lacking in managed forest landscapes (*addressing Principle 2*).
7. Substantially longer intervals between regeneration harvesting are likely to be the norm rather than the exception under an ecological model (*addressing Principle 3—timing*), reflecting a desire to develop structures and conditions that depend on old trees.
8. Variability and heterogeneity, of harvesting and legacy creation, thinning, decadence creation, etc., are concepts to keep in mind when applying silviculture in Archetype 4 ecosystems. Put simply, if a goal is to promote native biodiversity and maintain more options for an uncertain future, not all stands should be treated alike (*addressing Principle 4*).

PART V

Additional Considerations

CHAPTER 15

Ecological Silviculture for Riparian Forests

Riparian areas that are forested provide unique ecological services compared to other forest settings by virtue of their landscape setting at the interface between land and water (Palik et al. 1999) (Figure 15.1). For instance, riparian forests provide recreational opportunities that are perhaps unmatched by other types of forest. Riparian forests are important for flood mitigation and filtering of sediments and nutrients from water. Not only do riparian forests provide habitat for a variety of terrestrial species that depend on the adjacency of water during some portion of their life, but they may serve as travel corridors and refuges for species that are occasional users. Riparian forests may sometimes be valued for timber resources either because they are productive or because they support unique timber species not found elsewhere in the landscape.

The ecological services provided by riparian forests are a result of functional attributes related to linkages between land and water. Briefly, these include regulation and movements of nutrients, energy, and materials between land and water, as well as structural features that provide these links and that provide habitat for a variety of species.

Various definitions of riparian areas exist, but the ones that reflect functionality are of greatest relevance to us, for example:

Figure 15.1 *A forest adjacent to an aquatic ecosystems is termed a riparian forest.*

> *"Riparian areas are the three-dimensional ecotones of interaction that include terrestrial and aquatic ecosystems that extend down into the groundwater, up above the canopy, outward across the floodplain, up the near-slopes that drain to the water, laterally into the terrestrial ecosystem, and along the water course at a variable width."* (Ilhardt et al. 1999.)

Inherent in this definition is the recognition that ecological functions of riparian forests occur at various scales and that they are ecotones of function. There are many similar definitions of riparian areas that acknowledge the functional importance of riparian forests as ecotones between land and water (e.g., Swanson et al. 1982; Gregory et al. 1991).

Given the importance of functional connections of riparian forests to aquatic ecosystems, managing these forests to sustain functionality is a unique challenge for foresters. Not only will timber management influence the riparian forest itself, it will also influence the linked aquatic ecosystem, potentially for large distances downstream in the case of riverine settings. As such, managing to maintain the functional integrity of the riparian forest continues to be a key concern of foresters working in this landscape setting. Pursuing this management in the context of our principles for ecological silviculture is a highly relevant approach for sustaining functional riparian forests.

That is the topic of this chapter. Specifically, we will review what a riparian forest is from a functional standpoint and we will consider the various types of riparian forests that you might encounter. Next we will consider riparian forests from the standpoint of ecotones and how management of functionality is influenced by the ecotone concept. Finally, we will address riparian forest management from an ecological perspective, with specific attention given to applying our principles for ecological silviculture in riparian settings.

What is a Riparian Forest, Functionally?

Riparian forests, regardless of their type or the type of aquatic ecosystem they are associated with, share certain functional features related to movements and transformation of energy, nutrients, sediments, and organic matter. Moreover, riparian forests provide biodiversity and habitat functions for organisms that depend on both terrestrial and aquatic environments. From the standpoint of our interest in ecological silviculture, suffice it to say that the ecological functions of riparian forests are very much dependent on and influenced by their composition and structure (Table 15.1).

Historically, riparian forest management has not fully appreciated the functional links between terrestrial and aquatic ecosystems (Swanson and Franklin 1992). More recently, managers are placing greater emphasis on incorporating a functional perspective into silviculture of riparian areas (Berg 1995), but not necessarily from the ecological perspective you have learned about in this book. As we will discuss later in this chapter, silviculture based on a natural models approach, and which incorporates our foundational principles, can have an important role in accomplishing a goal of functional sustainability of riparian forests. We summarize the major riparian functions below, but much more thorough treatments of riparian areas and their ecological functions can be found elsewhere (e.g., Gregory et al. 1991; Naiman and Decamps 1997).

Figure 15.2 *Large deadwood in a stream.*

Deadwood Recruitment

Large deadwood input into streams is perhaps the most studied functional link between riparian forests and aquatic ecosystems (Figure 15.2). In general, the source area for deadwood into the aquatic ecosystem is the lateral distance away from the bank as defined by the average maximum height of trees (Fetherston et al. 1995; Gregory et al. 1991). In actuality, most deadwood comes from much nearer the water, e.g., nearly half may come from within a meter (3 ft) of the water's edge (Murphy and Koski 1989). This distance will vary with factors that influence tree height, such as productivity, species composition, and age. The source distance for deadwood may exceed tree height, sometimes greatly, if secondary transport occurs, as with floods or mass soil movement.

Tree and forest characteristics that influence deadwood once it gets into the aquatic ecosystem include sizes and species composition. For instance, the longevity of deadwood in streams is directly proportional to bole diameter (Murphy and Koski 1989), because large logs decay at slower rates than smaller logs (Harmon et al. 1986) and large logs are less prone to movement and breakage once in the water. Reducing tree diameters in the deadwood recruitment zone of a riparian forest, through shortened rotation lengths or conversion to smaller statured or short-lived

Table 15.1 *Riparian functions and their relationship to forest structure and composition.*

Functional Attribute	Forest Characteristic
Recruitment of large deadwood into the aquatic ecosystem	Tree species composition; live and dead tree density; tree age and size
Recruitment of particulate organic matter into the aquatic ecosystem	Plant composition; tree density
Light transmittance and water temperature regulation	Leaf area, canopy heterogeneity; tree density
Bank stability	Tree age and size; tree density; understory vegetation
Regulation of sediment, nutrient, and organic matter movement or uptake	Tree density; large deadwood abundance; tree species composition; forest floor and understory vegetation
Habitat for riparian species	Tree species composition; tree age and size; large deadwood; understory vegetation

Figure 15.3 *Leaf litter input to a stream. Photo credit: "stream" by muffinman71xx, licensed under CC BY-SA 2.0 (www.creativecommons.org/licenses/by-sa/2.0).*

species, will directly impact deadwood dynamics in the aquatic ecosystem. Moreover, log decomposition rates will also differ among species (Harmon et al. 1986), as will the susceptibility of riparian trees to mortality (and recruitment into the deadwood pool) from disturbance (Palik et al. 1998). Together, these factors influence addition and depletion rates of deadwood.

Particulate Organic Matter Flux

Riparian trees are the primary source of terrestrial particulate organic matter in aquatic ecosystems. Larger particulate pieces (> 1 mm in size), often termed coarse particulate organic matter, include leaves (and needles), twigs, bark, and reproductive parts. Leaves in particular are an important substrate and source of food for organisms in aquatic ecosystems (Figure 15.3). Leaf litter is particularly important in shaded settings where aquatic primary production is low, such as small streams and wetlands (Vannote et al. 1980; Barlocher et al. 1978; Palik et al. 2006). Terrestrial-origin coarse particulate organic matter can also be an important nutrient and energy source in large, unshaded rivers through deposition in their floodplains followed by re-translocation into the active channel (Johnson et al. 1995).

Most coarse particulate organic matter may come from vegetation near the water, say within a distance of one-half tree height (e.g., FEMAT 1993), generally within 20 m or less. There are not many studies that have actually tried to quantify this distance, but there is some evidence that it can be greater than the one-half tree height rule of thumb. For example, in some headwater streams in Minnesota, USA, up to one-third of the biomass of terrestrial coarse particulate organic matter comes from distances beyond 30 m, depending on the year and mode of input (Figure 15.4). This distance well exceeds the mean maximum tree height of 20 m for these forests (Palik et al. 1999). This same study found that coarse particulate organic matter input to the stream within the first few years after partial cutting was often reduced from the uncut forest (Figure 15.4). The point is that characteristics of the forest within at least 30 m of the water and even beyond will influence the quantity of coarse particulate organic matter input into the aquatic ecosystem. In particular, tree density, size, and age will influence the amount of litter reaching the water. Reducing tree density through silvicultural treatments, favoring smaller statured tree species, or shortening recovering times between harvests, could all potentially reduce the quantity of coarse particulate organic matter input into the aquatic ecosystem.

Figure 15.4 *Coarse particulate organic matter (mostly tree leaves) input into headwater streams in northern Minnesota, USA under varying riparian forest management conditions. The streams were surrounded by a) uncut forests (control), b) a 30 m wide uncut buffer with the adjacent forest outside of the buffer clearcut (uncut buffer), and c) a 30 m wide buffer (with adjacent clearcut) that was partially harvested to a basal area of 8 m^2/ha (partially cut buffer). Overhead input refers to material entering the stream from above, while lateral input consists of material moving horizontally from the forest to the stream. Values are means of three replicates (+/- 1 standard error). Data for overhead litter input in 1997 were redrawn from Palik et al. 1999, while other data are unpublished.*

The quality of particulate organic matter, particularly leaf litter, entering the water is also important for aquatic food web functioning; riparian forest composition has profound effects on the quality of this litter. For instance, litter of alder, species of which are abundant near streams in many regions (e.g., Swanson et a. 1982), have been found to increase turnover of particulate organic matter due to high nitrogen content and faster decomposition rates compared to many other species (Cole et al. 1978). Differences in litter processing rates that result from differences in quality can affect aquatic invertebrate communities (Webster and Benfield 1986), largely by determining the timing of resource availability to litter processors (Cummins et al. 1989). All other factors being equal, poorer nutrient quality litter decays slower and becomes bioavailable (to aquatic organisms that consume it) at a slower rate than higher quality litter.

Consequently, silvicultural activities that alter composition of the riparian forest, as well as litter quality, may feedback to make processed litter either more or less available at a particular time, depending on the species of litter. For example, in northern Minnesota riparian forests, harvesting black ash, a species with relatively high quality litter, in favor of trembling aspen, a species with relatively low quality litter, can alter the timing of litter becoming bioavailable to aquatic invertebrates (Palik et al. 2006).

Forest composition also determines the phenology of litter additions to water. Seasonal inputs of litter occur when deciduous species dominate the riparian forest. In contrast, evergreen species provide a more constant litter supply. This suggests that altering mixtures of evergreens and deciduous species over what occurs naturally in a riparian forest may influence aquatic functioning through these litter differences.

Regulation of Solar Radiation

The canopies of riparian forests regulate the amount and quality of light reaching the water surface (Figure 15.5), which in turn influences water temperatures (Moore et al. 2005) and **primary productivity**. Changes in light environments through forest alteration are of greatest interest in smaller aquatic ecosystems, such as **headwater streams**, where shading is usually high. This light influencing function of the riparian forest can have important consequences for aquatic primary production and some aquatic organisms (Tschaplinski and Hartman 1983; Gregory et al. 1991). For example, **macrophyte** production in the aquatic ecosystem increases with decreased shading by adjacent trees. Riparian canopies also regulate water temperature (Barton et al. 1985), which in turn has important consequences for fish populations through effects on primary productivity of food sources (Swanson et al. 1982), as well as dissolved oxygen concentrations (Ringler and Hall 1975).

Figure 15.5 *Varying degrees of shading by riparian forest along a small stream.*

Models based on stream orientation, riparian slope, vegetation height, and solar angle and direction, provide insight into the influence of riparian forests on shade and water temperature. For example, modeling has been used to identify riparian forest width requirements needed to maintain stream temperatures at pre-harvest levels. One such study found that maximum shading of small streams occurs when the width of riparian forest on the south side of a stream approaches 46 m (Steinblums et al. 1984). Keep in mind that these distances may buffer the stream ecosystem from temperature alteration, they do not necessarily provide temperature and other microclimatic buffering in the riparian forest itself, by protecting the latter from adjacent influences, such as large adjacent clearcuts (Brosofske et al. 1997).

The solar radiation regulating effect of riparian forests points to factors that may be manipulated through silviculture, potentially in ways that disrupt this regulation. In particular, reductions in tree density in the zone of influence can increase water temperatures (Moore et al. 2005), although factors other than tree shading also influence this. Alterations in species composition, such as changes from dense to sparse canopy trees, may do likewise.

Bank Stability

An important function of riparian trees and other vegetation is the stability that root systems of these plants impart to near bank environments. Root systems bind soil, which reduces surface erosion and sediment runoff to surface waters. Riparian plants also buffer banks from the erosive force of flowing water (Smith 1976). Larger trees, and more of them, as influenced by species, age, and density, will have expansive root systems and are likely better at preventing erosion than smaller trees. However, bank stability is not only a function of woody vegetation. The root systems of herbaceous vegetation may also help to reduce erosion rates (Dunaway et al. 1994). Thus, the entire riparian plant community, including trees, shrubs, and herbaceous plants, should be considered important for this function when developing silvicultural prescriptions for riparian forests.

Regulating the Movement of Nutrients, Sediment, and Organic Matter

Riparian forests influence the movement of water between terrestrial and aquatic ecosystems, and along with it, dissolved nutrients and suspended sediment and particulate organic matter. Forest structural characteristics influence these movements in several ways. Notably, large pieces of deadwood in a stream channel trap suspended sediments and particulate organic matter behind the dams they create (Figure 15.6). Managing forests to maintain abundant sources of large deadwood for the aquatic ecosystem should be a key concern of the ecological forester.

Trees growing on floodplains and large deadwood on the forest floor also trap sediments and particulate organic matter outside the active channel during over-bank flows (Swanson et al. 1982; Grubaugh and Anderson 1989; Harmon et al. 1986), well outside of the near-bank environment. This sediment trapping function of flooded forests points to the need to consider the entire flood-prone geomorphic setting as part of the functional riparian area and to manage these forest accordingly by maintaining proper tree densities and fostering the development of large trees that are ultimately sources of deadwood.

Even in geomorphic settings not prone to flooding, riparian forests trap sediment in overland flow as it moves downslope to the aquatic ecosystem. Tree boles, large deadwood, shrubs, herbaceous vegetation, and litter can all act as sediment traps (Figure 15.7). Most suspended material is likely trapped within the first 60 to 90 m of entering a

Figure 15.6 *Debris dam in a small stream in New Hampshire, USA. Sediment and organic matter will accumulate behind the dam.*

Figure 15.7 *The sediment trapping influence of riparian vegetation is evident from situations like this where vegetation is removed on a trail that runs perpendicular to a stream located at its base. Photo credit: David Foster.*

forest (O'Laughlin and Belt 1995), however this distance increases with the increasing slope of the forest. Regardless, managing forest structure to maintain its sediment trapping function is important.

Tree composition of riparian forests also influences the cycling and transformation of dissolved nutrients. Riparian trees contribute dissolved nutrients to aquatic ecosystems through leachates from litter and roots (Fisher and Likens 1973; Fiebig et al. 1990). As discussed previously, tree species vary in the nutrient content of their litter, so compositional changes may affect dissolved nutrient concentrations. Additionally, forest composition may influence **denitrification** in wet soils. For example, denitrification rates are higher in riparian forest dominated by alder rather than conifers (Gregory et al. 1991).

An interesting aspect of riparian nutrient fluxes is that transport is not all in the direction of land to water. Significant amounts of nitrogen, derived from fish carcasses, can find its way into riparian forests from the water. Much of the early research on this pathway of nutrient movement comes from riparian forests in Alaska, USA. For example, one study found that up to 24 percent of foliar nitrogen in Sitka spruce was derived from stream-spawning salmon (Helfield and Naiman 2001). This fish-derived nitrogen substantially enhanced growth of trees within 25 m of the stream. Where nitrogen inputs were highest, mean annual basal area increment of the trees was more than tripled, compared to reference sites. The pathway for this transfer of nitrogen from the stream to riparian trees may include summer floods that deposit salmon carcasses on streambanks. Also, dissolved nitrogen from fish carcasses may get transported to the rooting zone of trees through hyporheic flow. The *hyporheic zone* is an area beneath and alongside a stream where shallow groundwater and surface water mix (Edwards 1998). Finally, in the Alaskan studies, fish-derived nitrogen and other nutrients may find their way into riparian forests through dissemination of feces, urine, and fish carcasses by brown bear and other animals that eat salmon (Quinn et al. 2009).

Habitat

Riparian forests function as habitat for a variety of species. Structural features of riparian forests, including many of the features we have discussed elsewhere in this book, such as large deadwood, snags, vertical canopy layering, large trees with cavities, etc., are important habitat elements for a variety of organisms, just as they are in non-riparian forests. However, the true uniqueness of riparian forests as habitat derives from their adjacency to water, making the riparian setting a particularly important habitat for many species.

There are animal species that are best thought of as true obligates of riparian forests—that is, they occur and survive nowhere else in the landscape. Wood ducks, and probably other forest dwelling waterfowl, are an example of what might be considered an obligate riparian species, in that they spend the majority of their life in a riparian setting, feeding in water and using large riparian trees for nesting cavities. There are other species that are dependent on open water for a portion of their life cycle, notably breeding, but then spend the majority of their time in forest habitat, including upland forest that may be well away from the water. Wood frogs are a good example of this type of species (Figure 15.8). There probably are few plant species that are as riparian dependent as these animal examples, as most plant species that might be found in a riparian forest are likely also to be found in other wet forest types (that are not riparian as such).

Figure 15.8 *Wood frogs are good examples of species that spend the majority of their life on the forest floor of upland forests, but depend on small bodies of open water for breeding. Photo credit: Melissa Youngquist.*

There are many species of animals and plants that may better be described as facultative riparian, in that they preferentially use, or are found in, this setting occasionally to frequently. Bald eagles, for instance, use water as a source of fish and use large trees in the riparian forest for nesting and observation, but they also roam widely across the landscape securing food from many different habitats. The eastern gray squirrel is another example of this type of riparian user. While found in many habitats, gray squirrels often are found in hardwood stringers along ephemeral streams in pine-hardwood forests in the central USA (Fischer and Holler 1991). The important point of this discussion about habitat functions of riparian forests, as with the discussion of the other riparian functions, is that maintaining composition and structure reflective of the natural model for the forest is key to maintaining habitat.

Riparian Forests in Space and Time

Riparian forests vary greatly in their biotic and abiotic characteristics and the landscape settings where they occur. As you will see, they are not simply floodplain forests next to perennial streams, although these are probably the best studied riparian forests. Rather, they occur adjacent to a variety of aquatic ecosystems and on a diversity of geomorphic surfaces, with riverine settings being the best studied (Figure 15.9). Riparian forests also occur adjacent to other types of open water ecosystems, such as lakes and open-water wetlands (Figure 15.9). In some regions, there is a large conservation effort around small seasonal wetlands or ponds with concerns about how forest management influences their structure and function. While not considered riparian in the traditional sense, some of the same issues related to impacts to functionality from forest management apply to these small wetlands (Box 15.1).

Even within a type of water body, such as a stream, riparian forests can differ greatly in structure and composition, even over short distances, due to differences in their geomorphic setting. For instance, in one stream segment, the riparian forest may have a well-defined floodplain supporting flood and hydric soil tolerant species that grades toward the upland into slope and terrace forest composed of upland species (Figure 15.11a). The next segment on this same stream may consist of a sandy bluff that lacks a well-defined floodplain and supports xeric-site tree species up to the stream bank (Figure 15.11b). This type of variation occurs, for example, along streams in the Coastal Plain of the southeastern USA (Figure 15.12).

Figure15.9 *Examples of different types of riparian forests: a) a high gradient, small-order stream running through a mixed conifer-hardwood forest; b) a small ephemeral channel in a hardwood forest; c) a lakeshore pine-conifer forest; d) a stream running through a lowland hardwood forest; e) a large river floodplain forest dominated by silver maple. Photo credits: a) "Vulcan Stream" by r-z; d) "Forest on the River" by wackybadger; licensed under CC BY-SA 2.0 (www.creativecommons.org/licenses/by-sa/2.0).*

Figure15.9 (*continued*)

Riparian forests also vary in composition and structure over time because of natural disturbances. The role of natural disturbance in regulating the amount and spatial pattern of successional stages in watersheds is an important management consideration. First, remember that various natural disturbances affect riparian forests to reset the successional clock, just as they do in other types of forests. Consequently, within and among watersheds, natural disturbances create a spectrum of stand age classes and compositions; not just old, late successional forest. For example, consider the situation in the US Lake States during the mid-nineteenth century. During this period, there were approximately 1.8 million hectares of river bottom forest in Michigan, Minnesota, and Wisconsin (Frelich 1995). However, old-growth apparently made up no more than 55 percent of the total, the rest was younger, earlier successional forest. Remember also that the disturbance regime (the type, intensity, and return interval) differs greatly within and among types of riparian forests, just as it does in other forest settings. Consequently, foresters must evaluate disturbance regimes on a site- and watershed-specific basis—that is, there is not a single riparian forest archetype that is generally applicable in all cases.

Having a diversity of riparian forest developmental stages in a watershed, while likely reflecting natural spatial patterns (Sibley et al. 2012), is also important functionally. For example, the nutrient uptake ability of riparian vegetation will vary with developmental stage. Periodic and dispersed disturbances in a watershed that foster new forest development (Young Forest Stage) will help to maintain nutrient uptake efficiency in the watershed (Lowrance et al. 1984; Barten et al. 1998).

Box 15.1 **Seasonal ponds and forest management.**

Small seasonally flooded forest ponds (hereafter termed seasonal ponds) are of special conservation concern in many regions (Tiner 2003; Zedler 2003; Colburn 2004; Calhoun and DeMaynadier 2007). These ecosystems by definition are small, palustrine (isolated basin) wetlands that have a seasonally varying water regime. They may contain some trees in their basins and are usually found in a matrix of upland forest (Figure 15.10). Beginning in the later twentieth century, there emerged a growing recognition that many forest landscapes contained an abundance of seasonal ponds (e.g., Calhoun et al. 2003; Palik et al. 2003; Tiner 2003). There was also recognition of their importance as habitat for amphibians, invertebrates, and other species (Brooks 2000; Batzer et al. 2004; Francl 2008). Given their conservation importance, there is concern about the potential impacts of adjacent forest management on seasonal pond ecology (Batzer et al. 2000; Palik et al. 2001).

Figure 15.10 *A seasonal pond in an upland hardwood forest, Minnesota, USA.*

The small size of seasonal ponds, often less than 0.5 ha (e.g., Calhoun et al. 2003; Calhoun and DeMaynadier 2007; Palik et al. 2007), results in high perimeter-to-area ratios, which increase the potential for edge effects and interaction with the adjacent upland forest (Palik et al. 2006). For example, seasonal ponds receive substantial amounts of particulate organic matter (mostly leaves) from trees in the adjacent forest (Oertli 1993; Palik et al. 2006). An adjacent tree canopy also influences light reaching a pond surface (Batzer et al. 2000; Palik et al. 2001; Hanson et al. 2009), which will influence primary productivity and water temperatures. It follows that structural changes to the adjacent forest from disturbance may alter radiation and organic matter input, as well as water chemistry and hydrology (Semlitsch and Bodie 1998; Skelly et al. 1999; Palik et al. 2001; Batzer et al. 2000; Hanson et al. 2009), which in turn can result in changes in biotic communities. This linkage between adjacent forest and seasonal pond functionality has been demonstrated in many studies (e.g., Batzer et al. 2000; Hanson et al. 2009; Palik et al. 2006; Palik and Kastendick 2010; Kolka et al. 2011).

The functional relationship between seasonal ponds and adjacent forest should sound familiar to you; it is very similar to the functional linkages between aquatic ecosystems and riparian forest. Most foresters would not describe the forest around a seasonal pond as a riparian forest, but functionally it is comparable and so similar precautions and approaches for maintaining functionality warrant consideration. These approaches are discussed in detail later in this chapter. Suffice it to say for now that one such approach is to retain a forested buffer around a seasonal pond (Dodd and Cade 1998; Semlitsch 1998). In fact, a number of studies have shown that buffers do have a mitigating influence on alterations to pond biotic communities and ecosystem function, including mitigating changes in wetland plant communities (Palik and Kastendick 2010), invertebrate communities (Hanson et al. 2010), hydrology (Kolka et al. 2011), and the community of birds that use the forest around ponds (Hanowski et al. 2006).

Management Implications

Despite the variation within and among riparian forests, management approaches have not always fully accounted for this variability (Moore and Richardson 2012). For example, many organizations adhere *to best management practices* (BMP) for riparian areas that consist of a set of guidelines designed to protect aquatic ecosystems from the impacts of forest management (Blinn and Kilgore 2001). Many BMPs use a one-size-fits-all delineation of a riparian management zone (RMZ), with boundaries based on a set distance (with some variation with slope or stream size) from the water that determines what occurs inside and outside of the boundary. Many organizations choose to use this approach to designate no-management buffers inside of the boundary. While this is a legitimate approach in many situations, it can also limit opportunities to pursue objectives that may depend on silvicultural actions, including restoration, habitat management, and timber management, particularly in settings with less potential for impact to the aquatic ecosystem (Figure 15.13). Such an approach also fails to capture the ecotonal nature of riparian forests, as we discuss next.

Riparian Forests as Ecotones

It may be convenient to think about riparian forests as stands, like forests in other settings, but in actuality they are ecotones between land and water (Naiman et al. 1988) and as such are not easily delineated. A riparian ecotone is

Figure 15.11 *Hypothetical riparian forests along similar sized streams, but that differ greatly in structure and composition due to differences in their geomorphic settings; a) floodplain forest; b) bluff forest.*

Figure 15.12 *A real world example of the type of variation depicted in Figure 15.11. a) A hardwood dominated, flood-prone forest. b) Oak stringer forest bordered by longleaf pine. Both of these forest settings occur along the same stream in southwest Georgia, USA.*

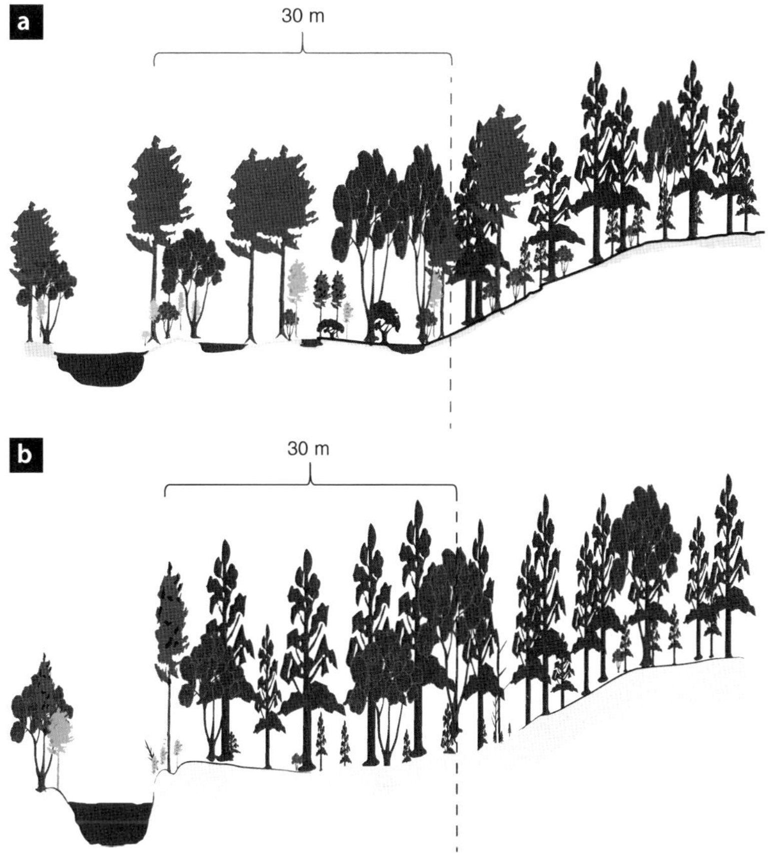

Figure 15.13 *Our stylized riparian forests from Figure 15.11 showing the placement of a riparian management boundary 30 m from the stream edge. a) Excluding entry within the first 30 m may be a legitimate approach because of high potential for impacts in the floodplain. b) There may be less potential for negative impacts to some riparian functions due to the different geomorphic condition.*

a transitional area from one ecosystem (forest) to another (aquatic) ecosystem. Ecologists have been studying riparian ecotones for decades, but foresters do not always consider the ecotone concept when managing riparian forests.

There are three types of ecotones of interest in riparian forests: physical ecotones, functional ecotones, and ecotones of impact. The physical ecotone reflects changes in abiotic characteristics from the water's edge to the upland forest. Various factors change across this ecotone, including geomorphology and landforms, soil texture and drainage, and water table depth and flooding frequency (Figure 15.14; Color Plate 15.1a). Concurrent with these physical changes will be changes in plant communities, including composition and structure (Figure 15.14; Color Plate 15.1a). Not all riparian ecotones will be structurally similar to the one shown in the figure, but most will consist of at least two intergrading physical and compositional conditions.

The second ecotone of interest is a functional one. This reflects the fact that the ecological linkages between riparian forests and water extend different distances along the water-to-upland gradient, but they all become increasingly less important with increasing distance from the water (Figure 15.15; Color Plate 15.1b). There are some rules of thumb for functional distances (Figure 15.15; Color Plate 15.1b) (adapted in part from O'Laughlin and Belt 1995), however, the exact extent of functional links is difficult to determine with certainty. Given this, it is helpful to think about the functional ecotone in a probabilistic sense. The probability that riparian functions are important is high immediately next to the water, but decreases away from the water, as various functions become less important. The point where the probability of the forest being riparian reaches zero is difficult to determine with certainty, but it is probably not so difficult to determine where the probability is low.

The third ecotone of interest is one of impacts to functions (Figure 15.16; Color Plate 15.1c). The idea is straightforward and relates directly to the functional probability concept discussed above. The importance of riparian functions increases with decreasing distance to the water, so the potential to negatively impact functions also increases closer to the water. For example, the probability of reducing organic matter flux to water with tree removal is high near the aquatic ecosystem, but decreases with distance away from the water.

It should be noted that impacts to riparian function are not always negative. They in fact may be positive because the potential to use silviculture to enhance riparian functions

Geomorphic Surface:	Floodplain	Hill slope	Upland Terrace
Water Table Depth:	Shallow	⟵⟶	Deep
Soil Drainage:	Poor	⟵⟶	Well
Soil Texture:	Organic	⟵⟶	Mineral
	Coarse Mineral		
Flood Frequency	High	⟵⟶	Low

Figure 15.14 *Stylized example of the physical riparian ecotone. Note the changes in dominant tree species from flood-tolerant deciduous species nearer the stream to xerophytic coniferous species on the hillslope and terrace.*

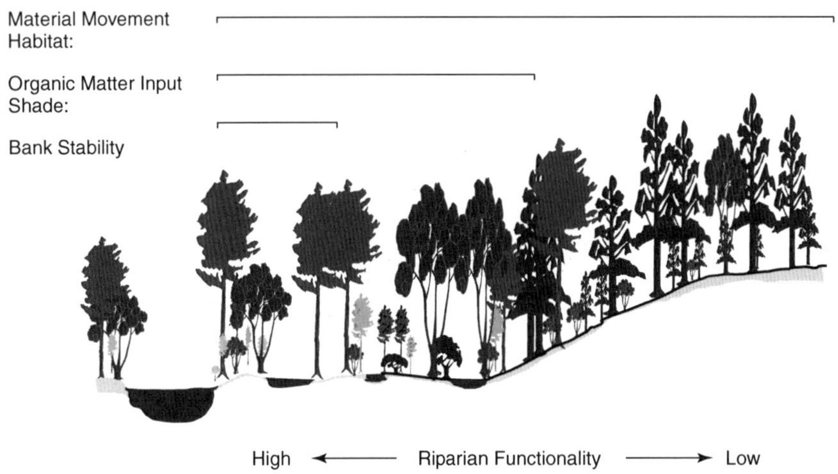

Figure 15.15 *Stylized example of the functional riparian ecotone. In the figure, material movement includes exchanges of water, sediment, and nutrients between the aquatic and terrestrial systems. Organic matter input includes large dead wood, particulate, and dissolved organic matter.*

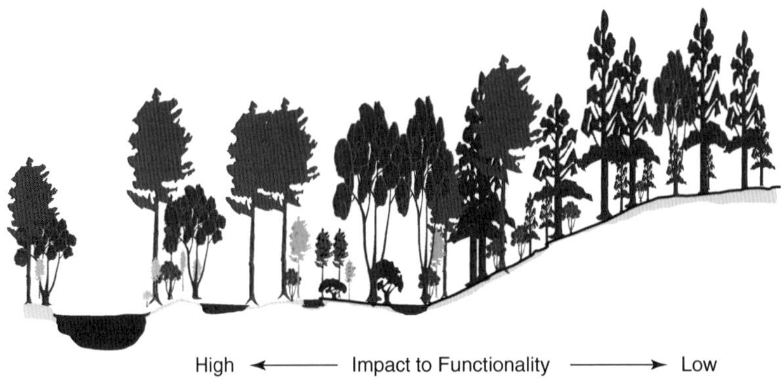

Figure 15.16 *Stylized depiction of an ecotone of impact (from management activities) to riparian functions.*

also increases closer to the water's edge. For instance, thinning in the riparian forest may increase growth of trees, resulting in larger individuals, and leading to increased stream shading and ultimately large dead wood contributions (Gregory 1997). Whether impacts are negative or positive, the covarying gradients of riparian function and impact provide us with the conceptual framework for integrating our fundamental ecological principles into silviculture for riparian forests, as we will explore in the next section.

Ecological Silviculture for the Riparian Ecotone

The first order of business for silvicultural planning in a riparian forest is to consider how to delineate a management unit that encompasses the riparian forest. Thinking back to the fixed-width approach for delineating a riparian management zone (RMZ), it should be clear that inappropriately placed boundaries may compromise the functional ecotone, depending on the nature of the riparian setting. For example, consider a stylized example where a boundary for an RMZ is placed at the location indicated (Figure 15.17a). Portions of this forest that contribute to riparian functionality likely fall outside of the RMZ. Depending on how the non-RMZ forest is managed, functionality could be impaired (Figure 15.17b). Also, this scenario (no management inside of the RMZ) results in only two potential direct impacts: low inside the boundary and high outside of the boundary.

The ecotonal nature of riparian forests suggests an alternative approach to delineating a management boundary,

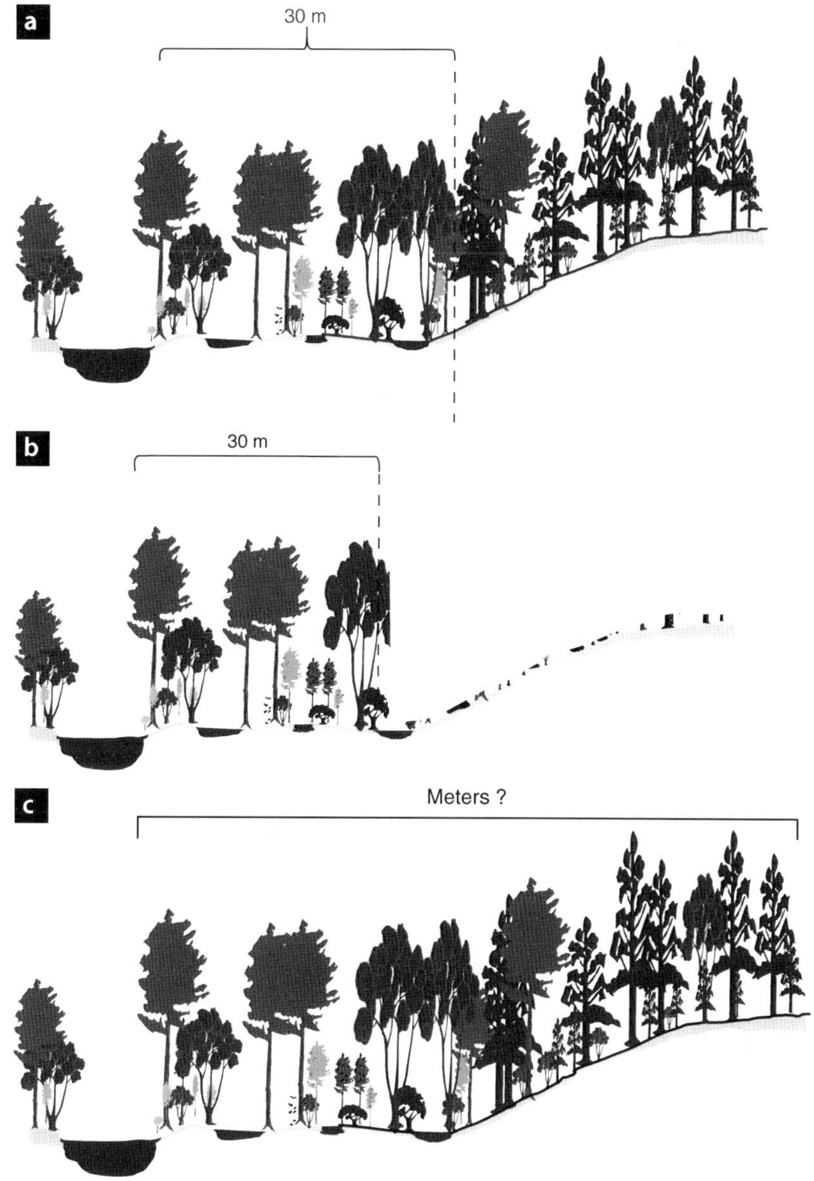

Figure 15.17 *a) Placement of an RMZ boundary at a distance that is within the functional riparian ecotone. b) Riparian functionality may be impaired from impacts to the forest outside of the boundary. c) A functionally delineated riparian management area.*

one that is integrative of physical and functional variation, and considers variation in impact. The alternative places the boundary farther landward along the ecotone, at a distance where the probability of riparian functionality is low (Figure 15.17c). This placement may well result in some "upland" forest falling within the RMZ, with the potential for limitation on management activities. However, this is only a problem if a hands-off approach to riparian management is the norm. It is not likely to be a problem, but instead an opportunity, if an ecological approach to riparian silviculture is adopted, one that recognizes the ecotonal nature of riparian forests and incorporates our fundamental principles of ecological silviculture.

Incorporating Continuity into Riparian Forest Silviculture

Sustaining structural continuity (*Principle 1*) in the riparian forest after harvest is important for all the reasons you have learned about previously (*lifeboating, enrichment, softening the matrix*), but it takes on added significance in the riparian setting because of important functional links that exist between forests and water. As we have already discussed, these links largely depend on forest structure. In some sense, sustaining continuity of structure may actually be easier in the riparian setting if functionality is being maintained as part of organizational policy. In fact, this is often the case when organizations follow voluntary or mandatory riparian best management practices that specify some minimum level of residual basal area after a harvest (Blinn and Kilgore 2001).

In riparian settings, meeting minimum basal area recommendations can be integrated with our understanding of live tree legacy retention and the concept of a riparian ecotone. For example, consider a stylized riparian forest (Figure 15.18a) where the residual stand structure after harvest reflects a dispersed retention spatial distribution (Figure 15.18b). While this structure may provide important habitat features in the forest, it may not adequately protect riparian functionality if there is insufficient retention near the water. Now consider the same forest after an aggregate retention harvest that more directly incorporates sustainability of riparian functions along an ecotone (Figure 15.18c). The total amount of retention is the same as in Figure 15.18b, but the amount increases from the upland to the water, acknowledging the increasing importance of riparian functions related to structure. Such a gradient would also reflect the changing importance of objectives, from more inclusive of timber objectives away from the water to more functionally focused objectives nearer the water.

A real world application of this approach comes from the southeastern USA, where mixtures of hardwood and pine species occupy terraces and slopes along streams. A forester may have the dual objectives of sustaining riparian functionality and regenerating shade-intolerant pines,

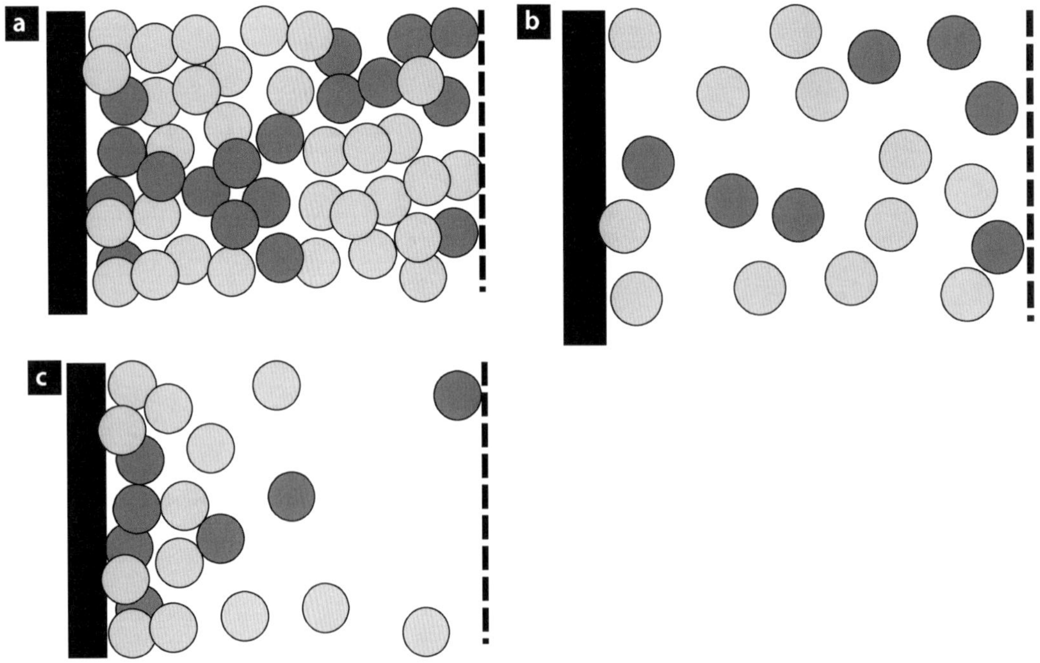

Figure 15.18 *Stylized riparian forest before harvest (a) and after a dispersed retention harvest (b) and an aggregate retention harvest (c). In the latter, retention is concentrated near the water, where functionality that is dependent on structure is highest.*

so they might concentrate retention near the stream, creating larger openings for regeneration of intolerant species on slopes and terraces (Palik et al. 1999).

Managing for Complexity and Diversity in Riparian Forests

Managing for structural complexity/diversity (*Principle 2*) in riparian forests can include the same objectives and activities as in any other forests that are structurally simplified (Chapter 8). This might include thinning to increase tree growth and development of large trees and ultimately large deadwood. In the riparian setting, thinning to increase growth and tree size may have added functional significance, such as increasing shading of the aquatic ecosystem and enhancing bank stability by developing large root systems. Managing for complexity in the riparian forest may also include deliberate decadence and deadwood creation to enhance forest habitat, but in the riparian setting this may be done to increase inputs of deadwood into the aquatic ecosystem.

As you learned previously, an important aspect of managing for complexity in established stands involves enhancing spatial heterogeneity of structural conditions. We have already discussed approaches for doing this in non-riparian forests with variable density thinning and modified selection systems (Chapter 8). In the riparian setting, use of these approaches would be guided by an understanding of the riparian ecotone and the need to provide greater protection of functionality nearer the water. For example, variable density thinning (VDT) that uses a skips and gaps approach (see Figure 8.14a) might be altered for a riparian setting by concentrating skips (unthinned areas) and avoiding larger gaps near the aquatic ecosystem (Figure 15.19).

The Timing of Silvicultural Activities

Appropriately timing silvicultural activities (*Principle 3*) by basing them on ecological cues rather than financial or timber growth metrics (Chapter 8) is as important in riparian forests as in other settings. In fact, allowing appropriate recovery between interventions, including regeneration harvests and thinning, may be more important in the riparian setting so as to minimize disruptions to riparian functionality. That is, overly short recovery periods not only impact the forest ecosystem itself, in terms of its complexity of structure and its composition, but the impact will extend to the linked aquatic ecosystem. For instance, short periods between interventions result in fewer (or no) large, old trees in the riparian forest, which in turn reduces shading potential and organic matter inputs, impacts habitat, and eventually limits large dead wood availability to the aquatic ecosystem (Figure 15.20; Color Plate 15.2).

The Context of Riparian Silviculture

Considering the larger scale context of silviculture (*Principle 4*) has added significance in riparian settings, due to the potential for cumulative effects on aquatic ecosystems. Cumulative effects are defined as *the impact on the environment that results from incremental impact of the action when added to the other past, present, and reasonably foreseeable future actions* (Gosselink et al. 1990). Cumulative effects can result from minor actions taking place over an extended period of time that collectively have significant negative effects. Cumulative effects have been most studied in the context of stream ecosystems and forest management and are taken to mean "the repeated, additive, or synergistic effects of forestry or other land use practices on various components of a stream's environment in time and space" (Burns 1991).

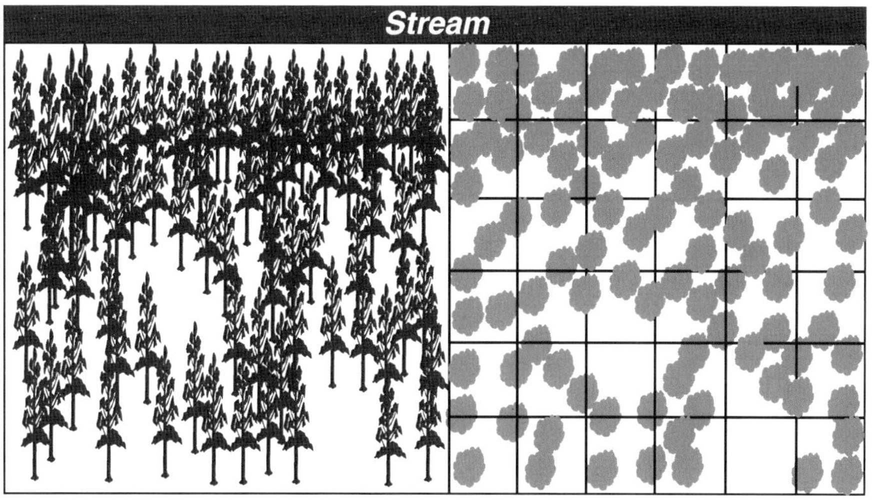

Figure 15.19 *Stylized depiction of variable density thinning using a skips and gaps approach in a riparian forest.*

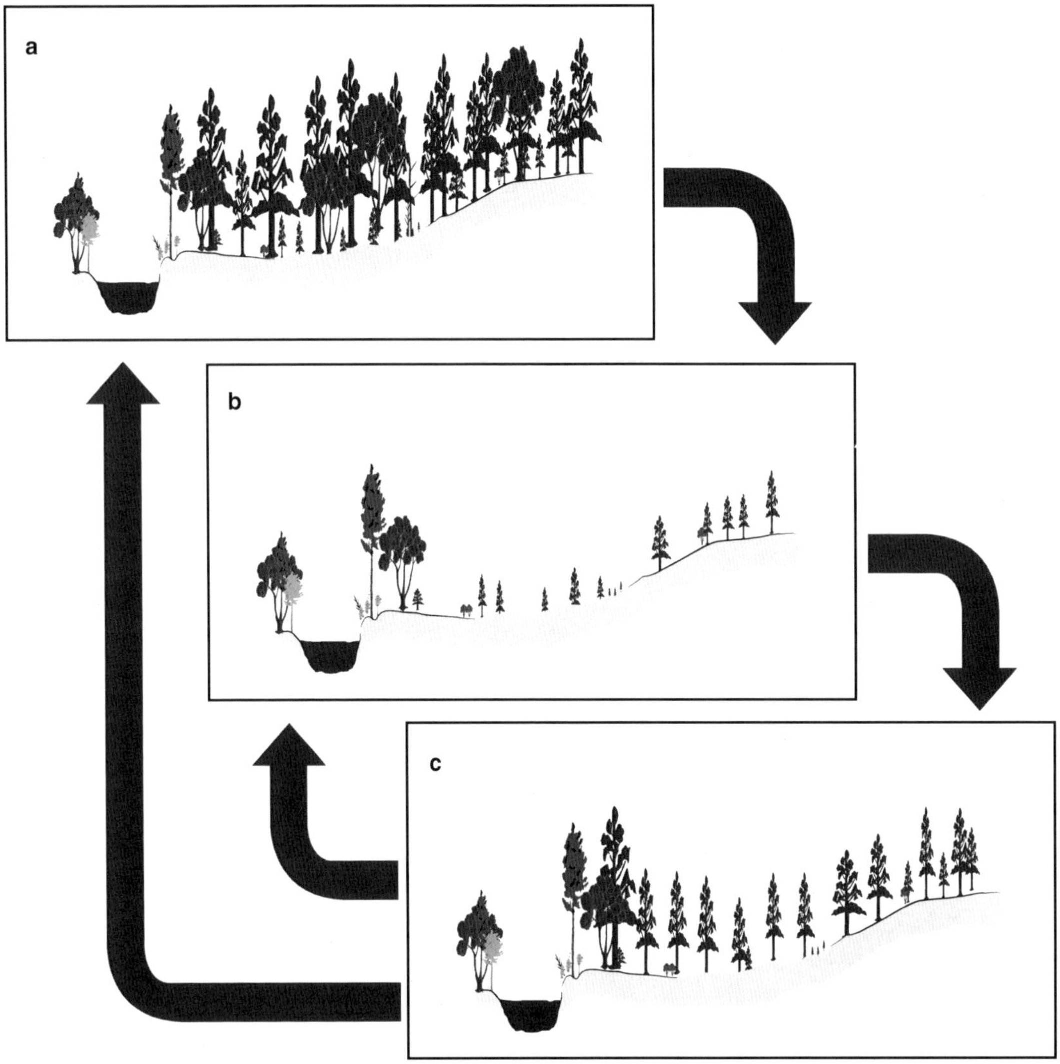

Figure 15.20 *Stylized depiction of multiple pathways of riparian forest development. a) The unharvested, Mature or Old Forest Stage ecosystem containing large old trees that eventually will provide large dead wood to the forest and stream. b) The same forest after harvesting of the large, old trees. c) The forest after development into the Young Forest Stage. Timing harvesting on financial or productivity cues would likely cycle this forest between b) and c), whereas timing based on ecological development would allow the forest to cycle back to a) before another harvest is considered.*

Given our consideration of riparian silviculture, the context issue becomes one of not impacting a watershed by treating too much of it (in a way that degrades riparian functionality) at any particular time. It is not that more intensive riparian management should never be considered, but rather that intensive riparian management should not be done in too large a percentage of a watershed in any given time period. Largely, concerns about cumulative effects have been addressed in watersheds drained by streams, but similar concerns apply to lake basins as well.

Ecological Guidelines for Riparian Silviculture

Riparian forests are the physical, biological, chemical, and functional nexus points of many forest landscapes, where they are integral for sustaining the health and productivity of both terrestrial and aquatic ecosystems. Society values riparian forests for the services they provide, such as flood mitigation, recreation, and wildlife habitat. In this chapter, we have pointed out how these services reflect the functional integrity of the riparian forest, which in turn is related to its structure and composition and how these are manipulated through silviculture.

Society may also value riparian forests for timber resources. Some riparian forest are highly productive because of their landscape positions, while others may contain tree species that are not found elsewhere in the landscape. Moreover, in some regions, a substantial proportion of the forest land base is riparian by virtue of an abundance of aquatic ecosystems in the landscape. Foresters charged with managing for timber resources will rarely ignore forests that are productive, abundant, or contain valuable tree species; riparian forests are no exception to this rule. Consequently, there is potential for impacts to aquatic ecosystems from adjacent forest management, impacts that degrade the functional linkages between forest and water.

Armed with an understanding of what a riparian forest is functionally, a fundamental question for a forester to ask is what kind of silviculture is appropriate in a riparian setting. In all cases, riparian silviculture must first and foremost protect or enhance functional linkages between forests and water. But this should not mean a de facto ban on timber management in the riparian setting. In fact, given our definition of the riparian ecotone based on functional attributes, an increased proportion of forest land that is part of the managed timber base would be considered riparian and thus some of this will likely be managed for timber. Moreover, as in other landscape settings with a long history of timber-focused management, restorative silviculture to enhance structural complexity and compositional diversity may be of high priority for riparian areas, particularly in regions such as northeastern North America, where long-lived conifer species were selectively removed from these areas by early harvesting activity.

Developing prescriptions for silviculture in riparian forests is something many foresters will need to do as part of their job. Our goal in this chapter has been to provide the tools needed to develop these prescriptions. A key aspect of doing this is incorporating our fundamental principles of ecological silviculture into the prescriptions, as they have as much relevance in riparian settings as they do in other types of forests. We summarize what you need to keep in mind to do this in the following guidelines:

Guideline 1: Understand that riparian forests are highly variable across and among landscapes. Just as with other types of forests, the structure and composition of a riparian forest will be different in different types of settings. As such, each riparian forest should be evaluated for its unique management needs and potentials. A one-size-fits-all approach to delineation of a boundary and determination of what to do in the riparian area (e.g., how much residual basal area to leave during a harvest) is not an ecologically prudent approach.

Guideline 2: Recognize what a riparian forest is, ecologically. Recognize riparian forests as ecotones of physical, compositional, and functional variation and not easily delineated stands. Determining the boundary of the riparian ecotone is difficult, but determining where the probability of riparian functionality being high or low is more straightforward. Riparian management zone boundary determination should be based on a functional probability determination.

Guideline 3: Sustaining or restoring riparian functions should be a primary silvicultural objective. When developing silvicultural prescriptions for riparian forests, keep the functional attributes that link land to water in mind, such as material uptake and interception, organic matter flux, shading and bank stability, and habitat. Regardless of how ecological a prescription might be, if it does not include support of riparian functions then it is ill-advised for the riparian setting.

Guideline 4: Consider a spatially variable application of a silviculture prescription in the riparian area. Variable retention harvesting and variable density thinning take on added meaning when applied to a riparian forest. While managing for continuity (*Principle 1*) and complexity/diversity (*Principle 2*) are just as important in riparian settings as elsewhere, how prescriptions incorporate the principles may differ in riparian forests. Specifically, level of retention may increase from the upland to the water's edge, paralleling the increasing importance of riparian functions that depend on continuity in structure.

Guideline 5: Remember that appropriate timing of silvicultural activities has added significance in riparian settings. Within a watershed, timing silvicultural interventions reflective of ecological cues related to forest development (*Principle 3*) has the added benefit of better sustaining functional connections between forest and water over time. Conversely, inappropriate timing of harvests and other activities can have the added impact (over and above the concerns we have discussed elsewhere in this book) of degrading functional linkages.

Guideline 6: Consider the potential for cumulative effects of riparian silviculture. The landscape context of silviculture (*Principle 4*) has special meaning in riparian settings due to concerns about cumulative effects that may degrade water quality and aquatic habitat. Even with proper attention given to *Principles 1–3* within individual riparian management zones, treating too much of the watershed at any particular time may result in unacceptable levels of accumulated impact.

CHAPTER 16

Silviculture in the Face of Climate Uncertainty

Perhaps the only real certainty about forests is that nothing is certain. No two forests are exactly alike in their structure and composition or in how they develop over time because the forces that act to shape them are not constant from place to place. There is nothing new about this type of uncertainty for a forester; most recognize that their prescriptions lead to somewhat general outcomes that are not predictable in the details. There are many reason that outcomes will differ even within what is considered the same type of forest. For instance, seed and propagule sources for tree species may be different or the species and amount of advance regeneration may differ, both leading to differing tree composition. The incidence of pest infestations may vary among locations, resulting in differences in composition and structure. Differing types, magnitudes, or durations of localized weather events, such as spring frosts or summer windstorms, will impact a host of factors including flowering and fruiting, tree growth, and stand structure. Again, foresters accept this uncertainty as part of doing business; they manage with the understanding that uncertainty of outcomes is a given with long-term management.

Another source of uncertainty that a forester must contend with is the unpredictability of markets for wood and wood products, as well as changing societal expectations about how forests should be managed. As above, these sources of uncertainty probably do not cause the forester who is developing silvicultural prescriptions too much heartburn since their prescriptions are part of a broader program of management for an organization or agency designed to provide some degree of flexibility to meet changing markets and social demands.

Now consider climate change as an additional source of uncertainty. In fact, climate change may well represent the greatest challenge and source of uncertainty that foresters have faced up and into the twenty-first century. This is reflected by growing urgency to develop strategies that increase the adaptive capacity of forests or, at the least, minimize ecosystem vulnerability to future climates (Spies et al. 2010). Broadly, strategies for addressing climate change may include manipulating forest structure and composition to reduce anticipated impacts of climate change or to increase the representation of desirable functional traits (e.g., drought tolerance) that may enhance ecosystem adaptation potential (Millar et al. 2007; Puettmann 2011). The efficacy of these and other approaches to minimizing climate change impacts are poorly understood, due in large part to the long timeframes necessary for empirical evaluation of management approaches. So, in addition to the uncertainty of climate change itself, there is the uncertainty about the effectiveness of adaptation strategies, presenting a daunting challenge for foresters tasked with developing long-term approaches for managing forests in the face of climate change. Moreover, climate change impacts are expected to interact synergistically with other stressors, such as invasive species, to create novel conditions for which we have little management experience to draw from to develop adaptation strategies.

The difference between climate change uncertainty and the other sources of uncertainty that we discussed above is that speculation and prediction about climate change is a relatively new concern, emerging only in the latter part of the twentieth century. Foresters are now told that they need to consider climate change adaptation in their prescription development, but they are also told that climate change predictions may have a high degree of uncertainty. Moreover, as mentioned above, there are likely difficult to predict interactions among climate change and other factors, including disturbance, pests, and invasive species. Understandably, the uncertainty of climate change predictions can lead to uncertainty about how to proceed with silvicultural prescriptions. Nonetheless, many of the ecological silvicultural principles and approaches we have discussed in previous chapters hold tremendous promise for addressing this uncertainty. Their application within the context of climate change is the subject of this chapter. In it we will review climate change in the context of forest management, including what we do and do not know about the issue. Then we will review the framework a forester uses to make decisions about silvicultural prescriptions in general, including how they incorporate uncertainty. We then provide a framework for including an uncertain climate future into silvicultural decision making. Finally, we discuss strategies and specific silvicultural approaches that can be included in prescriptions to incorporate climate change adaptation.

A Background on Climate Change and Forests

A certainty of nature is that climates are dynamic and always have been. However, there is an accumulating body of evidence that suggests that climate changes occurring since the mid-twentieth century are leading to conditions that have not occurred in the past several centuries to millennia (IPCC 2014). That is, climates look to be more dynamic than at any time in the past. These emerging novel climates, combined with additional pressures associated with human population growth, fragmentation of habitats and landscapes, increased pollutants, invasive species, and altered fire regimes, are resulting in a rapidly changing global environment that will pose challenges to the sustainability of forest ecosystems and the services they provide, as well as how to manage forests in the face of these challenges.

What We Know

Perhaps the most reliable signal associated with global climate change is the well-characterized increase in atmospheric carbon dioxide concentration (CO_2) generated from long-term observational records. This monitoring indicates that CO_2 concentrations have increased from about 315 ppm in the mid-twentieth century to about 400 ppm by the early twenty-first century (Figure 16.1a). While projections for the future vary widely based on different assumptions for carbon emissions and mitigation measures, carbon dioxide concentrations well above 400 ppm are expected by the end of the twenty-first century.

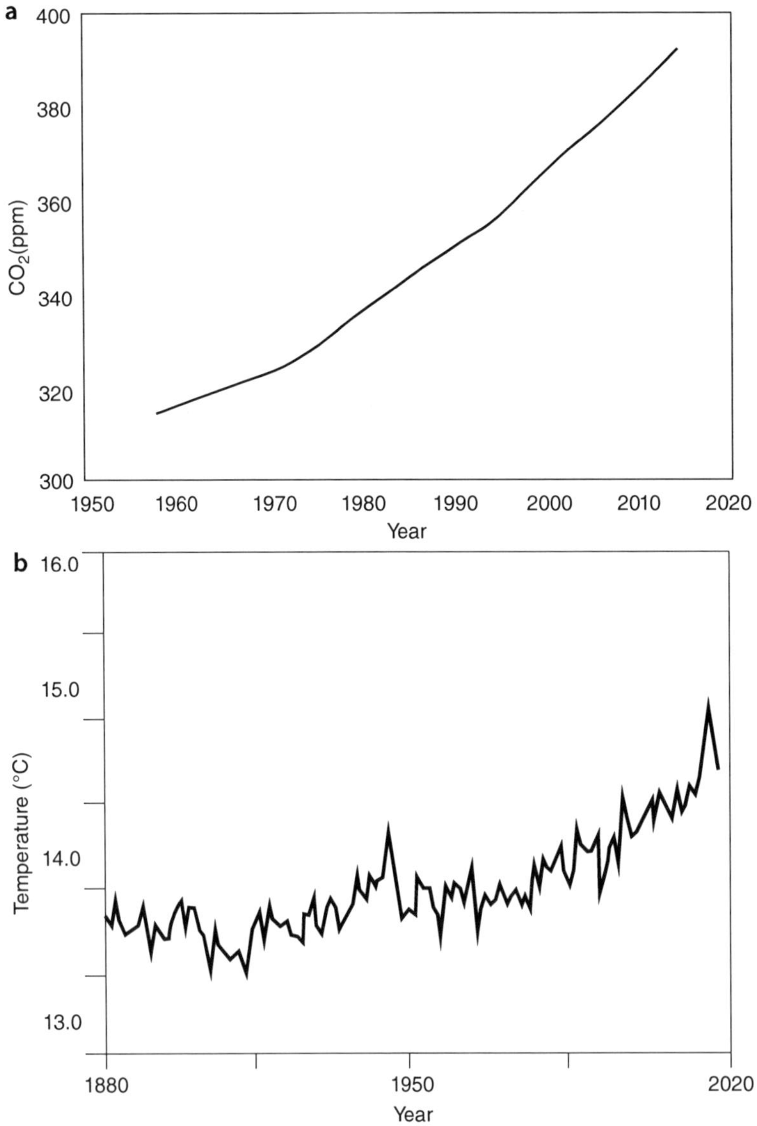

Figure 16.1 *Global mean atmospheric carbon dioxide concentrations and mean global temperatures over time. (a) Atmospheric carbon dioxide concentrations generalized from measured values at Mauna Loa, Hawaii, USA and the South Pole. (b) Mean global temperature (source data from NOAA National Climatic Data Center, http://www.ncdc.gov/cag).*

In association with the global increase in atmospheric carbon dioxide (and other greenhouse gases, notably methane) there has been a general increase in mean global temperature of about 1.0 °C since the mid-nineteenth century (Figure 16.1b), with varying levels of increases in different regions of the globe. Given projections for continued increases in carbon dioxide, it is likely that global mean temperatures will also continue to rise at rates and to levels that have not been experienced in the evolutionary timeframe that has shaped current forest ecosystems and their adaptive characteristics, particularly at higher latitudes.

So what do these changes mean for forests, particularly at higher latitudes and elevations where climate changes are likely to be the most severe? Among other things, conditions for establishment, growth, and survival of tree species in many regions are going to be different than they have been over the last several centuries. Given that the physiological and reproductive processes of each tree species have unique temperature and moisture optima, there is concern that at some point the environmental conditions in a particular region will change to the extent that the current forest species may no longer be able to cope with the new conditions during some phase of their development. Thus, a key challenge for the sustainability of forest ecosystems will be to develop silvicultural strategies and approaches that promote forest adaptation to changing climate conditions. In the near term, these strategies may focus on maintaining or restoring the vigor and composition of current species. Over longer time periods, strategies may focus more on adapting forests to new environmental conditions by increasing structural complexity and compositional diversity so as to maximize pathways for recovery and by introducing new species and promoting stand structural conditions that allow these species to be successful.

What We Don't Know with Certainty

Although we can be relatively certain that mean temperatures are increasing globally, and will continue to do so, the exact magnitude and timing of warming in different regions is less certain. Even less certain is how the abundance and seasonality of precipitation is changing, or how the frequency and intensity of wind storms and other extreme weather events is being altered (IPCC 2013). Another uncertainty is how drought may be expressed in the future. Temperature increases and changes in seasonal precipitation are predicted to lead to increasing frequency and severity of droughts in forests across the globe (Allen et al. 2010), but given the uncertainty of precipitation predictions, there is corresponding uncertainty in predictions about the magnitude, duration, and frequency of future droughts. Much of this uncertainty arises from regional variations in vegetation and topography. At even finer scales, complex local topography and microsite variability may greatly influence the local expression of climate change, including drought events.

Silviculture in the Face of Climate Uncertainty

The overarching challenge for coping with climate change from a forest management perspective, is dealing with uncertainty in the specifics of future climate. Despite this uncertainty, there is a growing consensus that management decisions at scales relevant to silviculture need to consider how actions either enhance or detract from a forest's potential to adapt to changing climate. In other words, the uncertainty of future climates, particularly at stand-scales, does not render consideration of climate change adaptation with silviculture a pointless exercise. Thus, a hallmark of the process by which silvicultural planning will incorporate a consideration of climate change is the acknowledgement and embracement of uncertain future conditions (Millar et al. 2007). Again, we have argued earlier that acknowledging and embracing uncertainty is not new to foresters, so considering climate change uncertainty should not be a completely foreign exercise. In fact, by considering potential forest ecosystem changes as a result of climate change, along with other sources of uncertainty, a forester encourages a deeper discussion of appropriate silvicultural actions across a range of potential changes and interactions. For example, a forester might ask if a given management action, taken for reasons other than climate change (e.g., growing a desired timber species or sustaining habitat for a particular species of songbird), will place an ecosystem at greater risk within the range of possible climate change outcomes.

We will go into more detail on specific adaptation strategies and silvicultural approaches to address those strategies. Suffice it to say for now that all strategies may involve actions that restore or sustain compositional, structural, and functional diversity to stands, so as to provide flexibility and the potential to shift stand development in different directions as warranted by evolving conditions. These are principles we have discussed in detail in earlier chapters in the context of sustaining biodiversity and other ecosystem services, yet they are also applicable to climate change adaptation. For instance, diversifying stands in composition and structure (within ranges of natural variation) may better ensure adaptation to changing climate conditions given that each species and structural condition has different climate responses. Moreover, restoring and sustaining diversity better safeguards the processes that promote adaptation in the face of all types of disturbance. In practice, managing for options and adaptability should not be an exercise associated strictly with climate change. Rather, a similar approach is an appropriate response when faced with all types of uncertainty that can impact forests, including for instance, changing markets or threats from invasive species. An adaptation approach driven by climate change is simply a smart way to manage forests in general.

Incorporating Climate Change Uncertainty into Silvicultural Considerations

General Considerations When Developing Silvicultural Prescriptions

What does a forester consider when developing silvicultural prescriptions? For all but the most industrialized settings, this process involves consideration of a number of factors at different spatial scales that influence the potential of a stand to respond to manipulation and achieve desired future conditions. These considerations help to refine management objectives (e.g., what should future composition of a stand include given the natural range of variability) and they also help to forecast the response of stands to manipulations designed to meet objectives. For instance, at the broadest spatial scales, sideboards for a prescription are provided by organizational or even multi-organizational plans that specify goals and objectives over some extended time period, say 10 to 15 years (e.g., land and resource management plans for USDA National Forests). These plans may include, among other things, broad goals related to desired forest composition and age structure. Multi-ownership regional landscape plans may reference goals of increasing landscape connectivity, the size of habitat patches, or the areas of certain forest conditions, such as interior or early successional forest. Such large-scale plans may also have goals of increasing habitat **refugia** and maintaining an array of conservation reserves that span a diversity of biophysical settings. Ultimately, silvicultural prescriptions will be influenced by these broad organizational plans.

Scaling down from regions and organizational landscapes to a local landscape context, a forester will ask a number of questions that will influence decisions about what to do in a stand. Local landscape issues to consider include proximity to water (riparian areas), viewscape concerns, **wildland-urban interface** proximity, or the location of recreational infrastructure (e.g., campgrounds, hiking trails). The ecological features of the local landscape will also be considered. For instance: what are the types, distributions, abundance, and disturbance history of the plant communities/forest types in the project area, what is the age distribution of forests, what is the proportion of natural origin versus plantation origin stands? Consideration will be given to the stand condition relative to other stands in the landscape. For instance: how does the stand's current and desired composition and structure compare to other stands and is the desired future condition conducive to local landscape objectives? Issues related to pests and other stressors will also be considered, such as the recent history of regeneration failures from overbrowsing by ungulates.

Finally, drilling down to the site or stand-scale, the forester will review a number of considerations that will influence goals and inform the actions needed to reach those goals. These include consideration of site characteristics, such as geomorphic setting, soil type, and productive potential, e.g., **site index**. Characteristics of the current forest will be assessed, including composition, size and age structures of trees, evidence of current and past pest issues, and occurrence of species of special concern. While a forester will approach this stand-scale assessment with objectives and desired future conditions in mind, the actions needed to meet those objectives and achieve desired future conditions will be informed by current conditions and history of the stand.

Our purpose in reviewing these considerations is to emphasize that the same concerns will still be of use in the context of developing silvicultural prescriptions for addressing climate change adaption. That is, addressing the latter should not be such a daunting task that the forester abandons what they already do when developing prescriptions. Rather, when incorporating climate change considerations into silvicultural planning, the forester begins by addressing the same suite of multiscale considerations, but now adds several in addition that reflect climate change uncertainty.

Climate Change Considerations

A key larger-scale consideration specific to climate change is whether habitat suitability for tree species of interest, either those currently in the stand or those desired to be in the stand, has the potential to change within a timeframe that encompasses the life span of these species. A forester knows through training and experience that a suite of factors interact to determine habitat suitability, including climate, physiography, soil biogeochemistry, disturbance, and biotic influences. Considering these relationships under climate change scenarios can help inform forest managers about the vulnerability of species to future climate and the ability for management to promote compositions that reduce stand-level vulnerability and enhance adaptability. Information on current and projected habitat suitability can be gained from resources like the Climate Change Tree Atlas (Prasad et al. 2007–ongoing) and other modeling efforts (e.g., Meier et al. 2012). These can be used to examine trends over regional landscapes and determine where particular species may be most at risk or likely to be successful. While these tools cannot tell the manager what will happen, they can provide an indication of the probability for habitat changes and thus the potential that tree populations and ecosystems will or won't adapt to climate change (Box 16.1).

In addition to changes in tree habitat suitability, a forester may also want to consider how *threats* such as wildfire,

Box 16.1 **Habitat suitability predictions: a) red pine forests in northern Minnesota, USA and b) northern hardwood forests in New Hampshire, USA.**

Recall that the model ecosystems we used to illustrate Archetypes 3 and 4, respectively (Chapter 6), are northern hardwood forests in northern New England, USA and red pine dominated mixed-species pine woodlands in the western Great Lakes region, USA. The distribution of these ecosystems may render both vulnerable to tree habitat shifts induced by climate change: New England hardwood forests occur at high latitudes and elevations, where temperature increases are likely to be significant, while mixed-species (pine) woodlands straddle multiple biogeographic regions, including boreal forests to the north, temperate hardwood forests to the south, and for Minnesota, USA at least, prairie parkland to the west (Aaseng et al. 2003).

Climate Predictions

Climate change predictions have been summarized in forest vulnerability assessments for northern Minnesota, USA (Handler et al. 2014) and northern New England and New York, USA (Janowiak et al. 2017). Although the assessments compared projections using several different emission scenarios, we focus here on results using a high CO_2 emission scenario (the GFDL A1FI emission model). Compared to a 1971 to 2000 baseline, the predicted average annual temperatures of northern Minnesota may be 4.9°C warmer by 2070 to 2099, while northern New England may be 4.2 °C warmer.

For Minnesota, fall (September–November) and winter (December–February) temperatures are predicted to increase by similar amounts (F 5.1 °C, W 5.4 °C), spring (March–May) is predicted to experience moderate warming (3.0 °C), while summer (June–August) temperature is predicted to warm by 6.3 °C. In northern New England, fall (4.4 °C), winter (4.3 °C), and summer (4.6 °C) may experience similar increases, while spring may have a somewhat lower increase (3.4°C).

In Minnesota, average annual precipitation is predicted to decrease marginally by only 1.0 cm, but with substantial seasonal variation. Average winter precipitation may increase by 1.8 cm, while summer precipitation is predicted to decrease by 12.2 cm. Spring and fall precipitation are predicted to increase by 8.1 cm and 1.5 cm, respectively. In northern New England, average annual precipitation may increase substantially, by 7.3 cm, with most of the increase coming in winter (5.8 cm). Summer precipitation is expected to decrease somewhat (4.6 cm), while spring (2.5 cm) and fall (3.6 cm) may have moderate increases.

In these two regions, average annual precipitation may not change appreciably, or it may even increase, but warmer temperatures combined with decreased growing season precipitation may result in increased intensity and duration of summer drought, with potential changes in tree habitat suitability.

Tree Habitat Suitability

These changes in climate may result in changes in habitat suitability for a number of tree species in both of our case study ecosystems (Table 16.1). These changes come from predictions using the Climate Change Tree Atlas (Prasad et al. 2007–ongoing) for the Northern Minnesota Drift and Lake Plains in Minnesota and White Mountains section in New Hampshire (ECOMAP 1993), with similar predictions

Table 16.1 *Predicted changes in importance value (reflecting changes in habitat suitability) between 2000 and 2100 for species currently or potentially associated with red pine (mixed-species woodland) ecosystem in northern Minnesota, USA and northern hardwood forests in northern New Hampshire and New York, USA.*

Minnesota Red Pine		N. New England Northern Hardwoods	
Species	Change in Importance Value[1]	Species	Change in Importance Value[2]
Trembling aspen	−19.57	Red spruce	−6.53
Balsam fir	−6.90	Yellow birch	−4.66
Paper birch	−6.14	American beech	−3.54
Black spruce	−5.36	Sugar maple	−3.29
Jack pine	−1.31	Quaking aspen	−1.11
Bigtooth aspen	− 1.00	Red maple	−1.05
Northern red oak	−0.84	Eastern white pine	0.25
White spruce	−0.69	Bitternut hickory	0.90
Red pine	−0.61	Eastern hemlock	1.07
Eastern white pine	0.30	American basswood	1.45
Red maple	1.83	Black birch	1.54
Black cherry	1.75	Northern red oak	3.84
White oak	3.01	—	—
Bur oak	3.05	—	—
Eastern red cedar	7.56	—	—

[1]Mean change within the northern Minnesota Drift and Lake Plains Section based on a Hadley high emission scenario model.
[2]Mean change within the Adirondack-New England Mixed Forest-Coniferous Forest-Alpine Meadow Province (White Mountains Section) based on a Hadley high emission scenario model.

coming from other independent modeling efforts (e.g., Ravenscroft et al. 2010). In brief, the Climate Change Tree Atlas predicts future habitat using regression tree analysis of importance values (relative abundance based on basal area and density) for trees on USDA Forest Service Inventory plots (https://www.fia.fs.fed.us), coupled with projected future climate variables. Importance values for plots are scaled to 20 km grid cells.

The model predicts declining habitat suitability for a number of characteristic tree species in each ecosystem (Table 16.1). Conversely, several species have projections of increasing importance. Eastern white pine, an iconic species in both regions, has predicted habitat suitability that is largely unchanged, reflecting the wide geographic and climatic distribution of the species across eastern North America (Burns and Honkala 1990). A number of species with distributions in the southern half of each of the regions are predicted to have increasing importance farther north by the end of the twenty-first century, including white oak, black cherry, and eastern red cedar in Minnesota, and northern red oak in New England.

extreme weather events, or pests and pathogens may change under a changing climate. For example, warmer, dryer growing seasons may reduce fuel moisture levels and increase the risk of catastrophic wildfire, warranting increased attention to fuel reduction treatments in forests prone to fire.

Altered climate, including milder winter temperatures and longer growing seasons, may increase the risk of attack from insect pests and diseases that were previously a minor problem. While climate change may influence the potential for insect outbreaks in many different ways, the direct relationship between seasonal temperature and population success suggests that climate warming may be a large and growing concern (Bale et al. 2002). The devastating impact of mountain pine beetle in western North America is an example of how warmer winter temperatures may greatly increase the threat to overcrowded, older-aged forests predisposed to an insect epidemic (Fettig et al. 2007).

Another new consideration that may affect silvicultural decisions is how societal expectations for ecosystem goods and services from forests will change in the face of climate alteration. There may be little change in the short term, with expectations still inclusive of merchantable timber species, game species populations, a full suite of native plants and animals, recreational opportunities, and in general, a composition and structure that the public perceives are normal. However, over time, there may be increasing demands on foresters to add new services to the mix, such as those reflecting carbon mitigation by increasing carbon stored as soil and plant biomass, or increasing long-term storage of carbon in various forest products (Box 16.2).

Ultimately, people will continue to want what they have come to expect from their forests in terms of ecosystem services, but they may gradually accept that the forests that provide these services are not their grandparents' forests. Providing for continuity of services will challenge the forester—that is, developing prescriptions that enhance adaptability to climate change, while still providing expected ecosystem services. In the next section, we discuss strategies to maintain continuity of function, as well as silvicultural approaches that can be used to implement these strategies.

Silvicultural Strategies and Approaches to Adapt Forests to Climate Change

Strategies

Despite the uncertainty of specific future climate conditions, particularly at finer spatial scales, silvicultural planning needs to embrace the concepts of managing forest for adaptation to new conditions. Adaptation strategies can be generally categorized into *resistance, resilience, and transition* (originally referred to as response) strategies (Millar et al. 2007; Table 16.2). Although presented as discrete categories, these strategies fall along a continuum, such that implementation of an adaptation approach may involve elements of two or three strategies.

A resistance strategy focuses on adapting the forest that is already there, in terms of its composition and structure, to expected climate changes, such as more frequent or severe growing season drought. In effect, a resistance strategy uses silviculture to make the forest you have now more robust to changing conditions in the near future, such as warmer, drier growing seasons. A resilience strategy works with the current ecosystem, but within a broader range of compositional and structural outcomes that are side-boarded by the **range of natural variation** for the ecosystem (Landres et al. 1999). The presumption is that a forest can be adapted to future climate by emphasizing native species and structures that have greater adaptability, but what the forest looks like will largely be within the range of conditions that occur (or occurred) naturally for that forest (assuming it is healthy and functioning normally). In fact, a forest managed using a resilience strategy is likely to fairly closely resemble a forest managed using the ecological principles we have presented in previous chapters. Note that although these conditions may fall within the natural range of variation, they may not currently be managed for on a regular basis as part of a classical timber-focused approach. Finally, a transition strategy assumes that habitat for current species, or at least some of them, will become less suitable or even unsuitable, due to climate change (e.g., Box 16.1) and that actions should be taken to transition forests to species and structural conditions that are predicted to be better adapted to future

Box 16.2 **Addressing climate change mitigation objectives through ecological silviculture.**

Forests are a large global **carbon sink** storing over 80% of global terrestrial aboveground carbon (Dixon et al. 1994). Given the critical role forests play in carbon mitigation, there is increased interest in managing forests for their carbon benefits to offset greenhouse gas emissions and increase the size and stability of terrestrial carbon stores (Malmsheimer et al. 2008). Forest carbon markets were developed in recognition of these benefits and landowners may receive payment for carbon offset projects that increase **carbon stocks** through: 1) establishing forests in areas currently in a different land use (*afforestation*), 2) reestablishing forests in recently harvested or understocked areas (*reforestation*), 3) modifying forest management practices to increase levels of carbon stored in the forest and in forest products (*improved forest management activities*), or 4) protecting forested areas from conversion to non-forest land use (*avoided conversion*) (Malmsheimer et al. 2011).

Of the abovementioned options for increasing carbon stocks, there is general consistency between many of the principles and practices associated with ecological silviculture and those recognized as *improved forest management activities* for carbon offset projects. In particular, ecological silviculture practices, including the use of extended rotation lengths and retention of live- and dead-tree legacies during harvests (Figure 16.2), are recognized as practices that increase on-site carbon stores relative to classical silvicultural approaches (Harmon et al. 2009; Nunery and Keeton 2010). Moreover, the larger live tree sizes associated with these management systems may create the potential for a greater level of off-site carbon storage in long-lived wood products, such as large **dimensional lumber**, utility poles, and construction logs.

It is important to recognize that carbon mitigation is one of many objectives being addressed on a given landscape and a singular focus on maximizing on-site carbon stores may conflict with the achievement of other ecological objectives (D'Amato et al. 2011). For example, carbon management often focuses on promoting and maintaining highly stocked forest conditions; however, such conditions may represent a significant deviation from historical, multiscale patterns in forest density and may increase vulnerability to disturbance events, including bark beetles and fire (Churchill et al. 2013). Furthermore, silvicultural approaches for climate adaptation often center on reducing ecosystem vulnerability by applying regeneration harvests and tree planting to increase the diversity of climate and disturbance responses contained in a given tree community (see sections below). Although these approaches may reduce levels of on-site carbon storage in the near term, they are critical for increasing options to sustain forest functioning in the face of changing climate conditions and emerging forest health threats. Potential options for balancing adaptation and mitigation objectives may include using multiaged regeneration methods that retain high levels of mature forest structure on site (Figure 16.2), in conjunction with planting future-adapted tree species that contain functional traits associated with high levels of carbon storage (e.g., high wood density; Jacobs et al. 2009).

Figure 16.2 *(a) A 143-year-old fire-origin red pine forest in north-central Minnesota thinned seven times since age 85 as part of an extended rotation management system and (b) large sugar maple reserve tree retained during a shelterwood overstory removal 30 years before the photo was taken in northern New York, USA.*

climate. These species likely are not currently present or abundant in the ecosystem, so *assisted migration* might be part of the approach needed for this strategy (Pedlar et al. 2012), as we discuss in more detail below. The three adaptation strategies generally translate to near term, mid-term, and longer term solutions for adapting to anticipated change. They also represent a gradient of increasing risk, effort, and (potentially) controversy as they reflect a progressively greater deviation away from accepted management approaches for a particular ecosystem (Table 16.2).

Silvicultural Approaches

The approaches (specific actions) discussed next are used to implement adaptation strategies; they are the on-the-ground elements of silviculture. These approaches may involve many

Table 16.2 *Climate change adaptation strategies (adapted from Millar et al. 2007).*

Strategy	Definition	Goal	Assumptions
No Action	No actions specific to climate change are taken	Allow response to climate change without direct intervention	High risk, low effort, good social acceptance (initially)
Resistance	Improve the defense of a forest to change	Maintain relatively unchanged conditions over time	Low risk and moderate effort, high social acceptance
Resilience	Accommodate some change but within the range of natural variation	Allow some change; encourage return to a reference condition	Medium risk and effort, good social acceptance
Transition	Accommodate change, allowing adaptive response to new conditions	Actively facilitate change to new condition to encourage adaptive responses	High risk and effort, low social acceptance (initially)

of the standard tools that foresters use to manage forest stands, including various tree harvesting, thinning, and regeneration techniques. The approaches do not necessarily correspond to only one adaptation strategy (i.e., resistance, resilience, response), but as you will see, they may be used alone or in combination to achieve objectives associated with multiple strategies. Broadly, the approaches we discuss next include density management, manipulation of tree composition, and reducing threats from extreme events.

Density Management

A pervasive question for foresters charged with managing forests in the face of climate change is how to minimize the negative impacts of increases in the frequency, intensity, and duration of growing season drought. Drought strongly affects tree growth and health and increases in drought events are expected to result in reduced productivity, heightened tree mortality, shifts in tree species composition, and reductions in carbon storage in many geographic regions (Klos et al. 2009; Allen et al. 2010).

Drought impacts can be direct, including reduced tree growth and vigor, and indirect, such as a predisposition of stressed trees to other disturbances, including wind, insects, or **pathogenic fungi** (discussed in more detail later). While the impacts of drought on forests are generally most severe on sites that are naturally water-limited, e.g., those with coarse textured soils (Allen and Breshears 1998), there are increasing incidences of drought-related tree mortality in a variety of forest ecosystems (Galiano et al. 2011; Michaelian et al. 2011; Granzow-de la Cerda et al. 2012), suggesting that drought related to climate change may impact a wide range of forests.

The direct impact of drought on forests is a density dependent phenomenon, that is, trees in higher density forests face greater moisture stress during drought than trees in lower density forests (Figure 16.3), all other conditions being equal. This is of concern because globally the area of high density forests (natural and planted) is extensive. These forests may be increasingly vulnerable to mortality and declines in productivity due to more impactful drought events (Allen et al. 2010).

Figure 16.3 *High density (on the right) and low density (on the left) ponderosa pine forest in Arizona, USA. Photo credit: Alessandra Bottero.*

One promising approach for minimizing the vulnerability of forest ecosystems to drought is the use of tree **stocking** control through thinning to increase resource availability to the remaining trees within a stand (Kohler et al. 2010; D'Amato et al. 2013; Bottero et al. 2017). As discussed in Chapter 8, foresters have long recognized the value of thinning and other forms of vegetation manipulation to reduce resource competition. Conventional application of thinning has commonly been used to maintain stand productivity above some minimum level by focusing resource allocation from a large number of smaller trees to a smaller number of larger trees. When applied over an extended period of time, repeated thinning has been used to maintain stand densities within bounds defined by minimums that do not sacrifice cumulative productivity and maximums that preclude periods of low stand vigor (e.g., the B- and A-lines respectively on a stocking diagram; Figure 16.4). Minimum and maximum density thresholds have been empirically characterized for many tree species and populations using conventional forestry metrics such as basal area or stem numbers per unit area.

For the purposes of understanding climate change adaptation, approaches to thinning that maintain stands at lower stocking levels, i.e., within the lower half of the stocking chart (Figure 16.4), are more likely to have lower vulnerability to drought compared to stands maintained at either high stocking over very low stocking (i.e., greater vulnerability to drought). A number of studies from different parts of the world have provided good evidence that density management can be used to minimize drought-induced stress and growth reductions in a variety of even-aged forest types, including ponderosa pine (McDowell et al. 2006; Kerhoulas et al. 2013; Bottero et al. 2017), Norway spruce (Kohler et al. 2010), Scots pine

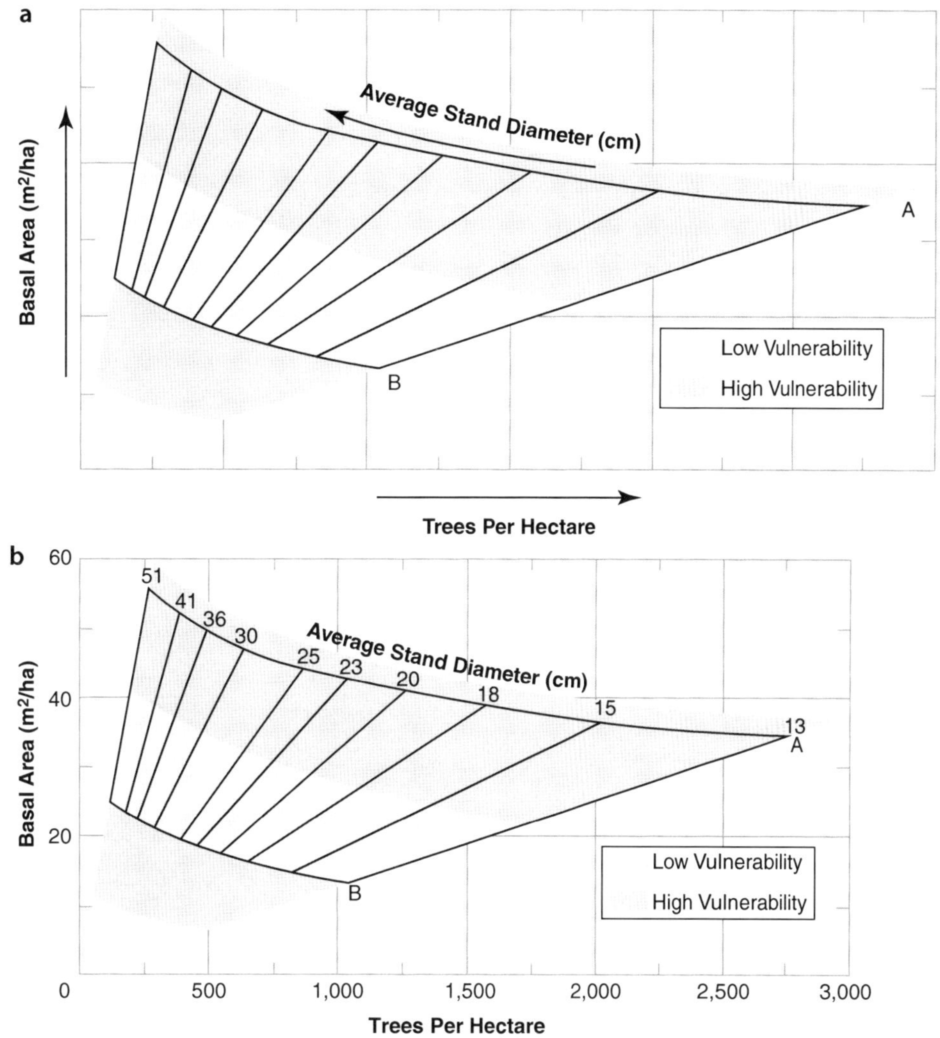

Figure 16.4 *(a) A generic (unitless) stocking diagram for a hypothetical stand. A fully stocked stand would be one that has a combination of average diameter, basal area, and density falling between the A-line and B-line on the diagram. Stands with stocking above the A-line are considered overstocked due to high tree densities and resultant resource limitations. Stands maintained below the B-line are considered understocked; that is, not fully utilizing all available resources. (b) An actual stocking guide for red pine in Minnesota, USA highlighting the zone of vulnerability for over- and understocked stands.*

(Sohn et al. 2016), and red pine (D'Amato et al. 2013; Bottero et al. 2017) (see Box 16.3).

Armed with this type of knowledge, foresters can use density management to enhance the ability of stand growth to be maintained during drought, or at least to minimize the growth reductions that might occur during or after drought. Thinning to control stocking levels should be viewed as a near-term approach for implementing a climate change *resistance strategy* in intact forests that are currently at high density (Box 16.4). However, this approach certainly can be used in conjunction with longer term approaches aimed at increasing the functional diversity of tree species, including a greater representation of drought tolerant species, as part of resilience or response strategies.

Box 16.3 **Density management to increase climate change resistance in ponderosa pine forests.**

Ponderosa pine woodlands of the semi-arid western United States are often overly dense due to longstanding fire suppression (Allen et al. 2002). There is accumulating evidence that this stand condition (high density) is sensitive to growth reduction during and after drought (Bottero et al. 2017). The relationships among tree density, growth, and drought can be quantified in several ways, including growth resistance to drought (this compares growth during a drought to growth before a drought) and growth resilience to drought (this compares growth after a drought to growth before a drought). The more resistant and resilient a stand is to drought, the less its growth declines during the drought and the faster it recovers after the drought. Growing stock experiments (e.g., see Box 8.8) have proven a useful way to relate stand growth to past drought events using tree ring analysis to quantify growth before, during, and after known drought events. One such experiment on the Fort Valley Experimental Forest in Arizona, USA is one of many from different regions and for different species that have been used to demonstrate that maintaining stands at lower relative density improves the resistance and resilience of stand growth to drought (Figures 16.3, 16.5). That is, trees growing at lower densities are less vulnerable to growth reduction from drought. Results from this and similar studies support the contention that managing stands at lower levels of stocking is a promising adaptive strategy (resistance) for reducing the adverse impacts of drought on forest growth in the coming decades.

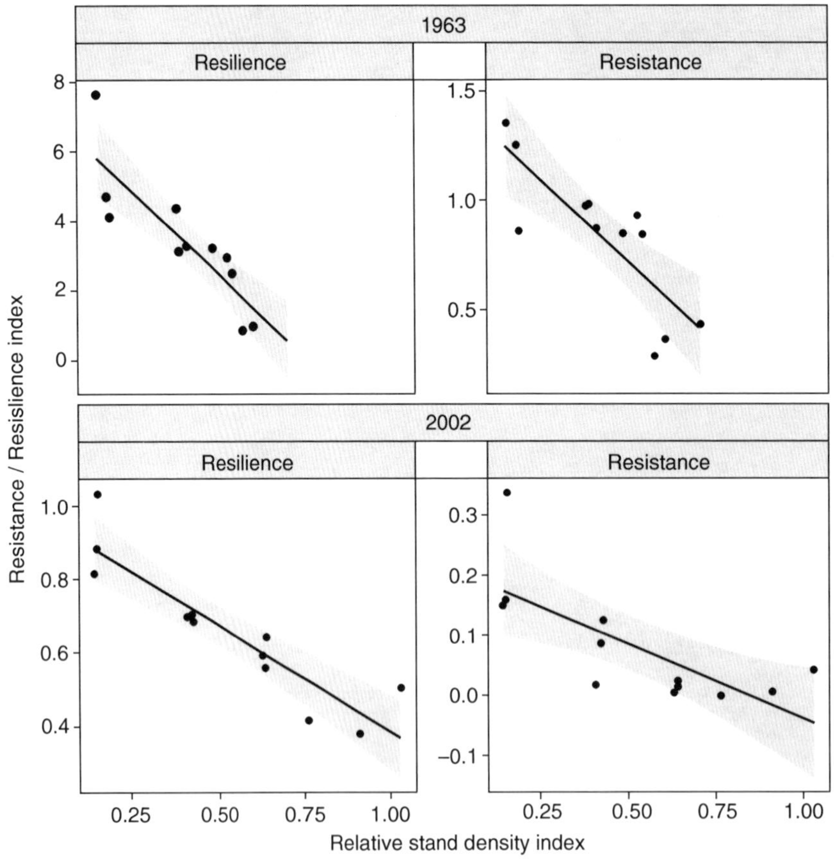

Figure 16.5 *Growth resistance and resilience to drought in ponderosa pine stands in northern Arizona, USA. Resistance (growth during a drought compared to growth before a drought) and resilience (growth after a drought compared to growth before a drought) are unitless measures related to relative stand density (as a proportion of maximum stand density index). Responses to two droughts are displayed: 1963 and 2002.*

Box 16.4 **Implementing a resistance adaptation strategy.**

Here we return to our northern hardwood forest and red pine woodland case studies (Archetypes 3 and 4, respectively) to consider examples of how to implement resistance adaptation strategies. For each forest, the objective for resistance is to maintain the currently dominant species for at least the near term; this includes sugar maple, American beech, yellow birch, and red spruce in New England, and red pine in the woodland ecosystem.

In both examples, the strategies are designed to minimize the impacts of future droughts and disturbances, such as ice storms, insects, and diseases, on current forest communities. A central premise of these strategies is that stand growth reductions during and after drought events, and likely mortality as well, can be mitigated by maintaining stocking at the lower end of vulnerability (e.g., see Figure 16.4). The higher tree-level vigor promoted by these stocking levels is also expected to increase resources available for resisting the impacts of insects and diseases. This strategy has been implemented on operational-scales for red pine forests on the Chippewa National Forest in northern Minnesota USA and for northern hardwoods on the Dartmouth College Second College Grant in New Hampshire USA. Both locations of part of the Adaptive Silviculture for Climate Change Experimental Network (Nagel et al. 2017).

For red pine, the resistance strategy involves periodic thinning to maintain stocking in the low vulnerability zone of stocking (Figure 16.4). Before thinning, stand basal areas averaged 32 m^2/ha (140 ft^2/ac) (Figure 16.6a), while an initial thinning reduced basal area to 23 m^2/ha (100 ft^2/ac) (Figure 16.6b), with an ultimate goal of maintaining basal area between 14 and 23 m^2/ha (60-100 ft^2/ac) after subsequent thinning.

In the northern hardwood installation, the resistance strategy involves single-tree selection, with removal across size classes and species to maintain basal area between 16 and 18 m^2/ha (70-80 ft^2/ac)—a residual basal area corresponding to the low vulnerability zone of stocking for these forests (Figure 16.7).

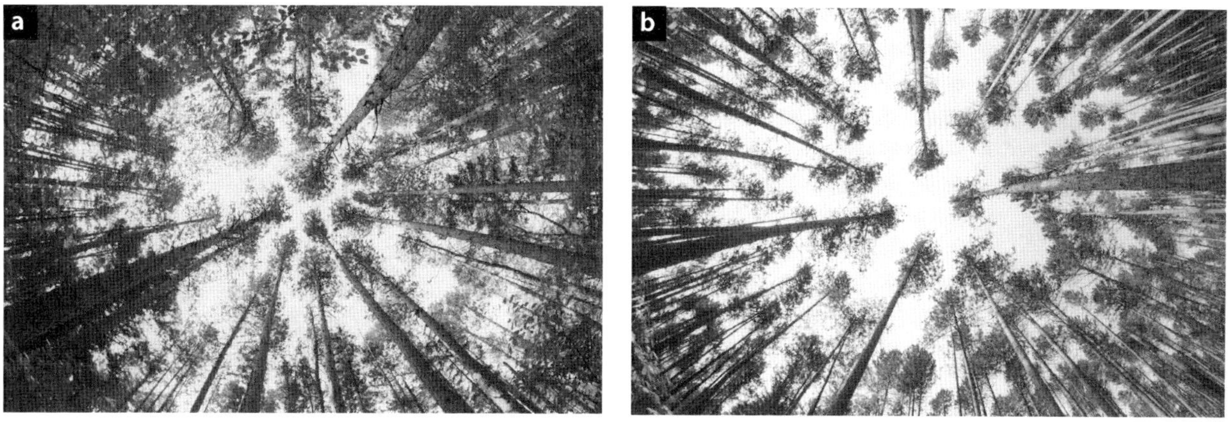

Figure 16.6 *Implementation of a resistance adaptation strategy in red pine forest in Minnesota, USA. (a) Hemispherical photograph of an unthinned, natural, fire-origin red pine stand dating to 1918 having a basal area of 32 m^2/ha (140 ft^2/ac). (b) The same stand after thinning to 23 m^2/ha (100 ft^2/ac). The stand condition on the right is measurably more resistant to growth reductions during and after drought. Photo credit: Jacob Muller.*

Figure 16.7 *Implementation of a resistance adaptation strategy in a northern hardwood forest in New Hampshire, USA. (a) Pre-harvest forest condition with basal area (25 m^2/ha, 110 ft^2/ac) approaching upper limit of acceptable stocking for promoting tree-level resistance to drought. (b) The same stand after single-tree selection harvest to 16–18 m^2/ha (70–80 ft^2/ac). The stand condition on the right is measurably more resistant to growth reductions during and after drought and also contains tree-level architecture more resistant to ice damage.*

This strategy is not only intended to reduce the impacts of future stressors on canopy trees, but was also selected to maintain an uneven-aged condition that included dominance of sugar maple in the regeneration layer. Occasional groups of two to three trees were also removed to encourage yellow birch and red spruce regeneration. In addition to conveying drought resistance, application in the northern hardwood installation also selects sugar maple with crown architecture more resistant to ice loading and American beech demonstrating resistance to beech bark disease (Figure 16.8), given the projected increases in ice storm events for northern New England and the current prevalence of beech bark disease in this region.

Should climate change predictions for the two regions prove accurate, the resistance approaches described here are unlikely to maintain the current dominant species indefinitely, as stresses on mature trees increase and as conditions become unfavorable for new seedling establishment.

Figure 16.8 *Individual tree-level conditions selected for a resistance treatment in a northern hardwood forest in New Hampshire, USA. (a) Sugar maple with U-shaped branch union. This branching pattern is more resistant to ice storm damage than V-shaped unions, which often have included bark and are weaker. (b) American beech demonstrating resistance to beech bark disease (stem on right has characteristic cankering associated with this disease complex).*

Manipulating Composition

Promoting resilient forests may include favoring species and genotypes that are adaptable to projected climate changes. Of particular concern is the need to favor species that are tolerant of anticipated warmer, dryer climates, and increased frequency of growing season droughts in both seedling and mature tree growth stages.

As discussed below, managing for adaptable species may, in some cases, involve species that are already part of the ecosystem, while in other cases this may require restoration of species that were abundant historically, but have been reduced in favor of other species (i.e., a resilience strategy). In addition, it is important to recognize there is tremendous variability within species in their climate response depending on life stage (e.g., seedling versus sapling versus overstory tree) and local growing environment (e.g., shallow versus deep soils). As such, management strategies aimed at promoting complex live-tree structures and factoring in underlying site conditions may enhance adaptation options. Finally, managing composition for climate change adaptation may involve the introduction of species that are new to the ecosystem (transition strategy).

Enhancing what you already have (or restoring what you should have)

Native forests are often composed of more than one tree species, sometimes several in abundance, even in ecosystems that we typically think of as being monospecific.

For example, Norway spruce forests in Europe naturally contain an admixture of European beech (Pretzsch et al. 2010) and several maple species, while northern hardwood forests in the Great Lakes region, USA often support four or five species in abundance and ten or more species in total (Curtis 1959), although many people think of these ecosystems as sugar maple forests. Even an ecosystem as seemingly monospecific as longleaf pine woodlands (Figure 16.9) in the southeastern coastal plain, USA, may contain 12 to 14 additional tree species, although none in high abundance (Goebel et al. 2001).

A key consideration for adaptation silviculture is to determine if any of the tree species that occur in the contemporary ecosystem, even in low abundance, may be future-climate adapted. This addresses the often cited first line of defense against climate change, specifically maintaining tree species diversity in managed forests (e.g., Brang et al. 2014). In this context, one or more species in the ecosystem may be future-climate adapted, so consideration should be given to increasing their abundance, or at least insuring that healthy populations are being maintained, so as to have the option to increase the abundance of these species in the future (as part of a resilience strategy). An interesting twist on the adaptation value of mixed-species stands is that survival of a focal tree may be higher during climate change stress when grown in mixtures with other species. This has been suggested for Norway spruce growing with European beech (Neuner et al. 2015), although the mechanism behind this is not clear.

With a goal of managing for native species mixtures, a simple approach to building resilience into a forest may be achieved by planting, seeding, or encouraging natural regeneration of a greater variety of species and genotypes following a harvest or natural disturbance event. The premise is to expand the gene pool, and therefore the probability of having adapted individuals in the forest. In this context, regeneration harvests and disturbances leading to regeneration may be viewed as opportunities to enhance the adaptive capacity of the forest to future climate, while at the same time addressing one of the primary goals of ecological forestry, which is to work as much as possible with the native species complement of the ecosystem. As such, understanding the natural developmental model for the ecosystem you are working with and the disturbance processes that historically encouraged the recruitment of the range of species will be critical to guiding regeneration methods employed to achieve an adaptation objective.

Note that this approach potentially could involve increasing the abundance of an adapted species beyond what typically is found in the ecosystem when managed primarily for timber, but likely still within the range of natural variation. For instance, deciduous species are often minor components of the tree layer in many conifer forests worldwide, but are often discouraged (Palik and Engstrom 1999). The reality is that some of these species may be better adapted to predicted climate change than the dominant species and thus may be candidates for increased abundance as part of an adaptation strategy (Figure 16.10).

An understandable concern associated with this approach is that future-adapted species may not be commercially desirable for timber products as currently defined. However, these species still likely have value for an array of other ecosystem services, including habitat, carbon sequestration, riparian protection, and aesthetic appeal. Moreover, history tells us that species with little or no commercial value can become important timber species in the future as technology and markets develop to use them; trembling aspen in the Great Lakes region of North America and loblolly pine in the southern USA are notable examples of commercially valuable tree species that were once considered weeds with little commercial value (Graham et al. 1963; Bennett 2015, 79–80).

Figure 16.9 *A longleaf pine woodland in Georgia, USA can be rich in associated tree species, although these typically occur in low abundance.*

Figure 16.10 *Red maple in a red pine forest. Red pine, the dominant species, may be vulnerable to climate change, while red maple may be better adapted to change. Photo credit: Jacob Muller.*

The need to favor future-adapted species may actually involve restoration of climate-adapted species when these have been eliminated in the forest. Timber-focused management, along with modification of natural fire regimes in many regions, has altered tree species composition such that some species that may be well adapted to projected climate change are not managed for (or are managed against), while other less-adapted species have increased in abundance. As an example, we refer again to our Archetype 4 case study of the Great Lakes red pine ecosystem. When managed for timber, the compositional goal for this forest is typically to maximize abundance of red pine, often in a plantation setting (Figure 16.11a), as it is a preferred timber species used for poles, dimensional lumber, and pulpwood. This is typically done at the expense of all other species that occur naturally in this ecosystem, several of which have projected future habitat suitability in the region that are more favorable than red pine (Table 16.1; Figure 16.11b). There are many examples of similar compositional losses for important commercial forest types globally, including Norway spruce, Scots pine, and aspen. As with increasing the abundance of a minor non-commercial species, restoring species that have been lost due to past management may run contrary to objectives of favoring species of high current commercial value. Yet the same arguments for increasing their abundance apply to restoration, that is, enhancing options as part of a resilience strategy.

Specific activities to tackle compositional goals as part of a resilience strategy may include common silvicultural tools, including site preparation followed by planting, direct

Figure 16.11 *(a) An industrial red pine plantation managed for pulp logs and sawtimber, with little attention given to maintaining a population of additional species. (b) A red pine plantation managed for timber, but also additional tree species diversity in created openings within the stand matrix. The latter approach helps to maintain species options for an uncertain climate future.*

seeding, capturing natural seed fall, competition control to favor advance regeneration of desired species, and retaining desired species that may be present in small numbers during both thinning and regeneration harvests so as to maintain a seed source (Box 16.5).

Adding new species and genotypes

A potential problem with relying on the inherent adaptability of current species is that the capacity for adaptation has evolved in response to past environmental conditions

Box 16.5 **Implementing a resilience adaptation strategy.**

A resilience adaptation strategy has been implemented in our two case study ecosystem (red pine-mixed-species woodlands and northern hardwood forests) as part of the Adaptive Silviculture for Climate Change Experimental Network (Nagel et al. 2017). For the red pine, the resilience strategy includes restoring or sustaining native tree species and structural conditions associated with reference conditions for the ecosystem, which included mixtures of up to 10–12 tree species and more open woodland structure (MN DNR 2003). For the northern hardwood forest, the resilience strategy has a goal of increasing the pathways for recovery following disturbance and extreme climate events by increasing representation of wind and ice resistant species and tree forms (e.g., yellow birch and red spruce) and increasing the abundance of native species with greater drought tolerance, including red maple and American beech. Note these species are projected to broadly decline in northern New England under a high emissions scenario (Table 16.1); however, fine scale modeling for the study landscape indicates little change in their abundance. As such, managers felt comfortable including these species as future options on this site, highlighting the importance of incorporating multiple projections where possible to inform management decisions for uncertain future conditions.

The red pine resilience strategy has been implemented on an operational scale on the Chippewa National Forest in northern Minnesota, USA (Figure 16.12). Specific actions include reducing basal area of red pine to 50–75% of total stocking and increasing the abundance of other native species that are predicted to have increased or stable habitat under a future climate. The silvicultural approach employs i) variable density thinning (see Chapter 8), ii) thinning the red pine dominated matrix to bring stocking into the low vulnerability zone (25 m^2/ha (110 ft^2/ac), and iii) maintaining mature canopy in unthinned patches (skips) to ameliorate understory microclimate conditions and sustain a diversity of habitats. The resilience strategy also includes planting of native future-adapted species in gaps, including eastern white pine, northern red oak, bur oak, and red maple.

The northern hardwood resilience strategy has been implemented on an operation-scale on the Dartmouth College Second College Grant, New Hampshire, USA (Figure 16.13a). The approach involves a modified group selection that resembles a variable density thinning, with 20% of the area in 0.04 to 0.1 ha (0.1 to 0.25 acre) gaps, 20% in no-cutting reserves the same dimensions as the gaps (skips), and light thinning in the remaining

Figure 16.12 *a) Post-harvest implementation of a resilience strategy in a red pine forest in Minnesota, USA showing a skip (background), gap (mid-ground), and thinned matrix (foreground). b) Native drought resistant northern red oak planted in resilience gaps. Photo credits: (a) Eli Sagor, (b) Jacob Muller.*

Figure 16.12 (*Continued*)

matrix to maintain basal area between 16 and 18 m^2/ha (70–80 ft^2/ac). When possible, gaps are centered on yellow birch or red spruce. Natural regeneration of yellow birch and red spruce is enhanced by incidental soil scarification in gaps due to harvests occurring in summer and fall (16.13b).

It should be noted that adaptive options tend to be cumulative, such that resistance strategies are also contained in these resilience approaches, with the matrix in both examples periodically thinned to maintain basal area within low-vulnerability zones of stocking. Moreover, implementation of a resilience strategy as described in these examples represents mid- to longer-term adaptation approaches for these regionally important forests that are actually well-addressed by using the ecological silviculture approaches we have already discussed (e.g., variable density thinning) for other purposes (e.g., biodiversity conservation). These approaches also create numerous pathways and mechanisms for the forests to respond to and recover from environmental changes broadly, such as invasive species.

Figure 16.13 *(a) View through thinned matrix toward 0.1 ha (0.25 acre) gap as part of resilience strategy in a northern hardwood forest in northern New Hampshire, USA. (b) 0.1 ha gap with incidental soil scarification by harvesting equipment to favor yellow birch and red spruce regeneration, two species with crown architecture that is more resistant to wind and ice damage, respectively.*

rather than the conditions projected for the future. What this means for our discussion is that there likely will not be enough time for new adaptations to evolve in place given the rapid changes in climate we are experiencing and that are projected through the twenty-first century. In other words, species that are currently believed to be future-adapted may become maladapted with further changes in climate.

The need then may be for foresters to take actions that encourage the addition of new species, or better adapted genotypes of current species, as part of a transition adaptation strategy in anticipation of future climate change. In some instances, future climate adapted species that are not thought to be in a region currently, may already have established populations for undetermined reasons (Figure 16.14). In such cases, actions might be taken to protect and encourage these species when encountered.

Often, actions to establish new species will need to be more deliberate; that is, getting the new species or genotypes into the forest will require active management, rather than passively waiting for this movement to happen. A reason for this is that there are limits to the dispersal capabilities of tree propagules. The pace of climate change

Figure 16.14 *An established and regenerating population of bald cypress in an Atlantic cedar swamp in Burlington County, New Jersey, USA, 160 km (100 miles) north of the species' established range. Photo credit: Bob Williams.*

may exceed natural dispersal capabilities of many species (Aitken et al. 2008; Meier et al. 2012). Moreover, barriers to dispersal caused by human alteration of the landscape, i.e., roads, cities, agriculture, etc. (Pitelka et al. 1997), make it even more questionable as to whether most tree species (and other forest organisms) will be able to migrate fast enough to keep pace with climate change.

An active approach to facilitate a response strategy will involve the intentional movement of species or genotypes with known adaptive characteristics to locations where these traits may be beneficial in the future environment. Current relationships between population variation in adaptive traits and climate can provide a basis for estimating the risk of population maladaptation to future climates. To the extent possible, these relationships are taken into account in habitat suitability models and are reflected in predictions for future changes in importance of species (e.g., Table 16.1) but still there is uncertainty in these predictions. For example, in our mixed-pine Archetype 4 example, eastern white pine, a species with a large geographic and climatic range in eastern North America, is predicted by some models (e.g., Prasad et al. 2007--ongoing; Ravenscroft et. al. 2010) to maintain its regional importance in northern Minnesota, USA through the twenty-first century, even under more extreme emissions scenarios, although other models predict severe reduction in habitat and eventual extirpation of the species through most of eastern North America (Joyce and Rehfeldt 2013).

The intentional movement of species to locations outside their natural range is termed *assisted migration* (McLachlan et al. 2007) and when discussed in the context of forestry, has been termed *forest assisted migration*, to distinguish it from the conservation biology concept of assisted migration to aid rescue of a threatened species (Pedlar et al. 2012). Forest assisted migration can be further divided into three subtypes, including *assisted population expansion, assisted range expansion, and translocation of exotics* (summarized in Leech et al. 2011). These three approaches represent increasing degrees of risk, cost, and controversy when applied to an adaptation response strategy in forests.

Assisted population expansion is the least controversial form of assisted migration, as it involves movement of species or genotypes over relatively short distances. In effect, foresters have been employing assisted population expansion for decades when they plant seedlings from populations of a species (from within its natural range) that are, for instance, more disease resistant than the local population. In the context of climate change adaptation, this practice is a bit different in that a forester might plant seedlings from a population of a species that comes from a warmer or dryer part of the regional range than the area of interest (although still within the natural range of the species).

Assisted range expansion is the movement of populations to locations adjacent to the species' current range from areas where the current climate is thought to be similar to the projected future climate of the new location. Caution must be used in undertaking assisted range expansion because in the near term, some transferred species may not be as well adapted to the current environment as local populations; that is, the benefit of enhanced adaptation to a future climate may come at a risk of maladaptation to the current climate.

Finally, translocation of exotics is the approach that presents the most risk and controversy in forestry. This can involve the interregional, transcontinental, or even intercontinental movement of species far outside their current geographic range. Such an approach may be warranted when there are no suitable candidate species for assisted range expansion in the region; that is, there are no species that would serve to fill important ecological or cultural roles of maladapted species.

A major difference between a response strategy (that involves assisted migration) and a resilience strategy (restoring/increasing abundance of native species) is that there has been less urgency to operationalize on-the-ground approaches for the former, perhaps because there is a much higher risk of failure in the near term. Nevertheless, there is still some urgency to conceptualize, test, and demonstrate potential response approaches at operational scales (e.g., Box 16.6). Part of this exploration involves

Box 16.6 **Implementation of a transition adaptation strategy.**

Once again we turn to two of our Archetype case study ecosystems: red pine woodlands and northern hardwood forests. For each, a transition adaptation strategy is based on the premise that managing for native species composition, done as part of a resilience strategy (Box 16.5), may meet with limited success in the long term with continued climate change. A second premise is that there will be a desire among stakeholders to maintain a forest that has some resemblance in structure to the current ecosystem, regardless of species composition. Thus transition adaptation strategies will likely need to bring new species into the ecosystem—species that may be better adapted to future climate conditions in the long term and provide a forest that is structurally and functionally similar to the current forest.

Examples of transition strategies with these goals in mind have been implemented on operational scales as part of the Adaptive Silviculture for Climate Change Network (Nagel et al. 2017) on both the Chippewa National Forest in northern Minnesota, USA (red pine) and the Dartmouth College Second College Grant in northern New Hampshire, USA (northern hardwoods). For red pine, specific objectives include: i) reducing basal area of dominant red pine to 20–50% of the total, ii) increasing abundance of species that are native to the ecosystem and are predicted to have increased habitat suitability under a future climate (i.e., the same species that are targeted in the resilience strategy in Box 16.5), and iii) introducing new tree species that currently are not part of the ecosystem, but are predicted to have increasing habitat suitability with climate change.

For the northern hardwood forest, objectives include: i) increasing the abundance of future-adapted species, favoring those better able to tolerate variable conditions and disturbance (drought, inundation, etc.), ii) shifting regeneration to an equal mixture of native future-adapted species and novel species that are future-adapted, and iii) increasing representation of shade-tolerant conifers that are future-adapted to maintain habitat and hydrologic functions.

For red pine, the silvicultural approach employs retention harvesting that resembles an expanding-gap irregular shelterwood, with 20% of stand area in 0.2 ha (0.5 ac) gaps initially. The matrix surrounding the gaps is thinned to 16 m^2/ha (70 ft^2/ac) and future-adapted species planted in both gaps and matrix (Figure 16.15). Future entries will expand existing gaps and maintain basal area in the low vulnerability range of stocking with periodic thinning (Figure 16.4). Regeneration activities include planting of eastern white pine, northern red oak, bur oak, and red maple, all native future adapted species, along with white oak, bitternut hickory, and black cherry, all species present in the next southern climate zone in Minnesota, USA. Planting also includes ponderosa pine, a species that is potentially an ecologically and socially acceptable replacement for red pine, but with greater drought tolerance (Figure 16.16).

Figure 16.15 *Post-harvest implementation of a variable retention harvest (expanding gap irregular shelterwood) used as part of a response adaptation strategy in a red pine forest in Minnesota, USA. Photo credit: Eli Sagor.*

The expanding gap irregular shelterwood system used in this strategy reflects the desire to progressively shift composition to greater representation of future-adapted species.

For the northern hardwood forest, the silvicultural approach employs a combination variable density thinning and retention harvesting and resembles a continuous cover irregular shelterwood, with 20% of the area in gaps that range from 0.1 to 0.4 ha (0.25–1.0 ac), 20% of the area in unharvested areas equivalent in size to gaps, and 60% of the matrix thinned to 16–18 m^2/ha

Figure 16.16 *Ponderosa pine planted in west-central Minnesota, USA. While a nonnative species, ponderosa pine is a widely planted tree in this region. Moreover, results from a long-term* ***provenance study*** *(Radsliff et al. 1981) suggest that seed sources from the eastern part of the range (Nebraska, western South Dakota, and eastern Montana, USA) may have reasonable levels of survival and growth in northern Minnesota.*

(70–80 ft^2/ac) (Figure 16.17). Regeneration efforts include fostering species that have growth forms and functional traits that convey heat and drought tolerance, including natural regeneration of American beech and red maple in gaps where possible, and planting future-adapted species. These species include northern hardwood species found in the region but not currently found on site, such as northern red oak, eastern white pine, eastern hemlock, American basswood, bigtooth aspen, and black cherry, as well as novel species that are currently out of range, including black birch, bitternut hickory, and American chestnut. Note planting has seldom been applied in northern hardwood forests given the historic reliability of securing natural regeneration of ecologically and economically desirable species; however, the vulnerability of many of these same species to future climate change and forest health impacts may require a rethinking of regeneration options and approaches in these forests, as reflected in this study.

Figure 16.17 *Post-harvest implementation of a response adaptation treatment in a northern hardwood forest in New Hampshire, USA using a hybrid variable density thinning/ retention harvest (resembling a continuous cover irregular shelterwood). The view is through a 0.4 ha (1 acre) gap to an adjacent reserved area of equivalent size centered on mature canopy red spruce.*

These examples of transition adaptation strategies may be high risk in the short-term, due to high potential for regeneration failures of the non-native species, and thus represent a longer-term adaptation approach for these ecosystems. Nonetheless, by progressively transitioning these forests with ecological silviculture, as opposed to clearcutting and planting, the forests retain structural legacies, maintain a range of regeneration niches, and have greater flexibility to maintain mature structure on site if regeneration is less successful in the near term.

developing the criteria that might be used to select species to target for assisted mitigation. These criteria might relate to morphology, function, or more broadly, the ecosystem services provided by the new species, compared to those that are predicted to decline. For instance, habitat characteristics for animals or wood properties of timber may be two of the many criteria that form the basis for selection. The criteria might involve assessing the social acceptability of new species; for example, the new species might provide acceptable functionality, but be unappealing to stakeholder groups based on other characteristics (e.g., appearance, attractiveness to pests).

A Reminder about Transition Strategies

When a transition strategy includes some form of assisted migration of new tree species, it is important to remember that the conditions needed for their success may be a long time coming. That is, the response strategy is likely to be high risk in the short term and better viewed as a longer-term approach for adaptation. It is also important to remember that the tree species currently in the overstory are not likely to rapidly decline in health any time soon, short of an extreme disturbance event that removes them. This is good, in that there will continue to be opportunities to husband these trees into the near future, for instance with thinning for drought resistance, and in the near-term they should continue to provide seed for new regeneration, if conditions allow. Moreover, in some cases these overstory trees may serve as a **nurse crop** that maintains favorable environmental conditions for the establishment of new species in the understory (Box 16.7). But it is also important to remember that the current overstory trees and shrub and herbaceous species in the understory may have negative interactions with the regeneration of the new species, via **competition** or **allelopathy**, and so shepherding the new species into the forest within the continued influence of the old forest may prove challenging.

Reducing Threats from Extreme Events

The previous two silvicultural approaches we have discussed—density management and manipulating composition—largely focused on adapting forests to increasing potential for drought by reducing vulnerabilities associated with high stand densities or inclusion of maladapted species. An additional concern to consider with climate change is the potential for increasing threats associated with catastrophic disturbances, including extreme weather events

Box 16.7 **Implementing an adaptation strategy to address the effects of invasive species and climate change in black ash wetland forest ecosystems.**

In addition to climate change, many forests around the globe are also threatened by the impacts of non-native, invasive insects and diseases (Lovett et al. 2016). The ability of many of these invasive species to effectively eliminate a given tree species from a region has already resulted in dramatic alternations to the structure, composition, and function of numerous forest types worldwide (e.g., Ellison et al. 2005). One such example is the emerald ash borer (EAB), which is native to northeastern Asia and was introduced to the Great Lakes region of North America in 2002. This wood-boring insect preferentially feeds on the phloem of ash spp. and has functionally eliminated ash trees from many urban and forested settings across the region (Herms and McCullough 2014). The threat of this insect spreading to the greatest concentrations of ash-dominated forests in the upper Lake States led to the development of adaptation strategies that could increase the resilience of black ash wetland forests to EAB impacts, while also considering future climate projections for the region.

Before developing adaptation strategies for black ash forests, it was important to account for natural developmental models and ecosystem properties of these systems. Black ash often constitutes 70–95 percent of overstory trees and plays a central role in regulating water table dynamics through its **transpiration**, so treatments need to consider potential implications of overstory mortality on increases in flooding duration. Reconstructions of disturbance dynamics and age structures for old-growth black forests indicate these forests are similar to Archetype 3 with gap-scale disturbance and multi-cohort conditions as defining features. These dynamics coupled with an understanding of hydrological processes highlight that overstory trees are generally present when recruitment events occur in these systems and likely facilitate cohort development by ameliorating increases in saturated conditions associated with tree mortality.

Based on this knowledge, adaptation strategies were developed that included the creation of small canopy gaps (0.04 ha) through group-selection treatments to provide opportunities for increasing the representation of non-EAB-host species in these forests (Figure 16.18a). As with the examples in Boxes 16.4 and 16.5, planting was used in these areas to increase the abundance of native, non-EAB-host species currently found in low abundance on site, as well as new tree species characteristic of wetland forests that are predicted to have increasing habitat suitability with climate change, but are not currently present. An alternative management response to the threat of EAB may have been preemptive salvage harvests of all ash to provide opportunities for planting non-host species (Figure 16.18b); however, this option ignores the ecological processes historically governing these systems, including the conditions under which recruitment occurs.

Figure 16.18 *Post-harvest conditions for (a) group selection harvest emulating historic gap-scale dynamics in black ash wetland forest systems, and (b) preemptive clearcut removing all ash in response to threat from the non-native, invasive emerald ash borer. Seedlings representing non-EAB-host species, including swamp white oak and red maple, were planted following harvests with higher levels of survival observed in group selection treatments due to maintenance of favorable hydrologic conditions and lower levels of graminoid competition. Trees in (b) are black ash stump sprouts surrounded by lake sedge. Photo credit for (a): Dave Hansen.*

As such, work with these forests has demonstrated lower non-host seedling survival, greater flooding, and loss of understory species assemblages in pre-salvage clearcuts relative to the adaptive approach described above (Looney et al. 2015; Looney et al. 2017; Slesak et al. 2014). This illustrates the importance of using natural developmental models to inform silvicultural approach, even in instances where the future existence of that species is threatened by an invasive species.

(Dale et al. 2001; Wuebbles et al. 2014). The disturbance agents may be biotic, such as heightened or new insects and disease pests, or abiotic, including wildfire and ice- and windstorms (Figure 16.19).

Figure 16.19 *Forest disturbances associated with extreme events such as (a) bark beetle outbreak in Norway spruce forests in the Czech Republic, (b) wildfire in the Boundary Waters Canoe Area Wildness, Superior National Forest, Minnesota USA, and (c) blowdown from a tornado near Brimfield Massachusetts, USA. Photo credit: (b) Doug Shinneman.*

Silviculture that follows an ecological forestry framework certainly can play a role in reducing damage from these threats. In fact, the approaches we have already discussed in this chapter and elsewhere in this book—density management and manipulating composition—can serve the multiple purposes of addressing ecological forestry goals, enhancing drought adaptation, and reducing vulnerability to disturbances associated with extreme events. For example, thinning overly dense stands to restore historical tree densities not only reduces soil moisture competition during drought, but by enhancing vigor of the remaining trees, may lessen mortality from insect pests that may become more of a threat with climate change (e.g., Neumann and Minko 1981; Clinton et al. 1993; Breshears et al. 2009).

The mountain pine beetle outbreak in western North America conifer forests provides an example of this interaction. Silvicultural objectives for these forests, which often have tree densities well above historical levels, may include thinning to maintain wide tree spacing and enhance tree vigor so as to reduce threats from the pine beetle (Whitehead et al., 2004; Whitehead and Russo, 2005). Another good example of an approach that addresses multiple interacting objectives comes from ponderosa pine forests in western North America. As mentioned previously, these forests tend to be overly dense due to longstanding fire exclusion. Restoration using ICO (Individual, Clump, Opening) thinning (see Box 8.5) accomplishes multiple objectives, including increasing the drought resistance of individual pines, making trees and stands more beetle- and fire-resistant, providing openings where regeneration of future adapted species can establish, and providing refugia in clumps that allow greater selection pressure for adaptation to stressful conditions (Churchill et al. 2013).

Manipulating composition to restore or maintain native tree species diversity may also be a good defense against climate change-related threats from insects and disease. For example, with the mountain pine beetle infestation in western North America, managing for multiple tree species that have differing levels of susceptibility has been suggested as a mitigation strategy (Fettig et al. 2007). Pest susceptibility, stand density, and tree composition can be interrelated when, for instance, monospecific stands are maintained at high density such that the tallest trees in the stand infect suppressed shorter trees through close proximity (Reich et al. 1991; Shaw et al. 2005). Maintaining native tree species diversity in mixed stands and reducing host tree density may reduce the potential for this threat (Puettmann et al. 2015). Moreover, when different species of trees are also of different heights, the shorter trees of one species may be less vulnerable to some pests when hidden by taller trees of another species (Castagneyrol et al. 2014), again pointing to the potential health value of maintaining mixed-species and multi-aged stands.

The potential for increased wildfires with warmer, dryer growing seasons is a threat that ecological silviculture can address as well. Dryer conditions during the growing season may result in reduced moisture of woody fuels more frequently in space and time, with a concurrent increased risk of wildfire. While many forests facing this threat are considered fire dependent, buildup of dangerous fuel levels due to long-term fire suppression poses a significant risk of catastrophic wildfire, particularly if fuels tend to be drier. Soil moisture deficits during the growing season will exacerbate the problem. As such, an adaptation approach related to climate change may include reducing unnatural fuel loads through various means (prescribed fire, mechanical treatments, thinning). This is also consistent with silviculture to restore forest structure that better emulates natural conditions.

Thinking about tree species composition in relation to wildfire, stands consisting of neighboring trees of different species may have lower probability of extensive crown fires if these species grow at different rates or establish at different times and thus are of different heights, resulting in discontinuous high density canopies (Puettmann et al. 2015). However, such multilayered stands, particularly if overly dense, may produce substantial **ladder fuels** (Schoennagel et al. 2004), greatly increasing potential destruction from fire. A combination of density management, manipulation of composition, and fuels reduction may be needed in many fire prone landscapes and likely will be even more of a need if climate change exacerbates fire risk.

Silvicultural approaches that maintain or restore mixed-species, multiaged, or multicanopied forests may also be useful for reducing threats from extreme wind events. Generally, extreme wind events have been predicted to increase in frequency and intensity in various regions with climate change (e.g. Dale et al. 2001; Knutson et al. 2010), although there is much uncertainty regarding these predictions depending on the models used, the region considered, and the type of storm (e.g., Brooks 2013; Handler et al. 2014). Should these increases occur, there is both empirical evidence and predictions from modeling that multi-aged stands, having vertical complexity of tree heights (irregular stands), are more stable to high wind events (Hanewinkel et al. 2014; Mason 2002; O'Hara and Ramage 2013). The potential for higher stability of multiaged forests may be due to greater stability of individual trees due to their lower height to diameter ratios (Mason 2002).

A Reminder about Natural Models and Reducing Threats

It is true that not all natural forests have mixed-species compositions, multi-aged structures, or irregular height distributions. Moreover, it is true that not all forests managed using ecological silviculture principles will have these structures. However, as we have discussed elsewhere in this book, it is often the case that natural forests of a particular type will be more rich in tree species and structurally complex than their managed counterparts, particularly when the latter have largely been managed for timber objectives. As such, there are good opportunities to manage most forests using the ecological principles we have discussed earlier (particularly, *continuity, complexity/diversity, timing*) in a way that brings them closer to a natural model of structure and composition and in so doing enhances their capability to resist to some degree the threats from extreme events associated with climate change.

What to Remember about Silviculture in the Face of Climate Change Uncertainty

The ecological forester needs to remember that uncertainty is always part of forest management. Good forestry should maintain options in the face of uncertainty whether this stems from changing markets, evolving societal expectation, or unpredictable natural disturbances. Climate change adds another layer of complexity to management, but addressing this can still use the same framework for maintaining options in the face of uncertainty that is used for other drivers of change.

Some specifics points to remember about silviculture and climate change adaptation:

1. Maintaining options to address climate change uncertainly can be as straightforward as not doing the same thing everywhere. For silviculture, working to apply the three adaptation strategies of resistance, resilience, and response in different places on the landscape will help to maintain options.
2. Worldwide, forests tend to be overly dense, often reflecting an increase over historical densities of native forests. This condition potentially renders forests more susceptible to productivity losses induced by climate change. The ecological forester may need to pursue aggressive thinning approaches to better maintain stands in the less vulnerable range of stocking (resistance strategy).
3. Many of the desired outcomes of ecological forestry, including restoration of stand structures to better reflect natural forests, maintenance of mixed-species stands, and encouraging structural complexity and heterogeneity, will be useful tools for mitigating risks associated with increased droughts and new disturbance regimes (resilience strategy).
4. The ecological forester may need to consider actions, such as assisted migration, that transition forests to new compositional and structural states; that is, states without clear natural analogs in the particular climate region of interest (response strategy). But even in such a new and different world, the foundational principles of ecological forestry will still have a place in the forester's toolbox.

CHAPTER 17

Summary and Conclusions: Why Ecological Silviculture Matters

Ecological silviculture is an approach for managing forest ecosystems, including trees, other organisms, abiotic components, and their interactions based on emulation of natural models of development. In doing this, ecological silviculture is used to sustain or restore the structure, composition, and function of ecosystems while managing for multiple services, inclusive of timber and other commodities. This book has delved into the nuts and bolts of ecological silviculture, thus defined. In this final chapter we summarize some of the major themes that emerge out of our treatment of ecological silviculture. We finish by considering the rightful place of ecological silviculture in the forest management landscape.

Five Major Themes of Ecological Silviculture

Here we summarize five themes that emerge out of our formulation of ecological silviculture. We do not review all that we have covered in this book. Rather, we reiterate the most important conceptual themes of our approach, often by presenting the particular theme in contrast to classical timber-focused silviculture. Our intent is not to suggest the shortcomings of the latter, but rather to help illustrate why ecological silviculture, as we present it, is something distinctly different from the classical approach.

Themes

1. Ecological Silviculture is Responsive to Changing Global Forces

At the onset of this book, we examined how various motivating forces shaping forests and forestry since World War II led to the conclusion that a new approach to silviculture is needed to meet contemporary objectives. These forces include an inability of the global forest to ever contain enough conservation reserves to fully sustain biodiversity, thus requiring that the broader forest landscape be managed more ecologically than a timber-focused approach allows. Another important force is the vast increase in knowledge of how forest ecosystems work and what it takes to sustain their services; timber-focused silviculture falls short of meeting this challenge. Finally, the entities that own and manage the forest, and what they want their management to accomplish, has changed considerably over time as a result of the rise of nontraditional forest owners and stewards, the increase in forests under conservation easements, the adoption of broader, ecosystem-focused missions among traditional forest stewards, and the advent of third-party sustainability certification. All of these changes dictate that silvicultural approaches must become more ecological in design and implementation than the classical (timber-focused) silvicultural approach allows.

2. Ecological Silviculture is Based on Natural Models

A central theme of this book is that ecological silviculture uses models of natural forest development as guides for silvicultural practices and prescriptions. We have presented such a model in a general way (i.e., Chapter 5; forest developmental stages), as well as specifics and variants of this model as exemplified by our disturbance archetypes (Chapters 11–14). The general idea is to use silviculture to emulate natural development of the forest type (archetype) in question. While ecological silviculture cannot fully emulate a natural model, the explicit goal is to reduce the disparity between a managed forest ecosystem and the natural model as much as feasible (Palik et al. 2002). This is an important distinction from classical silviculture. While the latter certainly pays attention to natural processes—e.g., "*silviculture involves methods to develop stand compositions and structures similar to those that characterize natural environments*" (Nyland 2016)—the classical approach is not explicitly intended to reduce the disparity between the natural and managed model. In fact, statements such as "*In silviculture, natural processes are deliberately guided to produce forests that are more useful than those of nature, and in less time*" (Ashton and Kelty 2017), suggest that classical silviculture can actually do it better (*develop forests*) than nature. While this certainly is true for some services, e.g., optimal production of bulk fiber or maximizing growth and yield of high value crop trees, this is not likely to be true for many other ecosystem services, including provision of wildlife habitat, conservation of biodiversity, sustainability of diverse tree species mixtures, adaptation to climate change and other novel stressors, and protection of riparian functionality.

3. Ecological Silviculture is Grounded in Ecological Principles

Classical silviculture is, in concept, "*conducted on the basis of ecological principles*" (Ashton and Kelty 2017). Yet in practice, these principles are rooted as much in the ecology of agricultural systems (e.g., resource supply and demand, inter- and intra-specific competition) as they are in natural models of forest development (Puettmann et al. 2009). While agricultural principles are important for understanding forests and how to manage them, they fall short of allowing the ecological forester to develop prescriptions that emulate models of natural forest development. In contrast, ecological silviculture is based directly on ecological (foundational) principles that follow from a natural model:

> *The principle of* continuity *recognizes the role of natural disturbances in forest ecosystems as manifest in legacies that survive from the pre-disturbance forest. The ecological forester uses an understanding of continuity to incorporate legacy management into silvicultural prescriptions.*
>
> *The principle of* complexity/diversity *acknowledges that natural forest ecosystems are more structurally complex and generally more species rich than forests managed using a timber-focused model. The ecological forester uses this understanding to devise intermediate treatments and regeneration methods that restore and sustain structural complexity and species richness.*
>
> The principle of timing *refers to the importance of allowing elements of old trees and stands to develop naturally. With this understanding, the forester learns to schedule treatments as much on ecological indicators as on the economic or productivity metrics. The ecological forester also recognizes that silvicultural treatments can be a partial substitute for time, for example, thinning to accelerate development of large trees.*
>
> *Finally, the principle of* context *allows the ecological forester to consider how stand-scale actions can accumulate to impact landscape function, e.g., cumulative effects on aquatic ecosystems and connectivity of habitat for wildlife species. Context also allows the ecological forester to understand natural landscape patterns and how to devise stand characteristics based on natural patterns.*

Consideration of the foundational principles within the context of the forest type being managed helps a forester become much more ecological in their approach than classical timber-focused silviculture allows.

4. Ecological Silviculture is Ecosystem Centric

A central theme of ecological silviculture is that it is ecosystem-focused. That is, prescriptions are designed and implemented with whole-ecosystem responses in mind, rather than focusing on the response of timber species as the framework upon which the response of other ecosystem components are hung. Classical silviculture is also cast as being ecosystem-focused: e.g., "*silviculture . . . has become a process for sustaining viable and productive ecosystems, rather than simply producing continuous supplies of commodities . . .*" (Nyland 2016) and "*The production of timber, though a common objective, is neither the only nor necessarily the dominant aim in silviculture*" (Aston and Kelty 2017). In reality, with classical silvicultural approaches, the ecosystem focus is often subverted due to a variety of overriding organizational goals or logistic constraints, such that silvicultural prescriptions are developed with a central focus on timber species. Consider these actual examples that illustrate our point.

> Example 1: *A forester has information on the natural forest type of a stand, including its native composition and dynamics, and has an organizational objective to restore natural forests to the extent possible, but still chooses to harvest and focus regeneration (through planting) on a single tree species that occurred only sporadically in the type. The justification is that the species being regenerated is the only one purchased by the regional mill. The forester states that there is no choice but to manage according to needs dictated by the current timber market, rather than the ecosystem or, importantly, potential future markets.*
>
> Example 2: *A public organization has a substantial area of single species conifer plantations managed at high density on sites that naturally were mixed-species woodlands. Establishment of these plantations dates to a time when the organization had a mandate to provide timber for dimensional lumber mills. The mandate for the organization has evolved considerably to be much more inclusive of sustainability of a broad array of ecosystem services as provided by natural forests (including timber). Still, foresters for this organization find it difficult to change their focus from timber to ecosystems, stating publically that since these forests have been plantations historically, the best course of action is to maintain them as plantations.*

The foresters involved in these real-world examples are likely well aware that classical silviculture, in concept, is meant to consider whole ecosystems as the entity being managed, with timber species as an important component, but not the only component. But the reality is that a timber-focused silvicultural approach often supersedes the ecosystem focus, due to economic constraints, outdated but entrenched organizational policies, or historical inertia. In effect, this is tail wagging the dog silviculture, in that timber interests end up having an overly large influence on how the forest ecosystem is managed, constraining the silvicultural creativity allowed at the level of the field forester.

The ecological forester attempts to overcome these constraints and the inertia of the past, and let the dog wag the tail, by emulating natural ecosystem models to the fullest extent possible, *while also meeting timber production objectives*. In this book, we have introduced you to several organizations that have a proven track record of doing this effectively, creatively, and economically.

5. Ecological Silviculture is Adaptation Silviculture

A nearly universal outcome of timber-focused silviculture is that it reduces native tree species richness (compared to the natural counterpart of the same forest type) and leads to a simplification of age and size structure. Sometime this simplification is deliberate, as in plantations, but often it is a consequence of silvicultural approaches that poorly emulate natural dynamics, thus limiting the establishment of the full range of species and inadvertently simplifying stand structure. This lack of diversity and complexity may limit adaptability of the forest in the face of unforeseen change and, in so doing, limit options for the forester.

In contrast, an explicit goal of ecological silviculture is to sustain (or often restore) the native tree species community of a stand (sustain *diversity*) and to treat stands in ways that allow development of complex (in a relative sense) age and size structures (enhance/restore *complexity*). Both conditions enhance adaptive capacity in the face of conditions that may impact a particular species or a certain tree size/age structure.

Thus, an important aspect of ecological silviculture is that it better positions forests to adapt to changing conditions, especially climate change, but also invasive pests, changing markets, and changing societal expectations; that is, ecological silviculture leads to forests that provide society with greater options in the face of uncertainty. Regarding climate change adaptation, we have argued that a wise strategy may be to include some actions to transition forests to new species that are likely to be better adapted to future climates. Even this should be done within the context of managing for species richness; that is, managing for a suite of species, native or new, rather than a single future-adapted species. Moreover, transition approaches should still account for the importance of legacies in affecting ecosystem recovery and response to disturbances and stressors by including retention as part of adaptation strategies.

Summary

This review of the major themes associated with ecological silviculture helps to illustrate why and how our approach emerges as a new silvicultural approach, rather than a tweaking of classical, timber-focused silviculture. While we are not the first to call for a new type of silviculture to address the contemporary forces and challenges shaping forests and forestry (Seymour and Hunter 1999; Puettmann et al. 2009; Franklin et al. 2018), we are the first to develop and articulate a comprehensive approach that is explicitly based on natural models of forest development. As we will summarize in the next section, our natural models approach is designed specifically to respond to the evolving needs of a wide range of forest owners and stewards and is thus applicable across virtually the entire forest landscape.

Ecological Silviculture's Place in the Forest Landscape

Ecological silviculture is the stand-scale toolbox used to meet the objectives of ecological forestry (Franklin et al. 2018). But where on the global forest landscape does ecological silviculture have a place? We submit that the rightful place is virtually everywhere that forests are actively managed for commodities and, potentially, in forest reserves as well, when the latter are in need of restoration and maintenance activities (see Figure 1.5). Where ecological silviculture likely has no place is in production forests primarily composed of single species plantations managed on short rotations to maximize short-term returns on investment; there will be little incentive to adopt an ecological approach in these production (agricultural) systems.

As we review below, the users of ecological silviculture include family forest owners, federal, state, and other public ownerships, tribal and other aboriginal entities, and various conservation organizations. While these owners and stewards are diverse in their missions and objectives, they share a desire to maintain or restore ecological conditions and functions in managed forests in ways that bring them closer to a natural model than is possible with a timber-focused approach.

Family Forest Owners

Family forest landowners, who, for example, own most of the private forest in the United States, generally have a complicated and diverse set of management objectives.

For instance, a study of family forest landowners in the United States found that beauty/scenery, protecting and improving wildlife habitat, passing land to their heirs, privacy, protecting biodiversity, and protecting water were the most important reasons for owning forest (Butler et al. 2016). Other studies, such as Birch (1996), have documented similar goals that families have for their forests. These studies substantiate that timber production is not the sole goal, or perhaps even a major goal, for many family forest landowners. Rather, they have a multiplicity of goals into which timber production and harvest must fit: timber production may pay the bills, but family forest landowners also have many other goals that direct their management in ways that argue for an explicitly ecological approach to silviculture.

Federal Forests

In the United States, the National Forest System encompasses the largest percentage of publicly owned forest land in the country. While US National Forests are managed for multiple use, including timber, they have increasingly shouldered responsibilities for conservation of biodiversity and watershed protection. Toward that end, silviculture for timber production as the driving priority has declined on many of these lands and instead is used to meet goals of broader ecosystem-focused projects, often with a bent toward ecological restoration. In fact, this is now codified in policy with the adoption of the 2012 planning rule for implementing the National Forest Management Act, whereby the US Forest Service has formally committed to managing the national forests to provide for ecological sustainability by maintaining or restoring the ecological integrity of terrestrial and aquatic ecosystems and watersheds in the plan area as the foundation of their management (USDA 2012). Our experience is that most managers of US National Forests are receptive students of an ecological approach to silviculture. Other countries that have significant national level forest ownership should be receptive as well. For instance, the Swedish state owns approximately 17 percent of the nation's managed forests, with conservation of biodiversity being a priority policy issue on these forests (https://www.skogsstyrelsen.se/globalassets/in-english/forests-and-forestry-in-sweden_2015.pdf). An ecological approach to silviculture is an appropriate model for meeting the objectives of this policy.

State Trust Forests

Most other public forests in the United States, for example, are state or county trust forests, such as those found in Washington, Pennsylvania, Minnesota, and Wisconsin. In many cases, public trust forests are managed for sustained yield with a primary purpose of producing revenue for schools and other purposes. However, often the stewards of these forests must balance their obligations to their direct beneficiaries (the trusts) with those of the broader general public. The dual obligations and greater visibility often result in management strategies that span those of private landowners and those of the national forests (Souder and Fairfax 1996). Moreover, many of these trust forests maintain third-party sustainability certification, so aspects of ecological forestry and ecological silviculture have to be used to meet certification standards.

Tribal Forests

Tribal forests on reservations in the United States are held in trust by the Bureau of Indian Affairs (BIA). Until the early 1990s, BIA foresters managed much of these forests to produce commercial timber revenue on a sustained yield basis for tribes within each reservation. Now, tribal objectives, which generally include managing for a variety of ecosystem services, much like family forests, play an important role in guiding management policy and practice (Gordon et al. 1993; Gordon et al. 2013). Furthermore, many tribes now have their own natural resource management agencies that have taken over responsibility for management of reservation forests from the BIA. Tribal managers are often more receptive to an ecological approach to silviculture than the previous stewards (Box 17.1). Similar ecological perspectives on forest management have emerged as part of aboriginal forestry initiatives in Canada (Parsons and Prest 2003).

Conservation Stewards and Forest Certification

Conservation organizations, notably The Nature Conservancy (TNC), increasingly own and manage forests, or are active partners in the management of public forests with other stakeholders (Cornett and White 2013). Not surprisingly, restoring ecosystem function and sustaining a broad range of ecosystem services is a major consideration of their management objectives. Silviculture for timber production is not generally a priority for these stewards, but treating forests with silviculture, including harvesting and thinning, is certainly part of what these entities do, but in ways designed to be more emulative of natural models, i.e., our approach, and ecological sustainability.

Finally, there can be a place for ecological silviculture even among owners and stewards that place a high priority on timber production, along with other ecosystem services, if these organizations pursue third party sustainability certification, notably through the Forest Stewardship Council or similar programs (Bennett 2015). Moreover, the advent of various forest sustainability laws, regulations, or guidelines administered by various government entities requires

Box 17.1 **The Leech Lake Band of Ojibwes' Forest Vision.**

The Leech Lake Band of Ojibwe (LLBO), located in north central Minnesota, USA, is part of the Minnesota Chippewa Tribe, a federally recognized tribal government that includes six bands and their reservations. The Leech Lake Indian Reservation encompasses 350,000 ha (864,158 ac) in the central pine-hardwood forest ecoregion, a zone of transition between boreal forest to the north and temperate forest to the south. Unlike all other Indian reservations in the US, most of the reservation lands fall within the administrative boundaries of the USDA Forest Service Chippewa National Forest (CNF), thus the LLBO actually only holds a small percentage of its reservation forest in direct ownership. However, the CNF has a federal trust responsibility to provide a homeland to the LLBO and to protect the reserved rights the LLBO retained under various federal treaties. Moreover, the 2004 Tribal Forest Protection Act authorizes the US Secretary of Agriculture (the parent agency of the US Forest Service) to work in agreement with Indian tribes to carry out projects that protect tribal forest interests. As such, LLBO foresters and natural resource managers work closely with CNF staff to better insure that the management direction of the National Forest corresponds with the Band's vision for their reservation forest. This vision is distinctly ecological and in line with silviculture that emulates a natural model of forest dynamics and that closely addresses the foundational principles of ecological silviculture. For example, management guidance from the LLBO to the CNF includes the following (grouped according to the foundational principles) in relation to transitioning conifer (red pine) plantations on the reservation forest to something closer to the native forest community:

Addressing Principle 1: Continuity:

"During harvest and any site prep, a more concerted effort would be needed to reserve the diversity of other tree species in the stand . . ."

Addressing Principle 2: Complexity/Diversity:

"Natural conifer stands often have gaps and openings. These openings are often the locations where other beneficial species of interest to Band members can grow. Allowing proper spacing between conifers will also increase the stand's capability to support forb and shrub species that are beneficial to the LLBO."

". . . naturally occurring conifer stands would not be simple communities. They would typically have a mix of conifer and hardwood species of various ages. This diversity is lost when plantation management is utilized."

Addressing Principle 3: Timing:

"Timely thinning will result in high live crown ratios and faster growing trees while further allowing other tree species and forbs located within the stand the opportunity to thrive, pushing the stand to a much more ecologically sound status."

". . . we further recommend extending the rotation of these stands to at/beyond 200 years to truly allow them to achieve full stand maturity and the ecological values that come with age."

Addressing Principle 4: Landscape Context:

". . . due to the extensive harvest that has occurred in the past, we are lacking in the amount of old conifer stands on the landscape and the values they provide for traditional activities and for wildlife. To remedy this situation there needs to be a very noticeable decrease in final harvest prescriptions in these stands."

Moreover, the LLBO vision recognizes the need to be proactively adaptive in the face of climate change uncertainty:

". . . work towards . . . creation of more diverse stands that provide habitat for wildlife and are better poised to withstand the unknowns of oncoming climate change."

The LLBO's forest vision, as exemplified in these excerpts, is about as in line with our ecological silviculture approach as any organization can get without having come up with our foundational principles themselves.

some consideration for and sustainability of natural forest composition and structure (e.g., leave tree guidelines), ecosystem productivity (e.g., minimizing soil compaction and rutting), and landscape function (e.g., riparian buffers).

Preventing the Forest Management Divergence

Adoption of an ecological approach to silviculture over much of the forest landscape by the diverse owners/stewards outlined above may be the only way to prevent what Bennett (2015) calls "*. . . the forest management divergence, which is characterized by the proliferation of protected areas and intensive plantations,*" with little to no management in native forests that occupy the middle ground between these bookends. Such a divergence arises at least in part from the public's distaste for clearcut (clearfell) regeneration systems that became so widespread in their use on state-controlled forests within many regions globally. In fact, this public reaction may ultimately be contributing to a decline in forestry as a trusted profession throughout much of the world (Bennett 2015).

Bennett points out that New Zealand and China are examples of the forest divergence occurring as a result of conscious policy decisions, while other counties (e.g., the United States, Australia) may be moving toward divergence due to market forces that make production of wood in plantations more profitable than in managed native forests. With such a framework in place, that is of separating forest production from protection with no middle ground of ecologically sustainable multiuse management, a host of new problems arise (Bennett 2015) including:

i. alienation of local people from forests, in favor of technical experts who determine what locals should think about and do with their forests;
ii. social unrest over a lack of forest-related jobs;
iii. declining health of protected forests from threats such as fire, climate change, and invasive species, due either to a lack of management or a lack of funds needed to mitigate these threats;
iv. devaluation of the cost of native forest timber, leading to use of overly intensive harvesting approaches such as clearcutting or biomass harvests.

Bennett (2015) argues that to prevent a forest divergence, stakeholders need to work together to develop policy that leads to the active management of native forests for resilience in the face of threats. He goes on to say that this includes "*Harvesting in native forests that have already been managed . . . but such harvesting should be directed by methods that increase or at a minimum sustain existing levels of biodiversity*." We submit that the ecological approach to silviculture presented in our book is uniquely situated to provide the methods needed to address the challenge of sustaining biodiversity, and all other aspects of ecosystems, over the bulk of the forest estate in ways that satisfy the public's desire for a wide range of ecosystem services. That is, our approach provides a means of practicing sustainable forestry in the middle ground (the vast majority of the forest estate) between production and protected areas, and in fact it serves to link these extremes into an integrated landscape framework that will prevent a *forest management divergence*. Moreover, by adopting our approach, foresters will remain or reemerge as the go-to professionals best suited to meet the challenge of managing the global forest sustainably.

The application of ecological silviculture will still need to be flexible in the details (i.e., in degree of emulation of a natural model), so as to recognize the often diverse needs of different landowners and stewards and account for the diversity of ecological conditions across any given ownership. Some of these stewards will focus on conservation of biological diversity and sustainability of ecological complexity, with wood production being a secondary and minor objective. Others will focus on timber production, but with much consideration given to sustaining biological diversity. Thus, the true landscape condition (as opposed to that depicted in Figure 1.5) will be something better represented by a landscape gradient (Figure 17.1), in which approaches (application of ecological silviculture) will vary with respect to their emulation of a natural models approach (Palik and D'Amato 2017). The commonality across this landscape gradient is that managers and stewards will strive to adhere to the *tenets of ecological silviculture*, which we first introduced in Chapter 1 and revisit here in the context of judging success at emulation of a natural model of silviculture (Box 17.2).

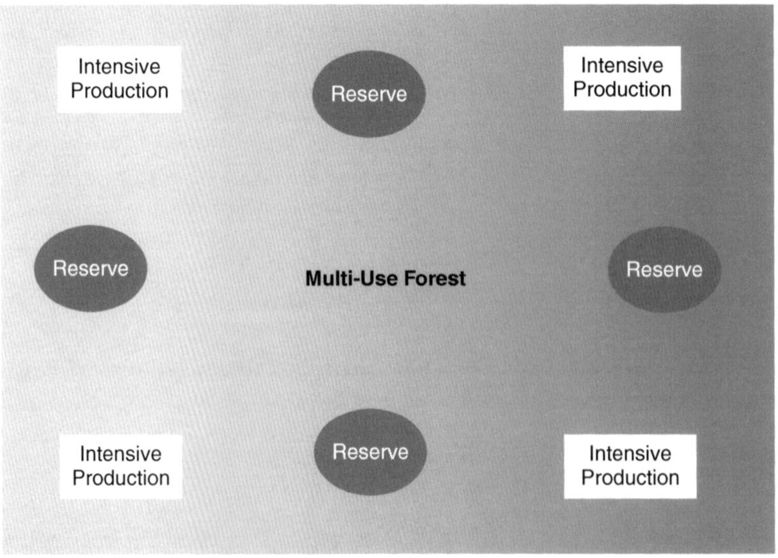

Figure 17.1 *Conceptual representations of a large forest landscape consisting of intensive production forests (white) and protected forests (dark gray) that are embedded in a multiuse forest landscape. In this representation, the multiuse forest is depicted as a gradient that reflects varying degrees of adherence to a natural model of silviculture.*

Box 17.2 **Revisiting the tenets of ecological silviculture. First presented in Chapter 1, the tenets are worth bearing in mind as you consider how to achieve the diverse management objectives that most landowners have for their forests.**

1. **Sustain ecosystems, rather than timber (or other singular objectives)**
   Even ecological foresters will generally maintain a strong commitment to producing commodities, but this will be done without compromising ecosystem productivity. If pursuit of landowner objectives compromises productivity, you likely have deviated too far from the natural model.
2. **Seek complexity and diversity**
   If your forest is not measurably more complex and species rich than with the outcome of a timber-focused model at all stages of development, you likely have deviated too far from the natural model.
3. **Seek forest continuity rather than discontinuity**
   New forests connect to the past through biological legacies. If your management does not have measurably more and varied types of legacies than a timber-focused model, you likely have deviated too far from the natural model.
4. **Maintain options for the future**
   Keeping all the parts of the forest is key to maintaining options for the future. If your management results in measurably fewer species or simplified structural conditions, you likely have deviated too far from the natural model.
5. **Practice silviculture in the context of landscapes**
   Natural forest landscapes often have different structure (and function) than timber-focused landscapes. If your management results in landscapes with measureable negative changes in landscape metrics (e.g., connectivity), you likely have deviated too far from the natural model in too much of the landscape.
6. **Let the ecosystem speak to you**
   Silviculture is an adaptive process that should rely on your intuitions and observations of how the structure, composition, and function of your forest is responding to treatments. If your gut tells you something is off relative to a natural model, you will want to reconsider and refine your approach.

Conclusions

There has been much rhetoric over recent years that speaks to the need for silvicultural approaches that better address ecosystem sustainability, that conserve biodiversity broadly, that manage forests as complex systems that are more adaptable to changing conditions, and that restore forests to more natural conditions. We contend that there has been less-than-sufficient action to move from rhetoric to change on the ground; that is, a clear blueprint for incorporating these ecological objectives into everyday practice is lacking. A driving goal for our book has been to provide this blueprint via our foundational principles of ecological silviculture and our examples of implementation using natural forest archetypes. Our belief is that armed with this knowledge, a forester is now in a position to integrate their skills with these principles to devise ecological silvicultural prescriptions for most of the forests and situations they encounter.

An ongoing challenge for foresters will be to implement ecological silvicultural prescriptions while still maintaining opportunities to meet multiple diverse objectives, often inclusive of timber management. The balance may shift toward one goal or the other at different times and different locations, depending upon priorities within and among management organizations. However, the ultimate objective should be to facilitate implementation of an ecological approach without compromising one objective for the sake of another, e.g., biodiversity to the exclusion of timber. In fact, as we have stressed earlier, if commodity production is ignored, then sustainability of biodiversity (or other ecosystem services) will never be widely implemented at scales needed to achieve significant change in the landscape, because the capital needed to pursue an ecological approach will be lacking.

We feel that there is (or there should be) a growing urgency to implement ecological silvicultural approaches on large portions of the forest landscape, due to, for example, increasing concerns related to climate change, globalization of timber markets, and threats from invasive species. We have intended to facilitate both development and implementation of such approaches by providing an overview of the scientific underpinnings of ecological silviculture and by synthesizing these underpinnings into generally applicable principles, i.e., the foundational principles. Our hope is that by distilling key concepts into practical principles and providing some examples of use we are helping to facilitate the development of practices that are adaptable to meet the varied needs and conditions in a wide array of forest ecosystems. In fact, the greatest gains in silviculture are made through its practice by creative foresters and we are excited to learn of the insights and ideas developed over time from the wider application of the principles and practices presented in this text.

Common and Latin Names of Plant and Animal Species Referenced in the Text

American basswood (*Tilia americana* L.)
American beech (*Fagus grandifolia* Ehrh.)
Atlantic white cedar (*Chamaecyparis thyoides* (L.) Britton, Sterns & Poggenb.)
bald cypress (*Taxodium distichum* (L.) Rich.)
bald eagle (*Haliaeetus leucocephalus* L.)
balsam fir (*Abies balsamea* (L.) Mill.)
beaked hazelnut (*Corylus cornuta* Marshall)
beargrass (*Xerophyllum tenax* (Pursh) Nutt.)
bigleaf maple (*Acer macrophyllum* Pursh)
bigtooth aspen (*Populus grandidentata* Michx.)
bitter cherry (*Prunus emarginata* (Douglas ex Hook.) D.Dietr.)
black ash (*Fraxinus nigra* Marshall)
black birch (*Betula lenta* L.)
black cherry (*Prunus serotina* Ehrh.)
black cottonwood (*Populus trichocarpa* Torr. & A. Gray ex Hook)
black spruce (*Picea mariana* (Mill.) Britton, Sterns & Poggenb.)
black-throated green warblers (*Setophaga virens*)
brown bear (*Ursus arctos* L.)
brown creeper (*Certhia americana* Bonaparte)
bur oak (*Quercus macrocarpa* Michx.)
California black oak (*Quercus kelloggii* Newb.)
canyon live oak (*Quercus chrysolepis* Liebm.)
cascara (*Rhamnus purshiana* (DC.) A. Gray)
coast redwood (*Sequoia sempervirens* (D.Don) Endl.)
Diplodia tip blight (*Diplodia pinea* [Desm.] J.J. Kickx)
Douglas-fir (*Pseudotsuga menziesii* (Mirb.) Franco)
Douglas-fir bark beetle (*Dendroctonus pseudotsugae* Hopkins)
dwarf mistletoe (*Arceuthobium spp.*)
eastern gray squirrel (*Sciurus carolinensis* Gmelin)
eastern hemlock (*Tsuga canadensis* (L.) Carr.)
eastern red cedar (*Juniperus virginiana* L.)
eastern white pine (*Pinus strobus* L.)
emerald ash borer (*Agrilus planipennis* Fairmare)
Engelmann spruce (*Picea engelmannii* Parry ex Engelm.)
Eucalyptus (*Eucalyptus* spp.)
European ash (*Fraxinus excelsior* L.)
European beech (*Fagus sylvatica* L.)
golden chinkapin (*Castanopsis chrysophylla* (Douglas ex Hook.) A.DC.)
gopher tortoise (*Gopherus Polyphemus*)
grand fir (*Abies grandis* (Douglas ex D.Don) Lindl.)
hazel (*Corylus* spp.)
huckleberries or blueberries (*Vaccinium* spp.)
incense-cedar (*Calocedrus decurrens* (Torr.) Florin)
ironwood (*Ostrya virginiana* (Mill.) K. Koch)
jack pine (*Pinus banksiana* Lamb.)
Japanese beech (*Fagus crenata* Blume)
Jeffrey pine (*Pinus jeffreyi* A.Murray bis)
juniper (*Juniperus* spp.)
lake sedge (*Carex lacustris* Wiild.)
Lenga (*Nothofagus pumilio* (Poepp. & Endl.) Krasser)
loblolly pine (*Pinus taeda* L.)
lodgepole pine (*Pinus contorta* Douglas ex Loudon)
longleaf pine (*Pinus palustris* Mill.)
marbled murrelet (*Brachyramphus marmoratus* Gmelin)
mountain ash (*Sorbus americana* Marshall)
mountain hemlock (*Tsuga mertensiana* (Bong.) Carr.)
mountain pine beetle (*Dendroctonus ponderosae* Hopkins)
noble fir (*Abies procera* Rehder)
northern bobwhite quail (*Colinus virginianus*)

northern California tanoak (*Lithocarpus densiflorus* (Hook. & Arn.) Rehder)
northern flying squirrels (*Glaucomys sabrinus* Shaw)
northern red oak (*Quercus rubra* L.)
northern spotted owl (*Strix occidentalis caurina* Xantus de Vesey)
Norway spruce (*Picea abies* (L.) H. Karst.)
oak (*Quercus* spp.)
Oregon white oak (*Quercus garryana* Douglas ex Hook.)
Pacific dogwood (*Cornus nuttallii* Audubon ex. Torr. and Gray)
Pacific madrone (*Arbutus menziesii* Pursh)
Pacific rhododendron (*Rhododendron macrophyllum* D. Don ex G. Don)
Pacific salmon (*Onchorhynchus* spp.)
Pacific silver fir (*Abies amabilis* (Douglas ex Loudon) J.Forbes)
Pacific yew (*Taxus brevifolia* Nutt.)
paper birch (*Betula papyrifera* Marshall)
Pennsylvania sedge (*Carex pensylvanica* Lam.)
pileated woodpecker (*Dryocopus pileatus*)
pin cherry (*Prunus pensylvanica* L.f.)
pine grass (*Calamagrostis rubescens* Buckley)
pine marten (*Martes americana* Turton)
pinyon pine (Pinus spp.)
pitch pine (*Pinus rigida* Mill.)
ponderosa pine (*Pinus ponderosa* Douglas ex C.Lawson).
radiata pine (*Pinus radiata* D.Don)
raspberry (*Rubus* spp.)
red alder (*Alnus rubra* Bong.)
red cockaded woodpecker (*Leuconotopicus borealis*)
red maple (*Acer rubrum* L.)
red pine (*Pinus resinosa* Ait.)
red spruce (*Picea rubens* Sarg.)
red squirrel (*Sciuris vulgaris* L.)
red-eyed vireo (*Vireo olivaceus* L.)
Roosevelt elk (*Cervus canadensis roosevelti*)
ruby-throated hummingbird (*Archilochus colubris* L.)
salmonberry (*Rubus spectabilis* Pursh)
Scotch elm (*Ulmus glabra* Huds.)
Scots pine (*Pinus sylvestris* L.).
shoe-string fungi (*Armillaria* genus)
silver fir (*Abies alba* Mill.)
silver maple (*Acer saccharinum* L.)
sitka spruce (*Picea sitchensis* (Bong.) Carriere)
snowbrush (*Ceanothus velutinus* Douglas ex Hook.)
southern pine beetle (*Dendroctonus frontalis* Zimmermann)
spruce beetle (*Dendroctonus rufipennis* Kirby)
subalpine fir (*Abies lasiocarpa* (Hook.) Nutt.
sugar maple (*Acer saccharum* Marsh.)
sugar pine (*Pinus lambertiana* Douglas)
swamp white oak (*Quercus bicolor* Wiild.)
sycamore maple (*Acer pseudoplatanus* L.)
table mountain pine (*Pinus pungens* Lamb.)
tanoak (*Lithocarpus densiflorus* (Hook. & Arn.) Rehd.)
teak (*Tectona grandis* L.f.)
trembling aspen (*Populus tremuloides* Michx.)
turkey oak (*Quercus laevis* Walter)
western hemlock (*Tsuga heterophylla* (Raf.) Sarg.)
western larch (*Larix occidentalis* Nutt.)
western redcedar (*Thuja plicata* Donn ex D. Don
western white pine (*Pinus monticola* Douglas ex D.Don)
white fir (*Abies concolor* (Gordon) Lindl. Ex Hildebr.)
white oak (*Quercus* alba L.)
white spruce (*Picea glauca* (Moench) Voss)
wood duck (*Aix sponsa* L.)
wood frog (*Lithobates sylvaticus* LeConte)
yellow birch (*Betula alleghaniensis* Britton)
yellow- or Alaska-cedar (*Callitropsis nootkatensis* (D. Don))

Glossary of Terms[a]

-3/2 power law of self-thinning dense populations that have reached a size at which mortality occurs demonstrate a negative relationship between log mean plant weight and log stand density; this generally has a slope of -3/2.

A-line pertains to the upper limit of stocking on a stocking chart. Stands above the A-line are considered overstocked.

Advance regeneration (reproduction) syn. Advance growth seedlings or saplings that develop or are present in the understory.

Adventitious a plant part that develops outside the usual order of time and/or position, e.g., an adventitious bud arises from any part of a stem, leaf, or root but lacks vascular connection with the pith; an adventitious shoot derives from an adventitious bud; an adventitious root arises from parts of the plant other than a preexisting root, e.g., from a stem or leaf.

Afforestation the establishment of a forest or stand in an area where the preceding vegetation or land use was not forest.

Aggregated retention the retention of structures or biological legacies as (typically) small, intact forest patches within the harvest unit.

Allelopathy the inhibitory or stimulatory effects of released organic chemicals by one plant on the germination, growth, or metabolism of a different plant.

Archetype pertaining to a model forest ecosystem in terms of natural disturbance, development/dynamics, and structure; examples of forests of the same archetype are similar in terms of these characteristics.

Assisted migration deliberate movement of a species to a different habitat.

Assisted population expansion within-range movement of species in response to changes in climate conditions.

Assisted range expansion deliberate movement of species to facilitate or mimic natural range expansion in the context of changing climate conditions.

Avoided conversion carbon project that consists of specific actions that prevent the conversion of privately owned forest land to a non-forest land use by dedicating the land to continuous forest cover through a conservation easement or transfer to public ownership.

B-line pertaining to a stocking chart or guide, the B-line is the lower limit of stocking that will allow the stand to develop naturally; stands below the B-line are considered understocked.

Bedding a site preparation procedure in which the soil is mounded mechanically to provide a well-drained ridge on which seedlings are planted or seeds distributed naturally or directly.

Below-ground competition two or more plants vying for the same physical soil space (interference competition) or soil-derived resources, including water or nutrients (exploitive competition).

Best management practices in forestry, best management practices or BMPs are proactive practical methods used during management to achieve desired goals related to protecting water quality, wildlife habitat, productivity, biodiversity, aesthetics, recreation, and heritage resources.

---

[a]Sources of defintions:
Silvicultural terms in Canada, 2nd edition, 1995. Canadian Forest Service, Natural Resources Canada. Used with permission.
Silviculture Instructors' Subgroup of the Society of American Foresters 1993.
Wisconsin Department of Natural Resources Silviculture Handbook 2431.5.

Biological diversity (biodiversity) the spectrum of life forms and ecological processes that support and sustain them. Biological diversity occurs at four interacting levels genetic, species, community, and ecosystem.

Biological legacy an organism, a reproductive portion of an organism, or a biologically derived structure or pattern inherited from a previous ecosystem. Biological legacies often include large trees, snags, and down logs left after harvesting to provide refugia and to structurally enrich the new stand.

Biomass the mass of living or formerly living organisms or structures derived from formerly living organisms per unit of area. It can be divided into above-ground biomass, below-ground biomass, living biomass, and dead biomass.

Biomass accumulation the increase in mass of living or dead organisms per unit area over a specified time.

Biotroph an organism that requires a living host.

Broadcast seeding to spread or apply seed more or less evenly over an entire area.

Buttressed stem a noticeable swelling and irregular shape of the lower trunk, more common in larger trees.

Cabin logs logs meeting minimum specifications for use in the construction of log homes.

Canopy gaps an opening in the forest canopy generally caused by death of large trees.

Carbon sink a reservoir having the capacity to accumulate or lose carbon over time (e.g., soils, aboveground biomass).

Carbon stocks measured, estimated, or modeled quantity of carbon held in a particular pool.

Chips wood chips created from logs, whole trees, logging residues, or stumps and used for generation of bioenergy, manufactured wood products, or pulp.

Clearcut (syn. Clearcutting) An area of forest land from which all merchantable trees have recently been harvested. To harvest all merchantable trees from an area of forest land.

Clearcutting method *see* regeneration method.

Clearcut with reserves *see* regeneration method.

Close-to-nature forestry a theory and practice of forestry having principles of maintaining continuous forest, avoidance of clearcutting, use of medium to long rotations, use of natural regeneration, avoidance of intensive soil cultivation, mineral fertilizers, and pesticides, increasing tree species diversity as appropriate.

Cohort (syn. Age class) a distinct aggregation of trees originating from a single natural event or regeneration activity, or a grouping of trees, e.g., 10-year age class, as used in inventory or management.

Cohort establishment pertaining to an event in stand development that resulted in an age class (cohort) of tree establishing.

Commercial thinning removal of trees to reduce stand density with the size and species of tree having economic value.

Competition the extent to which each organism maximizes fitness by both appropriating contested resources from a pool not sufficient for all, and adapting to the environment altered by all participants.

Competition-based mortality in forest stands, tree death that results from competition for limiting resources.

Complex adaptive system as pertains to forestry, considering a forest as composed of heterogeneous assemblages of individual agents, including plants, animals, and humans, closely interacting through flows involving markets, goods, and various other ecosystem services, with a capacity to self-organize following disturbance, and that displays nonlinear behaviors.

Complexity in reference to forest stands, refers to a measure of the number of different structural attributes present, the relative abundance of each of these attributes.

Complexity/diversity foundational principle of ecological silviculture that reflects the need to create and maintain structural complexity and species diversity, including spatial heterogeneity of conditions at multiple spatial scales through all silvicultural interventions.

Conservation forest forest land with a primary purpose of protecting biodiversity and ecosystem function. Usually formally declared by law or policy.

Context foundational principle of ecological silviculture that underscores the importance of planning and implementing silvicultural interventions in the context of how these actions accumulate to influence the structure and function of landscapes.

Continuity foundational principle of ecological silviculture emphasizing the provision for continuity in forest structure, function, and biota between pre- and post-harvest ecosystems during regeneration harvests.

Coppice natural regeneration originating from stump sprouts, stool shoots, or root suckers.

Coppice method *see* regeneration method.

Coppice system *see* coppice method.

Coppice with standards *see* regeneration method coppice.

Crop-tree a term to describe a tree that is commercially desirable, with the potential to grow straight, tall, and vigorously. Any tree selected to become or forming a component of the final crop. A crop tree can also be one selected for non-timber objectives, such as mast production or habitat potential.

Crop-tree thinning *see* thinning crown.

Crown Class a class of tree based on crown position relative to the crowns of adjacent trees.

Emergent Trees with crowns completely above the general level of the main canopy receiving full light from above and from all sides.

Dominant Trees with crowns extending above the general level of the main canopy of even-aged stands or, in uneven-aged stands, above the crowns of the tree's immediate neighbors, and receiving full light from above and partly from the sides.

Codominant Trees with crowns forming the general level of the main canopy in even-aged stands or, in uneven-aged stands, the main canopy of the tree's immediate neighbors, receiving full light from above and comparatively little from the sides.

Intermediate Trees with crowns extending into the lower portion of the main canopy of even-aged stands or, in uneven-aged stands, into the lower portion of the canopy formed by the tree's immediate neighbors, but shorter in height than the codominants. They receive little direct light from above and none from the sides.

Overtopped (Suppressed) Trees of varying levels of vigor that have their crowns completely covered by the crowns of one or more neighboring trees.

Crown fire a fire that spreads across the tops of trees or shrubs more or less independently of a surface fire.

Crown thinning *see* thinning crown.

Cull trees or logs or portions thereof that are of merchantable size but are rendered unmerchantable by defects.

Culmination of mean annual increment the age in the growth cycle of a tree or stand at which the mean annual increment (MAI) for height, diameter, basal area, or volume is at a maximum.

Cutting Cycle The planned interval between partial harvests in an uneven-aged stand. *See also* thinning interval.

Dauerwald German term often used to describe "close-to-nature forestry" with emphasis of management on the selection and promotion of individual trees, of all sizes and multiple species, rather than the creation of areas with trees of uniform size.

Deadwood dead trees and tree components (boles, branches, roots) in various state of decay.

Decadence dead and decaying components of trees, including portions of living stems.

Dendroecology the science that uses dendrochronology (the study of tree rings) to analyze past ecological processes.

Denitrification reduction of nitrogen oxides (nitrate and nitrite) to molecular nitrogen or nitrogen oxides with lower oxidation state of nitrogen by bacterial activity.

Density management diagrams form of stocking chart based on the natural patterns of development for self-thinning, even-aged stands assumed under the -3/2 self-thinning rule. Diagrams often use relative density or stand density index to establish zones related to different levels of resources available for individual tree and stand growth and associated patterns in mortality.

Detritus small pieces of dead and decomposing plants and animals.

Diameter distribution frequency distribution of tree diameters represented in a given forest stand, often expressed in terms of density (trees per unit area in a given diameter class).

Dimensional lumber timber cut to specific widths, thicknesses, and lengths.

Discing scarification technique using disks to break small slash and the organic layer and to cut vegetation, loosening and incorporating these into the soil.

Discounted present net value current worth of an investment in the future based on a specified rate of return or discount rate.

Dispersed retention the retention of structures or biological legacies in a dispersed or uniform pattern.

Dominant thinning *see* thinning selection.

Dry-mesic of sites or habitats characterized by moisture conditions falling between xeric and mesic.

Duff the partially decomposed organic material of the forest floor beneath the litter of freshly fallen twigs, needles, and leaves.

Early successional forest communities characterizing early stages of ecosystem development following a disturbance.

Ecological forestry management approach that applies an understanding of the structure, function, and dynamics of natural forest ecosystems to achieve integrated environmental, economic, and social outcomes.

Ecological restoration management approach that attempts to recovered degraded forests and landscapes to a prior reference condition.

Ecosystem a spatially explicit, relatively homogeneous unit of the earth that includes all interacting organisms and components of the abiotic environment within its boundaries.

Ecosystem function refers to the various ecological processes, such as nutrient cycling, primary productivity, and decomposition, which occur in an ecosystem.

Ecosystem structure refers to the biotic (e.g., plants, animals, primary producers, decomposers) and abiotic components (e.g., water, nutrients) of an ecosystem.

Epicormic shoot a shoot arising from a dormant or adventitious bud on the stem or branch of a woody plant.

Even-aged methods *see* regeneration method.

Even-aged system a planned sequence of treatments designed to maintain and regenerate a stand with one age class. The range of tree ages is usually less than 20 percent of the rotation (*see also* regeneration method clearcutting, seed tree, shelterwood, coppice).

Even-aged of a forest, stand, or forest type in which relatively small age differences exist between individual trees. The differences in age permitted are usually 10 to 20 years; if the stand will not be harvested until it is 100 to 200 years old, larger differences up to 25 percent of the rotation age may be allowed.

Foodweb the interlocking pattern of feeding relationships in a community.

Forest assisted migration general term for movement of widespread commercially and ecologically important species to ensure populations are established that are climatically adapted for the duration of a rotation.

Forest certification nongovernmental effort to use market mechanisms to ensure sustainability. Processes vary by organizations, but generally involve a written certification provide by an independent party that attests to the sustainable and responsible production of wood products based on social and environmental standards.

Free thinning *see* thinning free.

Function *see* ecosystem function.

Gap-level retention retention of living and dead canopy trees within harvest gaps created as part of uneven-aged regeneration methods, such as group selection or continuous cover irregular shelterwoods.

Geometric thinning *see* thinning mechanical.

Girdling making more or less continuous incisions around a living stem, through at least both bark and cambium, generally with the object of killing the tree. Sometimes termed mechanical girdling, to distinguish it from herbicide girdling when herbicide is added. Making a series of close downward and upward, i.e., V-shaped, incisions into the sapwood is termed notch-girdling.

Gross periodic annual increment growth increment of individual or stand of trees over a period of years divided by the period length. Includes all growth components, including mortality and harvest removals.

Growing space the physical area available to and utilized by a tree.

Growing stock all the trees growing in a forest or in a specified part of it, usually commercial species, meeting specified standards of size, quality, and vigor, and generally expressed in terms of number or volume.

Headwater stream first-order stream representing upper reaches of given watershed and stream network.

Heartwood the inner, nonliving part of a tree stem that is altered to a protective state as a result of normal, genetically controlled aging processes as cells die, and that provides mechanical support.

Heterogeneity the state of being not identical in some or all parameters in one or more samples or populations. Variation in the environment over space and time.

High severity disturbance that kills the majority of vegetation, such as a crown fire.

Ice contact partially stratified or unstratified glacial drift usually consisting of sand and gravel. This material was originally water-deposited (stratified) on or in glacial ice, then slumped when ice melted, thus mixing the material and destroying stratification.

Improved forest management activities forest management activities that result in increased carbon stocks within forests and/or reduce greenhouse gas emissions from forestry activities when compared to business-as-usual forestry practices.

Improvement cutting a cutting made in a stand past the sapling stage, primarily to improve composition and quality through the removal of less desirable trees of any species.

Intermediate treatments (*syn.* tending) a collective term for any treatment designed to enhance growth, quality, vigor, and composition of the stand after establishment or regeneration and prior to final harvest.

Irregular shelterwood system harvest cutting in which opening of the canopy is irregular and gradual; generally in groups, with the final cutting often in strips; regeneration is natural; regeneration interval is long, often up to half the rotation, and the resultant crop is considerably uneven-aged and irregular.

Ladder fuels combustible material that provides vertical continuity between vegetation strata and allows fire to climb into the crowns of trees or shrubs with relative ease.

Large deadwood (*syn.* coarse woody debris) the standing and downed dead wood in a forest.

Leaf area measure of ground surface covered by leaves or needles.

Legacy trees trees, usually mature or old-growth, that are retained on a site after harvesting or natural disturbance to provide a biological legacy.

Logging slash the residue left on the ground after logging *or* the unused portion of trees cut or killed during logging and left in the woods.

Low severity disturbance that only partially disturbs a given ecosystem, such as a downburst.

Low thinning *see* thinning low.

Macrophyte plant growing in or near the surface of water

Mast seed crop a seed year in which much more seed than usual is produced. Mast seeding or masting is a phenomenon exhibited by some species of plants, which can be defined as synchronous production of seed at long intervals.

Mechanical thinning *see* thinning mechanical.

Mesic of sites or habitats characterized by intermediate moisture conditions, i.e., neither decidedly wet nor dry.

Meso-scale moderate severity disturbances that affect areas greater than an individual stand, but are not experienced across an entire landscape.

Mid-tolerant general term for species intermediate in level of shade tolerance.

Mineral soil a soil consisting predominantly of, and having its properties determined predominantly by, mineral matter.

Mixed-severity disturbance regime characterized by variation in disturbance severity that can be observed at the intermediate or meso-scale.

Multi-aged *see* multi-cohort.

Multi-cohort forest composed of three or more cohorts of trees.

Multiple-use *Federal Land Policy and Management Act of 1976.* The management of the public lands and their various resource values so that they are utilized in the combination that will best meet the present and future needs of the American people.

Mycorrhizae the symbiotic association between certain fungi and plant roots which enhances the uptake of water and nutrients.

Natural disturbance disturbance associated with ecological or meteorological agents, such as fires, flooding, windstorms, or insects.

Natural origin forest originating from natural disturbance and natural regeneration.

Natural forest *see* natural origin.

Niche the ultimate unit of the habitat, i.e., the specific spot occupied by an individual organism.

Nitrogen fixation the conversion of elemental nitrogen (N_2) from the atmosphere to organic combinations or to forms readily utilizable in biological processes.

Noncommercial tree that is not merchantable either due to species or inadequate size or quality.

Normal stand a forest stand organized so that each age class occupies an equal area.

Nurse crop plant species that assists in the establishment of another species

Nutrient cycling the exchange or transformation of elements among the living (organic and biotic) and nonliving (inorganic and abiotic) components of an ecosystem.

Outwash stratified detritus (chiefly sand and gravel) removed or "washed out" from a glacier by meltwater streams and deposited in front of or beyond the end moraine or the margin of an active glacier.

Overmature 1) a tree or even-aged stand that has reached that stage of development when it is declining in vigor and health and reaching the end of its natural life span, or 2) a tree or even-aged stand that has begun to lessen in commercial value because of size, age, decay, or other factors.

Patch 1) a small area distinct from that about it, 2) a small part of a stand or forest, or 3) an ecosystem element (e.g. an area of vegetation that is relatively homogeneous internally and differs from surrounding elements).

Patch scarification (*syn.* patch scarifier) a mechanized implement used to expose patches of mineral soil in a systematic pattern.

Pathogenic fungi fungal species capable of causing disease.

Periodic annual increment (PAI) the growth of a tree or stand observed over a specific time period divided by the length of the period.

Pioneer a plant capable of invading bare sites and persisting there or colonizing them until supplanted by successional species.

Pits-and-mounds small, microtopographical variation in a forest floor caused over time by uprooting of trees via windthrow. Pits represent area formerly occupied by root mass, whereas mounds represent area to which soil roots and soils were displaced by wind event.

Plantation a stand composed primarily of trees established by planting or artificial seeding.

Planted forest forest originating primarily from artificial regeneration.

Pole a tree between a sapling and small sawtimber size. Size varies by region, e.g., for boreal and eastern forests (Canadian) 12–20 cm dbh.

Precommercial thinning a thinning that does not yield trees of commercial value, usually designed to reduce stocking in order to concentrate growth on the more desirable trees.

Primary production process whereby inorganic carbon is fixed and converted to organic matter by primary productions, such as trees and other plants.

Production forests forests managed according to agronomic and economic models with emphasis on economically efficient production of wood.

Protected forests an area, wholly or partly covered with trees, managed primarily to regulate stream flow, maintain water quality, minimize erosion, stabilize drifting sand, conserve ecosystems, or provide other benefits via protection.

Provenance study a planting in which population samples from stands of known geographic origins are grown together in one or more locations.

Pulpwood wood used in making pulp for paper.

Raking site preparation technique using a bulldozer or similar equipment with a blade having teeth instead of a plain edge, for pushing large, coarse woody debris and rocks off a site and leaving smaller stones, soil, small finer slash, and woody debris in place.

Range of natural variation spectrum of natural conditions possible in ecosystem structure, composition, and function, when considering both temporal and spatial scales.

Real Estate Investment Trust (REIT) a real estate investment trust is a company that owns, operates or finances income-producing forestland real estate.

Recovery periods interval between harvests in an ecological silvicultural system.

Reforestation the reestablishment of forest cover either naturally (by seeding, coppice, or root suckers) or artificially (by direct seeding or planting).

Refugia locations and habitats that support populations of organisms that are limited to small fragments of their previous geographic range.

Regeneration (*syn.* reproduction) method a cutting method by which a new age class is created. The major methods are Clearcutting, Seed Tree, Shelterwood, Selection, and Coppice (*see* harvesting method).

Coppice methods methods of regenerating a stand in which the majority of regeneration is from stump sprouts or root suckers.

> ***Coppice*** a method of regenerating a stand in which all trees in the previous stand are cut and the majority of regeneration is from sprouts or root suckers.
>
> ***Coppice with standards*** a coppice method in which reserve trees are retained to attain goals other than regeneration. The method normally creates a two-aged stand.

Even-aged methods methods to regenerate a stand with a single age class.

> ***Clearcutting*** a method of regenerating an even-aged stand in which a new age class

develops in a fully exposed microclimate after removal, in a single cubing, of all trees in the previous stand. Regeneration is from natural seeding, direct seeding, planted seedlings, and/or advance reproduction. Cutting may be done in groups or patches (Group or Patch Clearcutting), or in strips (Strip Clearcutting). In the Clearcutting System, the management unit or stand in which regeneration, growth, and yield are regulated consists of the individual clearcut stand (see Group Selection). When the primary source of regeneration is advance reproduction, the preferred term is Overstory Removal.

Clearcutting with reserves (*see* two-aged methods).

Seed tree an even-aged regeneration method in which a new age class develops from seeds that germinate in fully exposed microenvironments after removal of all the previous stand except a small number of trees left to provide seed. Seed trees are removed after regeneration is established.

Seed tree with reserves (*see* two-aged methods).

Shelterwood a method of regenerating an even-aged stand in which a new age class develops beneath the moderated microenvironment provided by the residual trees. The sequence of treatments can include three distinct types of cuttings: (1) an optional preparatory cut to enhance conditions for seed production; (2) an establishment cut to prepare the seed bed and to create a new age class; and (3) a removal cut to release established regeneration from competition with the overwood. Cutting may be done uniformly throughout the stand (Uniform Shelterwood), in groups or patches (Group Shelterwood), or in strips (Strip Shelterwood).

Shelterwood with reserves (*see* two-aged methods).

Two-aged methods methods designed to maintain and regenerate a stand with two age classes. In each case the resulting stand may be two-aged or tend toward an uneven-aged condition as a consequence of both an extended period of regeneration establishment and the retention of reserve trees that may represent one or more age classes.

Clearcutting with reserves a clearcutting method in which varying numbers of reserve trees are not harvested to attain goals other than regeneration.

Seed tree with reserves a seed tree method in which some or all of the seed trees are retained after regeneration has become established to attain goals other than regeneration.

Shelterwood with reserves a variant of the Shelterwood Method in which some or all of the shelter trees are retained, well beyond the normal period of retention, to attain goals other than regeneration.

Uneven-aged (selection) methods methods of regenerating a forest stand, and maintaining an uneven-aged structure, by removing some trees in all size classes either singly, in small groups, or in steps.

Group selection a method of regenerating uneven-aged stands in which trees are removed, and new age classes are established, in small groups. The maximum width of groups is approximately twice the height of the mature trees, with small openings providing microenvironment suitable for tolerant regeneration and the larger openings providing conditions suitable for more intolerant regeneration. In the Group Selection System, the management unit or stand in which regeneration, growth, and yield are regulated consists of a landscape containing an aggregation of groups (*see* clearcutting).

Group selection with reserves a variant of the Group Selection Method in which some trees within the group are not cut to attain goals other than regeneration within the group.

Single tree selection a method of creating new age classes in uneven-aged stands in which individual trees of all size classes are removed more-or-less uniformly throughout the stand to achieve desired stand structural characteristics.

Release a treatment designed to free young trees from undesirable, usually overtopping, competing vegetation. Treatments include cleaning, liberating, and weeding.

Resilience the capacity of a plant community or ecosystem to maintain or regain normal function and development following disturbance.

Resistance the ability of a plant community or ecosystem to avoid alteration of its present state by a disturbance.

Resource availability amount of key resources for normal growth, maintenance and reproduction in a given community or local microsite.

Response range of reactions to environmental change among species that contribute to the same ecosystem function.

Retention harvests silvicultural method that retains forest structural elements, such as large living and dead trees, at the time of harvest to serve as biological legacies in the resulting forest or cohort.

Return interval the time between disturbances in a defined area.

Roller chopping site preparation technique that is designed to knock down and chop smaller trees and slash using a large drum mounted with grader blades.

Rotation in even-aged silvicultural systems, the period between regeneration establishment and final cutting. Rotation may be based on many criteria including culmination of mean annual increment, mean size, age, attainment of particular minimum physical or value growth rate, and biological condition.

Rotation age the planned number of years between the formation or regeneration of a crop or stand and its final cutting at a specified stage or maturity.

Row thinning a thinning generally in plantations in which the trees are cut out in lines or narrow strips at fixed intervals throughout a stand.

Salvage cutting the removal of dead trees or trees being damaged or dying due to injurious agents other than competition, to recover value that would otherwise be lost.

Sanitation cutting the removal of trees to improve stand health by stopping or reducing actual or anticipated spread of insects and disease (*see* stand improvement).

Sapling a usually young tree larger than a seedling but smaller than a pole.

Saprotroph an organism that feeds on or derives nourishment from decaying organic matter.

Sawtimber trees that will yield logs suitable in size and quality for the production of lumber.

Scarification mechanical removal of competing vegetation or interfering debris, or disturbance of the soil surface, to enhance reforestation.

Seed-tree method the cutting of all trees except for a small number of widely dispersed trees retained for seed production and to produce a new age class in fully exposed microenvironment.

Selection methods regenerate and maintain a multiaged structure by removing some trees in all size classes either singly, in small groups, or in strips.

Selection thinning (*see* thinning selection).

Self-thinning tree mortality from the effect of the competition arising between trees on the same site.

Sere the individual sequential stages in forest succession.

Serotinous particularly applied to plant species or individuals with cones that remain on the tree without opening for one or more years (e.g., *Pinus contorta* and *Pinus banksiana*).

Shade intolerant having the capacity to compete for survival under direct sunlight conditions.

Shade tolerant having the capacity to compete for survival under shaded conditions.

Shearing a method of site preparation in which all standing material is removed at ground level using a shear blade attached to a large tractor.

Shelterwood method the cutting of most trees, leaving those needed to produce sufficient shade to produce a new age class in a moderated microenvironment.

Silvicultural prescription a planned series of treatments designed to change current stand structure to one that meets management goals and objectives. The prescription normally considers ecological, economic, and societal constraints.

Silvicultural system a planned program of vegetation treatment during the entire life of a stand. The three basic components are tending, harvesting, and regeneration. Named after the stand age class structure and the regeneration method employed.

Single-cohort forest composed of a single age class of trees.

Site index a species-specific measure of actual or potential forest productivity (site quality, usually for even-aged stands), expressed in terms of the average heigh of trees included in a specified stand component (defined as a certain number of dominants, codominants, or the largest and tallest trees per unit area) at a specified index or base age.

Site occupancy probability that a habitat patch is occupied by a given species.

Site preparation a hand or mechanized manipulation of a site designed to enhance the success of regeneration. Treatments may include bedding, burning, chemical spraying, chopping, disking, drainage, raking, and scarifying. All treatments are designed to modify the soil, litter, and vegetation and to create microclimate conditions conducive to the establishment and growth of desired species.

Size-density relationships relationship between maximum average size and associated density found in self-thinning populations.

Skips and gaps used to refer to variable density thinning due to the creation of harvest gaps and associated designation of unharvested areas (skips).

Soil seed bank the soil seed bank is the natural storage of seeds, often dormant, within the soil of most forest ecosystems.

Species richness a measure of the number of species present in a community, ecosystem, landscape, region, etc.

Stand as defined in classic silviculture, a contiguous group of trees sufficiently uniform in age class distribution, composition, and structure, and growing on a site of sufficiently uniform quality, to be a distinguishable unit. In ecological silviculture, forest areas that incorporate the structure, age, composition, and spatial pattern of trees of natural ecosystems, as determined by an underlying geomorphic, soil, or disturbance templates.

Stand development the growth of a stand through its various developmental stages - from seedling or coppice through thicket, sapling, and pole to the tree stage, i.e. to maturity, and finally to overmaturity.

Stand initiation the initial stage of stand structural development, lasting from the time of stand-replacing disturbance until the new cohort forms a continuous canopy and trees begin competing with each other for light and canopy space.

Stand-replacing disturbance that kills the majority of vegetation on a given site.

Stocking an indication of growing-space occupancy relative to a preestablished standard. Common indices of stocking are based on percent occupancy, basal area, relative density, and crown competition factor.

Stocking chart a chart used to guide thinning of a stand based on the relationship among basal area, tree density, and average tree diameter. A stocking chart is used to determine whether the stand is overstocked, fully stocked, or understocked (*see* A-line, B-line).

Stocking control the regulation and, more particularly, the limitation of seedling populations by natural, or direct or indirect artificial factors.

Stocking guide *see* stocking chart.

Stocking level *see* stocking.

Structural complexity degree of heterogeneity in living and dead components in a forest.

Structure the horizontal and vertical distribution of components of a forest stand including the height, diameter, crown layers and stems of trees, shrubs, herbaceous understory, snags, and down woody debris.

Surface fire a fire that burns only surface fuels such as litter, loose debris, and small vegetation.

Tending *see* intermediate treatments.

Thinning a cultural treatment made to reduce stand density of trees primarily to improve growth, enhance forest health, or to recover potential mortality.

Crown thinning (thinning from above, high thinning) the removal of trees from the dominant and codominant crown classes in order to favor the best trees of those same crown classes.

Free Thinning the removal of trees to control stand spacing and favor desired trees using a combination of thinning criteria without regard to crown position.

Low thinning (thinning from below) the removal of trees from the lower crown classes to favor those in the upper crown classes.

Mechanical thinning (geometric thinning) the thinning of trees in either even- or uneven-aged stands involving removal of trees in rows, strips, or by using fixed spacing intervals.

Selection thinning (dominant thinning) the removal of trees in the dominant crown class in order to favor the lower crown classes.

Thinning from above *see* thinning: crown.

Thinning from below *see* thinning: low.

Three-cohort *see* multi-cohort.

Timber Investment Management Organization (TIMO) is a management group that aids institutional investors in managing their timberland investment portfolios.

Timing foundation principle of ecological silviculture that refers to the importance of applying silvicultural interventions at ecologically appropriate time intervals.

Translocation of exotics movement of species to an area outside of its native range.

Transpiration the process by which water vapor passes from the foliage or other parts of a living plant to the atmosphere.

Two-aged stand a stand composed of two distinct age classes separated in age by more than 20 percent of rotation.

Two-aged system a planned sequence of treatments designed to maintain and regenerate a stand with two age classes.

Two-cohort forest composed of two age classes of trees.

Uneven-aged methods regenerate and maintain a multiaged structure by removing some trees in all size classes either singly, in small groups, or in strips.

Uneven-aged system a planned sequence of treatments designed to maintain and regenerate a stand with three or more age classes (*see* single tree selection, group selection).

Utility poles logs meeting minimum specification for use as utility poles.

Variable density thinning thinning used to promote greater heterogeneity in ecological conditions by varying intensity of removal of overstory trees across a stand. This includes retention of dense areas, as well as creation of harvest gaps.

Variable retention harvest system an approach to harvesting based on the retention of structural elements or biological legacies (trees, snags, logs, etc.) from the harvested stand for integration into the new stand to achieve various ecological objectives.

Veneer a thin sheet of wood of uniform thickness, produced by rotary cutting (peeling) or slicing, and sometimes by sawing.

Weeding a release treatment in stands not past the sapling stage that eliminates or suppresses undesirable vegetation regardless of crown position.

Wildland-urban interface area where houses and other developments meet or intermingle with wildland vegetation.

Woodland a low-density forest forming open habitats with more sunlight and less shade than closed-canopy forest. Woodlands support an understory with greater shrub and/or herbaceous plants, including grasses, than a typical closed-canopy forest.

Xeric pertaining to sites or habitats characterized by decidedly dry conditions.

References

Aaseng, N., J. Almendinger, K. Rusterholz, D. Wovcha, and T. R. Klein. 2003. *Field guide to the native plant communities of Minnesota: The Laurentian Mixed Forest Province.* St. Paul, MN: *Minnesota Department of Natural Resources.*

Abrams, M. D. 1984. Uneven-aged jack pine in Michigan. *Journal of Forestry* 82:306–309.

Adams, D. M., and A. R. Ek. 1974. Optimizing the management of uneven-aged forest stands. *Canadian Journal of Forest Research* 4:274–287.

Addington, R. N., G. H. Aplet, M. A. Battaglia, J. S. Briggs, P. M. Brown, A. S. Cheng, Y. Dickinson, J. A. Feinstein, K. A. Pelz, and C. M. Regan. 2018. Principles and practices for the restoration of ponderosa pine and dry mixed-conifer forests of the Colorado Front Range. *USDA Forest Service General Technical Report RMRS* 373:121.

Agee, J. K. 1993. *Fire ecology of Pacific Northwest forests.* Washington, DC: Island Press.

Agee, J. K., and C. N. Skinner. 2005. Basic principles of forest fuel reduction treatments. *Forest Ecology and Management* 211:83–96.

Aitken, S. N., S. Yeaman, J. A. Holliday, T. Wang, and S. Curtis-McLane. 2008. Adaptation, migration or extirpation: Climate change outcomes for tree populations. *Evolutionary Applications* 1:95–111.

Allen, C. D. 2007. Interactions across spatial scales among forest dieback, fire, and erosion in northern New Mexico landscapes. *Ecosystems* 10:797–808.

Allen, C. D., and D. D. Breshears. 1998. Drought-induced shift of a forest-woodland ecotone: Rapid landscape response to climate variation. *Proceedings of the National Academy of Sciences* 95:14839–14842.

Allen, C. D., A. K. Macalady, H. Chenchouni, D. Bachelet, N. McDowell, M. Vennetier, T. Kitzberger, A. Rigling, D. D. Breshears, E. H. Hogg, P. Gonzalez, et al. 2010. A global overview of drought and heat-induced tree mortality reveals emerging climate change risks for forests. *Forest Ecology and Management* 259:660–684.

Allen, C. D., M. Savage, D. A. Falk, K. F. Suckling, T. W. Swetnam, T. Schulke, P. B. Stacey, P. Morgan, M. Hoffman, and J. T. Klingel. 2002. Ecological restoration of Southwestern ponderosa pine ecosystems: A broad perspective. *Ecological Applications* 12:1418–1433.

Anderson, H. W., E. P. Boysen, D. C. Dey, and J. A. Rice. 2001. Natural regeneration of hardwoods. In *Regenerating the Canadian forest: Principles and practice for Ontario* R. G. Wagner and S. J. Colombo, 393–421. Markham, ON, CA: Fitzhenry and Whiteside Ltd.

Aravena, J. C., M. R. Carmona, C. A. Pérez, and J. J. Armesto. 2002. Changes in tree species richness, stand structure and soil properties in a successional chronosequence in northern Chiloé Island, Chile. *Revista Chilena de Historia Natural* 75:339–360.

Arbogast, C., Jr. 1957. *Marking guides for northern hardwoods under the selection system.* Paper LS-56. St. Paul, Minn.: USDA Forest Service, Lake States Forest Experiment Station. 20 p.

Ashton, M. S., and M. J. Kelty. 2017. *The practice of silviculture: Applied forest ecology.* 10th ed. New York: John Wiley and Sons.

Assman, E. 1970. *The principles of forest yield study.* Oxford, UK: Pergamon Press.

Aubry, K. B., C. B. Halpern, D. A. Maguire. 2004. Ecological effects of variable retention harvests in the northwestern United States: The DEMO study. *Forest, Snow, and Landscape Research* 78:119–137.

Aust, W. M., and C. R. Blinn. 2004. Forestry best management practices for timber harvesting and site preparation in the eastern United States: An overview of water quality and productivity research during the past 20 years (1982–2002). *Water, Air and Soil Pollution: Focus* 4:5–36.

Bailey, J. D., and W. W. Covington. 2002. Evaluating ponderosa pine regeneration rates following ecological restoration treatments in northern Arizona, USA. *Forest Ecology and Management* 155:271–278.

Bailey, J. D., and J. C. Tappeiner. 1998. Effects of thinning on structural development in 40- to 100-year-old Douglas-fir stands in western Oregon. *Forest Ecology and Management* 108:99–113.

Baker, J. B., M. D. Cain, J. M. Guldin, P. A. Murphy, and M. G. Shelton. 1996. *Uneven-aged silviculture for the loblolly and shortleaf pine forest cover types*. USDA Forest Service, General Technical Report SO-118. 65 p.

Bale, J. S., G. J. Masters, I. D. Hodkinson, C. Awmack, T. M. Bezemer, V. K. Brown, J. Butterfield, A. Buse, J. C. Coulson, J. Farrar, and J. E. Good. 2002. Herbivory in global climate change research: Direct effects of rising temperature on insect herbivores. *Global Change Biology* 8:1–16.

Barbour, M. G., and W. D. Billings. 2000. *North American terrestrial vegetation*. 2nd ed. New York: Cambridge University Press. 708 p.

Bare, B. B., and D. Opalach.1988. Determining investment-efficient diameter distributions for uneven-aged northern hardwoods. *Forest Science* 34:243–249.

Barlocher, F., R. J. Mackay, and G. B. Wiggins. 1978. Detritus processing in a temporary vernal pool in southern Ontario. *Archiv für Hydrobiologie* 81:269–295.

Barten, P. K., T. Kyker-Snowman, P. J. Lyons, T. Mahlstedt, R. O'Connor, and B. A. Spencer. 1998. Massachusetts: Managing a watershed protection forest. *Journal of Forestry* 96:10–15.

Barton, D. R., W. D. Taylor, and R. M. Biette. 1985. Dimensions of riparian buffer strips required to maintain trout habitat in southern Ontario streams. *North American Journal of Fisheries Management* 5:364–378.

Batzer, D. P., C. R. Jackson, and M. Mosner. 2000. Influences of riparian logging on plants and invertebrates in small, depressional wetlands of Georgia, USA. *Hydrobiologia* 441:123–132.

Batzer, D. P., B. J. Palik, and R. Buech. 2004. Relationships between environmental characteristics and macroinvertebrate communities in seasonal woodland ponds of Minnesota. *Journal of the North American Benthological Society* 23:50–68.

Bauhus, J., K. J. Puettmann, and C. Kühne. 2013. Close-to-nature forest management in Europe: Does it support complexity and adaptability of forest ecosystems? In *Managing forest as complex adaptive systems*, ed. C. Messier, K. J. Puettmann, and K. D. Coates, 187–213. New York: Routledge Press.

Bauhus, J., P. van der Meer, and M. Kanninen. 2010. *Ecosystem goods and services from plantation forests*. London, UK: Earthscan Forestry Library.

Bebber, D. P., S. C. Thomas, W. G. Cole, and D. Balsillie. 2004. Diameter increment in mature eastern white pine *Pinus strobus* L. following partial harvest of old-growth stands in Ontario, Canada. *Trees* 18:29–34.

Bédard, S. 2016. Introduction to hardwood forest ecology, exploitation, and silviculture in multi-aged silviculture of northern hardwood and mixedwood forests. Proceedings: Field tour in Québec. Québec Ministry of Forests, Wildlife, and Parks 38.

Bédard, S., F. Guillemette, P. Raymond, S. Tremblay, C. Larouche, and J. DeBlois. 2014. Rehabilitation of northern hardwood stands using multicohort silvicultural scenarios in Québec. *Journal of Forestry* 112:276–286.

Benda, L. E., N. L. Poff, D. Miller, T. Dunne, G. Reeves, G. Pess, and M. Pollack. 2004. The network dynamics hypothesis: How channel networks structure riverine habitats. *BioScience* 54:413–427.

Bennett, B. M. 2015. *Plantations and protected areas: A global history of forest management*. Cambridge, MA: MIT Press.

Benzie, J. W. 1977. *Manager's handbook for red pine in the North Central States*. USDA Forest Service, General Technical Report NC-33.

Berg, D. R. 1995. Riparian silvicultural system design and assessment in the Pacific Northwest Cascade Mountains, USA. *Ecological Applications* 5:87–96.

Bergeron, Y., and J. Brisson. 1990. Fire regime in red pine stands at the northern limit of the species' range. *Ecology* 71:1352–1364.

Bergeron, Y., B. Harvey, A. Leduc, and S. Gauthier. 1999. Forest management guidelines based on natural disturbance dynamics: Stand- and forest-level considerations. *The Forestry Chronicle* 75:49–54.

Bergman, H. F. 1924. The composition of climax plant formations in Minnesota. *Papers of the Michigan Academy of Science, Arts and Letters* 3:51–60.

Bettinger, P., K. Boston, J. P. Siry, and D. L. Grebner. 2009. *Forest management and planning*. Amsterdam, Netherlands: Academic Press/Elsevier.

Bingham, B. B., and J. O. Sawyer. 1992. Canopy structure and tree condition of young, mature, and old-growth Douglas-fir/hardwood forests. In *Proceedings of the symposium on biodiversity of northwestern California*, technical coordinators R. R. Harris and D. E. Erman, 141–149. University of California, Wildland Resources Center Report No. 29.

Birch, T. W. 1996. *Private forest-land owners of the United States*. USDA Forest Service, Research Bulletin NE-134, Radnor, PA.

Blinn, C. R., and M. A. Kilgore. 2001. Riparian management practices: A summary of state guidelines. *Journal of Forestry* 99:11–17.

Bolton, N. W., and A. W. D'Amato. 2011. Regeneration responses to gap size and coarse woody debris within natural disturbance-based silvicultural systems in northeastern Minnesota, USA. *Forest Ecology and Management* 262:1215–1222.

Bormann, F. H., and G. E. Likens. 1994. *Pattern and process in a forested ecosystem*. New York: Springer-Verlag.

Bottero, A., A. W. D'Amato, B. J. Palik, J. B. Bradford, S. Fraver, M. A. Battaglia, and L. A. Asherin. 2017. Density-dependent vulnerability of forest ecosystems to drought. *Journal of Applied Ecology*. doi:10.1111/1365-2664.12847

Boyd, R. 1999. *Indians, fire, and the land in the Pacific Northwest*. Corvallis, OR: Oregon State University Press.

Boyle, J. R., J. C. Tappeiner, R. H. Waring, and C. Tattersall Smith. 2016. Sustainable forestry: Ecology and silviculture for resilient forests. *Reference Module in Earth Systems and Environmental Sciences*. doi: 10.1016/B978-0-12-409548-9.09761-X.

Bradford, J. B., S. Fraver, A. M. Milo, A. W. D'Amato, B. Palik, and D. J. Shinneman. 2012. Effects of multiple interacting disturbances and salvage logging on forest carbon stocks. *Forest Ecology and Management* 267:209–214.

Brang, P., P. Spathelf, J. B. Larsen, J. Bauhus, A. Bonččina, C. Chauvin, L. Drössler, C. García-Güemes, C. Heiri, G. Kerr, and M. J. Lexer. 2014. Suitability of close-to-nature silviculture for adapting temperate European forests to climate change. *Forestry* 87:492–503.

Breshears, D. D., O. B. Myers, C. W. Meyer, F. J. Barnes, C. B. Zou, C. D. Allen, N. G. McDowell, and W. T. Pockman. 2009. Tree die-off in response to global change-type drought: Mortality insights from a decade of plant water potential measurements. *Frontiers in Ecology and the Environment* 7:185–189.

Briggs, D., D. DeBell, and W. Atkinson, eds. *Utilization and management of alder*. USDA Forest Service, General Technical Report PNW-70.

Brooks, H. E. 2013. Severe thunderstorms and climate change. *Atmospheric Research* 123:129–138.

Brooks, R. T. 2000. Annual and seasonal variation and the effects of hydroperiod on benthic macroinvertebrates of seasonal forest ("vernal") ponds in central Massachusetts, USA. *Wetlands* 20:707–715.

Brose, P. H., and T. A. Waldrop. 2006. Fire and the origin of Table Mountain pine pitch pine communities in the southern Appalachian Mountains, USA. *Canadian Journal of Forest Research* 36:710–718.

Brosofske, K. D., J. Chen, R. J. Naiman, and J. F. Franklin. 1997. Harvesting effects on microclimatic gradients from small streams to uplands in western Washington. *Ecological Applications* 7:1188–1200.

Brown, C. 2000. *The global outlook for future wood supply from forest plantations*. Food and Agriculture Organization of the United Nations.

Brown, R. T., J. K. Agee, and J. F. Franklin. 2004. Forest restoration and fire: Principles and the context of place. *Conservation Biology* 18:903–912.

Buckman, R. E. 2006. *Growth and yield of red pine in the Lake States*. USDA Forest Service, General Technical Report NC-271.

Buermeyer, K. R., and C. A. Harrington. 2002. Fate of overstory trees and patterns of regeneration 12 years after clearcutting with reserve trees in southwest Washington. *Western Journal of Applied Forestry* 17:78–85.

Burger, T. L., and J. Kotar. 2003. *A guide to forest communities and habitat types of Michigan*. Madison, WI: Department of Forest Ecology and Management, University of Wisconsin.

Burns, D. C. 1991. Cumulative effects of small modifications to habitat. *Fisheries* 16:12–17.

Burns, R. M., and B. H. Honkala. 1990. *Silvics of North America, Volume 2: Conifers*. Agriculture Handbook 654, USDA Forest Service, Washington, DC, USA. 877 p.

Butler, B. J., J. H. Hewes, B. J. Dickinson, K. Andrejczyk, S. M. Butler, and M. Markowski-Lindsay. 2016. Family forest ownerships of the United States, 2013: Findings from the USDA Forest Service's National Woodland Owner Survey. *Journal of Forestry* 114:638–647.

Butler, B. J., M. Markowski-Lindsay, S. Snyder, P. Catanzaro, D. B. Kittredge, K. Andreiczyk, B. J. Dickinson, D. Eryilmaz, J. H. Hewes, P. R. Ler, D. Ladle, and M. A. Kilgore. 2014. Effectiveness of landowner assistance activities: An examination of the USDA Forest Service's Forest Stewardship Program. *Journal of Forestry* 112:187–197.

Calhoun, A. J. K., and P. G. DeMaynadier. 2007. *Science and conservation of vernal pools in northeastern North America*. Boca Raton, FL: CRC Press.

Calhoun, A. J. K., R. E. Walls, S. S. Stockwell, and M. McCollough. 2003. Evaluating vernal pools as a basis for conservation strategies: A Maine case study. *Wetlands* 23:70–81.

Camp, A., C. D. Oliver, P. F. Hessburg, and R. L. Everett. 1997. Predicting late-successional fire refugia pre-dating European settlement in the Wenatchee Mountains. *Forest Ecology and Management* 95:63–77.

Canham, C. D., M. J. Papaik, and E. F. Latty. 2001. Interspecific variation in susceptibility to windthrow as a function of tree size and storm severity for northern temperate tree species. *Canadian Journal of Forest Research* 31:1–10.

Cao, S. 2008. Why large-scale afforestation efforts in China has failed to solve the desertification problem. *Environmental Science and Technology* 42:1826–1831.

Carey, A. B. 1995. Sciurids in Pacific Northwest managed and old-growth forests. *Ecological Applications* 5:648–661.

Carey, A. B. 2003. Biocomplexity and restoration of biodiversity in temperate coniferous forest: inducing spatial heterogeneity with variable-density thinning. *Forestry* 76:127–136.

Carey, A. B. 2009. Maintaining biodiversity in managed forests. In *Old growth in a new world: A Pacific Northwest icon reexamined*, ed. T. A. Spies, and S. L. Duncan, 58–69. Washington, DC: Island Press.

Castagneyrol, B., H. Jactel, C. Vacher, E. G. Brockerhoff, and J. Koricheva. 2014. Effects of plant phylogenetic diversity on herbivory depend on herbivore specialization. *Journal of Applied Ecology* 51:134–141.

Chambers, M. E., P. J. Fornwalt, S. L. Malone, and M. A. Battaglia. 2016. Patterns of conifer regeneration following high severity wildfire in ponderosa pine–dominated forests of the Colorado Front Range. *Forest Ecology and Management* 378:57–67.

Chen, J., J. F. Franklin, and T. A. Spies. 1992. Vegetation responses to edge environments in old-growth Douglas-fir forests. *Ecological Applications* 2:387–396.

Cheyney, E. G. 1942. *American silvics and silviculture.* Minneapolis, MN: University of Minnesota Press.

Christensen, N. L., J. K. Agee, P. F. Brusssard, J. Hughes, D. H. Knight, G.W. Minshall, J. M. Peek, S. J. Pyne, F. W. Swanson, J. W. Thomas, S. Wells, S. E. Williams, and H. A. Wright. 1989. Interpreting the Yellowstone fires of 1988. *BioScience* 39:678–685.

Churchill, D. J., G. C. Carnwath, A. J. Larson, and S. A. Jeronimo. 2017. *Historical forest structure, composition, and spatial pattern in dry conifer forests of the Western Blue Mountains, Oregon, USA.* USDA Forest Service, General Technical Report PNW-GTR-956.

Churchill, D. J., A. J. Larson, M. C. Dahlgreen, J. F. Franklin, P. F. Hessburg, and J. A. Lutz. 2013. Restoring forest resilience: From reference spatial patterns to silvicultural prescriptions and monitoring. *Forest Ecology and Management* 291:442–457.

Clark, J. S., L. Iverson, C. W. Woodall, C. D. Allen, D. M. Bell, D. C. Bragg, A. W. D'amato, F. W. Davis, M. H. Hersh, and I. Ibanez. 2016. The impacts of increasing drought on forest dynamics, structure, and biodiversity in the United States. *Global Change Biology* 22:2329–2352.

Clayoquot Scientific Panel. 1995. Scientific panel for sustainable forest practices in Clayoquot Sound, Report 5. *Sustainable ecosystem management in Clayoquot Sound: Planning and practices*, Victoria, BC, Canada. p. 296.

Clements, F. E. 1916. *Plant succession: An analysis of the development of vegetation*. Publication No. 242. Washington, DC: Carnegie Institution of Washington.

Cline, A. C., and S. H. Spurr. 1942. The virgin upland forest of central New England: A study of old growth stands in the Pisgah mountain section of southwestern New Hampshire. *Harvard Forest Bulletin* 21:58.

Clinton, B. D., L. R. Boring, and W. T. Swank. 1993. Canopy gap characteristics and drought influences in oak forests of the Coweeta Basin. *Ecology* 74:1551–1558.

Colburn, E. A. 2004. *Vernal pools: Natural history and conservation*. Granville, OH: The McDonald and Woodward Publishing Company.

Cole, D., S. Gessel, and J. Turner. 1978. Comparative mineral cycling in red alder and Douglas-fir. In *Utilization and Management of Alder*, eds. D. G. Briggs, D. S. DeBel, and W. A. Atkinson, 327–336. USDA Forest Service, General Technical Report PNW-70.

Cooper, C. F. 1960. Changes in vegetation, structure, and growth of southwestern pine forests since white settlement. *Ecological Monographs* 30:130–164.

Cooper, C. F. 1961. Pattern in ponderosa pine forests. *Ecology* 42:493–499.

Corace, R. G., III, P. C. Goebel, D. M. Hix, T. Casselman, and N. E. Seefelt. 2009. Ecological forestry at National Wildlife Refuges: Experiences from Seney National Wildlife Refuge and Kirtland's Warbler Wildlife Management Area, USA. *The Forestry Chronicle* 85:695-701.

Corace, R. G., III, L. M. Shartell, P. C. Goebel, D. M. Kashian, I. Drobyshev, and D. M. Hix. 2014. Bird communities of reference and altered mixed-pine forests: Implications for restoring fire-dependent ecosystems. Forest Ecology and Management 318:183-193.

Cornett, M. W., and M. A. White. 2013. Forest restoration in a changing world: Complexity and adaptation examples from the Great Lakes region of North America. In *Managing forests as complex adaptive systems: building resilience to the challenge of global change*, ed. C. Messier, K. J. Puettmann, and K. D. Coates, 113–133. New York: Routledge Press.

Covington, W. W., and J. D. Aber. 1980. Leaf production during secondary succession in northern hardwoods. *Ecology* 61:200–204.

Covington, W. W., and M. M. Moore. 1994. Southwestern ponderosa forest structure: Changes since Euro-American settlement. *Journal of Forestry* 92:39–47.

Crockett, J. L., and A. L. Westerling. 2018. Greater temperature and precipitation extremes intensify Western US droughts, wildfire severity, and Sierra Nevada tree mortality. *Journal of Climate* 31:341–354.

Cummins, K. W., M. A. Wilzbach, D. M. Gates, J. B. Perry, and W. B. Taliaferro. 1989. Shredders and riparian vegetation. *BioScience* 39:24–30.

Curtis, J. T. 1959. *The vegetation of Wisconsin*. Madison, WI: University of Wisconsin Press.

Curtis, R. O. 1995. *Extended rotations and culmination age of coastal Douglas-fir: old studies speak to current issues*. USDA Forest Service, Pacific Northwest Research Station, Research Paper PNW-485.

Curtis, R. O. 1997. The role of extended rotations. In *Creating a forestry for the 21st century*, ed. K. A. Kohm and Franklin, J. F., 165–170. Washington, DC: Island Press.

Curtis, R. O., D. S. DeBell, C. A. Harrington, D. P. Lavender, B. J. St. Clair, J. C. Tappeiner, and J. D. Walstad. 1998. *Silviculture for multiple objectives in the Douglas-fir region*. USDA Forest Service, General Technical Report PNW-GTR-435. 123 p.

Curtis, R. O., and D. D. Marshall. 1993. Douglas-fir rotations-time for reappraisal? *Western Journal of Applied Forestry* 8:81–85.

Dahir, S. E., and C. G. Lorimer. 1996. Variation in canopy gap formation among developmental stages of northern hardwood stands. *Canadian Journal of Forest Research* 26:1875–1892.

Dale, V. H., L. A. Joyce, S. McNulty, R. P. Neilson, M. P. Ayres, M. D. Flannigan, P. J. Hanson, L. C. Irland, A. E. Lugo, C. J. Peterson, and D. Simberloff. 2001. Climate change and forest disturbances: Climate change can affect forests by altering the frequency, intensity, duration, and timing of fire, drought, introduced species, insect and pathogen outbreaks, hurricanes, windstorms, ice storms, or landslides. *BioScience* 51:723–734.

Dale, V. H., F. J. Swanson, and C.M. Crisafulli, eds. 2005. *Ecological responses to the 1980 eruption of Mount St. Helens*. New York: Springer.

D'Amato, A. W., N. W. Bolton, C. Blinn, and A. R. Ek. 2009. *Current status and long-term trends of silvicultural practices in Minnesota: A 2008 assessment*. University of Minnesota, Department of Forest Resources Staff Paper Series No. 205:58.

D'Amato, A. W., J. B. Bradford, S. Fraver, and B. J. Palik. 2013. Effects of thinning on drought vulnerability and climate response in north temperate forest ecosystems. *Ecological Applications* 23:1735–1742.

D'Amato, A. W., and P. Catanzaro. 2007. *Restoring old-growth characteristics*. UMass Cooperative Extension Landowner Outreach Pamphlet. Amherst, MA: University of Massachusetts. 18 p.

D'Amato, A. W., P. F. Catanzaro, and L. S. Fletcher. 2015. Early regeneration and structural responses to patch selection and structural retention in second-growth northern hardwoods. *Forest Science* 61:183–189.

D'Amato, A. W., B. J. Palik, and C. C. Kern. 2010. Growth, yield, and structure of extended rotation *Pinus resinosa* stands in Minnesota, USA. *Canadian Journal of Forest Research* 40:1000–1010.

Davis, L. S., K. N. Johnson, P. Bettinger, and T. E. Howard. 2001. *Forest management: To sustain ecological, economic, and social values*. Long Grove, IL: Waveland Press.

Després, T., H. Asselin, F. Doyon, and Y. Bergeron. 2014. Structural and spatial characteristics of old-growth temperate deciduous forests at their northern distribution limit. *Forest Science* 60:871–880.

Dixon, R. K., S. Brown, R. A. Houghton, A. M. Solomon, M. C. Trexler, and J. Wisniewski. 1994. Carbon pools and flux of global forest ecosystems. *Science* 263:185–190.

Dodd, C. K., Jr., and B. S. Cade. 1998. Movement patterns and the conservation of amphibians breeding in small, temporary wetlands. *Conservation Biology* 12:331–339.

Donato, D. C., J. L. Campbell, and J. F. Franklin. 2012. Multiple successional pathways and precocity in forest development: Can some forests be born complex? *Journal of Vegetation Science* 23:576–584.

Donato, D. C., J. B. Fontaine, J. L. Campbell, W. B. Robinson, J. B. Kauffman, and B. E. Law. 2006. Post-wildfire logging hinders regeneration and increases fire risk. *Science* 311:352–352.

Donoso, P. J., and R. D. Nyland. 2006. Interference to hardwood regeneration in northeastern North America: The effects of raspberries (*Rubus* spp.) following clearcutting and shelterwood methods. *Northern Journal of Applied Forestry* 23:288–296.

Drobyshev, I., P. C. Goebel, D. M. Hix, R. G. Corace, and M. E. Semko-Duncan. 2008a. Pre-and post-European settlement fire history of red pine dominated forest ecosystems of Seney National

Wildlife Refuge, Upper Michigan. *Canadian Journal of Forest Research* 38:2497–2514.

Drobyshev, I., P. C. Goebel, D. M. Hix, R. G. Corace, and M. E. Semko-Duncan. 2008b. Interactions among forest composition, structure, fuel loadings and fire history: A case study of red pine-dominated forests of Seney National Wildlife Refuge, Upper Michigan. *Forest Ecology and Management* 256:1723–1733.

Ducey, M., J. Gunn, and A. Whitman. 2013. Late-successional and old-growth forests in the northeastern United States: Structure, dynamics, and prospects for restoration. *Forests* 4:1055.

Dunaway, D., S. R. Swanson, J. Wendel, and W. Clary. 1994. The effect of herbaceous plant communities and soil textures on particle erosion of alluvial streambanks. *Geomorphology* 9:47–56.

Dunning, D. 1928. A tree classification for the selection forests of the Sierra Nevada. *Journal of Agricultural Research* 36:755–771.

Dyck, B. 2003. Precision forestry: The path to increased profitability! In *Precision forestry: Proceedings of the second international precision forestry symposium*, 3–8. University of Washington, College of Forest Resources.

Eberhart, K. E., and P. M. Woodard. 1987. Distribution of residual vegetation associated with large fires in Alberta. *Canadian Journal of Forest Research* 117:1207–1212.

ECOMAP. 1993. "National hierarchical framework of ecological units." Unpublished administrative paper, USDA Forest Service.

Edwards, D. P., T. H. Larsen, T. D. S. Docherty, F. A. Ansell, W. W. Hsu, M. A. Derhe, K. C. Hamer, and D. S. Wilcove. 2011. Degraded lands worth protecting: The biological importance of Southeast Asia's repeatedly logged forests. *Proceedings of the Royal Society* B 278:82–90.

Edwards, R. T. 1998. The Hyporheic zone. In *River ecology and management: Lessons from the pacific coastal ecoregion*, ed. R. J. Naiman and R. E. Bilby, 399–429. New York: Springer-Verlag.

Ellison, A. M., M. S. Bank, B. D. Clinton, E. A. Colburn, K. Elliott, C. R. Ford, D. R. Foster, B. D. Kloeppel, J. D. Knoepp, G. M. Lovett, J. Mohan, D. A. Orwig, N. L. Rodenhouse, W. V. Sobczak, K. A. Stinson, J. K. Stone, C. M. Swan, J. Thompson, B. Von Holle, and J. R. Webster. 2005. Loss of foundation species: Consequences for the structure and dynamics of forested ecosystems. *Frontiers in Ecology and the Environment* 3(9):479–486.

Emmingham, W. H., P. T. Oester, S. A. Fitzgerald, G. M. Filip, and W. D. Edge. 2005. *Ecology and management of eastern Oregon forests: A comprehensive manual for forest managers*. Corvallis, OR: Oregon State University Extension Service.

Erickson, M. D., D. D. Reed, and G. D. Mroz. 1990. Stand development and economic analysis of alternative cutting methods in northern hardwoods: 32-year results. *Northern Journal of Applied Forestry* 7:153–158.

Eyre, F. H., and P. Zehngraff. 1948. *Red pine management in Minnesota* (Circular 778). United States Department of Agriculture.

Eyre, F. H., and W. M. Zillgitt. 1953. Partial cuttings in northern hardwoods of the Lake States: Twenty-year results. *USDA Forest Service Lake States Forest Experiment Station Technical Bulletin* 1076:124.

Fahey, R. T., and C. G. Lorimer. 2014. Persistence of pine species in late-successional forests: Evidence from habitat-related variation in stand age structure. *Journal of Vegetation Science* 25:584–600.

Fajardo, A., J. M. Graham, J. M. Goodburn, and C. E. Fiedler. 2007. Ten-year responses of ponderosa pine growth, vigor, and recruitment to restoration treatments in the Bitterroot Mountains, Montana, USA. *Forest Ecology and Management* 243:50–60.

Fajvan, M. A., and R. S. Seymour. 1993. Canopy stratification, age structure, and development of multicohort stands of eastern white pine, eastern hemlock, and red spruce. *Canadian Journal of Forest Research* 23:1799–1809.

Farrell, M. L., and B. F. Chabot. 2012. Assessing the growth potential and economic impact of the U.S. maple syrup industry. *Journal of Agriculture, Food Systems, and Community Development* 2:11–27.

Fassnacht, K. S., D. R. Bronson, B. J. Palik, A. W. D'Amato, C. G. Lorimer, and K. J. Martin. 2014. *Accelerating the development of old-growth characteristics through active management: Applications to northern hardwood forests in the Lake States*. USDA Forest Service, General Technical Report NRS-144.

Fassnacht, K. S., D. R. Bronson, B. J. Palik, A. W. D'Amato, C. G. Lorimer, and K. J. Martin. 2015. *Accelerating the development of old-growth characteristics in second-growth northern hardwoods*. USDA Forest Service, General Technical Report NRS-144. Newtown Square, PA. 33 p.

Fassnacht, K. S., E. A. Padley, T. W. Steele, C. G. Lorimer, B. J. Palik, A. W. D'Amato, and K. J. Martin. 2013. *Development and implementation of treatments in a long-term experiment to enhance forest structural and compositional complexity in second-growth northern hardwoods forests*. Technical Bulletin 198.

Madison, WI: Wisconsin Department of Natural Resources.

Fedrowitz, K., J. Koricheva, S. C. Baker, D. B. Lindenmayer, B. Palik, R. Rosenvald, W. Beese, J. F. Franklin, J. Kouki, E. Macdonald, C. Messier, A. Sverdrup-Thygeson, and L. Gustafsson. 2014. Can retention forestry help conserve biodiversity? A meta-analysis. *Journal of Applied Ecology* 51:1669–1679.

FEMAT: Forest Ecosystem Management Assessment Team. 1993. *Forest ecosystem management: An ecological, economic, and social assessment.* Washington, DC: US Government Printing Office.

Fetherston, K. L., R. J. Naiman, and R. E. Bilby. 1995. Large woody debris, physical process, and riparian forest development in montane river networks of the Pacific Northwest. *Geomorphology* 13:133–144.

Fettig C. J., K. D. Klepzig, R. F. Billings, A. S. Munson, T. E. Nebeker, J. F. Negrón, and J. T. Nowak. 2007. The effectiveness of vegetation management practices for prevention and control of bark beetle infestations in coniferous forests of the western and southern United States. *Forest Ecology and Management* 238: 24–53.

Fiebig, D. M., and M. A. Lock. 1991. Immobilization of dissolved organic matter from groundwater discharging through the stream bed. *Freshwater Biology* 26:45–55.

Fiedler, C. E., and S. F. Arno. 2015. *Ponderosa: People, fire, and the West's most iconic tree.* Missoula, MT: Mountain Press.

Fiedler, C. E., R. R. Becker, and S. A. Haglund. 1988. Preliminary guidelines for uneven-aged silvicultural prescriptions in ponderosa pine. In *Ponderosa pine: The species and its management*, eds. D. M. Baumgartner and J. E. Lotan, 235–241. Pullman, WA: Washington State University.

Finney, M. A., R. Bartlette, L. Bradshaw, K. Close, B. M. Collins, P. Gleason, W. M. Hao, P. Langowski, J. McGinely, C. W. McHugh, E. Martinson, P. N. Omi, W. Shepperd, and K. Zeller. 2003. Fire behavior, fuel treatments, and fire suppression on the Hayman Fire. In *Hayman fire case study*, ed. R. T. Graham, 33–35. USDA Forest Service, General Technical Report RMRS-144.

Fischer, R. A., and N. R. Holler. 1991. Habitat use and relative abundance of gray squirrels in southern Alabama. *The Journal of Wildlife Management* 55:52–59.

Fisher, S. G., and G. E. Likens. 1973. Energy flow in Bear Brook, New Hampshire: An integrative approach to stream ecosystem metabolism. *Ecological Monographs* 43:421–439.

Flannigan, M. D., M. A. Krawchuk, W. J. de Groot, B. M. Wotton, and L. M. Gowman. 2009. Implications of changing climate for global wildland fire. *International Journal of Wildland Fire* 18:483–507.

Food and Agriculture Organization of the United Nations. 2015. *Global forest resources assessment 2015: How are the world's forests changing?* Rome, Italy: FAO.

Forest Stewardship Council. 2009. "Our vision and mission." https://us.fsc.org/mission-and-vision.187.htm.

Forest Stewardship Council. 2010. "FSC-US forest management standard (v1.0)." https://us.fsc.org/forest-management-certification.225.htm.

Forman, R. T. 1995. Some general principles of landscape and regional ecology. *Landscape Ecology* 10:133–142.

Foster, D. R., J. D. Aber, J. M. Melillo, R. Bowden, and F. Bazzaz. 1997. Forest response to disturbance and anthropogenic stress: Rethinking the 1938 Hurricane and the impact of physical disturbance vs. chemical and climate stress on forest ecosystems. *BioScience* 47:437–445.

Foster, D. R., and E. R. Boose. 1992. Patterns of forest damage resulting from catastrophic wind in central New England, USA. *Journal of Ecology* 80:79–98.

Foster, D. R., K. F. Lambert, D. B. Kittredge, B. M. Donahue, C. M. Hart, W. G. Labich, S. Meyer, J. Thompson, M. Buchanan, J. N. Levitt, R. Pershel, K. Ross, G. Elkins, C. Daigle, B. Hall, E. K. Faison, A. W. D'Amato, R. T. T. Forman, P. Del Tredici, L. C. Irland, B. A. Colburn, D. A. Orwig, J. D. Aber, A. Berger, C. T. Driscoll, W. S. Keeton, R. J. Lilieholm, N. Pederson, A. M. Ellison, M. L. Hunter, and T. J. Fahey. 2017. *Wildlands and woodlands, farmlands and communities: Broadening the vision for New England.* Petersham, MA: Harvard Forest.

Fox, T. R., E. J. Jokela, and H. L. Allen. 2007. The development of pine plantation silviculture in the southern United States. *Journal of Forestry* 105:337-347.

Francl, K. E. 2008. Summer bat activity at woodland seasonal pools in the northern Great Lakes region. *Wetlands* 28:117–124.

Franklin, J. F. 1963. *Natural regeneration of Douglas-fir and associated species using modified clear-cutting systems in the Oregon Cascades.* USDA Forest Service, Research Paper PNW-3, 14 p.

Franklin, J. F. 1989. Toward a new forestry. *American Forests* 95:37–44.

Franklin, J. F., and J. K. Agee. 2003. Forging a science-based national forest fire policy. *Issues in Science and Technology* 20:59–66.

Franklin, J. F., D. R. Berg, D. A. Thornburgh, and J. C. Tappeiner. 1997. Alternative silvicultural approaches to timber harvesting: Variable retention harvest systems. In *Creating a forestry for the 21st century*, ed. K. A. Kohm and J. F. Franklin, 111–139. Washington, DC: Island Press.

Franklin, J. F., and D. S. DeBell. 1988. Thirty-six years of tree population change in an old-growth *Pseudotsuga-Tsuga* forest. *Canadian Journal of Forest Research* 18:633–639.

Franklin, J. F., and C. T. Dyrness. 1973. *Natural vegetation of Oregon and Washington*. USDA Forest Service, General Technical Report PNW-8.

Franklin, J. F., and C. T. Dyrness. 1988. *Natural vegetation of Oregon and Washington*. Corvallis, OR: Oregon State University Press.

Franklin, J. F., and R. T. T. Forman. 1987. Creating landscape patterns by forest cutting: Ecological consequences and principles. *Landscape Ecology* 1:5–18.

Franklin, J. F., and K. N. Johnson. 2004. Forests face new threat: Global market changes. *Issues in Science and Technology* 20:41–48.

Franklin, J. F., and K. N. Johnson. 2012. A restoration framework for federal forests in the Pacific Northwest. *Journal of Forestry* 110:429–439.

Franklin, J. F., and K. N. Johnson. 2013. Ecologically based management: a future for federal forestry in the Pacific Northwest. *Journal of Forestry* 111:429–432.

Franklin, J. F., K. N. Johnson, D. J. Churchill, K. Hagmann, D. Johnson, and J. Johnston. 2013. *Restoration of dry forests in eastern Oregon: A field guide*. Portland, OR: The Nature Conservancy of Oregon.

Franklin, J. F., K. N. Johnson, and D. L. Johnson. 2018. *Ecological forest management*. Long Grove, IL: Waveland Press.

Franklin, J. F., and D. B. Lindenmayer. 2002. *Conserving forest biodiversity: A comprehensive multiscaled approach*. Island Press, Washington, D C.

Franklin, J. F., D. B. Lindenmayer, J. A. MacMahon, A. McKee, J. Magnusson, D. A. Perry, R. Waide, and D. R. Foster. 2000. Threads of continuity: Ecosystem disturbances, biological legacies and ecosystem recovery. *Conservation Biology in Practice* 1:8–16.

Franklin J. F., and J. A. MacMahon. 2000. Messages from a mountain. *Science* 288:1183–1185.

Franklin, J. F., R. J. Mitchell, and B. J. Palik. 2007. *Natural disturbance and stand development principles for ecological forestry.* USDA Forest Service, Northern Research Station, General Technical Report NRS-19.

Franklin, J. F., H. H. Shugart, and M. E. Harmon. 1987. Tree death as an ecological process. *BioScience* 37:550–556.

Franklin, J. F., T. A. Spies, R. V. Pelt, A. B. Carey, D. A. Thornburgh, D. R. Berg, D. B. Lindenmayer, M. E. Harmon, W. S. Keeton, D. C. Shaw, K. Bible, and J. Chen. 2002. Disturbances and structural development of natural forest ecosystems with silvicultural implications, using Douglas-fir forests as an example. *Forest Ecology and Management* 155:399–423.

Franklin, J. F., and M. E. Swanson. 2008. Forest landscape structure, degradation and condition: Some commentary and fundamental principles. In *Managing and designing landscapes for conservation: Moving from perspectives to principles*, ed. D. B. Lindenmayer and R. J. Hobbs, 131–145. Hoboken, NJ: John Wiley and Sons.

Franklin, J. F., and R. Van Pelt. 2004. Spatial aspects of structural complexity in old-growth forests. *Journal of Forestry* 102:22–28.

Fraver, S., and B. J. Palik. 2012. Stand and cohort structures of old-growth *Pinus resinosa*-dominated forests of northern Minnesota, USA. *Journal of Vegetation Science* 23:249–259.

Frelich, L. E. 1995. Old forest in the Lake States today and before European settlement. *Natural Areas Journal* 15:157–167.

Frelich, L. E. 2002. *Forest dynamics and disturbance regimes: Studies from temperate evergreen-deciduous forests*. Cambridge, UK: Cambridge University Press.

Frelich, L. E., and L. J. Graumlich. 1994. Age-class distribution and spatial patterns in an old-growth hemlock-hardwood forest. *Canadian Journal of Forest Research* 24:1939–1947.

Frelich, L. E., and C. G. Lorimer. 1991. Natural disturbance regimes in hemlock-hardwood forests of the upper Great Lakes region. *Ecological Monographs* 61:145–164.

Frelich, L. E., and P. B. Reich. 1995. Spatial patterns and succession in a Minnesota southern-boreal forest. *Ecological monographs* 65:325–346.

Freund, J. A., J. F. Franklin, A. J. Larson, and J. A. Lutz. 2014. Multi-decadal establishment for single-cohort Douglas-fir forests. *Canadian Journal of Forest Research* 44:1068–1078.

Fulé, P. Z., and W. W. Covington. 1997. Fire regimes and forest structure in the Sierra Madre Occidental, Durango, Mexico. *Acta Botánica Mexicana* 41:43–79.

Fulé, P. Z., J. E. Crouse, J. P. Roccaforte, and E. L. Kalies. 2012. Do thinning and/or burning treatments in western USA ponderosa or Jeffrey pine-dominated forests help restore natural fire behavior? *Forest Ecology and Management* 269:68–81.

Galiano, L., J. Martínez-Vilalta, and F. Lloret. 2011. Carbon reserves and canopy defoliation determine the recovery of Scots pine 4 years after a drought episode. *New Phytologist* 190:750–759.

Gandhi, K. J., and D. A. Herms. 2010. Direct and indirect effects of alien insect herbivores on ecological processes and interactions in forests of eastern North America. *Biological Invasions* 12:389–405.

Gayer, K. 1886. *Der gemischte Wald: seine begründung und Pflege, insbesondere durch Horst: Und Gruppenwirtschaft*. Berlin, Germany: Paul Parey.

Gilbert, A. M., and V. S. Jensen. 1958. *A management guide for northern hardwoods in New England*. USDA Forest Service, Northeastern Forest Experiment Station Paper NE-112. 22 p.

Gillson, L., T. P. Dawson, S. Jack, and M. A. McGeoch. 2013. Accommodating climate change contingencies in conservation strategy. *Trends in Ecology and Evolution* 28:135–142.

Gilmore, D. W., and B. Palik. 2006. *A revised manager's handbook for red pine in the north central region*. USDA Forest Service, General Technical Report NC-264.

Goebel, P. C., B. J. Palik, L. K. Kirkman, M. B. Drew, L. West, and D. C. Pederson. 2001. Forest ecosystems of a Lower Gulf Coastal Plain landscape: Multifactor classification and analysis. *Journal of the Torrey Botanical Society* 128:47–75.

Goff, F. G., and D. West. 1975. Canopy-understory interaction effects on forest population structure. *Forest Science* 21:98–108.

Goodburn, J. M., and C. G. Lorimer. 1998. Cavity trees and coarse woody debris in old-growth and managed northern hardwood forests in Wisconsin and Michigan. *Canadian Journal of Forest Research* 28:427–438.

Goodburn, J. M., and C. G. Lorimer. 1999. Population structure in old-growth and managed northern hardwoods: An examination of the balanced diameter distribution concept. *Forest Ecology and Management* 118:11–29.

Gordon, J., J. F. Franklin, K. N. Johnson, D. Patton, J. Sedell, J. Sessions, and E. Williston. 1993. *An assessment of Indian forests and forest management in the United States*. Report for the Indian Forest Management Team. Portland Oregon: Intertribal Timber Council.

Gordon J., J. Sessions, J. Bailey, D. Cleaves, V. Corrao, A. Leighton, L. Mason, M. Rasmussen, H. Salwasser, and M. Sterner. 2013. "Assessment of Indian forests and forest management in the United States." Report for the Intertribal Timber Council, Portland Oregon, USA. Accessed August 1, 2018. https://www.itcnet.org/issues_projects/issues_2/forest_management/assessment.html

Gore, J. A., and W. A. Patterson III. 1986. Mass of downed wood in northern hardwood forests in New Hampshire: Potential effects of forest management. *Canadian Journal of Forest Research* 16:335–339.

Gosselink, J. G., G. P. Shaffer, L. C. Lee, D. M. Burdick, D. L. Childers, N. C. Leibowitz, S. C. Hamilton, R. Boumans, D. Cushman, S. Fields, M. Koch, and J. M. Visser. 1990. Landscape conservation in a forested wetland watershed: Can we manage cumulative impacts? *BioScience* 40:588–600.

Graham, R. T., A. E. Harvey, T. B. Jain, and J. R. Tonn. 1999. *Effects of thinning and similar stand treatments on fire behavior in western forests*. USDA Forest Service, General Technical Report PNW-GTR-463.

Graham, R. T., and T. B. Jain. 2005. Overview of ponderosa pine ecosystems. In *Proceedings of the symposium on ponderosa pine: Issues, trends, and management*, ed. M. W. Ritchie, D. A. Maguire, and A. Youngblood, 1–33. USDA Forest Service, Forest General Technical Report PSW-198.

Graham, R. T., and T. B. Jain. 2005a. Application of free selection in mixed forests of the inland northwestern United States. *Forest Ecology and Management* 209:131–145.

Graham, S. A., R. P. Harrison Jr., and C. E. Westell Jr. 1963. *Aspens: Phoenix trees of the Great Lakes region*. Ann Arbor, MI: University of Michigan Press.

Granzow-de la Cerda, Í., F. Lloret, J. E. Ruiz, and J. H. Vandermeer. 2012. Tree mortality following ENSO-associated fires and drought in lowland rain forests of Eastern Nicaragua. *Forest Ecology and Management* 265:248–257.

Graves, A. T., M. A. Fajvan, and G. W. Miller. 2000. The effects of thinning intensity on snag and cavity tree abundance in an Appalachian hardwood stand. *Canadian Journal of Forest Research* 30:1214–1220.

Greenberg, C., B. Collins, and F. Thompson, III, eds. 2011. *Sustaining young forest communities: Ecology and management of early successional habitats in the central hardwood region, USA*. Berlin, Germany: Springer Science & Business Media.

Gregory, S. V. 1997. Riparian management in the 21st century. In *Creating a forestry for the 21st century: The science of ecosystem management*, ed. J. F. Franklin and K. A. Kohm, 69–85. Washington, DC: Island Press.

Gregory, S. V., *F. J.* Swanson, *W. A.* McKee, *and* K. W. Cummins. 1991. An ecosystem perspective of riparian zones. *BioScience* 41:540–551.

Gronewold, C. A., A. W. D'Amato, and B. J. Palik. 2010. The influence of cutting cycle and stocking level on the structure and composition of managed old-growth northern hardwoods. *Forest Ecology and Management* 259:1151–1160.

Grubaugh, J. W., and R. V. Anderson. 1989. Upper Mississippi River: Seasonal and floodplain forest influences on organic matter transport. *Hydrobiologia* 174:235–244.

Gustafsson, L., S. C. Baker, J. Bauhus, W. J. Beese, A. Brodie, J. Kouki, D. B. Lindenmayer, A. Lõhmus, G. M. Pastur, C. Messier, M. Neyland, B. Palik, A. Sverdrup-Thygeson, W. J. A. Volney, A. Wayne, and J. F. Franklin. 2012. Retention forestry to maintain multifunctional forests: A world perspective. *Bioscience* 62:633–664.

Guyette, R., T. Gallagher, B. Palik, D. Dey, and M. Stambaugh. 2016. "Early fire history at the Cutfoot Experimental Forest." Unpublished report, Missouri Tree-Ring Laboratory.

Haffey, C., T. D. Sisk, C. D. Allen, A. E. Thode, and E. Q. Margolis. 2018. Limits to ponderosa pine regeneration following large high-severity forest fires in the United States Southwest. *Fire Ecology* 14.

Hagan, J. M., and A. A. Whitman. 2004. Late successional forest: A disappearing age class and implications for biodiversity. *Forest Mosaic Science Notes* 2(4).

Hagenbuch, S., K. Manaras, J. Shallow, K. Sharpless, and M. Snyder. 2011. *Silviculture with birds in mind: Options for integrating timber and songbird habitat management in northern hardwood stands in Vermont.* Montpelier, VT: Audubon Vermont and Vermont Forests, Parks, and Recreation, 24 p.

Hagmann, K., J. F. Franklin, and K. N. Johnson. 2013. Historical structure and composition of ponderosa pine and mixed-conifer forests in south-central Oregon. *Forest Ecology and Management* 304:492–504.

Hagmann, R. K., J. F. Franklin, and K. N. Johnson. 2014. Historical conditions in mixed-conifer forests on the eastern slopes of the northern Oregon Cascade Range, USA. *Forest Ecology and Management* 330:158–170.

Hale, C. M., J. Pastor, and K. A. Rusterholz. 1999. Comparison of structural and compositional characteristics in old-growth and mature, managed hardwood forests of Minnesota, U.S.A. *Canadian Journal of Forest Research* 29:1479–1489.

Hallin, W. E. 1954. Unit area control: Its development and application. USDA Forest Service, California Forest and Range Experimental Station Miscellaneous Paper 16. 10 p.

Halpern, C. B., and T. A. Spies. 1995. Plant species diversity in natural and managed forests of the Pacific Northwest. *Ecological Applications* 5:913–934.

Halpin, C. R. 2014. "Long-term stand development and demographic sustainability of tree populations in northern hardwood forests." PhD diss., University of Wisconsin-Madison.

Handler, S., M. J. Duveneck, L. Iverson, E. Peters, R. M. Scheller, K. R. Wythers, L. Brandt, P. Butler, M. Janowiak, P. D. Shannon, C. Swanston, K. Barrett, R. Kolka, C. McQuiston, B. Palik, P. B. Reich, C. Turner, M. White, C. Adams, A. D'Amato, S. Hagell, P. Johnson, R. Johnson, M. Larson, S. Matthews, R. Montgomery, S. Olson, M. Peters, A. Prasad, J. Rajala, J. Daley, M. Davenport, M. R. Emery, D. Fehringer, C. L. Hoving, G. Johnson, L. Johnson, D. Neitzel, A. Rissman, C. Rittenhouse, and R. Ziel. 2014. *Minnesota forest ecosystem vulnerability assessment and synthesis: A report from the Northwoods Climate Change Response Framework project.* USDA Forest Service, General Technical Report NRS-133. Newtown Square, PA, USA. 228 p.

Hanewinkel, M., T. Kuhn, H. Bugmann, A. Lanz, and P. Brang. 2014. Vulnerability of uneven-aged forests to storm damage. *Forestry* 87:525–534.

Hann, S. W., and B. B. Bare. 1979. *Uneven-aged forest management: State of the art (or science?).* USDA Forest Service, General Technical Report. INT-50.

Hanowski, J., N. Danz, and J. Lind. 2006. Response of breeding bird communities to forest harvest around seasonal ponds in northern forests, USA. *Forest Ecology and Management* 229:63–72.

Hansen, A. J., T. A. Spies, F. J. Swanson, and J. L. Ohmann. 1991. Conserving biodiversity in managed forests. *BioScience* 41:382–392.

Hanson, J. J., and C. G. Lorimer. 2007. Forest structure and light regimes following moderate wind storms: Implications for multi-cohort management. *Ecological Applications* 17:1325–1340.

Hanson, J. J., C. G. Lorimer, C. R. Halpin, and B. J. Palik. 2012. Ecological forestry in an uneven-aged, late-successional forest: Simulated effects of contrasting treatments on structure and yield. *Forest Ecology and Management* 270:94–107.

Hanson, M. A., S. E. Bowe, F. G. Ossman, J. Fieberg, M. G. Butler, and R. Koch. 2009. Influences of forest harvest and environmental gradients on aquatic invertebrate communities of seasonal ponds. *Wetlands* 29:884–895.

Hanson, M. A., B. J. Palik, J. O. Church, and A. T. Miller. 2010. Influences of upland timber harvest on aquatic invertebrate communities in seasonal ponds: Efficacy of forested buffers. *Wetlands Ecology and Management* 18:255–267.

Harmon, M. E. 2001. Moving towards a new paradigm for woody detritus management. *Ecological Bulletins* 49:269–278.

Harmon, M. E., J. F. Franklin, F. J. Swanson, P. Sollins, S. V. Gregory, J. D. Lattin, N. H. Anderson, S. P. Cline, N. G. Aumen, J. R. Sedell, and G. W. Lienkaemper. 1986. Ecology of coarse woody debris in temperate ecosystems. *Advances in Ecological Research* 15:133–302.

Harmon, M. E., J. F. Franklin, F. J. Swanson, P. Sollins, S. V. Gregory, J. D. Lattin, N. H. Anderson, S. P. Cline, N. G. Aumen, J. R. Sedell, G. W. Lienkaemper, K. Cromack Jr., and K. W. Cummins. 2004. Ecology of coarse woody debris in temperate ecosystems. *Advances in Ecological Research* 34:59–234.

Harmon, M. E., A. Moreno, and J. B. Domingo. 2009. Effects of partial harvest on the carbon stores in Douglas-fir/western hemlock forests: A simulation study. *Ecosystems* 12:777–791.

Harrington, C. A., S. D. Roberts, and L. C. Brodie. 2005. Tree and understory responses to variable-density thinning in western Washington. In *Balancing ecosystem values: innovative experiments for sustainable forestry*, eds. C. E. Peterson and D. A. Maguire, 97–106. USDA Forest Service, General Technical Report PNW-GTR-635.

Harvey, B. J., D. C. Donato, and M. G. Turner. 2016. Drivers and trends in landscape patterns of stand-replacing fire in forests of the US Northern Rocky Mountains (1984–2010). *Landscape ecology* 31:2367–2383.

Hatcher, W., S. Rondeau, D. L. Johnson, K. N. Johnson, and J. F. Franklin. 2017. Klamath Tribes: Managing their homeland forests in partnership with the USDA Forest Service. *Journal of Forestry* 115:447–455.

Hawley, R. C. 1921. *The practice of silviculture*. New York: John Wiley & Sons, Inc.

Hendricks, J. J., C. A. Wilson, and L. R. Boring. 2002. Foliar litter position and decomposition in a fire-maintained longleaf pine wiregrass ecosystem. *Canadian Journal of Forest Research* 32:928–941.

Heinselman, M. L. 1981. Fire intensity and frequency as factors in the distribution and structure of northern ecosystems. In *Proceedings of fire regimes and ecosystem properties*, 7–57. USDA Forest Service, General Technical Report WO-26.

Heinselman, M. L., 1981a. Fire and succession in the conifer forests of northern North America. In *Forest succession*, ed. D. C. West, H. H. Shugart, and D. B. Botkin, 374–405. New York: Springer.

Heinselman, M. L. 1996. *The Boundary Waters wilderness ecosystem*. Minneapolis, MN: University of Minnesota Press.

Helfield, J. M., and R. J. Naiman. 2001. Effects of salmon-derived nitrogen on riparian forest growth and implications for stream productivity. *Ecology* 82:2403–2409.

Helms, J. A. 1998. *Dictionary of forestry*. Cambridge, MA: Society of American Foresters and CABI Publishing. 210 p.

Henry, J. D., and J. M. A. Swan. 1974. Reconstructing forest history from live and dead plant material—an approach to the study of forest succession in southwest New Hampshire. *Ecology* 55:772–783.

Herms, D. A., and D. G. McCullough. 2014. Emerald ash borer invasion of North America: History, biology, ecology, impacts, and management. *Annual Review of Entomology* 59:13–30.

Hessburg, P. F., and J. K. Agee. 2003. An environmental narrative of inland Northwest United States forests, 1800-2000. *Forest Ecology and Management* 178:23–59.

Hessburg, P. F., D. J. Churchill, A. J. Larson, R. D. Haugo, C. Miller, T. A. Spies, M. P. North, N. A. Povak, R. T. Belote, P. H. Singleton, W. L. Gaines, R. E. Keane, G. H. Aplet, S. L. Stephens, P. Morgan, P. A. Bisson, B. E. Rieman, R. B. Salter, and G. H. Reeves. 2015. Restoring fire-prone landscapes: Seven core principles. *Landscape Ecology* 30:1805–1835.

Hessburg, P. F., R. B. Salter, and K. M. James. 2007. Re-examining fire severity relations in pre-management era mixed conifer forests: Inferences from landscape patterns of forest structure. *Landscape Ecology* 22:5–24.

Hessburg, P. F., T. A. Spies, D. A. Perry, C. N. Skinner, A. H. Taylor, P. M. Brown, S. L. Stephens, A. J. Larson, D. J. Churchill, N. A. Povak, et al. 2016. Tamm Review: Management of mixed-severity fire regime forests in Oregon, Washington, and Northern California. *Forest Ecology and Management* 366:221–250.

Heyerdahl, E. K., L. B. Brubaker, and J. K. Agee. 2002. Annual and decadal climate forcing of historical fire regimes in the interior Pacific Northwest. *The Holocene* 12:597–604.

Heyerdahl, E. K., D. McKenzie, L. D. Daniels, A. E. Hessl, J. S. Littell, and N. J. Mantua. 2008. Climate drivers of regionally synchronous fires in the inland Northwest (1651–1900). *International Journal of Wildland Fire* 17:40–49.

HFRA. 2003. *Healthy Forest Restoration Act of 2003*. Public Law 108-148, Statutes at Large 117: 1887.

Hickey, J., M. Neyland, A. Rothe, and J. Bauhus. 2015. Is continuous-cover silviculture, as practiced in

Bavaria, suitable for use in wet eucalypt forests in Tasmania, Australia? *Australian Forestry* 78:29–44.

Holling, C. S., and G. K. Meffe. 1996. Command and control and the pathology of natural resource management. *Conservation Biology* 10:328–337.

Holm, S. O. 2015. A management strategy for multiple ecosystem services in boreal forests. *Journal of Sustainable Forestry* 34:358–379.

Holmes, S. B., and D. G. Pitt. 2007. Response of bird communities to selection harvesting in a northern tolerant hardwood forest. *Forest Ecology and Management* 238:280–292.

Horsley, S. B., R. P. Long, S. W. Bailey, R. A. Hallett, and T. J. Hall. 2000. Factors associated with the decline and disease of sugar maple on the Allegheny Plateau. *Canadian Journal of Forest Research* 30:1365–1378.

Howard, J., and R. Westby. 2013. *US timber production, trade, consumption and price statistics 1965–2011*. USDA Forest Service, Research Paper FPL-RP–676.

Hunter, M. L. 2005. A mesofilter conservation strategy to complement fine and coarse filters. *Conservation Biology* 19:1025–1029.

Hunter, M. L., and F. K. Schmiegelow. 2011. *Wildlife, forests, and forestry: Principles of managing forests for biological diversity*. 2nd ed. Saddle River, NJ: Prentice-Hall.

Hutnick, R. J. 1952. Reproduction on windfalls in a northern hardwood stand. *Journal of Forestry* 50:693–694.

Ilhardt, B., E. S. Verry, and B. Palik. 1999. Defining riparian areas. In *Riparian management in riparian forests of the continental eastern United States*, ed. E. S. Verry, J. W. Hornbeck, and C. A. Dolloff, 23–42. Boca Raton, FL: Lewis Press.

IPCC. 2013. Climate change 2013: The physical science basis. Contribution of Working Group I to the Fifth Assessment Report of the Intergovernmental Panel on Climate Change, eds. T. F. Stocker, D. Qin, G.-K. Plattner, M. Tignor, S. K. Allen, J. Boschung, A. Nauels, Y. Xia, V. Bex and P. M. Midgley. Cambridge, United Kingdom and New York, NY: Cambridge University Press.

IPCC. 2014. Climate change 2014: Synthesis report. Contribution of Working Groups I, II, and III to the Fifth Assessment Report of the Intergovernmental Panel on Climate Change, eds. R.K. Pachauri and L.A. Meyer, 151. Geneva, Switzerland: IPCC.

Isaac, L. A. 1943. *Reproductive habits of Douglas-fir*. Washington, DC: Charles Lathrop Pack Forestry Foundation.

Jackson, K. T. 1985. *Crabgrass frontier: The suburbanization of the United States*. New York: Oxford University Press.

Jacobs, D. F., M. F. Selig, and L. R. Severeid. 2009. Aboveground carbon biomass of plantation-grown American chestnut (*Castanea dentata*) in absence of blight. *Forest Ecology and Management* 258:288–294.

Jacqmain, E. I., R. H. Jones, and R. J. Mitchell. 1999. Influences of frequent cool-season burning across a soil moisture gradient on oak community structure in longleaf pine ecosystems. *American Midland Naturalist* 141:85–100.

Jain, T. B., M. A. Battaglia, H.-S. Han, R. T. Graham, C. R. Keyes, J. S. Fried, and J. E. Sandquist. 2012. *A comprehensive guide to fuel management practices for dry mixed conifer forests in the northwestern United States*. USDA Forest Service, General Technical Report RMRS-292.

James, F. C., P. M. Richards, C. A. Hess, K. E. McCluney, E. L. Walters, and M. S. Schrader. 2004. Sustainable forestry for the red-cockaded woodpecker's ecosystem. In *Red-cockaded woodpecker: Road to recovery*, ed. R. Costa and S. J. Daniels, 60–69. Blaine, WA: Hancock House.

Janowiak, M. K., A. W. D'Amato, C. W. Swanston, L. Iverson, F. Thompson. III, W. Dijak, S. Matthews, M. Peters, A. Prasad, J. S. Fraser, L. A. Brandt, P. R. Butler, S. D. Handler, P. D. Shannon, D. Burbank, J. Campbell, C. Cogbill, M. J. Duveneck, M. Emery, N. Fisichelli, J. Foster, J. Hushaw, L. Kenefic, A. Mahaffey, T. L. Morelli, N. Reo, P. Schaberg, K. R. Simmons, A. Weiskittel, S. Wilmot, D. Hollinger, E. Lane, L. Rustad, and P. Templer. 2017. *New England and New York forest ecosystem vulnerability assessment and synthesis: A report from the New England Climate Change Response Framework*. USDA Forest Service General Technical Report NRS-XX.

Janowiak, M. K., L. M. Nagel, and C. R. Webster. 2008. Spatial scale and stand structure in northern hardwood forests: Implications for quantifying diameter distributions. *Forest Science* 54:497–506.

Johnson, B. L., W. B. Richardson, and T. J. Naimo. 1995. Past, present, and future concepts in large river ecology. *BioScience* 45:134–141.

Johnson, C. E., R. B. Romanowicz, and T. G. Siccama. 1997. Conservation of exchangeable cations after clear-cutting of a northern hardwood forest. *Canadian Journal of Forest Research* 27:859–868.

Johnson, K. N., and J. F. Franklin. 2008. *A plan for the Klamath Tribes' management of the Klamath Reservation forest*. Klamath Indian Nation.

Johnson, K. N., J. F. Franklin, and D. L. Johnson. 2003. *A plan for the Klamath Tribe's management of the Klamath Reservation forest*. Report 184. Corvalis, OR: Oregon State University. 120 p.

Johnson, M. C., D. L. Peterson, and C. L. Raymond. 2007. *Guide to fuel treatments in dry forests of the Western United States: Assessing forest structure and fire hazard.* USDA Forest Service, General Technical Report PNW-686.

Johnson, P. S., S. R. Shifley, and R. Rogers. 2009. *The ecology and silviculture of oaks.* New York: CABI Publishing.

Johnstone, J. F., C. D. Allen, J. F. Franklin, L. E. Frelich, B. J. Harvey, P. E. Higuera, M. C. Mack, R. K. Meentemeyer, M. R. Metz, and G. L. Perry. 2016. Changing disturbance regimes, ecological memory, and forest resilience. *Frontiers in Ecology and the Environment* 14:369–378.

Jones, C. N., D. L. McLaughlin, K. Henson, C. A. Haas, and D. A. Kaplan. 2018. From salamanders to greenhouse gases: Does upland management affect wetland functions? *Frontiers in Ecology and the Environment* 16:14–19.

Jönsson, M. T., S. Fraver, B. G. Jonsson, M. Dynesius, M. Rydgård, P.-A. Esseen. 2007. Eighteen years of tree mortality and structural change in an experimentally fragmented Norway spruce forest. *Forest Ecology and Management* 242, 306–313.

Joyce, D. G., and G. E. Rehfeldt. 2013. Climatic niche, ecological genetics, and impact of climate change on eastern white pine (*Pinus strobus* L.): Guidelines for land managers. *Forest Ecology and Management* 295:173–192.

Jürgensen, C., W. Kollert, and A. Lebedys. 2014. *Assessment of industrial roundwood production from planted forests.* Rome, Italy: Food and Agriculture Organization of the United Nations.

Kabrick, J. M., D. C. Dey, and D. Gwaze. 2007. *Shortleaf pine restoration and ecology in the Ozarks: Proceedings of a symposium.* USDA Forest Service, General Technical Report NRS-15.

Kaufmann, M. R., D. Binkley, P. Z. Fule, M. Johnson, S. L. Stephens, and T. W. Swetnam. 2007. Defining old growth for fire-adapted forests of the Western United States. *Ecology and Society* 12.

Kaya, J., and J. Buongiorno. 1989. A harvesting guide for uneven-aged northern hardwood stands. *Northern Journal of Applied Forestry* 6:9–14.

Keane, R. E., M. F. Mahalovich, B. L. Bollenbacher, M. E. Manning, R. A. Loehman, T. B. Jain, L. M. Holsinger, and A. J. Larson. 2018. Effects of climate change on forest vegetation in the Northern Rockies. In *Climate change and Rocky Mountain ecosystems,* 59–95. New York: Springer.

Keen, F. P. 1943. Ponderosa pine tree classes redefined. *Journal of Forestry* 41:249–253.

Keeton, W. S. 2005. Managing for old-growth structure in northern hardwood forests. In *Balancing ecosystem values: Innovative experiments for sustainable forestry,* eds. C. E. Peterson and D. A. Maguire, 107–118. USDA Forest Service, General Technical Report PNW-GTR-635.

Kelty, M. J., and A. W. D'Amato. 2006. Historical perspective on diameter-limit cutting in northeastern forests. In *Proceedings of the conference on diameter-limit cutting in northeastern forests,* eds. L. S. Kenefic and R. D. Nyland, 3–15. USDA Forest Service General Technical Report NE-342.

Kenderes, K., B. Mihók, and T. Standovár. 2008. Thirty years of gap dynamics in a central European beech forest reserve. *Forestry* 81:111–123.

Kenefic, L. S., and R. D. Nyland. 2007. Cavity trees, snags, and selection cutting: A northern hardwood case study. *Northern Journal of Applied Forestry* 24:192–196.

Kerhoulas, L. P., T. E. Kolb, M. D. Huteau, and G. W. Koch. 2013. Managing climate change adaptation in forests: A case study from the US Southwest. *Journal of Applied Ecology* 50:1311–1320.

Kern, C. C., P. B. Reich, R. A. Montgomery, and T. F. Strong. 2012. Do deer and shrubs override canopy gap size effects on growth and survival of yellow birch, northern red oak, eastern white pine, and eastern hemlock seedlings? *Forest Ecology and Management* 267: 134–143.

Kern, C. C., A. W. D'Amato, and T. F. Strong. 2013. Diversifying the composition and structure of managed, late-successional forests with harvest gaps: what is the optimal gap size? *Forest Ecology and Management* 304: 110–120.

Kern, C. C., J. I. Burton, P. Raymond, A. W. D'Amato, W. S. Keeton, A. A. Royo, M. B. Walters, C. R. Webster, and J. L. Willis. 2017. Challenges facing gap-based silviculture and possible solutions for mesic northern forests in North America. *Forestry: An International Journal of Forest Research* 90: 4–17.

Khakimulina, T., S. Fraver, and I. Drobyshev. 2016. Mixed-severity natural disturbance regime dominates in an old-growth Norway spruce forest of northwest Russia. *Journal of Vegetation Science* 27:400–413.

Kirkman, L. K, K. L. Coffey, R. J. Mitchell, and B. Moser. 2004. Ground cover recovery patterns and life-history traits: Implications for restoration obstacles and opportunities in a species rich savanna. *Journal of Ecology* 92:409–421.

Kirkman, L. K., and S. B. Jack. 2017. *Ecological restoration and management of longleaf pine forests.* Boca Raton, FL: CRC Press.

Kirkman, L. K., R. J. Mitchell, M. J. Kaeser, S. D. Pecot, and K. L. Coffey. 2007. The perpetual forest: Using undesirable species to bridge restoration. *Journal of Applied Ecology* 44:604–614.

Klos, R. J., G. G. Wang, W. L. Bauerle, and J. R. Rieck. 2009. Drought impact on forest growth and mortality in the southeast USA: An analysis using forest health and monitoring data. *Ecological Applications* 19:699–708.

Knapp, A. K., M. D. Smith, S. E. Hobbie, S. L. Collins, T. J. Fahey, G. J. A. Hansen, D. A. Landis, K. J. La Pierre, J. M. Melillo, T. R. Seastedt, G. R. Shaver, and J. R. Webster. 2012. Past, present, and future roles of long-term experiments in the LTER network. *BioScience* 62:377–389.

Kneeshaw, D. D., and Y. Bergeron. 1998. Canopy gap characteristics and tree replacement in the southeastern boreal forest. *Ecology* 79:782–795.

Knight, F. R. 1997. "A spatial analysis of a Sierra Nevada old-growth mixed-conifer forest." MS thesis, University of Washington.

Knutson, T. R., J. L. McBride, J. Chan, K. Emanuel, G. Holland, C. Landsea, I. Held, J. P. Kossin, A. K. Srivastava, and M. Sugi. 2010. Tropical cyclones and climate change. *Nature Geoscience* 3:157–163.

Kohler, M., J. Sohn, G. Nägele, and J. Bauhus. 2010. Can drought tolerance of Norway spruce (*Picea abies* (L.) Karst.) be increased through thinning? *European Journal of Forest Research* 129:1109–1118.

Kohm, K. A., and J. F. Franklin. 1997. *Creating forestry for the 21st century: The science of ecosystem management.* Washington, DC: Island Press.

Kolb, T. E., J. K. Agee, P. Z. Fule, N. G. McDowell, K. Pearson, A. Sala, and R. H. Waring. 2007. Perpetuating old ponderosa pine. *Forest Ecology and Management* 249:141–157.

Kolka, R. K., B. J. Palik, D. P. Tersteeg, and J. C. Bell. 2011. Effects of riparian buffers on hydrology of northern seasonal ponds. *Transactions of the ASABE* 54:2111–2116.

Kruckeberg, A. R. 1967. Ecotypic response to ultramafic soils by some plant species of northwestern United States. *Brittonia* 19:133–151.

Kuuluvainen, T. 2009. Forest management and biodiversity conservation based on natural ecosystem dynamics in northern Europe: The complexity challenge. *AMBIO: A Journal of the Human Environment* 38:309-315.

Kuuluvainen, T., J. Mäki, L. Karjalainen, and H. Lehtonen. 2002. Tree age distributions in old-growth forest sites in Vienansalo wilderness, eastern Fennoscandia. *Silva Fennica* 36:169–184.

Kuuluvainen, T., and J. Siitonen. 2013. Fennoscandian boreal forests as complex adaptive systems: Properties, management challenges and opportunities. In *Managing forests as complex adaptive systems: Building resilience to the challenge of global change*, ed. C. Messier, K. J. Puettman, and K. D. Coates, 244-268. Abingdon, UK: Routledge Press.

Lance, A. N., and M. Phinney. 2001. Bird responses to partial retention timber harvesting in central interior British Columbia. *Forest Ecology and Management* 142:267–280.

Landres, P. B., P. Morgan, and F. J. Swanson. 1999. Overview of the use of natural variability concepts in managing ecological systems. *Ecological Applications* 9:1179-1188.

Lang, K. D., L. A. Schulte, and G. R. Guntenspergen. 2009. Windthrow and salvage logging in an old-growth hemlock-northern hardwoods forest. *Forest Ecology and Management* 259:56–64.

Langston, N. 1995. *Forest dreams, forest nightmares: The paradox of old growth in the inland West.* Seattle, WA: University of Washington Press.

Larson, A. J., R. T. Belote, C. A. Cansler, S. A. Parks, and M. S. Dietz. 2013. Latent resilience in ponderosa pine forest: Effects of resumed frequent fire. *Ecological Applications* 23:1243–1249.

Larson, A. J., and D. Churchill. 2008. Spatial patterns of overstory trees in late-successional conifer forests. *Canadian Journal of Forest Research* 38:2814–2825.

Larson, A. J., and D. J. Churchill. 2012. Tree spatial patterns in fire-frequent forests of western North America, including mechanisms of pattern formation and implications for designing fuel reduction and restoration treatments. *Forest Ecology and Management* 267:74–92.

Larson, A. J., and J. F. Franklin. 2005. Patterns of conifer tree regeneration following an autumn wildfire event in the western Oregon Cascade Range, USA. *Forest Ecology and Management* 218:25–36.

Larson, A. J., J. A. Lutz, R. F. Gersonde, J. F. Franklin, and F. F. Hietpas. 2008. Potential site productivity influences the rate of forest structural development. *Ecological Applications* 18:899–910.

Latham, P., and J. Tappeiner. 2002. Response of old-growth conifers to reduction in stand density in western Oregon forests. *Tree Physiology* 22:137–146.

Leak, W. B. 1964. An expression of diameter distribution for unbalanced, uneven-aged stands and forests. *Forest Science* 10:39–50.

Leak, W. B. 1975. Age distribution in virgin red spruce and northern hardwoods. *Ecology* 56:1451–1454.

Leak, W. B. 1996. Long-term structural change in uneven-aged northern hardwoods. *Forest Science* 42:160–165.

Leak, W. B. 2002. *Origin of sigmoid diameter distributions.* USDA Forest Service, Research Paper NE-718.

Leak, W. B. 2003. Regeneration of patch harvests in even-aged northern hardwoods in New England. *Northern Journal of Applied Forestry* 20:188–189.

Leak, W. B., and S. M. Filip. 1977. 38 years of group selection in New England northern hardwoods. *Journal of Forestry* 75:641–643.

Leak, W. B., D. S. Solomon, and P. S. DeBald. 1987. *Silvicultural guide for northern hardwood types in the Northeast (revised)*. USDA Forest Service, Research Paper NE-603.

Leak, W. B., M. Yamasaki, and R. Holleran. 2014. *Silvicultural guide for northern hardwoods in the northeast*. USDA Forest Service, General Technical Report NRS-132. 46 p.

Leech, S. M., P. Lara Almuedo, and G. O'Neil. 2011. Assisted migration: Adapting forest management to a changing climate. *BC Journal of Ecosystems and Management* 12:18–34.

Leopold, A. 1949. *A Sand County almanac and sketches here and there*. New York: Oxford University Press.

Lindenmayer, D. B. 2009. Forest wildlife management and conservation. *Annals of the New York Academy of Sciences* 1162:284–310.

Lindenmayer, D. B., B. J. Burton, and J. F. Franklin. 2008. *Salvage logging and its ecological consequences*. Washington, DC: Island Press.

Lindenmayer D. B., and J. F. Franklin. 2002. *Conserving forest biodiversity: A comprehensive multiscaled approach*. Washington, DC: Island Press.

Lindenmayer, D. B., J. F. Franklin, and J. Fischer. 2006. General management principles and a checklist of strategies to guide forest biodiversity conservation. *Biological Conservation* 131:433–445.

Lindenmayer, D. B., J. F. Franklin, A. Lõhmus, S. C. Baker, J. Bauhus, W. Beese, A. Brodie, B. Kiehl, J. Kouki, G. Martínez Pastur, C. Messier, M. Neyland, B. Palik, A. Sverdrup-Thygeson, J. Volney, A. Wayne, and L. Gustafsson. 2012. A major shift to the retention approach for forestry can help resolve some global forest sustainability issues. *Conservation Letters* 5:421–431.

Lindenmayer, D. B., W. F. Laurance, and J. F. Franklin. 2012. Global decline in large old trees. *Science* 338:1305–1306.

Lindenmayer, D. B., and R. F. Noss. 2006. Salvage logging, ecosystem processes, and biodiversity conservation. *Conservation Biology* 20:949–958.

Littell, J. S., D. L. Peterson, K. L. Riley, Y. Liu, and C. H. Luce. 2016. A review of the relationships between drought and forest fire in the United States. *Global Change Biology* 22:2353–2369.

Long, J. N. 1985. A practical approach to density management. *Forestry Chronicle* 61:23–27.

Long, J. N. 1995. Using stand density index to regulate stocking in uneven-aged stands. In *Uneven-aged management: Opportunities, constraints and methodologies*, ed. K. L. O'Hara, 110-122. Montana Forest and Conservation Experiment Station Miscellaneous Publication No. 56.

Looney, C. E., A. W. D'Amato, B. J. Palik, and R. A. Slesak. 2015. Overstory treatment and planting season affect survival of replacement tree species in emerald ash borer threatened Fraxinus nigra forests in Minnesota, USA. *Canadian Journal of Forest Research* 45:1728–1738.

Looney, C. E., A. W. D'Amato, B. J. Palik, R. A. Slesak, and M. A. Slater. 2017. The response of Fraxinus nigra forest ground-layer vegetation to emulated emerald ash borer mortality and management strategies in northern Minnesota, USA. *Forest Ecology and Management* 389:352–363.

Loope, W. L., and J. B. Anderton. 1998. Human vs. lightning ignition of presettlement surface fires in coastal pine forests of the upper Great Lakes. *The American Midland Naturalist* 140:206–218.

Lorimer, C. G., S. E. Dahir, and E. V. Nordheim. 2001. Tree mortality rates and longevity in mature and old-growth hemlock-hardwood forests. *Journal of Ecology* 89:960–971.

Lorimer, C. G., and L. E. Frelich. 1984. A simulation of equilibrium diameter distributions of sugar maple (*Acer saccharum*). *Bulletin of the Torrey Botanical Club* 111:193–199.

Lorimer, C. G., and L. E. Frelich. 1994. Natural disturbance regimes in old-growth northern hardwoods: Implications for restoration efforts. *Journal of Forestry* 92:33–38.

Lorimer, C. G., and C. R. Halpin. 2014. Classification and dynamics of developmental stages in late-successional temperate forests. *Forest Ecology and Management* 334:344–357.

Louma, D. L., C. A. Stockdale, R. Molina, and J. L. Eberhart. 2006. The spatial influence of *Pseudotsuga menziessi* retention trees on ectomycorrhiza diversity. *Canadian Journal of Forest Research* 36:2561–2573.

Lovett, G., K. Weathers, M. Arthur, and J. Schultz. 2004. Nitrogen cycling in a northern hardwood forest: Do species matter? *Biogeochemistry* 67:289–308.

Lovett, G. M., M. Weiss, A. M. Liebhold, T. P. Holmes, B. Leung, K. F. Lambert, D. A. Orwig, F. T. Campbell, J. Rosenthal, D. G. McCullough, R. Wildova, M. P. Ayres, C. D. Canham, D. R. Foster, S. L. LaDeau, and T. Weldy. 2016. Nonnative forest insects and pathogens in the United States: Impacts and policy options. *Ecological Applications* 26:1437–1455.

Lowrance, R., R. Todd, J. Fail Jr., O. Hendrickson Jr., R. Leonard, and L. Asmussen. 1984. Riparian forests as nutrient filters in agricultural watersheds. *BioScience* 34:374–377.

Malmsheimer, R. W., J. L. Bowyer, J. S. Fried, E. Gee, R. L. Izlar, R. A. Miner, I. A. Munn, E. Oneil, and W. C. Stewart. 2011. Managing forests because carbon matters: Integrating energy, products, and land management policy. *Journal of Forestry* 109:S7-S50.

Malmsheimer, R. W., P. Heffernan, S. Brink, D. Crandall, F. Deneke, C. Galik, E. Gee, J. A. Helms, N. McClure, and M. Mortimer. 2008. Forest management solutions for mitigating climate change in the United States. *Journal of Forestry* 106:115–173.

Markkanen, A., and P. Halme. 2012. Polypore communities in broadleaved boreal forests. *Silva Fennica* 46:317–331.

Marshall, R. 1927. The growth of hemlock before and after the release from suppression. *Harvard Forest Bulletin* 11:43.

Mascia, M. B., and S. Pailler. 2011. Protected area downgrading, downsizing, and degazettement (PADD) and its conservation implications. *Conservation Letters* 4:9–20.

Maser, C., R. F. Tarrant, J. M. Trappe, and J. F. Franklin. 1988. *From the forest to the sea: A story of fallen trees.* USDA Forest Service, General Technical Report PNW-GTR-229.

Mason, W. L. 2002. Are irregular stands more windfirm? *Forestry* 75:347-355.

Matthews, J. D. 1989. *Silvicultural systems.* Oxford, UK: Oxford University Press.

Marx, L., and M. B. Walters. 2008. Survival of tree seedlings on different species of decaying wood maintains tree distribution in Michigan hemlock-hardwood forests. *Journal of Ecology* 96:505–513.

Matonis, M. S., and D. Binkley. 2018. Not just about the trees: Key role of mosaic-meadows in restoration of ponderosa pine ecosystems. *Forest Ecology and Management* 411:120–131.

McArdle, R. E., M. H. Meyer, and D. Bruce. 1949. *The yield of Douglas-fir in the Pacific Northwest.* USDA Forest Service Technical Bulletin 201.

McComb, W., and D. Lindenmayer. 1999. Dying, dead, and down trees. In *Maintaining biodiversity in forest ecosystems*, ed. M. L. Hunter, 335–372. Cambridge, UK: Cambridge University Press.

McCune, B. 1993. Gradients in epiphyte biomass in three *Pseudotsuga-Tsuga* forests of different ages in western Oregon and Washington. *Bryologist* 96:405–411.

McDowell, N. G., H. D. Adams, J. D. Bailey, M. Hess, and T. E. Kolb. 2006. Homeostatic maintenance of ponderosa pine gas exchange in response to stand density changes. *Ecological Applications* 16:1164–1182.

McElhinny, C., P. Gibbons, C. Brack, and J. Bauhus. 2005. Forest and woodland stand structural complexity: Its definition and measurement. *Forest Ecology and Management* 218:1–24.

McGee, G. G. 2000. The contribution of beech bark disease-induced mortality to coarse woody debris loads in northern hardwood stands of Adirondack Park, New York, USA. *Canadian Journal of Forest Research* 30:1453–1462.

McGee, G. G., D. J. Leopold, and R. D. Nyland. 1999. Structural characteristics of old-growth, maturing, and partially cut northern hardwood forests. *Ecological Applications* 9:1316–1329.

McGuire J. P., R. J. Mitchell, E. B. Moser, S. D. Pecot, D. H. Gjerstad, and C. W. Hedman. 2001. Gaps in a gappy forest: Plant resources, pine regeneration, and understory response to tree removal in longleaf pine savannas. *Canadian Journal of Forest Research* 31:765–778.

McIntyre, R. K., S. B. Jack, R. J. Mitchell, and J. K. Hiers. 2008. Multiple value management: The Stoddard-Neel approach to ecological forestry in longleaf pine grasslands. Newton, GA: JW Jones Ecological Research Center.

McLachlan, J. S., J. J. Hellmann, M. W. Schwartz. 2007. A framework for debate of assisted migration in an era of climate change. *Conservation Biology* 21:297–302.

McLaughlin, J. A. 2001. Impact of *Armillaria* root disease on succession in red pine plantations in southern Ontario. *Forestry Chronicle* 77:519–524.

Meier, E. S., H. Lischke, D. R. Schmatz, and N. E. Zimmermann. 2012. Climate, competition and connectivity affect future migration and ranges of European trees. *Global Ecology and Biogeography* 21:164–178.

Mendell, B. 2016. From cigar tax to timberland trusts. *Forest History* (Spring/Fall) 32–35.

Merrens, E. J., and D. R. Peart. 1992. Effects of hurricane damage on individual growth and stand structure in a hardwood forest in New Hampshire, USA. *Journal of Ecology* 80:787–795.

Messier, C., K. J. Puettmann, and K. D. Coates, eds. 2013. *Managing forests as complex adaptive systems: Building resilience to the challenge of global change.* London, UK: Routledge.

Messier, C., R. Tittler, D. D. Kneeshaw, N. Gélinas, A. Paquette, K. Berninger, H. Rheault, P. Meek, and N. Beaulieu. 2009. TRIAD zoning in Quebec: Experiences and results after 5 years. *The Forestry Chronicle* 85:885–896.

Meyer, H. A. 1952. Structure, growth, and drain in balanced uneven-aged forests. *Journal of Forestry* 50:85–92.

Meyer, W. H. 1934. *Growth in selectively cut ponderosa pine forests of the Pacific Northwest*. USDA Forest Service, Pacific Northwest Research Station: Technical Bulletin No 407.

Michaelian, M., E. H. Hogg, R. J. Hall, and E. Arsenault. 2011. Massive mortality of aspen following severe drought along the southern edge of the Canadian boreal forest. *Global Change Biology* 17:2084–2094.

Millar, C. I., N. L. Stephenson, and S. L. Stephens. 2007. Climate change and forests of the future: Managing in the face of uncertainty. *Ecological Applications* 17:2145–2151.

Miller, J. D., and H. D. Safford. 2012. Trends in wildfire severity: 1984 to 2010 in the Sierra Nevada, Modoc Plateau, and southern Cascades, California, USA. *Fire Ecology* 8:41–57.

Miller, T. F., D. J. Mladenoff, and M. K. Clayton. 2002. Old-growth northern hardwood forests: Spatial autocorrelation and patterns of understory vegetation. *Ecological Monographs* 72:487–503.

Minnesota Forest Resources Council. 2013. *Sustaining Minnesota forest resources: Voluntary site-level forest management guidelines for landowners, loggers and resource managers*. St. Paul, MN: Minnesota Forest Resources Council.

Mitchell, R. J., J. K. Hiers, J. J. O'Brien, S. B. Jack, and R. T. Engstrom. 2006. Silviculture that sustains: The nexus between silviculture, frequent prescribed fire, and conservation of biodiversity in longleaf pine forests of the southeastern United States. *Canadian Journal of Forest Research* 36:2724–2736.

Mitchell, R. J., J. K. Hiers, J. J. O'Brien, S. B. Jack, and R. T. Engstrom. 2006a. Silviculture that sustains: The nexus between silviculture, frequent prescribed fire, and conservation of biodiversity in longleaf pine forests of the southeastern United States. *Canadian Journal of Forest Research* 40:596–596.

Mitchell, R. J., W. L. Neel, J. K. Hiers, F. T. Cole, and J. B. Atkinson. 2000. *A model management plan for conservation easements in longleaf pine-dominated landscapes*. Newton, GA: Joseph W. Jones Ecological Research Center.

MN DNR. 2003. *Field guide to the native plant communities of Minnesota: The Laurentian Mixed Forest Province*. St. Paul, MN: Minnesota Department of Natural Resources.

Möller, A. 1922. *Der Dauerwaldgedanke-Sein Sinn und seine Bedeutung*. Berlin, Germany: Verl. Julius Springer.

Monserud, R. A., and H. Sterba. 1999. Modeling individual tree mortality for Austrian forest species. *Forest Ecology and Management* 113:109–123.

Montgomery, R. A., B. J. Palik, S. B. Boyden, and P. B. Reich. 2013. New cohort growth and survival in variable retention harvests of a pine ecosystem in Minnesota, USA. *Forest Ecology and Management* 310:327–335.

Moore, R. D., and J. S. Richardson. 2012. Natural disturbance and forest management in riparian zones: Comparison of effects at reach, catchment, and landscape scales. *Freshwater Science* 31:239–247.

Moore, R., D. L. Spittlehouse, and A. Story. 2005. Riparian microclimate and stream temperature response to forest harvesting: A review. *Journal of the American Water Resources Association* 41:813–834.

Mori, A. S., and R. Kitagawa. 2014. Retention forestry as a major paradigm for safeguarding forest biodiversity in productive landscapes: A global meta-analysis. *Biological Conservation* 175:65–73.

Moser, W. K., S. M. Jackson, V. Podrazsky, and D. R. Larsen. 2002. Examination of stand structure on quail plantations in the Red Hills region of Georgia and Florida managed by the Stoddard-Neel system: An example for forest managers. *Forestry* 75:443–449.

Muir, P. S., R. L. Mattingly, J. C. Tappeiner II, J. D. Bailey, W. E. Elliott, J. C. Hagar, J. C. Miller, E. B. Peterson, and E. E. Starkey. 2002. Managing for biodiversity in young Douglas-fir forests of western Oregon. US Geological Survey, Biological Resources Division, Biological Science Report USGS/BRD/BSR-2002-0006. Corvallis, OR: US Geological Survey. 76 p.

Murphy, M. L., and K. V. Koski. 1989. Input and depletion of woody debris in Alaska streams and implications for streamside management. *North American Journal of Fisheries Management* 9:427–436.

Myers, R. K., and D. H. van Lear. 1998. Hurricane-fire interactions in coastal forests of the south: A review and hypothesis. *Forest Ecology and Management* 103: 265–276.

Myers, R. L. 1990. *Ecosystems of Florida*. Orlando, FL: University of Central Florida Press.

Myers, R. L., and D. A. Rodríguez-Trejo. 2009. Fire in tropical pine ecosystems. *Tropical Fire Ecology*, 557–605.

Nadkarni, N. M. 1984. Epiphyte biomass and nutrient capital of a neotropical elfin forest. *Biotropica* 6:249–256.

Nagel, L. M., B. J. Palik, M. A. Battaglia, A. W. D'Amato, J. M. Guldin, C. W. Swanston, M. K. Janowiak, M. P. Powers, L. A. Joyce, C. I. Millar, and D. L. Peterson. 2017. Adaptive silviculture for climate change: A national experiment in manager-scientist partnerships to apply an adaptation framework. *Journal of Forestry* 115: 167-178.

Nagel, T. A., M. Svoboda, and M. Kobal. 2014. Disturbance, life history traits, and dynamics in an old-growth forest landscape of southeastern Europe. *Ecological Applications* 24:663–679.

Naiman, R. J., and H. Décamps. 1997. The ecology of interfaces: Riparian zones. *Annual Review of Ecology and Systematics* 28:621–658.

Naiman, R. J., H. Décamps, J. Pastor, and C. A. Johnston. 1988. The potential importance of boundaries of fluvial ecosystems. *Journal of the North American Benthological Society* 7:289–306.

Neel, L. 2010. *The art of managing longleaf: A personal history of the Stoddard-Neel approach.* Athens, GA: University of Georgia Press.

Neumann, F. G., and G. Minko. 1981. The sirex wood wasp in Australian radiata pine plantations. *Australian Forestry* 44:46–63.

Neuner, S., A. Albrecht, D. Cullmann, F. Engels, V. C. Griess, W. A. Hahn, M. Hanewinkel, F. Härtl, C. Kölling, K. Staupendahl, and T. Knoke. 2015. Survival of Norway spruce remains higher in mixed stands under a dryer and warmer climate. *Global Change Biology* 21:935–946.

Newland, J. A., and T. H. DeLuca. 2000. Influence of fire on native nitrogen-fixing plants and soil nitrogen status in ponderosa pine-Douglas-fir forests in western Montana. *Canadian Journal of Forest Research* 30:274–282.

Newman, D. H. 1988. *The optimal forest rotation a discussion and annotated bibliography.* USDA Forest Service, General Technical Report SE-48.

Nicholls, T. H., and M. E. Ostry. 1990. *Sphaeropsis sapinea* cankers on stressed red and jack pines in Minnesota and Wisconsin. *Plant Diseases* 74:54–56.

Noel, J. M., W. J. Platt, and E. B. Moser 1998. Structural characteristics of old- and second-growth stands of longleaf pine (*Pinus palustris*) in the Gulf Coastal Region of the U.S.A. *Conservation Biology* 12:533–548.

North, M. P., B. M. Collins, H. D. Safford, and N. L. Stephenson. 2016. Montane forests. In *Ecosystems of California*, ed. H. Mooney, and E. Zavaleta, 553–577. Berkeley, CA: University of California Press.

North, M. P., S. L. Stephens, B. M. Collins, J. K. Agee, G. Aplet, J. F. Franklin, and P. Z. Fule. 2015. Reform forest fire management. *Science* 349:1280–1281.

North, M. P., P. Stine, K. L. O'Hara, W. J. Zielinski, and S. L. Stephens. 2009. *An ecosystem management strategy for sierran mixed- conifer forests.* USDA Forest Service, General Technical Report PSW-220.

Norton, M. R., and S. J. Hannon. 1997. Songbird response to partial-cut logging in the boreal mixedwood forest of Alberta. *Canadian Journal of Forest Research* 27:44–53.

Nowacki, G. J., M. D. Abrams, and C. G. Lorimer. 1990. Composition, structure, and historical development of northern red oak stands along an edaphic gradient in north-central Wisconsin. *Forest Science* 36:276–292.

Nunery, J. S., and W. S. Keeton. 2010. Forest carbon storage in the northeastern United States: Net effects of harvesting frequency, post-harvesting retention, and wood products. *Forest Ecology and Management* 259:1363–1375.

Nyamai, P. A. 2013. "Factors affecting regeneration-layer dynamics in mixed-pine forest ecosystems of eastern Upper Michigan and implications for forest ecosystem restoration." PhD diss., Ohio State University.

Nyland, R. D. 2007. *Silviculture: Concepts and applications.* Long Grove, IL: Waveland Press.

Oertli, B. 1993. Leaf litter processing and energy flow through macroinvertebrates in a woodland pond (Switzerland). *Oecologia* 96:466–477.

O'Hara, K. L. 1996. Dynamics and stocking-level relationships of multi-aged ponderosa pine stands. *Forest Science Monograph* 42.

O'Hara, K. L. 1998. Silviculture for structural diversity: A new look at multiaged systems. *Journal of Forestry* 96:4–10.

O'Hara, K. L. 2001. The silviculture transformation: A commentary. *Forest Ecology and Management* 151: 81–86.

O'Hara, K. L. 2005. Multiaged silviculture of ponderosa pine. In *Proceedings of the symposium on ponderosa pine: Issues, trends, and management*, eds. M. W. Ritchie, D. A. Maguire, and A. Youngblood, 59–70. USDA Forest Service Forest PSW-198.

O'Hara, K. L. 2014. *Multiaged silviculture: Managing for complex forest stand structures.* Oxford, UK: Oxford University Press.

O'Hara, K. L., L. P. Leonard, and C. R. Keyes. 2012. Variable-density thinning and a marking paradox: Comparing prescription protocols to attain stand variability in coast redwood. *Western Journal of Applied Forestry* 27:143–149.

O'Hara, K. L., and L. M. Nagel. 2013. The stand: Revisiting a central concept in forestry. *Journal of Forestry* 111:335–340.

O'Hara, K. L., and C. D. Oliver. 1999. A decision system for assessing stand differentiation potential and prioritizing precommercial thinning treatments. *Western Journal of Applied Forestry* 14:7–13.

O'Hara, K. L., and B. S. Ramage. 2013. Silviculture in an uncertain world: Utilizing multi-aged management systems to integrate disturbance. *Forestry* 86: 401–410.

O'Hara, K. L., N. I. Valappil, and L. M. Nagel. 2003. Stocking control procedures for multiaged ponderosa pine stands in the inland Northwest. *Western Journal of Applied Forestry* 18:5–14.

O'Laughlin, J., and G. H. Belt. 1995. Functional approaches to riparian buffer strip design. *Journal of Forestry* 93:29–32.

Oldén, A., O. Ovaskainen, J. S. Kotiaho, S. Laaka-Lindberg, and P. Halme. 2014. Bryophyte species richness on retention aspens recovers in time but community structure does not. *PloS one* 9(4):e93786.

Oliver, C. D., and B. C. Larson. 1996. *Forest stand dynamics*. New York: Wiley.

Ostry, M. E., M. J. Moore, C. C. Kern, R. C. Venette, and B. J. Palik. 2012. Multiple diseases impact survival of pine species planted in red pine stands harvested in spatially variable retention patterns. *Forest Ecology and Management* 286:66-72.

Ostry, M. E., T. H. Nicholls, and D. D. Skilling. 1990. *Biology and control of* Sirococcus *shoot blight on red pine*. USDA Forest Service, North Central Research Station Research Paper NC-295.

Oswalt, S. N, B. W. Smith, P. D. Miles, and S. A. Pugh. 2014. *Forest resources of the United States, 2012: A technical document supporting the Forest Service 2010 update of the RPA Assessment*. USDA Forest Service General Technical Report WO-91. 218 p.

Palik, B. J., D. P. Batzer, R. Buech, D. Nichols, K. Cease, L. Egeland, and D. E. Streblow. 2001. Seasonal pond characteristics across a chronosequence of adjacent forest ages in Northern Minnesota, USA. *Wetlands* 21:532–542.

Palik, B., D. Batzer, and C. Kern. 2006. Upland forest linkages to seasonal wetlands: Litter flux, processing, and food quality. *Ecosystems* 9:142–151.

Palik, B. J., and A. W. D'Amato. 2017. Ecological forestry: Much more than retention harvesting. *Journal of Forestry* 115:51–53.

Palik, B. J., and A. W. D'Amato. 2019. Variable retention harvesting in Great Lakes mixed-pine forests: emulating a natural model in managed ecosystems. *Ecological Processes* 8:16.

Palik, B. J., and R. T. Engstrom. 1999. Species composition. In *Maintaining biodiversity in forest ecosystems*, ed. M. L. Hunter Jr., 65–94. Cambridge, UK: Cambridge University Press.

Palik, B. J., S. W. Golladay, P. C. Goebel, and B. W. Taylor. 1998. Riparian geomorphology and coarse woody debris recruitment in a Georgia coastal plain stream. *EcoScience* 5:551–560.

Palik, B. J., and D. Kastendick. 2010. Response of seasonal pond plant communities to upland forest harvest in northern Minnesota forests, USA. *Forest Ecology and Management* 260:628–637.

Palik, B. J., R. J. Mitchell, and J. K. Hiers. 2002a. Modeling silviculture after natural disturbance to sustain biodiversity in the longleaf pine (*Pinus palustris*) ecosystem: Balancing complexity and implementation. *Forest Ecololgy and Management* 155:347–356.

Palik, B. J., R. J. Mitchell, G. Houseal, and N. Pederson. 1997. Effects of canopy structure on resource availability and seedling responses in a longleaf pine ecosystem. *Canadian Journal of Forest Research* 27:1458–1464.

Palik, B. J., R. J. Mitchell, S. D. Pecot, M. A. Battaglia, and M. Pu. 2003. Spatial distribution of overstory retention influences resources and growth of longleaf pine seedlings. *Ecological Applications* 13:674–686.

Palik, B. J., R. A. Montgomery, P. B. Reich, and S. B. Boyden. 2014. Biomass growth response to spatial pattern of variable-retention harvesting in a northern Minnesota pine ecosystem. *Ecological Applications* 24:2078-2088.

Palik, B. J., and N. Pederson. 1996. Natural disturbance and overstory mortality in longleaf pine ecosystems. *Canadian Journal of Forest Research* 26:2035–2047.

Palik, B. J., and K. S. Pregitzer. 1992. A comparison of presettlement and present-day forests on two bigtooth aspen-dominated landscapes in northern Lower Michigan. *The American Midland Naturalist* 127:327–338.

Palik, B. J., and K. S. Pregitzer. 1993. The repeatability of stem exclusion during even-aged development of bigtooth aspen-dominated forests. *Canadian Journal of Forest Research* 23:1156–1168.

Palik, B. J., and K. S. Pregitzer. 1993a. The vertical development of early successional forests in northern Lower Michigan, USA. *Journal of Ecology* 81:271–285.

Palik, B. J., and K. S. Pregitzer. 1994. White pine seed tree legacies in an aspen landscape: Influences on post-disturbance white pine population structure. *Forest Ecology and Management* 67:191–202.

Palik, B. J., and J. Robl. 1999. *Structural legacies of catastrophic windstorm in a mature Great Lakes aspen forest*. USDA Forest Service, Research Paper NC-337.

Palik, B., D. Streblow, L. Egeland, and R. Buech. 2007. Landscape variation of seasonal pool plant communities in forests of northern Minnesota, USA. *Wetlands* 27:12–23.

Palik, B., J. Zasada, and C. Hedman. 1999. Ecological considerations for riparian silviculture. In *Riparian management in riparian forests of the continental eastern United States*, ed. E. S. Verry, J. W. Hornbeck, and C. A. Dolloff, 233–254. New York: Lewis Press.

Parsons, R., and G. Prest. 2003. Aboriginal forestry in Canada. *The Forestry Chronicle* 79:779–784.

Parsons, W. F. J., D. H. Knight, and S. L. Miller. 1994. Root gap dynamics in a lodgepole pine forest, nitrogen transformations in gaps of different size. *Ecological Applications* 4:354–362.

Payn, T., J.-M. Carnus, P. Freer-Smith, M. Kimberley, W. Kollert, S. Liu, C. Orazio, L. Rodriguez, L. N. Silva, and M. J. Wingfield. 2015. Changes in planted forests and future global implications. *Forest Ecology and Management* 352:57–67.

Pearson, G. A. 1950. *Management of ponderosa pine in the Southwest as developed by research and experimental practice*. Agriculture Monograph No. 6. Washington, DC: USDA Forest Service. 218 p.

Pedlar, J. H., D. W. McKenney, I. Aubin, T. Beardmore, J. Beaulieu, L. Iverson, G. A. O'Neill, R. S. Winder, and C. Ste-Marie. 2012. Placing forestry in the assisted migration debate. *BioScience* 62:835–842.

Peterson, C. J., and S. T. A. Pickett. 1995. Forest reorganization: A case study in an old-growth forest catastrophic blowdown. *Ecology* 76:763–774.

Peterson, D. L., J. K. Agee, G. H. Aplet, and D. P. Dykstra. 2009. *Effects of timber harvest following wildfire in western North America*. USDA Forest Service, General Technical Report PNW-776.

Peterson, D. L., M. C. Johnson, D. McKenzie, J. K. Agee, T. Jain, and E. D. Reinhardt. 2005. *Forest structure and fire hazard in dry forests of the western United States*. USDA Forest Service, General Technical Report PNW-628.

Peterson, D. W., E. K. Dodson, and R. J. Harrod. 2015. Post-fire logging reduces surface woody fuels up to four decades following wildfire. *Forest Ecology and Management* 338:84–91.

Pickett, S. T. A., and P. S. White, eds. 1985. *The ecology of natural disturbance and patch dynamics*. New York: Academic Press.

Pickles, B. J., and S. W. Simard. 2017. Mycorrhizal networks and forest resilience to drought. *In Mycorrhizal mediation of soil*, eds. N. Collins Johnson, C. Gehring, and J. Jansa, 319–339. Amsterdam, Netherlands: Elsevier.

Piha, A., T. Kuuluvainen, H. Lindberg, and I. Vanha-Majamaa. 2013. Can scar-based fire history reconstructions be biased? An experimental study in boreal Scots pine. *Canadian Journal of Forest Research* 43:669–675.

Pimm, S. L., G. E. Davis, L. Loope, and C. T. Roman. 1994. Hurricane Andrew. *BioScience* 44:224–229.

Pirard, R., L. Dal Secco, and R. Warman. 2016. Do timber plantations contribute to forest conservation? *Environmental Science and Policy* 57:122–130.

Pitelka, L., and Plant Migration Workshop Group. 1997. Plant migration and climate change: A more realistic portrait of plant migration is essential to predicting biological responses to global warming in a world drastically altered by human activity. *American Scientist* 85:464–473.

Platt, W. J., G. W. Evans, and S. L. Rathbun. 1988. The population dynamics of a long-lived conifer (*Pinus palustris*). *American Naturalist* 131:491–525.

Plotkin, A. B., D. Foster, J. Carlson, and A. Magill. 2013. Survivors, not invaders, control forest development following simulated hurricane. *Ecology* 94:414–423.

Plotkin, J. B., J. Chave, and P. S. Ashton. 2002. Cluster analysis of spatial patterns in Malaysian tree species. *The American Naturalist* 160:629–644.

Powers, M. D., C. R. Webster, K. S. Pregitzer, and B. J. Palik. 2009. Spatial dynamics of radial growth and growth efficiency in residual *Pinus resinosa* following aggregated retention harvesting. *Canadian Journal of Forest Research* 39:109–117.

Prasad, A. M., L. R. Iverson, S. Matthews, and M. Peters. 2007-ongoing. "A climate change atlas for 134 forest tree species of the eastern United States [database]." https://www.nrs.fs.fed.us/atlas/tree, Northern Research Station, USDA Forest Service, Delaware, OH, USA.

Pretzsch, H. 2005. Stand density and growth of Norway spruce (*Picea abies* (L.) Karst.) and European beech (*Fagus sylvatica* L.): Evidence from long-term experimental plots. *European Journal of Forest Research* 124:193–205.

Pretzsch, H., J. Block, J. Dieler, P. H. Dong, U. Kohnle, J. Nagel, H. Spellmann, and A. Zingg. 2010. Comparison between the productivity of pure and mixed stands of Norway spruce and European beech along an ecological gradient. *Annals of Forest Science* 67:712. doi: 10.1051/forest/2010037.

Prevost, M., P. Raymond, and J. M. Lussier. 2010. Regeneration dynamics after patch cutting and scarification in yellow birch conifer stands. *Canadian Journal of Forest Research* 40:357–369.

Prugh, L. R., K. E. Hodges, A. R. Sinclair, and J. S. Bradshares. 2008. Effect of habitat area and isolation on fragmented animal populations. *Proceedings of the National Academic of Sciences* 105:20770–20775.

Puettmann, K. J. 2011. Silvicultural challenges and options in the context of global change: Simple fixes and opportunities for new management approaches. *Journal of Forestry* 109:321–331.

Puettmann, K. J., K. D. Coates, and C. C. Messier. 2008. *A critique of silviculture: Managing for complexity.* Washington, DC: Island Press.

Puettmann, K. J., C. Messier, and K. D. Coates. 2013. Managing forests as complex adaptive systems. In *Managing forest as complex adaptive systems*, ed. C. Messier, K. J. Puettmann, and K. D. Coates, 3–16. New York: Routledge Press.

Puettmann, K. J., S. M. Wilson, S. C. Baker, P. J. Donoso, L. Drössler, G. Amente, B. D. Harvey, T. Knoke, Y. Lu, S. Nocentini, and F. E. Putz. 2015. Silvicultural alternatives to conventional even-aged forest management: What limits global adoption? *Forest Ecosystems* 2:8.

Pyšek, P., and D. M. Richardson. 2010. Invasive species, environmental change and management, and health. *Annual Review of Environment and Resources* 35:25–55.

Quinn, T. P., S. M. Carlson, S. M. Gende, and H. B. Rich Jr. 2009. Transportation of Pacific salmon carcasses from streams to riparian forests by bears. *Canadian Journal of Zoology* 87:195–203.

Radsliff, W. A., C. A. Mohn, W. H. Cromell, and W. H. Gray. 1981. *Ponderosa pine provenance tests in Minnesota*. Minnesota Forestry Research Note No. 277. St. Paul, MN: University of Minnesota. 4 p.

Rapp, V., and T. A. Spies. 2003. *New findings about old-growth forests*. Science Update 4. USDA Forest Service, Pacific Northwest Research Station, Portland, OR.

Ravenscroft, C., R. M. Scheller, D. J. Mladenoff, and M. A. White. 2010. Forest restoration in a mixed-ownership landscape under climate change. *Ecological Applications* 20:327–346.

Raymond, P., S. Bédard, V. Roy, C. Larouche, and S. Tremblay. 2009. The irregular shelterwood system: Review, classification, and potential application to forests affected by partial disturbances. *Journal of Forestry* 107:405–413.

Rebertus, A. J., T. Kitzberger, T. T. Veblen, and L. M. Roovers. 1997. Blowdown history and landscape patterns in the Andes of Tierra del Fuego, Argentina. *Ecology* 78:678–692.

Reeves, G. H., L. E. Benda, K. M. Burnett, P. A. Bisson, and J. R. Sedell. 1995. A disturbance-based ecosystem approach to maintaining and restoring freshwater habitats of evolutionarily significant units of anadromous salmonids in the Pacific Northwest. *American Fisheries Society Symposium* 17:334–339.

Reeves, G. H., K. M. Burnett, and E. V. McGarry. 2003. Sources of large woody debris in the main stem of a fourth-order watershed in coastal Oregon. *Canadian Journal of Forest Research* 33:1363–1370.

Reeves, G. H., and T. A. Spies. 2017. Watersheds and landscapes. In *People, forests, and change: Lessons from the Pacific Northwest*, ed. D. H. Olson and B. Van Horne, 207–22. Washington, DC: Island Press.

Reich, R. M., P. W. Mielke Jr., and F. G. Hawksworth. 1991. Spatial analysis of ponderosa pine trees infected with dwarf mistletoe. *Canadian Journal of Forest Research* 21:1808–1815.

Reynolds, R. T., A. J. S. Meador, J. A. Youtz, T. Nicolet, M. S. Matonis, P. L. Jackson, D. G. DeLorenzo, and A. D. Graves. 2013. *Restoring composition and structure in southwestern frequent-fire forests: A science-based framework for improving ecosystem resiliency*. USDA Forest Service, General Technical Report RMRS-310.

Ricker, D. P., and Z. A. Zasada. 1958. *Timber management guide for the National Forests of the North Central states-red pine type*. USDA Forest Service, North Central Region, Milwaukee, WI, USA. 13 p.

Ringler, N. H., and J. D. Hall. 1975. Effects of logging on water temperature, and dissolved oxygen in spawning beds. *Transactions of the American Fisheries Society* 104:111–121.

Rodman, K. C., A. J. S. Sanchez-Meador, M. M. Moore, and D. W. Huffman. 2017. Reference conditions are influenced by the physical template and vary by forest type: A synthesis of Pinus ponderosa-dominated sites in the southwestern United States. *Forest Ecology and Management* 404:316–329.

Roe, A. L., R. R. Alexander, and M. D. Andrews. 1970. *Engelmann spruce regeneration practices in the Rocky Mountains*. USDA Forest Service, Research Paper No. 115.

Rugani, T., J. Diaci, and D. Hladnik. 2013. Gap dynamics and structure of two old-growth beech forest remnants in Slovenia. *PLoS ONE* 8:e52641.

Rummel, R. S. 1951. Some effects of livestock grazing on ponderosa pine forest and range in central Washington. *Ecology* 32:594–607.

Runkle, J. R. 1982. Patterns of disturbance in some old-growth mesic forests of eastern North America. *Ecology* 63:1533–1546.

Ryan, M. G., D. Binkley, and J. H. Fownes. 1997. Age-related decline in forest productivity: Pattern and process. *Advances in Ecological Research* 27:213–262.

Sanchez-Meador, A. J., A. Azpeleta, M. Stoddard, B. Bagdon, and S. Nepal. 2017. Using Climate-FVS to inform management decisions: Three case studies from the American Southwest. In 2017 Forest Vegetation Simulator (FVS) e-Conference, 126.

Sarr, D., K. Puettmann, R. Pabst, M. Cornett, and L. Arguello. 2004. Restoration ecology: New perspectives and opportunities for forestry. *Journal of Forestry* 102:20–24.

Sawyer, J. O., and T. Keeler-Wolf. 1995. *A manual of California vegetation*. Sacramento, CA: California Native Plant Society.

Schmitt, C. B., N. D. Burgess, L. Coad, A. Belokurov, C. Besançon, L. Boisrobert, A. Campbell, L. Fish, D. Gliddon, K. Humphries, V. Kapos, C. Loucks, I. Lysenko, L. Miles, C. Mills, S. Minnemeyer, T. Pistorius, C. Ravilious, M. Steininger, and G. Winkel. 2009. Global analysis of the protection status of the world's forests. *Biological Conservation* 142:2122–2130.

Schoennagel, T., J. K. Balch, H. Brenkert-Smith, P. E. Dennison, B. J. Harvey, M. A. Krawchuk, N. Mietkiewicz, P. Morgan, M. A. Moritz, and R. Rasker. 2017. Adapt to more wildfire in western North American forests as climate changes. *Proceedings of the National Academy of Sciences* 114:4582–4590.

Schoennagel, T., and C. R. Nelson. 2011. Restoration relevance of recent National Fire Plan treatments in forests of the western United States. *Frontiers in Ecology and the Environment* 9:271–277.

Schoennagel, T., T. T. Veblen, and W. H. Romme. 2004. The interaction of fire, fuels, and climate across Rocky Mountain forests. *Bioscience* 54:661–676.

Semlitsch, R. D. 1998. Biological delineation of terrestrial buffer zones for pond- breeding salamanders. *Conservation Biology* 12:1113–1119.

Semlitsch, R. D., and J. R. Bodie. 1998. Are small, isolated wetlands expendable? *Conservation Biology* 12:1129–1133.

Seymour, R. S., and M. L. Hunter. 1992. *New forestry in eastern spruce-fir forests: Principles and applications to Maine*. Orono, ME: College of Forest Resources, University of Maine.

Seymour, R. S., and M. L. Hunter Jr. 1999. Principles of ecological forestry. In *Maintaining biodiversity in forest ecosystems*, ed. M.L. Hunter Jr., 22–61. Cambridge, UK: Cambridge University Press.

Seymour, R. S., A. S. White, and P. G. deMaynadier. 2002. Natural disturbance regimes in northeastern North America: Evaluating silvicultural systems using natural scales and frequencies. *Forest Ecology and Management* 155:357–367.

Sharpe, G. W. 1956. "A taxonomical-ecological study of vegetation by habitats in eight forest types of the Olympic Rain Forest, Olympic National Park, Washington." PhD diss., University of Washington.

Shaw, D. C., J. Chen, E. A. Freeman, and D. M. Braun. 2005. Spatial and population characteristics of dwarf mistletoe infected trees in an old-growth Douglas-fir–western hemlock forest. *Canadian Journal of Forest Research* 35:990–1001.

Shirley, H. L. 1932. Light intensity in relation to plant growth in a virgin Norway pine forest. *Journal of Agricultural Research* 44:227–244.

Shive, K. L., H. K. Preisler, K. R. Welch, H. D. Safford, R. J. Butz, K. L. O'Hara, and S. L. Stephens. 2018. From the stand-scale to the landscape-scale: Predicting the spatial patterns of forest regeneration after disturbance. *Ecological Applications*.

Show, S. B., and E. I. Kotok. 1924. *The role of fire in the California pine forests*. Washington, DC: USDA.

Sibley, P. K., D. P. Kreutzweiser, B. J. Naylor, J. S. Richardson, and A. M. Gordon. 2012. Emulation of natural disturbance (END) for riparian forest management: Synthesis and recommendations. *Freshwater Science* 31:258–264.

Silver, E. J., S. Fraver, A. W. D'amato, T. Aakala, and B. J. Palik. 2013. Long-term mortality rates and spatial patterns in an old-growth *Pinus resinosa* forest. *Canadian Journal of Forest Research* 43:809–816.

Silvertown, J. W., and J. Lovett Doust. 1993. *Introduction to plant population biology*. Oxford, UK: Blackwell Scientific Publications.

Simard, S. W., K. J. Beiler, M. A. Bingham, J. R. Deslippe, L. J. Philip, and F. P. Teste. 2012. Mycorrhizal networks: Mechanisms, ecology and modelling. *Fungal Biology Reviews* 26:39–60.

Singer, M. T., and C. G. Lorimer. 1997. Crown release as a potential old-growth restoration approach in northern hardwoods. *Canadian Journal of Forest Research* 27:1222–1232.

Skelly, D. K., E. E. Warner, and S. A. Cortwright. 1999. Long-term distributional dynamic of a Michigan amphibian assemblage. *Ecology* 80:2326–2337.

Slesak, R. A., C. F. Lenhart, K. N. Brooks, A. W. D'Amato, and B. J. Palik. 2014. Water table response to harvesting and simulated emerald ash borer mortality in black ash wetlands in Minnesota, USA. *Canadian Journal of Forest Research* 44(8):961–968.

Smith, D. G. 1976. Effect of vegetation on lateral migration of anastomosed channels of a glacier meltwater river. *Geological Society of America Bulletin* 87:857–860.

Smith, D. M. 1962. *The practice of silviculture*. 7th ed. New York: John Wiley and Sons.

Smith, D. M., B. C. Larson, M. J. Kelty, and P. M. S. Ashton. 1996. *The practice of silviculture: Applied forest ecology*. 9th ed. New York: John Wiley and Sons.

Smith, F. W., and J. N. Long. 2001. Age-related decline in forest growth: An emergent property. *Forest Ecology and Management* 144:175–181.

Society of American Foresters (SAF). 2008. *The dictionary of forestry*. Washington, DC: The Society of American Foresters.

Sohn, J. A., F. Hartig, M. Kohler, J. Huss, and J. Bauhus. 2016. Heavy and frequent thinning promotes drought adaptation in *Pinus sylvestris* forests. *Ecological Applications* 26:1030–1046.

Souder, J. A., and S. K. Fairfax. 1996. *State trust lands: History, management, and sustainable use*. Lawrence, KS: University Press of Kansas.

Spies, T. A. 2009. Science of old growth, or a journey into wonderland. In *Old growth in a new world: A Pacific Northwest icon reexamined*, ed. T. A. Spies and S. L. Duncan, 31–43. Washington, DC: Island Press.

Spies, T. A., and J. F. Franklin. 1991. The structure of natural young, mature, and old-growth Douglas-fir forests in Oregon and Washington. In *Wildlife and vegetation of unmanaged Douglas-fir forests*, eds. L. F. Ruggiero, K. B. Aubry, A. B. Carey, and M. H. Huff, 91–109. USDA Forest Service, General Technical Report PNW-285.

Spies, T. A., and J. F. Franklin. 1996. The diversity and maintenance of old-growth forests. In *Biodiversity in managed landscapes: Theory and practice*, ed. R. C. Szaro, and D. W. Johnston, 296–314. New York: Oxford University Press.

Spies, T. A., J. F. Franklin, and T. B. Thomas. 1988. Coarse woody debris in Douglas-fir forests of western Oregon and Washington. *Ecology* 69:1689–1702.

Spies, T. A., T. W. Giesen, F. J. Swanson, J. F. Franklin, D. Lach, and K. N. Johnson. 2010. Climate change adaptation strategies for federal forests of the Pacific Northwest, USA: Ecological, policy, and socio-economic perspectives. *Landscape Ecology* 25:1185–1199.

Spurr, S. H., and A. C. Cline. 1942. Ecological forestry in central New England. *Journal of Forestry* 40:418–420.

Stambaugh, M. C., R. P. Guyette, and J. Marschall. 2013. Fire history in the Cherokee nation of Oklahoma. *Human Ecology* 41:749–758.

Steen, H. K. 1990. *David M. Smith and the history of silviculture*. Durham, NC: Forest History Society.

Steinblums, I. J., H. A. Froehlich, and J. K. Lyons. 1984. Designing stable buffer strips for stream protection. *Journal of Forestry* 82:49–52.

Stephens, E. P. 1955. "The historical-developmental method of determining forest trends." PhD diss., Harvard University.

Stephens, S. L., B. M. Collins, C. J. Fettig, M. A. Finney, C. M. Hoffman, E. E. Knapp, M. P. North, H. Safford, and R. B. Wayman. 2018. Drought, tree mortality, and wildfire in forests adapted to frequent fire. *BioScience* 68:77–88.

Stephens, S. L., B. M. Collins, and G. Roller. 2012. Fuel treatment longevity in a Sierra Nevada mixed conifer forest. *Forest Ecology and Management* 285:204–212.

Stephens, S. L., D. L. Fry, and E. Franco-Vizcaíno. 2008. Wildfire and spatial patterns in forests in northwestern Mexico: The United States wishes it had similar fire problems. *Ecology and Society* 13:1538–1547.

Stephens, S. L., J. M. Lydersen, B. M. Collins, D. L. Fry, and M. D. Meyer. 2015. Historical and current landscape-scale ponderosa pine and mixed conifer forest structure in the Southern Sierra Nevada. *Ecosphere* 6:1–63.

Stephens, S. L., C. I. Millar, and B. M. Collins. 2010. Operational approaches to managing forests of the future in Mediterranean regions within a context of changing climates. *Environmental Research Letters* 5:1–9.

Stevens-Rumann, C. S., K. B. Kemp, P. E. Higuera, B. J. Harvey, M. T. Rother, D. C. Donato, P. Morgan, T. T. Veblen, and F. Lloret. 2018. Evidence for declining forest resilience to wildfires under climate change. *Ecology Letters* 21:243–252.

Stine, P., P. F. Hessburg, T. A. Spies, M. G. Kramer, C. J. Fettig, A. J. Hansen, J. F. Lehmkuhl, K. L. O'Hara, K. Polivka, P. H. Singleton, S. Charnley, and A. Merschel. 2014. *The ecology and management of moist mixed-conifer forests in eastern Oregon and Washington: A synthesis of the relevant biophysical science and implications for future land management.* USDA Forest Service, General Technical Report PNW-897.

Strong, T. F., G. G. Erdmann, and J. N. Niese. 1995. Forty years of alternative management practices in second-growth, pole-size northern hardwoods I: Tree quality development. *Canadian Journal of Forest Research* 25:1173–1179.

Swanson, F. J., and J. F. Franklin. 1992. New forestry principles from ecosystem analysis of Pacific Northwest forests. *Ecological Applications* 2:262–274.

Swanson, F. J., S. V. Gregory, J. R. Sedell, and A. G. Campbell. 1982. *Land-water interactions: The riparian zone*. Stroudsburg, PA: Hutchinson Ross Publishing Company.

Swanson, M. E., J. F. Franklin, R. L. Beschta, C. M. Crisafulli, D. A. DellaSala, R. L. Hutto, D. B. Lindenmayer, and F. J. Swanson. 2011. The forgotten stage of forest succession: Early-successional ecosystems on forest sites. *Frontiers in Ecology and the Environment* 9:117–125. Safford, H. D., J. T. Stevens, K. Merriam, M. D. Meyer, and

A. M. Latimer. 2012. Fuel treatment effectiveness in California yellow pine and mixed conifer forests. *Forest Ecology and Management* 274:17–28.

Tappeiner, J. C., D. A. Maguire, T. B. Harrington, and J. D. Bailey. 2015. Silviculture and ecology of western U. S. Forests. 2nd ed. 446 p. Corvallis, OR: Oregon State University Press.

Tappeiner, J. C., D. Huffman, D. Marshall, T. A. Spies, and J. D. Bailey. 1997. Density, ages, and growth rates in old-growth and young-growth forests in coastal Oregon. *Canadian Journal of Forest Research* 27:638–648.

Tepley, A. J., F. J. Swanson, and T. A. Spies. 2013. Fire-mediated pathways of stand development in Douglas-fir/western hemlock forests of the Pacific Northwest, USA. *Ecology* 94:1729-1743.

Tepley, A. J., F. J. Swanson, and T. A. Spies. 2014. Post-fire tree establishment and early cohort development in conifer forests of the western Cascades of Oregon, USA. *Ecosphere* 5:1–23.

Thomas, J. W., E. D. Forsman, J. B. Lint, E. C. Meslow, B. R. Noon, and J. Verner. 1990. A conservation strategy for the northern spotted owl. Interagency committee to address the conservation of the northern spotted owl. Washington, DC: USDA Forest Service, USDI Bureau of Land Management, USDI Fish and Wildlife Service, and USDI National Park Service.

Tiner, R. W. 2003. Spatially isolated wetlands of the United States. *Wetlands* 23:494–516.

Toumey, J. W. 1928. *Foundations of silviculture upon an ecological basis.* New York: John Wiley & Sons.

Tritton, L. M. 1980. "Deadwood in the northern hardwood forest ecosystem." PhD diss., Yale University.

Trotsiuk, V., M. Svoboda, P. Janda, M. Mikolas, R. Bace, J. Rejzek, P. Samonil, O. Chaskovskyy, M. Korol, and S. Myklush. 2014. A mixed severity disturbance regime in the primary *Picea abies* (L.) Karst. forests of the Ukrainian Carpathians. *Forest Ecology and Management* 334:144–153.

Troup, R. S. 1927. Dauerwald. *Forestry* 1:78–81.

Troup, R. S. 1928. *Silvicultural Systems.* Oxford, UK: Clarendon Press.

Tschaplinski, P. J., and G. F. Hartman. 1983. Winter distribution of juvenile coho salmon (*Oncorhynchus kisutch*) before and after logging in Carnation Creek, British Columbia, and some implications for overwinter survival. *Canadian Journal of Fisheries and Aquatic Sciences* 40:452–461.

Turner, D. P., and E. H. Franz. 1985. The influence of western hemlock and western redcedar on microbial numbers, nitrogen mineralization, and nitrification. *Plant and Soil* 88:259–267.

USDA. 2012. "Increasing the pace of restoration and job creation on our national forests." https://www.fs.fed.us/publications/restoration/restoration.pdf.

USDA. 2012. National forest system land management planning. *Federal Register* 77(68): 21162-21276.

USDA Forest Service. 2004. Chequamegon-Nicolet National Forest Final Forest Plan (2004 Land and Resource Management Plan). 670 p. Milwaukee, WI.

Van Pelt, R. 2008. *Identifying old trees and forests in Eastern Washington.* Olympia, WA: Washington State Department of Natural Resources.

Van Pelt, R., and S. C. Sillett. 2008. Crown development of coastal *Pseudotsuga menziesii*, including a conceptual model for tall conifers. *Ecological Monographs* 78:283–311.

Vannote, R. L., G. W. Minshall, K. W. Cummins, J. R. Sedell, and C. E. Cushing. 1980. The river continuum concept. *Canadian Journal of Fisheries and Aquatic Sciences* 37:130–137.

Verner, J., K. S. McKelvey, B. R. Noon, R. J. Gutierrez, G. I. Gould Jr., and T. W. Beck. 1992. *The California spotted owl: A technical assessment of its current status.* USDA Forest Service, General Technical Report PSW-GTR-133. 285 p.

Vilsack, T. 2009. "Vision for the forest service: Remarks as prepared for delivery Seattle, Washington." August 14, 2009. https://www.fs.fed.us/video/tidwell/vilsack.pdf.

Volkov, A. D., A. N. Gromtsev, and V. I. Sakovets. 1997. *Climax forests in the north-western taiga zone of Russia: natural characteristic, present state and conservation problems.* Preprint of report for the meeting of the Learned Council. Forest Research Institute, Karelian Research Centre, Russian Academy of Sciences. 32 p.

Walker, B., C. S. Hollin, S. R. Carpenter, and A. Kinzig. 2004. Resilience, adaptability and transformability in social-ecological systems. *Ecology and Society* 9.

Walker, L. R., N. V. L. Brokaw, D. J. Lodge, and R. B. Waide, eds. 1991. Special issue: Ecosystem, plant, and animal responses to hurricanes in the Caribbean. *Biotropica* 23:313–521.

Wallace, L. L., ed. 2004. *After the fires: The ecology of change in Yellowstone National Park.* New Haven, CT: Yale University Press.

Wallenius, T., T. Kuuluvainen, R. Heikkila, and T. Lindholm. 2002. Spatial tree age structure and fire history in two old-growth forests in eastern Fennoscandia. *Silva Fennica* 36:185-199.

Watkins, L. 2011. *The forest resources of Ontario 2011.* Ontario Ministry of Natural Resources, Sault Ste. Marie Ontario, Forest Evaluation and Standards Section, Forests Branch. 307 pp.

Way, A. G. 2011. *Conserving southern longleaf: Herbert Stoddard and the rise of ecological land management.* Athens, GA: University of Georgia Press.

Weaver, H. 1943. Fire as an ecological and silvicultural factor in the ponderosa pine region of the Pacific slope. *Journal of Forestry* 41:7–15.

Weaver, H. 1961. Ecological changes in the ponderosa pine forest of cedar valley in southern Washington. *Ecology* 42:416–420.

Webster, C. R., and N. R. Jensen. 2007. A shift in the gap dynamics of *Betula alleghaniensis* in response to single-tree selection. *Canadian Journal of Forest Research* 37:682–689.

Webster, C. R., and C. G. Lorimer. 2005. Minimum opening sizes for canopy recruitment of midtolerant tree species: A retrospective approach. *Ecological Applications* 15:1245–1262.

Webster, J. R., and E. F. Benfield. 1986. Vascular plant breakdown in freshwater ecosystems. *Annual Review of Ecology and Systematics* 17:567–594.

Westerling, A. L., H. G. Hidalgo, D. R. Cayan, and T. W. Swetnam. 2006. Warming and earlier spring increase western US forest wildfire activity. *Science* 313:940–943.

Westerling, A. L., and T. W. Swetnam. 2003. Interannual to decadal drought and wildfire in the western United States. *EOS, Transactions American Geophysical Union* 84:545–555.

Westphal, C., N. Tremer, G. V. Oheimb, J. Hansen, K. V. Gadow, and W. Härdtle. 2006. Is the reverse J-shaped diameter distribution universally applicable in European virgin beech forests? *Forest Ecology and Management* 223: 75–83.

White, A. S. 1985. Presettlement regeneration patterns in a southwestern ponderosa pine stand. *Ecology* 66:589–594.

Whitehead, R. J., and G. L. Russo. 2005. *"Beetle-proofed" lodgepole pine stands in interior British Columbia have less damage from mountain pine beetle.* Report BC-X-402. Natural Resources Canada, Canadian Forest Service, Pacific Forestry Centre, Victoria, BC, 17 pp.

Whitehead, R. J., L. Safranyik, G. Russo, T. L. Shore, and A. L. Carroll. 2004. Silviculture to reduce landscape and stand susceptibility to the mountain pine beetle. In *Mountain pine beetle symposium: challenges and solutions*, eds. T. L. Shore, J. E. Brooks, and J. E. Stone, 233–244. Report BC-X-399, Natural Resources Canada, Canadian Forest Service, Pacific Forestry Centre, Victoria, BC.

Whitman, A. A., and J. M. Hagan. 2007. An index to identify late-successional forest in temperate and boreal zones. *Forest Ecology and Management* 246:144–154.

Whitney, G. G. 1986. Relation of Michigan's presettlement pine forests to substrate and disturbance history. *Ecology* 67:1548-1559.

Williamson, G. B., and E. M. Black. 1981. High temperature of forest fires under pines as a selective advantage over oaks. *Nature* 293:643–644.

Williamson, R. L. 1982. *Response to commercial thinning in a 110-year-old Douglas-fir stand.* USDA Forest Service, Research Paper PNW-296. Portland, OR. 16 p.

Wilson, B. T., A. J. Lister, R. I. Riemann, and D. M. Griffith. 2013. *Live tree species basal area of the contiguous United States (2000-2009).* USDA Forest Service.

Wisconsin Department of Natural Resources. 2004a. *Wisconsin's statewide forest plan.* Madison, WI: Wisconsin Department of Natural Resources, 76 p.

Wisconsin Department of Natural Resources. 2004b. *Old-growth and old-forests handbook.* Madison, WI: Wisconsin Department of Natural Resources, 76 p.

Woods, K. D. 2004. Intermediate disturbance in a late-successional hemlock-northern hardwood forest. *Journal of Ecology* 92:464–476.

Wuebbles, D. J., K. Kunkel, M. Wehner, and Z. Zobel. 2014. Severe weather in United States under a changing climate. *Eos, Transactions American Geophysical Union* 95:149–150.

York, R., and J. Battles. 2008. Growth response of mature trees versus seedlings to gaps associated with group selection management in the Sierra Nevada, California. *Western Journal of Applied Forestry* 23:94–98.

Youngblood, A. 2005. Silvicultural systems for managing ponderosa pine. In *Proceedings of the symposium on ponderosa pine: issues, trends, and management*, eds. M. W. Ritchie, D. A. Maguire, and A. Youngblood, 49–58. USDA Forest Service Forest, General Technical Report PSW-198.

Youtz, J. A., R. T. Graham, R. T. Reynolds, and J. Simon. 2008. Implementing northern goshawk habitat management in Southwestern forests: A template for restoring fire-adapted forest ecosystems. In *Integrated restoration of forested ecosystems to achieve multiresource benefits: Proceedings of the 2007 national silviculture workshop*, 173–191. USDA Forest Service, General Technical Report PNW-733.

Zedler, P. H. 2003. Vernal pools and the concept of isolated wetlands. *Wetlands* 23:597–607.

Zhang, Q., K. S. Pregitzer, and D. D. Reed. 2000. Historical changes in the forests of the Luce District of the Upper Peninsula of Michigan. *The American Midland Naturalist* 143:94–110.

Zinke, P. J. 1962. The pattern of influence of individual forest trees on soil properties. *Ecology* 43:130–133.

Zinke, P. J., and R. L. Crocker. 1962. The influence of Giant Sequoia on soil properties. *Forest Science* 8:2–11.

Index